Handbook of Heterocyclic Corrosion Inhibitors

Handbook of Heterocyclic Corrosion Inhibitors presents a comprehensive overview of corrosion inhibition using heterocyclic compounds. It covers numerous, emerging heterocyclic compound-based industrial corrosion inhibitors that are oriented toward minimizing corrosive damages and prevention methods.

Describing the fundamentals of heterocycles, corrosion, and corrosion inhibition, the book considers the potential of different series of *N*-heterocycles, such as acridine and acridone-based, carbazole-based, imidazole and imidazoline-based, indole and indoline-based, melamine-based, etc. It presents the corrosion inhibition potential of oxygen- and sulfur-based heterocycle compounds. The book also explores issues with corrosion as a result of improper design with descaling, acidification, refinery, and transport processes.

The book will be of interest to researchers and graduate students studying corrosion science, heterocyclic chemistry, material science and engineering, energy, chemistry, and colloid science. It will also be a valuable reference for corrosion scientists and R&D engineers working in industrial corrosion and industry-based corrosion protection systems.

Handbook of Heterocyclic Corrosion Inhibitors
Principles and Applications

Edited by Chandrabhan Verma

CRC Press
Taylor & Francis Group
Boca Raton London New York

CRC Press is an imprint of the
Taylor & Francis Group, an **informa** business

Designed cover image: Shutterstock

First edition published 2024
by CRC Press
2385 NW Executive Center Drive, Suite 320, Boca Raton FL 33431

and by CRC Press
4 Park Square, Milton Park, Abingdon, Oxon, OX14 4RN

CRC Press is an imprint of Taylor & Francis Group, LLC

Library of Congress Cataloging-in-Publication Data

Names: Verma, Chandrabhan, editor.
Title: Handbook of heterocyclic corrosion inhibitors : principles and applications / edited by Chandrabhan Verma.
Description: Boca Raton : CRC Press, 2024. | Includes bibliographical references and index.
Identifiers: LCCN 2023029563 (print) | LCCN 2023029564 (ebook) | ISBN 9781032454399 (hardback) | ISBN 9781032454405 (paperback) | ISBN 9781003377016 (ebook)
Subjects: LCSH: Corrosion resistant materials. | Heterocyclic compounds.
Classification: LCC TA418.75 .H36 2024 (print) | LCC TA418.75 (ebook) | DDC 661/.8--dc23/eng/20231002
LC record available at https://lccn.loc.gov/2023029563
LC ebook record available at https://lccn.loc.gov/2023029564

ISBN: 978-1-032-45439-9 (hbk)
ISBN: 978-1-032-45440-5 (pbk)
ISBN: 978-1-003-37701-6 (ebk)

DOI: 10.1201/9781003377016

Typeset in Times LT Std
by KnowledgeWorks Global Ltd.

Contents

Preface

Corrosion is one of the most significant issues of the twenty-first century as it is associated with severe safety concerns and monetary losses. The cost of corrosion globally is estimated by NACE to be over US$2.5 trillion, or roughly 3.4% of world GDP. Costs associated with corrosion can be reduced by 15% (US$375) to 35% (US$875) annually using previously developed solutions. Many industrial procedures, including pickling, acid cleaning, and oil-well acidification processes, result in significant corrosion-related losses. The corrosion problem significantly impacts a wide range of sectors, notably those reliant on petroleum. The gases and organic acids in petroleum fluids can easily corrode and damage refinery equipment, storage tanks, and transport pipelines. Surface corrosion may also result from the buildup of scale and deposits. The accumulation of inorganic salts in solid sediments causes corrosion in cooling water systems. Along with oil-well acidification, acid descaling, and acid pickling, corrosion is a significant issue.

The use of heterocyclic compounds is one of the most popular methods of corrosion mitigation. Heterocycles utilize their nonbonding electrons and electrons from side chains or aromatic ring(s) to adsorb on the metal surface and form a corrosion-resistance layer. Numerous heterocyclic compounds containing N, O, and S prevent corrosion in various applications. Because of their adsorption, they impede cathodic and anodic processes. This book compiles the most recent findings on heterocyclic compounds, including pyridine-, imidazole-, quinoline-, melamine-, naphthyridine-, indole-, nitrogen-, sulfur-, oxygen-, and pyridine-based heterocycles as corrosion inhibitors in the real world.

Overall, this book contains 19 chapters. Chapter 1 describes the heterocyclic compounds, their properties, nomenclature, and applications, particularly corrosion inhibition. Their ability to inhibit corrosion is influenced by the factors of heteroatoms, including basicity, electronegativity, position, relative orientation, and number, which are also described in this chapter. Chapter 2 reports the corrosion inhibition mechanism of different heterocyclic compounds in various corrosive electrolytes. Chapters 3–14 provide a collection of different heterocyclic compounds' corrosion inhibition potential and their mechanism of action. Chapters 15–17 report on using heterocyclic compounds in corrosion inhibition for various industries, including petrochemical, oil-well acidization, and acid pickling. Chapters 18 and 19 report corrosion inhibition's green and sustainable aspects with recent trends and future opportunities.

I, Dr. Chandrabhan Verma (the editor), thank all contributors for their outstanding efforts. On behalf of the Chemical Rubber Company (CRC, Taylor & Francis), I am very much thankful to the authors of all chapters for their outstanding and passionate efforts in making this book. Special thanks to Dr. Kyra Lindholm (acquisitions editor) Dr. Tam, and Sonia (Editorial Assistant at CRC) for their dedicated support and help during this project. In the end, all thanks to the CRC for publishing the book.

Chandrabhan Verma
Ph.D.
Editor

About the Editor

Chandrabhan Verma works in the Department of Chemical Engineering, Khalifa University of Science and Technology, Abu Dhabi, United Arab Emirates. Dr. Verma obtained B.Sc. and M.Sc. degrees from Udai Pratap Autonomous College, Varanasi (UP), India. He received his Ph.D. from the Department of Chemistry, Indian Institute of Technology (Banaras Hindu University) Varanasi, under the supervision of Prof. Mumtaz A. Quraishi in Corrosion Science and Engineering. He is a member of the American Chemical Society (ACS) and a lifetime member of the World Research Council (WRC). He is a reviewer and editorial board member of various internationally recognized platforms such as ACS, RSC, Elsevier, Wiley, and Springer. Dr. Verma has to his credit numerous research and review articles in different areas of science and engineering published by ACS, Elsevier, RSC, Wiley, Springer, etc. His current research focuses on designing and developing industrially applicable corrosion inhibitors. Dr. Verma has also edited a few books for the ACS, Elsevier, RSC, Springer, and Wiley. He has received several awards for his academic achievements, including a Gold Medal in M.Sc. (Organic Chemistry; 2010) and Best Publication Awards from the Global Alumni Association of IIT-BHU (Second Prize 2013).

1 Heterocyclic Compounds
Fundamental and Corrosion Inhibition

Reema Sahu[1], Dakeshwar Kumar Verma[1], and Chandrabhan Verma[2]
[1]Department of Chemistry, Govt. Digvijay Autonomous Postgraduate College, Rajnandgaon, Chhattisgarh, India
[2]Department of Chemical Engineering, Khalifa University of Science and Technology, Abu Dhabi, United Arab Emirates

1.1 FUNDAMENTALS OF HETEROCYCLIC COMPOUNDS

Heterocyclic compounds, also known as ring structures, are cyclic compounds with at least two different elemental atoms as constituents of their rings. The area of organic chemistry that deals with these heterocycles' synthesis, characteristics, and uses is known as heterocyclic chemistry. The ring of a heterocyclic compound contains at least two unique components. The heteroatoms nitrogen (N), oxygen (O), and sulfur (S) are most frequently found on a cyclic ring [1–3]. Examples of heterocyclic compounds include vitamins, critical micronutrients, and nucleic acids. Heterocyclic compounds are present in most medicines, insecticides, dyes, and plastics. Additionally, since heterocyclic rings of a particular size share many characteristics, classification by ring size is practical. Certain broad generalizations can be applied to heterocyclic rings, just as they can be applied to carbocyclic rings. Due to their tiny size, three- and four-membered rings are geometrically stressed and easily opened. They are also quickly produced. These heterocycles are common types of reactive intermediates. Building five- and six-membered rings is simple and likewise relatively stable. Their diameters also permit the emergence of aromatic character. Larger rings with seven or more members are stable, although they are less easily created and have received less research [4–6].

1.1.1 AROMATICITY AND REACTIVITY

The idea of aromaticity first emerged in mid-nineteenth century to distinguish between formally unsaturated benzene and unsaturated hydrocarbons. Low reactivity was the critical characteristic of benzene-like compounds, frequently associated with aromatic molecules. As a result, resonance energy was used to determine thermodynamic stability and provide the first quantitative indicator of aromaticity. Later, various theoretical methods were developed to estimate this amount, and now the criterion is frequently regarded as the most fundamental. As a result, multiple notions based on magnetism were created, with the nucleus-independent chemical shift being the most useful in determining aromaticity. In the heterocyclic five-membered rings of pyrrole, furan, and thiophene, the heteroatom has at least one pair of nonbonding valence shell electrons. A p-orbital inhabited by two electrons and oriented parallel to the carbon p-orbitals is produced by hybridizing this heteroatom to the sp^2 state. The resulting planar ring satisfies the first criterion for aromaticity. The Hückel Rule is satisfied since the π-system comprises six electrons, four from the two double bonds and two from the heteroatom [7, 8].

The electrophilic substitution reaction is a significant synthetic process for aromatic molecules. A reaction known as electrophilic aromatic substitution occurs when a substituent replaces a hydrogen atom on a heterocyclic ring. The electrophilic substitution reaction is more likely to take place when the aromatic ring's electron density is high. This is because it is simpler to attack the electrophiles with higher aromatic ring's electron density [9].

There are two reaction sites for five-membered heterocyclic compounds, such as furan, thiophene, and pyrrole: the second and third. The electrophilic substitution reaction occurs at the second position of the electrophile. If a substituent exists at the 2-position, more resonant structures can be written when an electrophilic agent combines with an aromatic ring. The six-membered ring has a lower electron density than the five-membered heterocyclic molecules. The carbon atoms in aromatic compounds have fewer electrons. When we write the resonance structure, the aromatic ring's double bond (electron) can be transferred to the nitrogen atom [10].

The atom of carbon is hence positively charged. The aromatic ring's electron density is reduced, making it less reactive due to the transfer of electrons from the aromatic ring to the nitrogen atom. An electrophilic substitution process involving pyridines requires extremely harsh conditions. Additionally, it is possible to see that the *ortho-* and *para-*locations are positively charged by looking at the resonance structure. Therefore, an electrophilic substitution always occurs at the *meta-*position for six-membered heterocyclic compounds like pyridine. The reaction occurs in the *meta-*position because the *ortho-* and *para-*locations have low electron densities [11].

1.1.2   CLASSIFICATION OF HETEROCYCLIC COMPOUNDS

Heterocyclic compounds can be divided into various types according to the following electronic arrangement.

1.1.2.1   Monocyclic Heterocyclic Compounds (Three- to Six-membered)

1.1.2.1.1   *Three-membered Heterocyclic Compounds*

Aziridine, oxirane, and thiirane are three-membered heterocyclic rings, each containing one nitrogen, one oxygen, or one sulfur element. These three-atom heterocyclic compounds can either be saturated or unsaturated. They can be further

FIGURE 1.1   Chemical structures of aziridine, oxirane, thiirane, diaziridine, and oxiaziridine.

divided into two groups based on the number of heteroatoms present: heterocyclic compounds with one heteroatom and heterocyclic compounds with more than one heteroatom [12, 13]. Some examples are aziridine, oxirane, thiirane, azirine, oxirene, thiirene, diaziridine, and oxiaziridine (Figure 1.1).

1.1.2.1.2   Four-membered Heterocyclic Rings

As with the comparable three-membered rings, nucleophilic displacement processes are employed to create the four-membered rings of azetidine, oxetane, and thietane, each containing one nitrogen, one oxygen, or one sulfur atom. These four-atom heterocyclic molecules can either be saturated or unsaturated. It can be further divided into two groups based on the number of heteroatoms present: heterocyclic compounds with one heteroatom and heterocyclic compounds with more than one heteroatom. Azetidine, oxetane, thietane, azete, oxete, thiete, diazetidine, and dioxetane are a few notable instances of this class (Figure 1.2) [12, 13].

Penicillin and cephalosporins, two related families of antibiotics, are the most significant heterocycles having four-membered rings. The azetidinone ring—the suffix -one designating a double-bonded oxygen atom joined to a ring carbon atom—is present in both series. The -lactam ring, which gives the -lactam antibiotics—the class to which the penicillin and cephalosporins belong—their name, is another frequent term for the azetidinone ring. During the extensive penicillin structure and synthesis research during World War II, the chemistry of azetidinones was intensively investigated.

1.1.2.1.3   Five-membered Heterocyclic Rings

By replacing a C=C bond for a heteroatom containing a lone pair of electrons, these heterocyclic compounds are synthesized from benzene. It is possible to further divide them into two groups based on the number of heteroatoms present: heterocyclic compounds with one heteroatom and heterocyclic compounds with more than

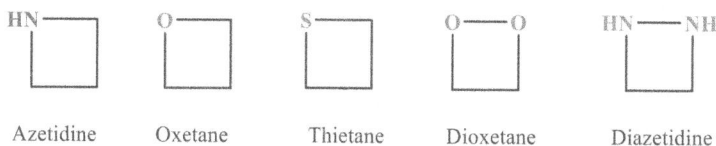

FIGURE 1.2   Chemical structures of some common four-membered heterocyclic rings.

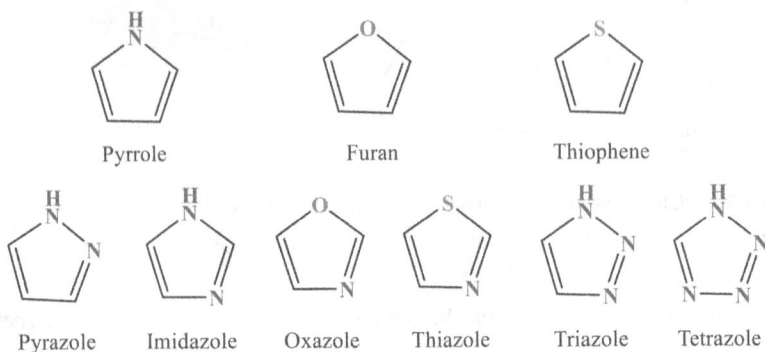

FIGURE 1.3 Chemical structures of some common five-membered heterocyclic rings.

one heteroatom [12, 13]. Furan, pyrrole, thiophene, pyrazole, imidazole, oxazole, thiazole, triazole, and tetrazole are a few notable members of this class (Figure 1.3).

Pyrrolidine, tetrahydrofuran, and thiophane represent the names of saturated derivatives of pyrrole, furan and thiophene, respectively. Indole, benzofuran, and benzothiophene are the three names for the bicyclic compounds formed when a pyrrole, furan, or thiophene ring is fused to a benzene ring.

1.1.2.1.4   Six-membered Heterocyclic Rings

By substituting an electron-bare heteroatom for one of the carbon atoms in benzene, several heterocyclic compounds are created. Heterocyclic compounds with more than one heteroatom can be further divided into two groups based on the number of present heteroatoms. Pyridine, pyran, thiopyran, pyridazine, pyrimidine, pyrazine, etc. are examples of this class (Figure 1.4).

Picoline, lutidine, and collidine denote the names for mono-, di-, and trimethylpyridines, respectively. Numbers indicate the positions of the methyl groups; for example, 2,4,6-collidine contains overall three methyl groups at 2-, 4-, and 6-positions. Additionally, the designations picolinic acid, nicotinic acid (which is produced

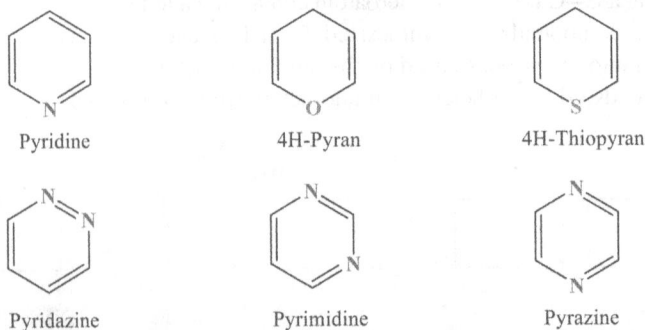

FIGURE 1.4 Chemical structures of some common six-membered heterocyclic rings.

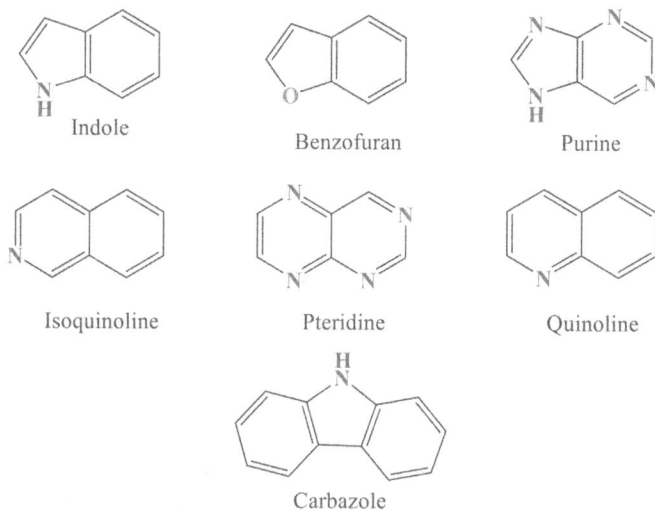

FIGURE 1.5 Chemical structures of some common fused heterocyclic compounds.

from nicotine and is an oxidation product), and isonicotinic acid are frequently used to refer to the pyridine derivatives containing $-COOH$ functional groups at 2-, 3-, and 4-positions, respectively. Coal tar and bone oil include pyridine, picoline, lutidine, and collidine.

1.1.2.2 Condensed or Fused Heterocyclic Compound

Two or more fused rings can be found in a condensed or fused heterocyclic compound. Condensed or fused heterocyclic compounds, such as indole, quinine, isoquinoline, and carbazole, can be partially carbocyclic or partially heterocyclic. Purine, pteridine, and other fully heterocyclic compounds are examples of condensed or fused heterocyclic compounds (Figure 1.5).

1.1.2.3 Nomenclature of Heterocyclic Compounds

There are three systems for naming heterocyclic compounds:

- The common nomenclature
- The replacement method
- The Hantzsch–Widman (IUPAC or systematic) method

1.1.2.3.1 Common Nomenclature

The trivial name that corresponds to each compound is used. This usually results from the presence of the substance, its initial preparation, or its unique features. When there are many heteroatoms of the same type, the counting begins with the saturated one (Figure 1.6) [14].

If there are many types of heteroatoms, the ring is assigned a number beginning with the heteroatom with the greatest value (O > S > N) and going in the direction of

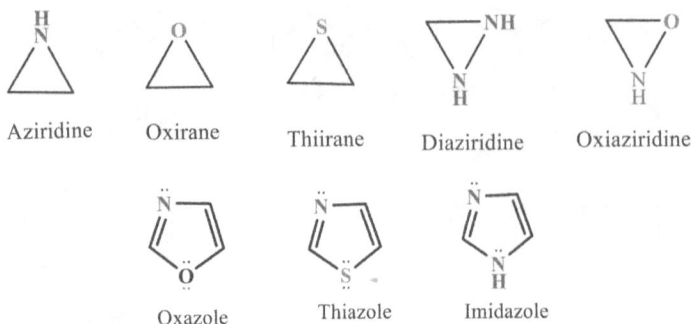

FIGURE 1.6 Common names of some heterocyclic compounds.

assigning the other heteroatoms as low numbers as possible. If there are substituents, their position should be determined by the number of atoms they are attached to, and they should then be listed alphabetically (Figure 1.7).

If two, three, or four atoms are saturated, the words dihydro, trihydro, or tetrahydro are employed. These words are followed by the matching fully unsaturated trivial name, with digits preceding them to show the position of saturated atoms as low as feasible (Figure 1.8).

Trivial names for certain significant heterocyclic chemical classes are provided in Figure 1.9).

1.1.2.3.2 Replacement Nomenclature

In replacement naming, the name of the heterocycle consists of the name of the equivalent carbocycle and an elemental prefix for the newly inserted heteroatom.

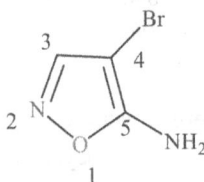

5-Amino-4-bromoisoxazole

FIGURE 1.7 Chemical structure of some common condensed or fused heterocyclic compounds.

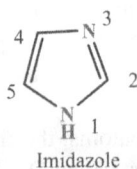

Imidazole

FIGURE 1.8 Chemical structure of imidazole.

(a) Five-membered heterocycles (1 & 2 heteroatoms)

Furan	Thiophene	Pyrrole	Pyrazole	Thiazole

(b) Six-membered heterocycles (1 & 2 heteroatoms)

Tetrahydropyran   Tetrahydrofuran   Pyrrolidine

Dioxane   Piperidine   Morpholine

(c) Common ring-fused heterocycles

Indole   Benzofuran   Benzothiophene

Isoindole   Purine   Indolizine

(d) Common ring fused azines

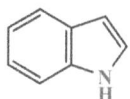 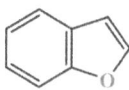 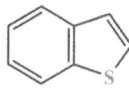

Quinoline   Isoquinoline   Quinozoline   Pteridine

(e) Saturated heterocyclic compounds

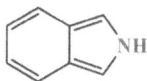 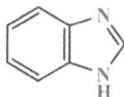 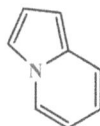

Tetrahydrofuran   Pyrrolidine   Piperidine   Tetrahydro-2H-pyran

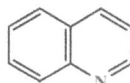 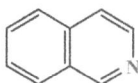 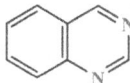 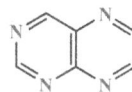

1,4-dioxane   Morpholine

FIGURE 1.9 Trivial names of some common heterocyclic compounds.

TABLE 1.1

The Priority Order of Heteroatoms in Nomenclature of Heterocyclic Compounds

Atom	O	Se	S	N	P
Prefix	Oxa	Selena	Thia	Aza	Phospha

Priority decreasing order ⟶

cyclopropane

Oxirane

Cyclopropene

1,2-oxazirene

1,3-cyclopentadine

Oxacyclopenta-2,4 diene

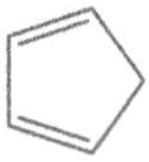

Cyclopenta-1,3-diene

Isothiazole

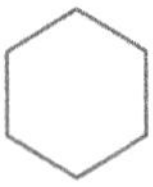

Cyclohexane

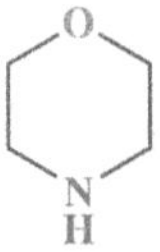

Morpholine

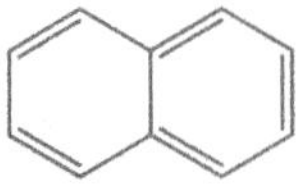

Naphthalene

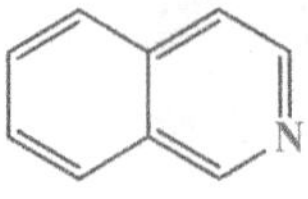

Isoquinoline

Benzene

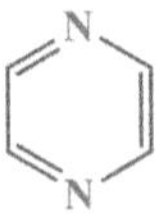

Pyrazine

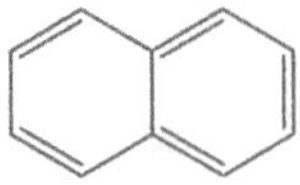

Naphthalene

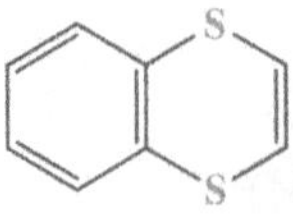

Benzo[b][1,4]dithiine

[15]. If there are multiple heteroatoms, they need to be listed in the order of priority displayed in Table 1.1.

1.1.2.3.3   Hantzsch–Widman Nomenclature

The Arthur Hantzsch and Oskar Widman proposed similar approaches for naming of heterocycles in 1887 and 1888, which are honored by the moniker Hantzsch–Widman nomenclature [16]. Combining the proper prefix, which indicates the nature/type and position (in the ring) of the heteroatom present in the ring, with the suffix, which establishes both the ring size and the level of unsaturation, three- to ten-membered rings can be identified using this technique. Additionally, the suffixes distinguish between heterocycles with nitrogen and those without nitrogen.

$$IUPAC\ name = locants + prefix + suffix$$

1.1.2.3.3.1   Hantzsch–Widman Rules for Fully Saturated and Fully Unsaturated Heterocycles   Determine which heteroatom is present in the ring, then select the appropriate prefix from the table (e.g., thia for sulfur, aza for nitrogen, and oxa for

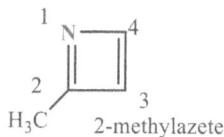

FIGURE 1.10   Chemical structure of 2-methylazete.

oxygen). The heteroatom is always at position 1, and if there are any substituents, they are given the lowest values (Figure 1.10).

When there are two or more comparable heteroatoms in the ring (two nitrogens are represented by diaza), a multiplicative prefix (di, tri, etc.) and locants are used, and the numbering preferably starts at a saturated rather than an unsaturated atom, as seen in Figure 1.11.

The name will have more than one prefix with locants to indicate the relative position of the heteroatoms if there are multiple types of heteroatoms in the ring. Atom prefixes must be listed in a specific sequence of importance. When combining the prefixes (e.g., oxa and aza), two vowels may end up together because "Oxa" (for oxygen) always comes before "aza" (for nitrogen), so the vowel at the end of the first part should be omitted. Starting with the highest-priority heteroatom, the other heteroatoms in the ring are assigned the smallest numbers possible (Figure 1.12).

Choose the appropriate suffix from Table 1.2, depending on the ring size.

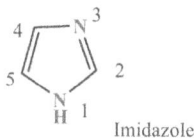

FIGURE 1.11   Hantzsch–Widman rules for 1,3-diazoles.

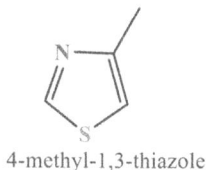

4-methyl-1,3-thiazole

FIGURE 1.12   Chemical structure of 4-methyl-1,3-thiazole.

TABLE 1.2
Ring Size and Suffix for Common Heterocyclic Compounds

Ring Size	Suffix	Ring Size	Suffix
3	ir	4	et
5	ol	6	in
7	ep	8	oc
9	on	10	ec

a   oxa+diaza+ole = 1,2,5-Oxadiazole   Perhyro+aza+ine = perhydrozine

FIGURE 1.13   Naming of 1,2,5-Oxadiazole and perhydrozine.

The endings indicate the size and degree of unsaturation of the ring.

Ring Size	With N		Without N	
	Unsaturated	Saturated	Unsaturated	Saturated
3	Irine	Iridine	irene	irane
4	ete	etidine	ete	etane
5	ole	olidine	ole	olane
6	ine	a	in	inane
7	Epine	a	epin	epane
8	Ocine	a	ocin	Ocane
9	Onine	a	onin	Onane
10	ecine	a	ecin	ecane

If two vowels come together in the prefix or suffix, combine them and omit the first vowel (Figure 1.13).

1.2   CORROSION INHIBITION POTENTIAL OF HETEROCYCLIC COMPOUNDS

1.2.1   EFFECT OF HETEROATOM

One of the most efficient, cost-effective, and valuable corrosion inhibitors is the utilization of organic molecules with P, N, O, and S in their molecular structures. The heteroatoms, particularly N, O, S, and P, transfer their electrons (charge) to the metallic d-orbitals and create a solid metal protective layer through coordinate bonding. The corrosion inhibition potential of heterocyclic compounds with one or more heteroatoms can correlate with their electronegativity. The tendency of an atom in a molecular structure to pull its neighboring pair of electrons toward itself is known as electronegativity. This trait has no dimensions because it is only a predisposition. It represents the final consequence of atoms' propensities to attract electron pairs forming bonds in various elements. Linus Pauling created the most widely used scale. The electronegativities of the typical heteroatoms followed the order: O (3.44 eV) > N (3.04 eV) > S (2.58 eV) > P (2.19 eV) [17, 18].

Considering the metal–corrosion inhibitor interaction as a charge transfer process, the corrosion inhibition potential of the inhibitor will follow the inverse order of

Electronegativity of O, N, S & P (Pauling scale)

O (3.44 eV) > N (3.04 eV) > S (2.58 eV) > P (2.19 eV)

H_2
C

O

H
N

S

H
P

| 1 | 2 | 3 | 4 | 5 |

Cycloprapane   Oxirane   Aziridine   Thiirane   Phosphirane

CH2

O

NH

S

PH

| 1 | 2 | 3 | 4 | 5 |

Cyclobutane   Oxetane   Azetidine   Thietane   Phosphetane

Order of IE: 5 > 4 > 3 > 2 > 1

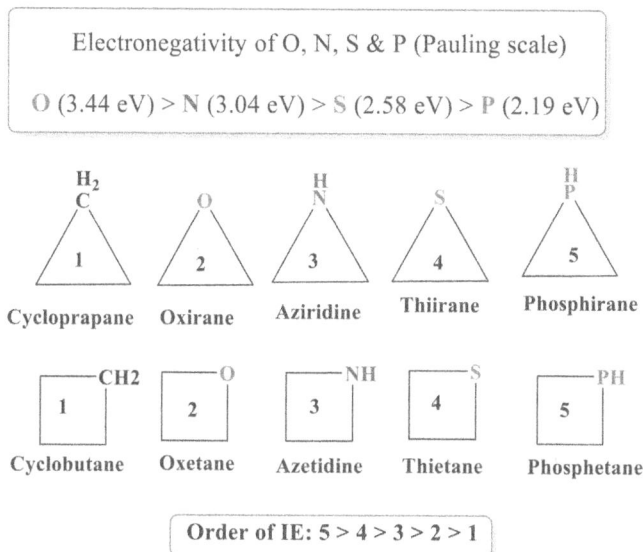

FIGURE 1.14 Expected corrosion inhibition efficiency (IE) order of some common three- and four-membered heterocyclic compounds with a single heteroatom.

electronegativity, i.e., the least electronegative element (P among the above) containing heterocyclic compound will be the most effective corrosion inhibitor. Therefore, the corrosion inhibition potential of heterocyclic compounds having different heteroatoms can be illustrated as per the following sequence (keeping other variables/parameters constant): P > S > N > O. Literature study suggests that heterocyclic compounds having N, O, S, and P in their chemical structures are widely used in corrosion mitigation. Obviously, in most of the cases, these heteroatoms serve as site for binding with the metallic surface. Noticeably, they also possess empty p- or d-orbital where they can accept electrons being transferred from metallic d-orbitals. This process is called retrodonation. Thus, both donation and back- (retro-) donation modes of chemical bonding are reinforced by these heteroatoms (Figure 1.14) [19, 20].

1.2.2 EFFECT OF POSITION OF HETEROATOM AND NATURE OF FUNCTIONAL GROUPS

The position or location of the heteroatom present in the form of functional group of is a very important parameter that determines the performance of heterocyclic corrosion inhibitors. Obviously, electron-donating groups, such as $-SH$, $-OCH_3$, $-OH$, $-NMe_2$, $-NH_2$, $-C(CH_3)_3$, $-CH_2CH_3$, $-CH_3$, etc., are expected to increase the corrosion inhibition performance. On the other hand, electron-donating functional groups, including $-CF_3$, $-NO_2$, $-CHO$, $-CN$, $-COOH$, etc., are expected to decrease the performance of heterocyclic inhibitors. The electron-donating substituents are expected to increase the participation of heteroatom(s) present in the ring

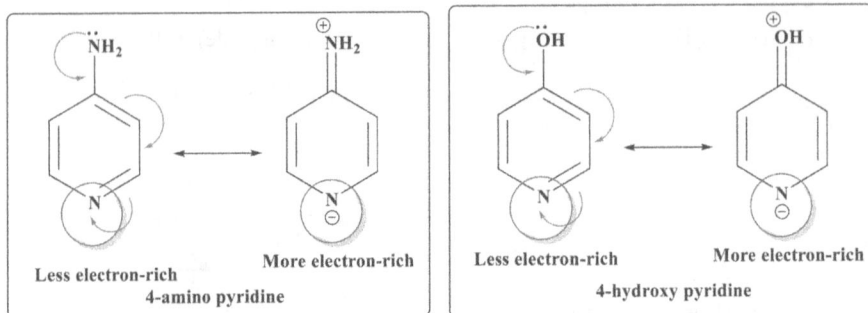

FIGURE 1.15 Schematic illustration of the effect of -NH2 and -OH substituents on electron density enhancement of donor site.

by increasing the electron density through their +I-effect and/or +R-effect, while electron-withdrawing substituents exert opposite effect (Figure 1.15).

Besides, inductive and resonance effect, polar functional groups irrespective of their electron-donating and electron-withdrawing nature favor the binding of inhibitor molecules through chelation if they present suitable position. For example, pyridine substituents having polar substituents at 2- and 6-positions are expected to favor the chelate formation with the metal and metal ions and therefore they are expected to exhibit better performance than nonsubstituted pyridine. Similarly, suitably substituted imidazole and other heterocyclic compounds can also behave as chelating ligands and manifest excellent anticorrosive activity (Figure 1.16).

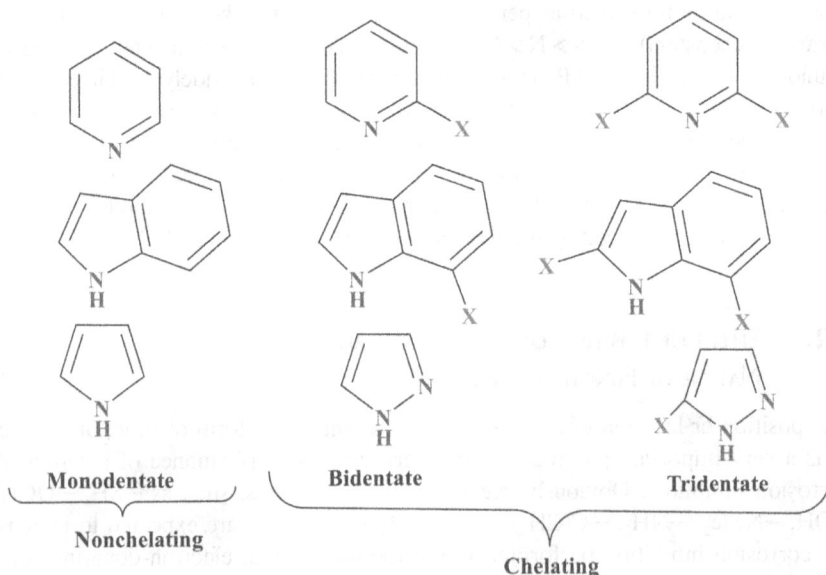

FIGURE 1.16 Common heterocyclic compounds heaving as mono-, bi- and polydentate ligands.

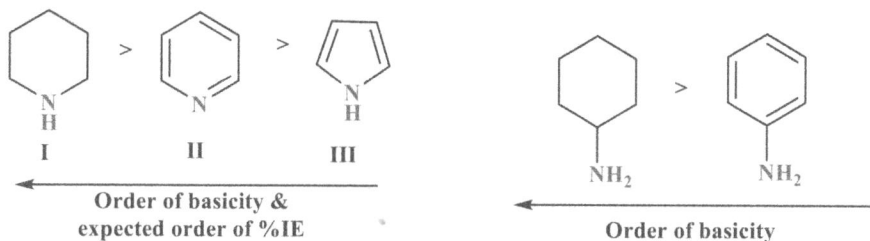

FIGURE 1.17 The basicity and expected order of %IE of some common six-membered heterocyclic compounds.

1.2.3 EFFECT OF HYBRIDIZATION AND BASICITY

Because heteroatoms donate electrons to develop new bonds with the metal surface, they are less likely to share them if they have a higher tendency to withdraw the electrons. It follows that basicity and electronegativity go hand in hand. The greater an atom's electronegativity, the less likely it is that it will share its electrons. The electronegativity of heterocyclic compounds is impacted by the heteroatom hybridization, which also influences their capacity to inhibit corrosion. sp^3-hybridized nitrogen (25% s-character) (e.g., N) is expected to have more basic and greater tendency to donate the electrons while bonding with the metallic surface as compared to the sp^2-hydrizided nitrogen (33.3% s-character). Therefore, the corrosion inhibition potential of heterocyclic compounds can be correlated with their electronegativity, i.e., inverse of basicity (Figure 1.17).

1.2.3.1 Effect of the Relative Number of Heteroatoms and Effect of Substituents

The corrosion inhibition potential of heterocyclic compounds can also be correlated with the relative number of heteroatoms and the nature of substituents present in their molecular structures. Obviously, heterocyclic compounds having greater number of heteroatoms are expected to have better corrosion inhibition performance as compared to the heterocyclic compounds having lesser number of heteroatoms. On this basis, it can be assumed that imidazole and imidazole derivatives would be more effective corrosion inhibitors as compared to the pyrrole and pyrrole derivatives, if they are substituted with identical substituents or functional groups. Similarly, pyrazine, pyrimidine, and pyrridazine are expected to have better corrosion inhibition performance than pyridine. Likewise, cinnoline, phthalazine, quinazoline, and quinoxaline and their derivatives are assumed to have better inhibition efficiency than quinolone and quinolone derivatives, if they are substituted with identical substituents or functional groups. By considering heteroatoms as adsorption center, it can be assumed that the presence of electron-donating functional groups increases the %IE and the presence of electron-withdrawing substituents decreases the %IE. Therefore, 4-aminopyridine is expected to have better inhibition performance than 4-nitropyridine (Figure 1.18).

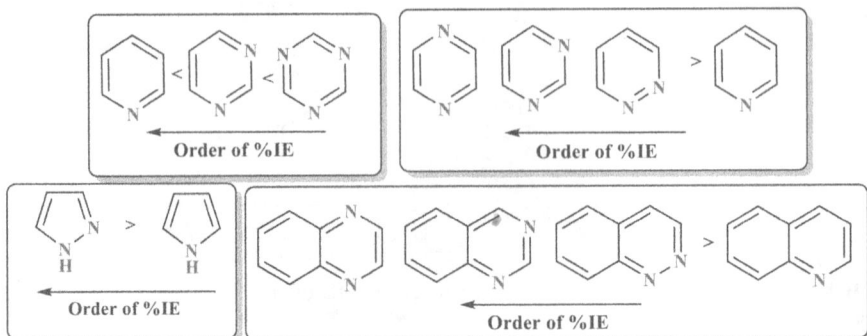

FIGURE 1.18 The expected order of %IE of some common six-membered heterocyclic compounds.

1.3 CONCLUSIONS

One of the biggest issues facing both industrialized and emerging countries is corrosion. As a result, several efforts are being made to reduce corrosion, particularly in industrial areas. Heterocyclic compounds have been shown to be effective inhibitors of metal corrosion. By creating a barrier of defense on the metal surface, the heterocyclic compounds prevent corrosion. Due to their remarkable capacity to coordinate and bind with the metallic surface as well as their high potential for corrosion inhibition, heterocyclic compounds are most frequently used in aqueous corrosion protection. The basics of heterocyclic compounds and how well they inhibit corrosion are covered in this chapter. This chapter provides a general overview of the characteristics, uses, and categorization of several series of heterocyclic compounds. This chapter also discusses how differentiating properties of heterocyclic compounds and heteroatoms affect their ability to prevent corrosion. The characteristics of heteroatoms, such as basicity, electronegativity, location, relative orientation, and number, have an impact on their capacity to prevent corrosion. By increasing the proximity of heteroatoms, basicity, and functional groups or substituents that donate electrons, their ability to suppress corrosion rises. The efficacy of heterocyclic compounds as inhibitors is adversely affected by increases in electronegativity, decreases in heteroatom basicity, and the presence of electron-withdrawing substituents or functional groups.

REFERENCES

1. A.S. Shawali, Chemical Reviews, 93 (1993) 2731–2777.
2. R.V. Orru, M. de Greef, Synthesis, 2003 (2003) 1471–1499.
3. A.Z. Halimehjani, I.N. Namboothiri, S.E. Hooshmand, RSC Advances, 4 (2014) 48022–48084.
4. C.-T. Ho, J.T. Carlin, in: Formation and aroma characteristics of heterocyclic compounds in foods, ACS Symposium Series, American Chemical Society (USA), 1989.
5. C. Rao, R. Venkataraghavan, Canadian Journal of Chemistry, 42 (1964) 43–49.
6. J. Tocher, D. Edwards, Free Radical Research Communications, 6 (1989) 39–45.

7. R.R. Gupta, M. Kumar, V. Gupta, Heterocyclic Chemistry: Volume II: Five-Membered Heterocycles, Springer Science+Business Media, 2013.

8. B.Y. Simkin, V. Minkin, M. Glukhovtsev, Advances in Heterocyclic Chemistry, 56 (1993) 303–428.

9. M. Kawase, T. Kitamura, Y. Kikugawa, The Journal of Organic Chemistry, 54 (1989) 3394–3403.

10. K. Smith, G.A. El-Hiti, Current Organic Synthesis, 1 (2004) 253–274.

11. J.A. Joule, K. Mills, Heterocyclic Chemistry at a Glance, John Wiley & Sons, 2012.

12. A.R. Katritzky, C.A. Ramsden, E.F. Scriven, R.J. Taylor, Comprehensive Heterocyclic Chemistry III, in: V1 3-Memb. Heterocycl., Together with All Fused Syst. Contain. a 3-Memb. Heterocycl. Ring. V2 4-Memb. Heterocycl. Together with All Fused Syst. Contain. a 4-Memb. Heterocycl. Ring. V3 Five-Memb. Rings with One Heteroat. Together with Their Benzo and Other Carbocycl.-Fused Deriv. V4 Five-Memb. Rings with Two Heteroat., Each with Their Fused Carbocycl. Deriv., Elsevier, 2008, pp. 1–13718.

13. A.R. Katritzky, C.W. Rees, Comprehensive Heterocyclic Chemistry, Pergamon Press, 1984.

14. R.R. Gupta, M. Kumar, V. Gupta, R.R. Gupta, M. Kumar, V. Gupta, Heterocyclic Chemistry: Volume I: Principles, Three-and Four-Membered Heterocycles, 1998, pp. 3–38.

15. A.R. Katritzky, C.A. Ramsden, J.A. Joule, V.V. Zhdankin, Handbook of Heterocyclic Chemistry, Elsevier, 2010.

16. W. Powell, Pure and Applied Chemistry, 55 (1983) 409–416.

17. L.R. Murphy, T.L. Meek, A.L. Allred, L.C. Allen, The Journal of Physical Chemistry A, 104 (2000) 5867–5871.

18. R.G. Pearson, Accounts of Chemical Research, 23 (1990) 1–2.

19. C. Verma, D.K. Verma, E.E. Ebenso, M.A. Quraishi, Heteroatom Chemistry, 29 (2018) e21437.

20. C. Verma, M. Quraishi, Coordination Chemistry Reviews, 446 (2021) 214105.

2 Corrosion Protection Using Heterocycles

Mechanism of Corrosion Inhibition in Different Electrolytes

Dheeraj Singh Chauhan[1] and
Mumtaz Ahmed Quraishi[2]
[1]Modern National Chemicals, Second
Industrial City, Dammam, Saudi Arabia
[2]Interdisciplinary Research Center for Advanced
Materials, King Fahd University of Petroleum
and Minerals, Dhahran, Saudi Arabia

2.1 INTRODUCTION

Heterocycles refer to organic compounds wherein one or more carbons are replaced by a heteroatom (e.g., N, S, P, O). Some of the common examples of such frameworks are pyridines, pyrimidines, thiazoles, azoles, etc. [1–9]. The heterocyclic molecules exist in a number of biological molecules, namely, amino acids, proteins, vitamins, carbohydrates, nucleic acids, etc. Several plant extracts contain complex mixtures of various heterocycles. For corrosion inhibition, mostly five- and six-membered heterocyclic rings have been reported. The lone pair of electrons on the heteroatoms of these molecules are involved in the adsorption of target metallic substrates. In addition, the presence of heteroatom-containing functional groups such as $-OH$, $-NH_2$, $-NO_2$, $-SH$, $-COOH$, and the π-electrons, heterocyclic, and phenyl rings are some of the important features that allow effective adsorption and corrosion inhibition from the surrounding aggressive electrolyte.

Heterocyclic corrosion inhibitors have been reported in a wide variety of corrosive environments, including strong acids, neutral and saline media, alkaline conditions, and sweet and sour media [10–13]. Several practical situations occurring in different industrial practices lead to the creation of such types of aggressive media, such as acid-pickling, oil-well stimulation, heat exchangers, batteries, storage and transport of oil, etc. [14–23]. This chapter discusses the mechanisms involved in the adsorption and protection behavior of metals and alloys in the presence of heterocyclic corrosion inhibitors. In the following sections, we have provided a glimpse of the

DOI: 10.1201/9781003377016-2

underlying mechanisms of adsorption and inhibition in acidic media, sweet and sour environments, neutral and saline environments, and alkaline environments.

2.2   CORROSION AND INHIBITION IN ACIDIC ENVIRONMENTS

2.2.1   BRIEF ACCOUNT OF ACID-PICKLING AND ACIDIZING

The technique of acid-pickling refers to treating a metal surface to remove contaminants, impurities, and scales. During pre-cleaning, grease, oils, and salts are removed; acid mixtures are used to remove oxide scales during pickling. Hydrochloric acid (recommended concentration 5–15%) is most commonly used for pickling due to several advantages. Faster pickling can be achieved at lower temperatures and acid concentrations using HCl. The pickling solutions make use of added corrosion inhibitors to prevent a direct corrosive attack on the metal surface [24–26]. The acidizing process is carried out in the oil wells to stimulate oil flow for recovery. Under high pressure, the acid formulation is pumped through the borehole into the rock formations' pores to enlarge the wellbore's flow channels. The acidizing process makes use of acids in the range of 5–28% [27]. This forms an aggressive medium against tubular steel structures [28–32]. Corrosion inhibitors are added to the acid solutions during acidizing to minimize the corrosion rate during the acid treatment. The choice of a suitable corrosion inhibitor depends on the type of acid being used. A number of research articles have been published on corrosion inhibitors for mild steel in 1 M HCl. This is a model system often used for testing new organic corrosion inhibitors. Several articles are also available on 15% HCl involving heterocyclic corrosion inhibitors. However, the inherent complexity of such highly concentrated acids, high temperatures, and the requirement of high inhibitor concentrations have restricted the number of publications in this research area.

2.2.2   MECHANISM OF CORROSION AND PROTECTION

Herein we have taken the example of Fe metal as a representative of steel alloys to explain the corrosion and protection behavior. As discussed above, the steel–acid interface is the most commonly studied system for corrosion and corrosion inhibitor testing. The corrosion and dissolution of Fe metal in the acid media depend on the nature of the adsorbed intermediate species formed. In the H_2SO_4 solution, the species is mainly $(FeOH)_{ads}$; in the HCl medium, it is $(FeCl)_{ads}$. The relevant chemical reactions for the H_2SO_4 solution are as follows [33, 34]:

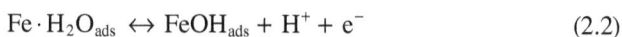

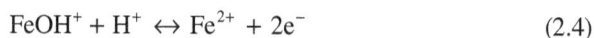

$$Fe + H_2O \leftrightarrow Fe \cdot H_2O_{ads} \tag{2.1}$$

$$Fe \cdot H_2O_{ads} \leftrightarrow FeOH_{ads} + H^+ + e^- \tag{2.2}$$

$$FeOH_{ads} \leftrightarrow FeOH^+ + e^- \tag{2.3}$$

$$FeOH^+ + H^+ \leftrightarrow Fe^{2+} + 2e^- \tag{2.4}$$

For HCl solutions, the process of Fe corrosion can be given as follows [35, 36]:

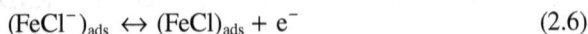

$$Fe + Cl^- \leftrightarrow (FeCl^-)_{ads} \tag{2.5}$$

$$(FeCl^-)_{ads} \leftrightarrow (FeCl)_{ads} + e^- \tag{2.6}$$

$$(FeCl)_{ads} \rightarrow FeCl^+ + e^- \tag{2.7}$$

$$FeCl^+ + e^- \rightarrow Fe^{2+} + Cl^- \tag{2.8}$$

The corresponding cathodic process is the evolution of hydrogen gas via the reduction of hydrogen ions [33, 34]:

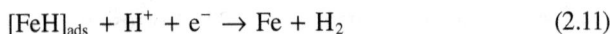

$$Fe + H^+ \leftrightarrow [FeH^+]_{ads} \tag{2.9}$$

$$[FeH^+]_{ads} + e^- \leftrightarrow [FeH]_{ads} \tag{2.10}$$

$$[FeH]_{ads} + H^+ + e^- \rightarrow Fe + H_2 \tag{2.11}$$

With a corrosion inhibitor, additional steps are involved in the corrosion and protection mechanism depending on the nature of the test electrolytic solution. For H_2SO_4, it is as follows [34, 37, 38]:

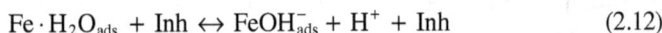

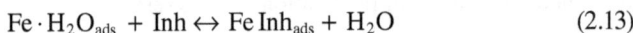

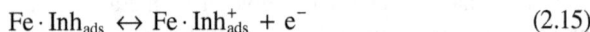

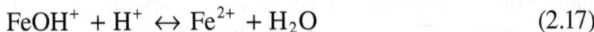

$$Fe \cdot H_2O_{ads} + Inh \leftrightarrow FeOH^-_{ads} + H^+ + Inh \tag{2.12}$$

$$Fe \cdot H_2O_{ads} + Inh \leftrightarrow Fe\,Inh_{ads} + H_2O \tag{2.13}$$

$$FeOH^-_{ads} \rightarrow FeOH_{ads} + e^- \left(\text{rate-determining step}\right) \tag{2.14}$$

$$Fe \cdot Inh_{ads} \leftrightarrow Fe \cdot Inh^+_{ads} + e^- \tag{2.15}$$

$$FeOH_{ads} + Fe \cdot Inh^+_{ads} \leftrightarrow FeOH^+ + Fe \cdot Inh_{ads} \tag{2.16}$$

$$FeOH^+ + H^+ \leftrightarrow Fe^{2+} + H_2O \tag{2.17}$$

The above mechanism shows that the corrosion inhibitor replaces some pre-adsorbed water molecules to provide the intermediate Fe·Inh$_{ads}$. In the rate-determining step, it reduces the amount of FeOH$^-_{ads}$ thereby retarding the Fe dissolution. For hydrochloric acid, the mechanism is as follows [34, 37]:

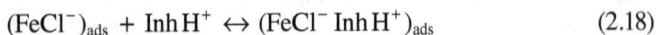

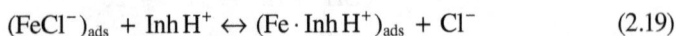

$$(FeCl^-)_{ads} + Inh\,H^+ \leftrightarrow (FeCl^-\,Inh\,H^+)_{ads} \tag{2.18}$$

$$(FeCl^-)_{ads} + Inh\,H^+ \leftrightarrow (Fe \cdot Inh\,H^+)_{ads} + Cl^- \tag{2.19}$$

In the presence of an inhibitor, the mitigation of the rate of cathodic corrosion is as follows [34, 39]:

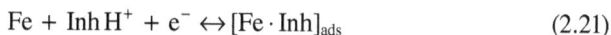

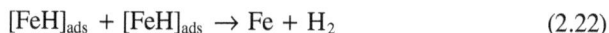

$$Fe + H^+ + e^- \leftrightarrow [FeH]_{ads} \tag{2.20}$$

$$Fe + Inh\,H^+ + e^- \leftrightarrow [Fe \cdot Inh]_{ads} \tag{2.21}$$

$$[FeH]_{ads} + [FeH]_{ads} \rightarrow Fe + H_2 \tag{2.22}$$

This shows that the H^+ and $InhH^+$ (protonated form of inhibitor) compete for the active site for adsorption. The heteroatoms can directly form coordinate bonds to the metallic surface and adsorb via chemical adsorption. The lone-pair electrons on the heteroatoms of the heterocycle and that of any functional groups can get donated to the inhibitor and facilitate the adsorption. In addition, the inhibitor can also gain electrons from the partly filled d-orbitals of the metal atoms. Schematically, the process of adsorption and protection behavior of a heterocyclic corrosion inhibitor is depicted in Figure 2.1.

Experimental and computational methods are generally applied for a proper elucidation of the adsorption and inhibition behavior of corrosion inhibitors. Gravimetric analysis (weight loss tests) allows information on the adsorption behavior in terms of a measured decrease in the weight loss (WL) of the metallic samples with the inhibitor. Electrochemical techniques via impedimetric measurements (EIS), polarization measurements (PDP, LPR), frequency modulations (EFM), etc. shed further light on the mechanism of adsorption and protection [41–43]. For acid solutions, computational studies have been reported using DFT-based reactivity indices to gain theoretical support for the experimentally obtained results. The results provide frontier molecular orbital (FMO) energies from which several parameters such as the molecular orbital energy gap (ΔE), electronegativity (χ), global hardness (η), and softness (σ) are obtained [17, 22, 44]. These parameters provide the trends in

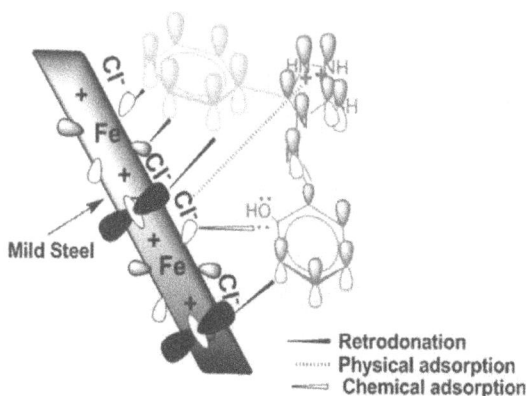

FIGURE 2.1   Schematic of adsorption and inhibition of Schiff base of pyridyl-substituted triazole on mild steel in 1 M HCl. (Reproduced from Ref. [40]. © Elsevier.)

the reactivity or propensity to undergo adsorption. Further support is gained via molecular simulation methods of Monte Carlo (MC) and molecular dynamics (MD) simulations. These parameters facilitate an understanding of the possible orientation of the corrosion inhibitor over an appropriately chosen plane of the desired metal surface. Further, the energy indices such as interaction energy, adsorption energy, binding energy, etc. allow the quantitative determination of the inclination toward adsorption.

In acidic environments, heterocyclic molecule-based corrosion inhibitors have been frequently studied (Figure 2.2). It should be noted that heterocyclic corrosion inhibitors refer to diverse molecules from synthetic organic compounds, natural extracts, amino acids, carbohydrates, pharmaceutical products, ionic liquids, etc. [44, 52, 53]. Quraishi et al. have investigated several drugs as inhibitors for steel surfaces in acid solutions [36, 54–59]. Cefotaxime sodium on mild steel surfaces showed a mixed-type behavior, and its adsorption did not modify the mechanism of anodic–cathodic reactions in the corrosion process [45]. Cefalexin drug, containing S and N together in a six-membered ring along with −NH$_2$, −COOH functional groups and phenyl ring, was investigated for mild steel [46]. The energy of activation in the inhibited sample was greater than for the blank steel, suggesting the creation of an energy barrier as opposed to the corrosion process. Adsorption of the drug was

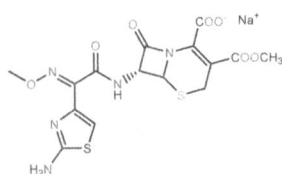

Cefotaxime sodium
Mild steel/ 1M HCl
IE 95.8%/ 300 mgL⁻¹ [45]

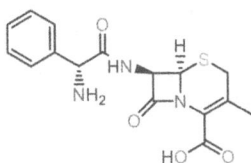

Cefalexin
Mild steel/ 1M HCl
IE 92.1%/ 400 mgL⁻¹ [46]

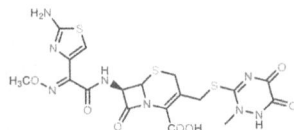

Ceftriaxone
Mild steel/ 1M HCl
IE 90.1%/ 400 mgL⁻¹ [47]

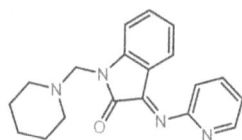

N-(Piperidinomethyl)-3-
[(pyridylidene)amino]isatin
Mild steel/ 1M HCl
IE 94.1%/ 300 mgL⁻¹ [48]

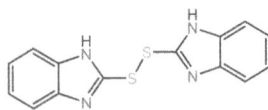

Bis (benzimidazol-2-yl) disulphide

Mild steel/ 1M HCl, 0.5M H₂SO₄
IE 98.2%/ 120 mgL⁻¹
IE 99.1%/ 100 mgL⁻¹ [49]

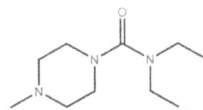

Diethylcarbamazine

Mild steel/ 1 M HCl
94.36/ 6.27 ×10⁻⁴ M [50]

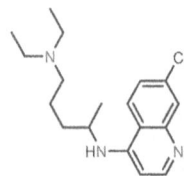

Chloroquine
Mild steel/ 1 M HCl
IE 99%/ 3.1 ×10⁻⁴ M [51]

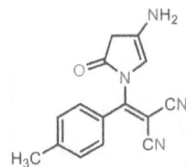

Pyrazolone
N80 steel/ 15 % HCl
IE 93.9%/ 150 mgL⁻¹ [52]

FIGURE 2.2 Heterocyclic molecules reported as corrosion inhibitors in acidic solutions.

aligned to Langmuir isotherm with a physical mode. Ceftriaxone drug, containing a 1,2,4-triazine ring and a thiazole ring, provided >90% protection at 400 mg L^{-1} for mild steel [47]. A mixed-type performance was observed via PDP analysis with a primarily physical mode of adsorption, as revealed by the WL technique. An isatin derivative containing pyridine and piperidine rings was evaluated for mild steel surface in 1 M HCl [48]. The energy of activation was found to be lower compared to the blank steel solution suggesting that the inhibitor is likely to undergo strong adsorption at elevated temperatures. A benzimidazole derivative containing two benzimidazole moieties connected with the disulfide linkage (−S−S−) was evaluated [49]. High inhibition performance of >95% was obtained in 1 M HCl and 0.5 M H$_2$SO$_4$ solutions. Diethylcarbamazine containing a piperazine group was evaluated as a corrosion inhibitor [50]. The inhibitor provided efficient adsorption and protection behavior with alignment to Langmuir isotherm. Chloroquine drug containing a pyridine ring adjacent to a chlorobenzene ring was evaluated for mild steel [51]. Excellent protection effect with 99% efficiency was obtained. Computational studies were undertaken to correlate the inhibitive action with the chemical structure of the inhibitor. Frontier orbital energy indices indicated that the inhibitor would likely donate and accept electrons from the metallic substrate.

2.3 MECHANISM OF CORROSION AND INHIBITION IN SWEET ENVIRONMENTS

2.3.1 BRIEF ACCOUNT OF CORROSION IN SWEET MEDIA

The phenomenon of sweet corrosion is mainly attributed to the corrosion of steel surfaces due to the dissolved CO$_2$. Concentrated brine along with the dissolved CO$_2$ aggravates the situation, causing severe pitting of the underlying steel surface [60–62]. Alloy steel is a preferred structural material for pipelines and oil-well equipment owing to its low cost and good mechanical properties [63–67]. However, the steel surface experiences severe corrosion in the presence of such an aggressive medium.

2.3.2 MECHANISM OF CORROSION AND INHIBITION

Dissolved CO$_2$, in the presence of aqueous medium, produces carbonic acid (H$_2$CO$_3$), which dissociates further to produce HCO$_3^-$ and CO$_3^{2-}$ [68–70]. A greater corrosion rate is produced by H$_2$CO$_3$, at a given pH, in comparison to that obtained in mineral acids such as H$_2$SO$_4$ and HCl that undergo complete dissociation in an aqueous medium [68, 71]. The major cathodic corrosion reactions leading to hydrogen evolution can be given as follows (Figure 2.3):

$$2H^+_{(aq)} + 2e^- \rightarrow H_{2(g)} \qquad (2.23)$$

$$2H_2CO_{3(aq)} + 2e^- \rightarrow H_{2(g)} + 2HCO^-_{3(aq)} \qquad (2.24)$$

$$2HCO^-_{3(aq)} + 2e^- \rightarrow H_{2(g)} + 2CO^{2-}_{3(aq)} \qquad (2.25)$$

FeCO$_3$ formation

Temperature ≥ 80 °C pH = 6
Temperature ≥ 120 °C pH = 4

Combined FeCO$_3$ and Fe$_3$O$_4$ formation

pCO$_2$ >3 bar
Temperature ≥ 150 °C

Fe$_3$O$_4$ formation

Temperature ≥ 200 °C
pCO$_2$ <1 bar

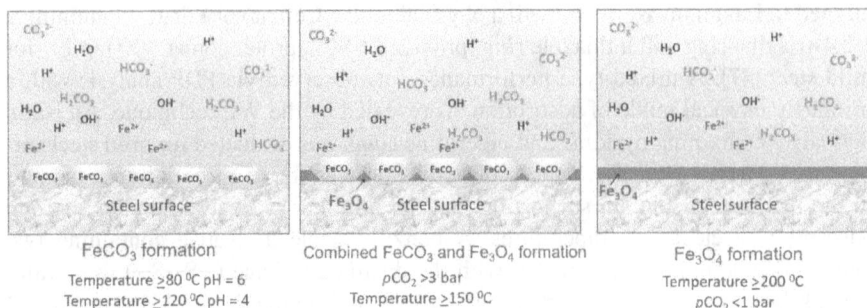

FIGURE 2.3 Influence of pH, temperature, and partial pressure of CO$_2$ (pCO$_2$) on corrosion production formation during sweet corrosion. (Reproduced from Ref. [72]. © Elsevier.)

The metal dissolution at the anode proceeds as follows:

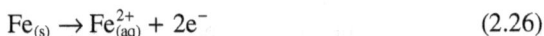

$$Fe_{(s)} \rightarrow Fe^{2+}_{(aq)} + 2e^- \qquad (2.26)$$

The imidazoline (IM)-based molecules are frequently studied as sweet corrosion inhibitors due to their ability to develop a protective film on the metallic substrate [68, 73–76]. The IM framework generally contains three substructures, as shown in Figure 2.4 [68, 69]. The five-membered IM ring acts as the hydrophilic, polar head group. Often a polar functional group such as —NH$_2$ or —OH is attached to the IM ring contributing toward the hydrophilicity of the inhibitor and aiding in the adsorption. The long alkyl chain acts as the hydrophobic part that hinders the attack of the aggressive aqueous electrolyte. Any additional alkyl chains present in the inhibitor can contribute toward the hydrophobicity of the inhibitor molecule.

The amidine N atom in IM (I) displayed in Figure 2.5 forms a frequently used IM-based corrosion inhibitor against CO$_2$ environment [77–79]. Such derivatives are weak nucleophiles but act as strong bases (pK_b ~ 1). In a CO$_2$/H$_2$O media, the IMs form bicarbonate salts III and IV and, upon partial hydrolysis, convert them into amide form V [69]. Besides forming carbamate salts II and bicarbonates III and IV, several neutral as well as ionic species, viz., CO$_2$, H$_2$O, HCO$_3^-$, CO$_3^{2-}$, OH$^-$, and H$_3$O$^+$, also exist in the medium (Figure 2.5) [69]. In the case of adsorption of

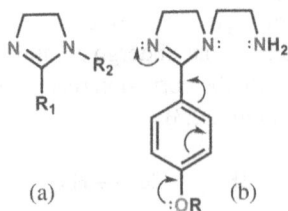

(a)   (b)

FIGURE 2.4 (a) Chemical structure of a typical imidazoline-based corrosion inhibitor. (b) Introduction of a hydrophobic species at R1 and a hydrophilic species at R2. (Reproduced from Ref. [69]. © Elsevier.)

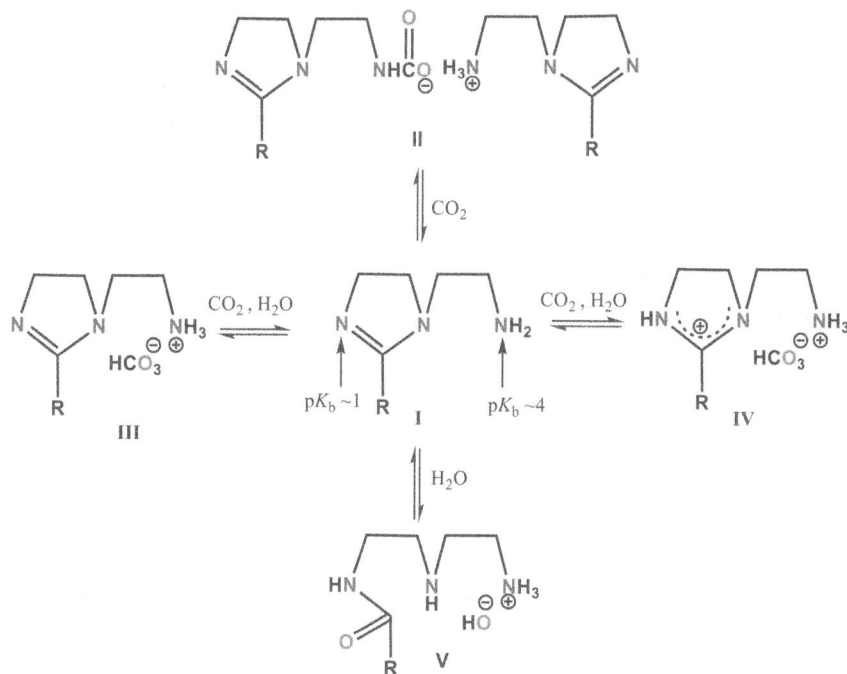

FIGURE 2.5 Various ionic species of IMs existing in CO_2 medium. (Reproduced from Ref. [69]. © Elsevier.)

a pyridine-containing benzimidazole-based inhibitor on the X60 steel (Figure 2.6) [80], the electrochemical study was undertaken using the rotating cylinder electrode (RCE). Herein, considering the acidic pH (3.83) of the NACE brine solution along with Cl⁻ ions in high concentration, the nitrogen of the pyridine ring and the benzimidazole moiety, having a basic character, can get protonated to afford N-H⁺ species. This protonated N can provide the active site for the inhibitor adsorption via Coulombic interaction with the preadsorbed Cl⁻ ions. Therefore, this can provide a synergistic corrosion inhibition action and enhancement in inhibitor adsorption.

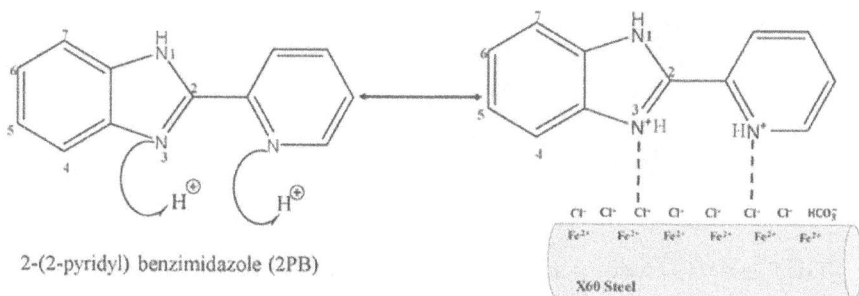

FIGURE 2.6 Schematic of the interaction between inhibitor 2PB and the surface of X60 steel during CO_2 corrosion inhibition in the NACE brine. (Reproduced from Ref. [80]. © Elsevier.)

A series of diphenyl imidazoles were analyzed for J55 steel in CO_2-saturated 3.5% NaCl [81]. The authors evaluated the influence of electron-donating and electron-withdrawing groups on inhibition. The $-OCH_3$ group–containing molecule showed better performance than that with $-CH_3$ and $-NO_2$ group–containing molecules. This order of efficiencies was in accordance with that of the Hammett substitution coefficients [82, 83]. Similar trends were noted for a series of three imidazolidine derivatives containing two phenyl rings and an imidazoline ring [84]. A detailed MD simulation study provided support to the experimentally obtained results. A macrocycle HPT was investigated for J55 steel surface [62]. The large macrocyclic ring contained six N atoms, four ketone groups, and two imine ($-C=N$) linkages. This allowed efficient adsorption on the steel surface. Three porphyrin derivatives P1 (with 4 pyridine rings), P2 (4 phenyl rings), and P3 (4 hydroxyphenyl rings) were studied for the same medium [85]. At an optimum concentration of 400 mg L^{-1}, the order of inhibition efficiencies was P1 > P2 > P3. For two more porphyrins (I with 4 pentafluorophenyl rings and Pd atom coordinated in the center to 4 imidazole rings; and II with 4 carboxylic acid-substituted phenyl rings) were tested for J55 steel [86]. Compound II provided better performance compared to I. A glucose derivative containing two glucose rings connected with a hexamethylene chain was analyzed for API X60 steel in the sweet medium [41]. The molecule's high dispersibility and performance of 91.82% was noted at a quite low dose. One quinoline derivative containing a cinnamyl moiety connected to the quinolone ring with $-NH_2$, $-OH$, and $-C\equiv N$ groups was studied for C1018 steel surface in sweet environment [11]. At temperatures of 25 °C and 60 °C, with 5 mM KI, the inhibitor provided a high performance. A triazine derivative containing 3 hydroxyphenyl rings was studied as inhibitor in hydrodynamic conditions and high efficiency of 97.05% was obtained [87]. Salient features of the inhibitor were the single-step synthesis assisted by ultrasonic irradiation in aqueous medium and high aqueous solubility. Some heterocyclic molecule-based inhibitors for sweet medium are given in Figure 2.7.

2.4   MECHANISM OF CORROSION AND INHIBITION IN NEUTRAL AND SALINE ENVIRONMENTS

2.4.1   BRIEF ACCOUNT OF CORROSION IN NEUTRAL AND SALINE MEDIA

Copper and its alloys have been considerably reported as model systems for studying heterocyclic corrosion inhibitors' adsorption and protective action. Copper metal and several of its alloys have shown considerable resistance to corrosion in saline media [5, 26, 88]. Cu corrosion and dissolution are suggested to take place in a medium containing Cl^- [89–91]:

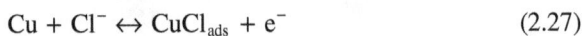

$$Cu + Cl^- \leftrightarrow CuCl_{ads} + e^- \tag{2.27}$$

Based on the Cl^- concentration, $CuCl_{ads}$ can further dissolve as follows:

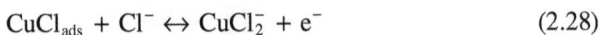

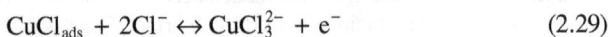

$$CuCl_{ads} + Cl^- \leftrightarrow CuCl_2^- + e^- \tag{2.28}$$

$$CuCl_{ads} + 2Cl^- \leftrightarrow CuCl_3^{2-} + e^- \tag{2.29}$$

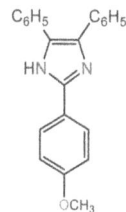

Diphenylimidazole
J55 steel/ CO₂-saturated 3.5% NaCl
IE 90.0%/ 400 mgL⁻¹ [81]

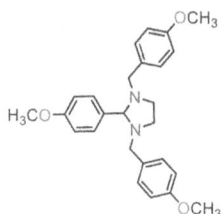

Imidazolidine
J55 steel/ CO₂-saturated 3.5% NaCl
IE 92.0/% 400 mgL⁻¹ [84]

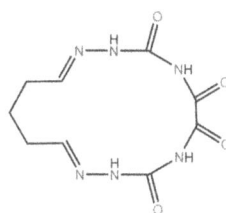

Hexaazacyclopentadecane
J55 steel/ CO₂-saturated 3.5% NaCl
IE 93%/ 400 mgL⁻¹ [62]

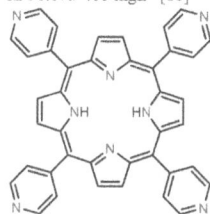

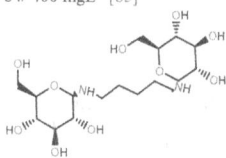

Porphine P1
J55 steel/ CO₂-saturated 3.5% NaCl
92/ 400 mgL⁻¹ [85]

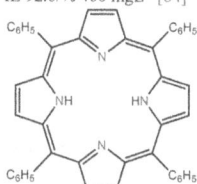

Porphine P2
J55 steel/ CO₂-saturated 3.5% NaCl
82/ 400 mgL⁻¹ [85]

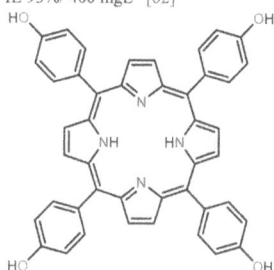

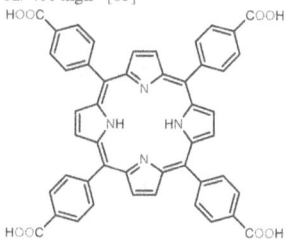

Porphine P3
J55 steel/ CO₂-saturated 3.5% NaCl
84/ 400 mgL⁻¹ [85]

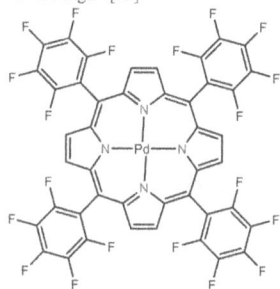

Porphyrin-F-Pd(II)

J55 steel/ CO₂-saturated 3.5% NaCl

IE 86%/ 400 mgL⁻¹ [86]

Porphyrin tetrabenzoic acid

J55 steel/ CO₂-saturated 3.5% NaCl

IE 93%/ 400 mgL⁻¹ [86]

Hexamethylene-1,6-bis(N-D-glucopyranosylamine)
API X60 steel/ CO₂-saturated 3.5% NaCl
IE 91.82%/ 2.27 × 10⁻⁴ M [41]

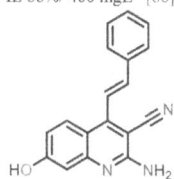

(E)-2-amino-7-hydroxy-4-styrylquinoline-3-carbonitrile
C1018 steel/ CO₂-saturated 3.5% NaCl
IE 89%/ 100 mgL⁻¹ [11]

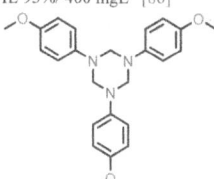

1,3,5-tris(4-methoxyphenyl)-1,3,5-triazinane
API X60 steel/ CO₂-saturated 3.5% NaCl
IE 90.19%/ 100 mgL⁻¹ [87]

FIGURE 2.7 Heterocyclic molecules reported as corrosion inhibitors in CO₂ environments.

Chloride solution
Atacamite ($Cu_2(OH)_3Cl$) or malachite ($CuCO_3 \cdot Cu(OH)_2$)
Cupric hydroxide ($Cu(OH)_2$) or oxide (CuO)
Cuprous oxide (Cu_2O)
Cuprous chloride ($CuCl$)
Copper metal

FIGURE 2.8 Generalized scheme of stratification of various species in the film of corrosion product on Cu surface in a saline environment. (Reproduced from Ref. [89]. © Elsevier.)

The oxide formation can follow the above in aerated condition:

$$2CuCl_2^- + 2OH^- \rightarrow Cu_2O + H_2O + 4Cl^- \tag{2.30}$$

Further oxidation could produce a less protective cupric hydroxyl chloride, also known as atacamite (Figure 2.8) [89, 92].

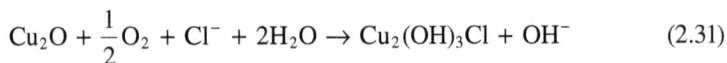

$$Cu_2O + \frac{1}{2}O_2 + Cl^- + 2H_2O \rightarrow Cu_2(OH)_3Cl + OH^- \tag{2.31}$$

Thus, due to the anodic dissolution, complex soluble species containing mainly $CuCl^{2-}$ and $CuCl_3^{2-}$ are generated in bulk. Contrariwise, the film of $Cu_2O/CuO/Cu_2(OH)_3Cl$ could afford a barrier against diffusion [89, 91].

2.4.2 MECHANISM OF CORROSION INHIBITION

In a chloride-containing environment, the oxidation of metal at the anode surface produces metal ions [26, 93]:

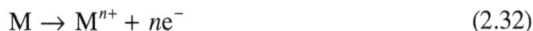

$$M \rightarrow M^{n+} + ne^- \tag{2.32}$$

The cationic form (M^{n+}) reacts with the Cl^- ions to yield $[M(Cl)_n]_{ads}$, which, in the presence of an inhibitor, forms a metal–inhibitor complex [26, 93, 94]:

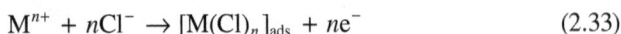

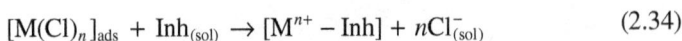

$$M^{n+} + nCl^- \rightarrow [M(Cl)_n]_{ads} + ne^- \tag{2.33}$$

$$[M(Cl)_n]_{ads} + Inh_{(sol)} \rightarrow [M^{n+} - Inh] + nCl^-_{(sol)} \tag{2.34}$$

The inhibitor having acidic hydrogen can lead to the formation of a neutral metal–inhibitor complex, for example [94, 95]:

$$CuCl_{(ads)} + C_5H_5N_5 \rightarrow [Cu - C_5N_5H_4] + Cl^- + H^+ \qquad (2.35)$$

A generalized equation for inhibitors containing active/acidic hydrogen can be given as follows:

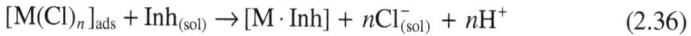

$$[M(Cl)_n]_{ads} + Inh_{(sol)} \rightarrow [M \cdot Inh] + nCl^-_{(sol)} + nH^+ \qquad (2.36)$$

The film of inhibitor deposited on the metallic substrate provides isolation to the metal surface from the solution containing Cl$^-$. The presence of metal cations facilitates the adsorption of inhibitors via synergistic action.

The 4- and 5-carboxybenzotriazole octyl esters (4-CBTAH-OE and 5-CBTAH-OE, respectively) were studied for copper substrate in 0.5 M sulfate solution under aerated condition [96]. Surface-enhanced Raman scattering (SERS) investigation revealed that at the neutral pH, the inhibitors afforded protection via the formation of a polymeric metal inhibitor (Figure 2.9a). Pyrazoles undergo adsorption on the Cu substrate via interaction with the lone-pair electrons of the N atoms on the pyrazole ring, via the six delocalized π-electrons of the azole ring. Another pyrazole derivative containing dithiocarboxylate, dodecyl, and hydroxyl groups was investigated for inhibition of Cu

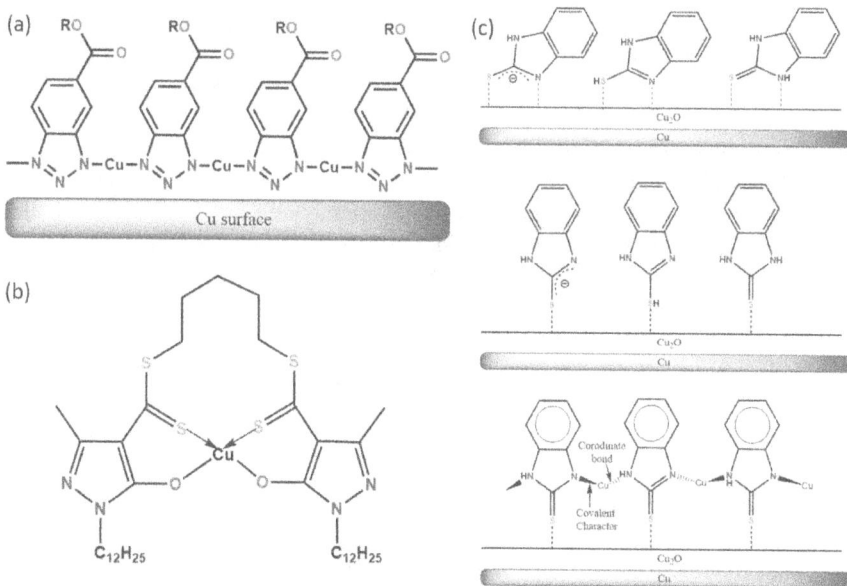

FIGURE 2.9   (a) Polymeric film deposited on Cu surface due to Cu-benzotriazole-octyl ester surface complex formation. (Reproduced from Ref. [96]. © Elsevier.) (b) Square–planar complex formed between tetradentate ligand 1,5-bis(4-dithiocarboxylate-1-dodecyl-5-hydroxy-3-methylpyrazolyl)-pentane and Cu(II). (Reproduced from Ref. [97]. © Elsevier.) (c) Possible alignments of MBIH on the Cu/Cu$_2$O in various forms. (Reproduced from Ref. [98]. © Elsevier.)

corrosion, revealing that the inhibitor first underwent physical adsorption and then chemical adsorption by forming a complex between Cu(II) and the inhibitor molecules [97]. A square planar complex was formed between the inhibitor and the Cu substrate, wherein the inhibitor acted as a tetradentate ligand (Figure 2.9b). The adsorption and protective action of 2-mercaptobenzimidazole (MBIH) on the copper substrate was studied for 180 days, followed by electrochemical analyses and surface examination [98]. It was noted that the protective film of the MBIH interacted with the Cu surface mainly via N and S heteroatoms (Figure 2.9c). In addition, the inhibitor–metal together formed a surface complex via coordinate bond formation.

Antidiabetic drug miglitol was evaluated for mild steel in 700 mg L^{-1} NaCl [99]. The introduction of Zn^{2+} ions improved the inhibition performance due to synergistic action. Two indazole derivatives containing benzimidazole connected to two long alkyl chains each were studied for Cu in 3% NaCl [100]. The inhibitors developed a self-assembled monolayer (SAM) on the Cu substrate. Computational studies suggested that the N atoms behaved as the active site for adsorption. Surfactant derivatives of bis-glucobenzimidazolone were prepared and studied for mild steel surface in a 200 mg L^{-1} NaCl environment [101]. High protection efficiencies were obtained at a quite low concentration of 10^{-5} M. The inhibitors showed mixed-type behavior, as revealed by the PDP analyses. Adsorption of neutral p-tolyl imidazole was comparatively better than that in the protonated form [102]. Purines contain an imidazole ring adjacent to one pyrimidine ring and are widely occurring N-containing heterocycles. 6-Benzylaminopurine exhibited a superior inhibition performance compared to imidazole, adenine, and purine on Cu surface in seawater [103]. The results were in accordance with the increase in molecular weight and the number of heteroatoms. 5-(Phenyl)-4H-1, 2, 4-triazole-3-thiol was investigated for Cu surface in 3.5% NaCl under stationary and stirred conditions [104]. The efficiency rose from 73% to 90% when the concentration was raised from 500 mg L^{-1} to 1500 mg L^{-1}. 2, 6-Diaminopyridine was evaluated for Al surface in 3.5% NaCl [105]. A mixed-type action with the anodic prevalence was noticed. 2,6-Dimethylpyridine was analyzed for Al surface in distilled water [106]. The PDP and adsorption evidenced a mixed nature obeyed the Langmuir isotherm. Figure 2.10 shows some heterocyclic corrosion inhibitors' chemical structures and corrosion inhibition performance for neutral and saline environments.

2.5   MECHANISM OF CORROSION AND INHIBITION IN ALKALINE ENVIRONMENTS

2.5.1   BRIEF ACCOUNT OF CORROSION IN ALKALINE MEDIA

Alkaline environments are generally encountered in industries such as fuel cells, batteries, alkaline etching, boiler feed water systems, etc., posing considerable corrosion concerns [107–111]. In an alkaline environment, the steel surface exists in the passive state. Nevertheless, contaminations can have a detrimental effect on the passive film. Cl^- ions act as one of the common contaminants that facilitate localized corrosion upon arriving at the metallic substrate [111]. The Cl^- effects the breakdown of passivity, which can be elucidated as a balance of competitive phenomena: (i) stabilization of the passive film due to the adsorption of OH^- and (ii) Cl^- ions leading to film disruption. With increase in the activity of the Cl^- above that of OH^- ions,

Miglitol
Mild steel/ 700 mgL^{-1} NaCl
IE 78.25%/ 100 mgL^{-1} [99]

Alkylimidazole
Copper/ 3% NaCl
IE 99.6/ 5 mM [100]

Bisglucobenzimidazolone
Mild steel/ 200 mgL^{-1} NaCl
IE 90.0%/ 10^{-5} M [101]

p-Tolylimidazole
Copper/ 0.5 M NaCl pH 5.6
IE 88.6% [102]

6-Benzylaminopurine
Copper/ seawater
IE 94.43%/ 5 × 10^{-3} M [103]

5-(Phenyl)-4H-1,2,4-triazole-3-thiol
Copper/ 3.5% NaCl
IE 90%/ 1500 mgL^{-1} M [104]

2.6-Diaminopyridine
Aluminium/ 3.5% NaCl
IE 98.4%/ 100 mgL^{-1} [105]

2,6-Dimethylpyridine
Aluminium/ distilled water
IE 86%/ 0.0187 M [106]

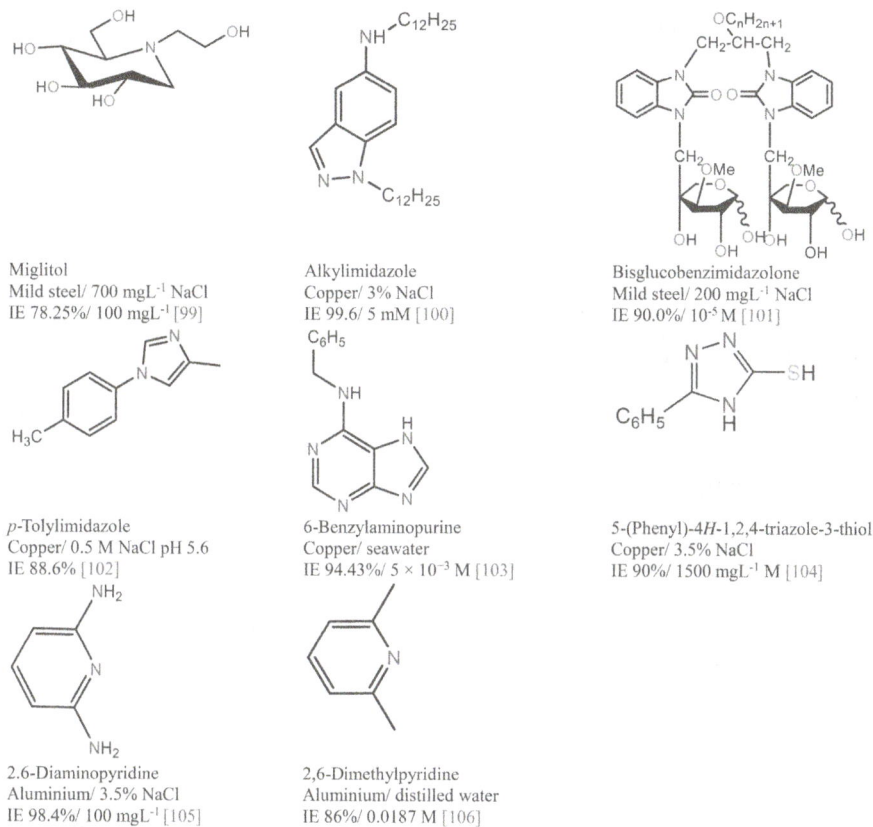

FIGURE 2.10 Heterocyclic molecules reported as corrosion inhibitors in neutral and saline environments.

corrosion proceeds [111, 112]. The formation of Fe(OH)$_{ads}$ can be taken as the first step during passivation in an alkaline medium. A thicker hydroxide film having a passive nature forms due to the oxidation of this layer [111, 113–115]:

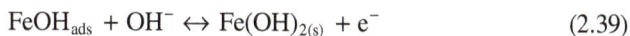

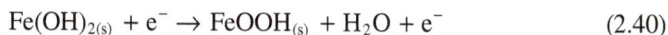

$$Fe + OH^- \leftrightarrow FeOH^-_{ads} \qquad (2.37)$$

$$FeOH^-_{ads} \leftrightarrow FeOH_{ads} + e^- \qquad (2.38)$$

$$FeOH_{ads} + OH^- \leftrightarrow Fe(OH)_{2(s)} + e^- \qquad (2.39)$$

$$Fe(OH)_{2(s)} + e^- \rightarrow FeOOH_{(s)} + H_2O + e^- \qquad (2.40)$$

The presence of Cl$^-$ ions effects the increased metal dissolution and a lowering of the surface coverage of the Fe(OH)$_{ads}$ as follows:

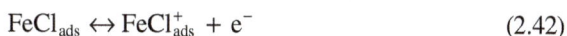

$$Fe + Cl^- \leftrightarrow FeCl^-_{ads} + e^- \qquad (2.41)$$

$$FeCl_{ads} \leftrightarrow FeCl^+_{ads} + e^- \qquad (2.42)$$

In alkaline medium, the hydrogen evolution reactions can be given as follows:

$$H_2O + e^- \rightarrow H_{ads} + OH^- \tag{2.43}$$

Here H_{ads} denotes hydrogen atom adsorbed on the surface of the cathode following the Tafel reaction [111, 116]:

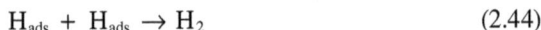

$$H_{ads} + H_{ads} \rightarrow H_2 \tag{2.44}$$

Instead of the above, following the Heyrovsky reaction, the removal of H_{ads} can take place [111, 117]:

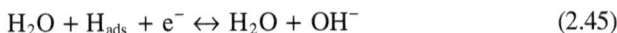

$$H_2O + H_{ads} + e^- \leftrightarrow H_2O + OH^- \tag{2.45}$$

2.5.2 MECHANISM OF CORROSION INHIBITION

Heterocyclic molecule-based corrosion inhibitors in the alkaline media for the protection of a metal/alloy have been extensively studied. Two pyridine-3-carbonitriles containing pyridine(I) and thiophene(II) rings were explored for mild steel in 1 M NaOH employing PDP, EIS, and hydrogen evolution techniques [111]. A greater protection efficiency was obtained for I, which was attributed to the availability of a greater electron density in I. A series of azoles were evaluated for Cu in pH 6.4, 8.4, and 10.4 sodium borate buffer solutions using electroanalytical techniques [118]. The inhibitors showed a mixed-type nature, as studied by PDP, and a capacitive control of the electrochemical process, as shown by the EIS results. Langmuir isotherm provided a suitable explanation to the inhibitor adsorption. Mercaptobenzothiazole (MBT) provided the best performance of 95% among all. The results of DFT-based computational analysis supported the experimentally obtained results. For 2-mercaptobenzothiazole (MBT), a notable decrease in both anodic and cathodic currents in comparison to the blank was observed [110]. The N and S atoms of the thiazole ring and the S atom of the thiol group (−SH) of MBT could allow three possible sites for coordination with Cu (Figure 2.11). MBT molecule is purported to present in a thione form in acid solution, and in a thiol form in an alkaline environment. Additionally, the surface of Cu bears a considerable affinity toward S atom. Azoles form a polymeric film on the Cu surface that can afford a strong protection. In the alkaline environment, Cu surface is covered with Cu_2O layer that could allow Cu^+ ions for interaction with the azoles. The chemical adsorption as in the case of benzotriazole (BTA) can form a polymeric film with the Cu ions as follows [110]:

$$n(BTA - H)_{ads} + nCu \rightleftarrows [Cu(BTA)]_n + nH^+ + ne^- \tag{2.46}$$

A rise in the pH of the environment favors the inhibitor adsorption and film formation of a corrosion inhibitor film occurs according to the above equilibrium.

Purine was analyzed for the copper surface in neutral (pH 7) and alkaline (pH 9) 0.5 M sodium sulfate via OCP, PDP, and chronoamperometric measurements [119]. Excellent adsorption performance was observed for purine in both environments.

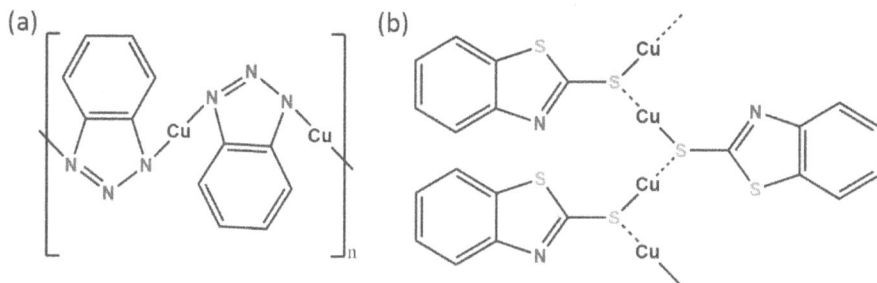

FIGURE 2.11 Structures of the polymeric complexes of (a) Cu-BTA and (b) Cu-MBT deposited in 0.1 M NaOH. (Reproduced from Ref. [110]. © Elsevier.)

Within the investigated pH range, purine exists in the neutral form. The purine adsorption on Cu substrate can be explained as follows:

$$4Cu + 4Pu_{ads} + O_2 \leftrightarrow 4Cu(I)Pu + 2H_2O \tag{2.47}$$

A triazole containing ethyl, $-SH$, and $-NH_2$ groups (EAMT), was studied for Al 6061 alloy in 0.5 M NaOH employing electrochemical techniques [120]. Adsorption of EAMT was in agreement with Langmuir isotherm at a varied temperature range from 30 °C to 60 °C and showed a mixed behavior. The protection performance increased with an increase in inhibitor dose, and decreased with an increase in temperature. In another study, a triazole containing methyl, $-SH$, and $-NH_2$ groups (MAMT) was investigated for the Al [121]. The synergistic corrosion inhibition effect of Zn^{2+} on the protection behavior of carboxymethylcellulose (CMC) was evaluated on Al surface in well water (pH 11) using the gravimetric, electrochemical, and surface analytical techniques. A complex formation takes place between CMC and Zn^{2+} ions in the medium. When Al surface is immersed, the diffusion of CMC-Zn^{2+} complex takes place from the solution bulk electrolyte to the Al substrate. A subsequent conversion to CMC-Al^{3+} complex takes place thereby releasing Zn^{2+} as follows [122]:

$$CMC\text{-}Zn^{2+} + Al^{3+} \rightarrow CMC\text{-}Al^{3+} + Zn^{2+} \tag{2.48}$$

$$Zn^{2+} + OH^- \rightarrow Zn(OH)_2 \tag{2.49}$$

Water-soluble polysaccharides alginates (ALG) and pectates (PEC) were evaluated as inhibitors for Al surface in NaOH medium [123]. These differ in the $-OH$ group position at C2 and C3, respectively [124]. At a 0.8% concentration, respective protection efficiencies of 68.93% and 78.49% were obtained from ALG and PEC. The greater extent of protection for PEC was attributed to the varying geometrical arrangement of the functional groups. The two polyelectrolytes underwent deprotonation in the alkaline environment to afford negatively charged alkoxides (Figure 2.12). The charged polymers formed a protective film on the metallic substrate that mitigated the corrosion and dissolution [123]. Figure 2.13 shows some heterocyclic corrosion inhibitors' chemical structures and corrosion inhibition performance for an alkaline environment.

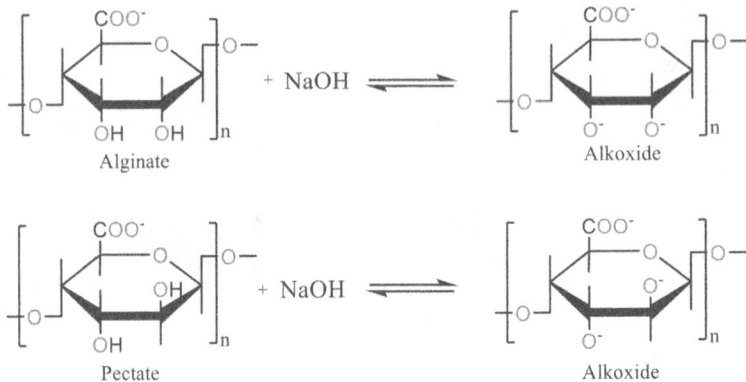

FIGURE 2.12 Formation of alkoxides from alginate and pectate due to ionization in NaOH. (Reproduced from Ref. [123]. © Portuguese Electrochemical Society.)

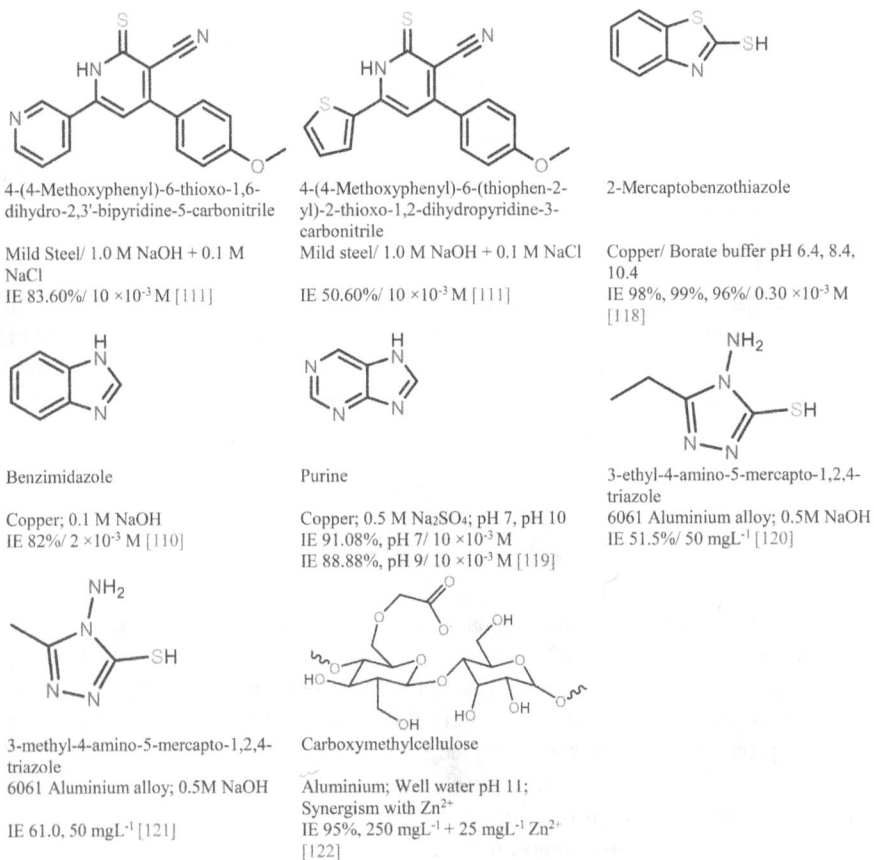

Alginate

Alkoxide

Pectate

Alkoxide

4-(4-Methoxyphenyl)-6-thioxo-1,6-dihydro-2,3'-bipyridine-5-carbonitrile

Mild Steel/ 1.0 M NaOH + 0.1 M NaCl
IE 83.60%/ 10 ×10⁻³ M [111]

4-(4-Methoxyphenyl)-6-(thiophen-2-yl)-2-thioxo-1,2-dihydropyridine-3-carbonitrile

Mild steel/ 1.0 M NaOH + 0.1 M NaCl

IE 50.60%/ 10 ×10⁻³ M [111]

2-Mercaptobenzothiazole

Copper/ Borate buffer pH 6.4, 8.4, 10.4
IE 98%, 99%, 96%/ 0.30 ×10⁻³ M [118]

Benzimidazole

Copper; 0.1 M NaOH
IE 82%/ 2 ×10⁻³ M [110]

Purine

Copper; 0.5 M Na₂SO₄; pH 7, pH 10
IE 91.08%, pH 7/ 10 ×10⁻³ M
IE 88.88%, pH 9/ 10 ×10⁻³ M [119]

3-ethyl-4-amino-5-mercapto-1,2,4-triazole
6061 Aluminium alloy; 0.5M NaOH
IE 51.5%/ 50 mgL⁻¹ [120]

3-methyl-4-amino-5-mercapto-1,2,4-triazole
6061 Aluminium alloy; 0.5M NaOH

IE 61.0, 50 mgL⁻¹ [121]

Carboxymethylcellulose

Aluminium; Well water pH 11;
Synergism with Zn²⁺
IE 95%, 250 mgL⁻¹ + 25 mgL⁻¹ Zn²⁺ [122]

FIGURE 2.13 Heterocyclic molecules reported as corrosion inhibitors in alkaline media.

2.6   CONCLUSIONS

The heterocyclic compounds refer to a vast variety of organic compounds, e.g., amino acids, drugs, natural extracts, proteins, vitamins, carbohydrates, etc. Many such molecules constitute various biomolecules. The presence of π-bonds, heteroatoms, heterocycles, and the functional groups containing these heteroatoms promote the adsorption and protective performance of these molecules on metals and alloys. In literature, several research articles appeared on the use of heterocyclic molecules as inhibitors against corrosion of various metals and alloys. Most of these organic molecules show a mixed type of inhibition behavior, with both physical and chemical adsorption. Several heterocycles form strong chemical adsorption, e.g., S-containing heterocycles, thiol SAMS, etc. on Cu surfaces, and some N-heterocycles on steel and other metallic surfaces. The organic heterocycles have been reported as inhibitors in diverse corrosive media, viz., concentrated acids, saline environments, CO_2 environments, alkaline media, etc. The effect of additives such as synergistic agents and surfactants have been studied on these molecules. Detailed thermodynamic and kinetic indices have been elucidated to explain the adsorption of the heterocyclic molecules on metallic materials in corrosive environments. Comprehensive computational data is available correlating the structures of these organic molecules with the protection performance.

2.7   PROSPECTS

Organic heterocycles have emerged in the literature as potential replacements to conventionally used toxic inorganic corrosion inhibitors. Some of the heterocycles form the framework of several biomolecules and hence are naturally green and biocompatible. It is desirable to evaluate the detailed toxicity, biodegradation, and bioaccumulation parameters of these heterocycles to gain information of their commercial applicability. In addition, a major issue associated with organic corrosion inhibitors is solubility. In this context, more studies are required to facilitate solubility using cosolvents and surfactants. Further, the prospects of synergistic action in corrosion inhibition using halide ions and other organic molecules need to be explored. Surfactant-containing and ionic liquid derivatives of heterocycles need to be studied in more detail. Also, it is apparent that a scarce amount of literature is available on heterocyclic molecule-based inhibitors in sweet media and in oil-well acidification. Therefore, the mechanistic details of the impact of corrosion products, high temperature, high pressure, hydrodynamic conditions, etc. need to be investigated.

REFERENCES

1. P. Dohare, M.A. Quraishi, C. Verma, H. Lgaz, R. Salghi, E.E. Ebenso, Results in Physics, 13 (2019) 102344.
2. J. Haque, K.R. Ansari, V. Srivastava, M.A. Quraishi, I.B. Obot, Journal of Industrial and Engineering Chemistry, 49 (2017) 176–188.
3. S.-H. Yoo, Y.-W. Kim, K. Chung, N.-K. Kim, J.-S. Kim, Industrial & Engineering Chemistry Research, 52 (2013) 10880–10889.

4. M. Abdallah, H. Al-Tass, B.A. Jahdaly, A. Fouda, Journal of Molecular Liquids, 216 (2016) 590–597.
5. D.S. Chauhan, M.A. Quraishi, C. Carrière, A. Seyeux, P. Marcus, A. Singh, Journal of Molecular Liquids, 289 (2019) 111113.
6. A. Suhasaria, M. Murmu, S. Satpati, P. Banerjee, D. Sukul, Journal of Molecular Liquids, 313 (2020) 113537.
7. Y. Qiang, H. Li, X. Lan, Journal of Materials Science & Technology, 52 (2020) 63–71.
8. R. Zhang, Y. Qin, L. Zhang, S. Luo, Organic Letters, 19 (2017) 5629–5632.
9. M. Lebrini, M. Lagrenée, M. Traisnel, L. Gengembre, H. Vezin, F. Bentiss, Applied Surface Science, 253 (2007) 9267–9276.
10. M.A. Quraishi, D.S. Chauhan, V.S. Saji, Journal of Molecular Liquids, 341 (2021) 117265.
11. I.B. Onyeachu, D.S. Chauhan, M.A. Quraishi, I.B. Obot, A. Singh, Journal of Adhesion Science and Technology, 36 (2022) 1858–1882.
12. D.S. Chauhan, M. Quraishi, A.A. Sorour, C. Verma, Journal of Petroleum Science and Engineering, 215 (2022) 110695.
13. M. Salman, V. Srivastava, M. Quraishi, D.S. Chauhan, K. Ansari, J. Haque, Russian Journal of Electrochemistry, 57 (2021) 228–244.
14. L.K. Goni, M.A.J. Mazumder, S.A. Ali, D.S. Chauhan, Water-Soluble Polymeric Corrosion Inhibitors, in: M.J. Mazumder, M. Quraishi, A. Al-Ahmed (Eds.) Polymeric Corrosion Inhibitors for Greening the Chemical and Petrochemical Industry, Wiley-VCH GmbH, 2023, pp. 97–123.
15. D.S. Chauhan, V. Srivastava, Y. Lin, M.A. Quraishi, Polymers as Corrosion Inhibitors for Sweet Environment, in: M.J. Mazumder, M. Quraishi, A. Al-Ahmed (Eds.) Polymeric Corrosion Inhibitors for Greening the Chemical and Petrochemical Industry, Wiley-VCH GmbH, 2023, pp. 193–220.
16. D.S. Chauhan, M.A. Quraishi, H. Al-Qahtani, M.A.J. Mazumder, Green Polymeric Corrosion Inhibitors: Design, Synthesis, and Characterization, in: M.J. Mazumder, M. Quraishi, A. Al-Ahmed (Eds.) Polymeric Corrosion Inhibitors for Greening the Chemical Petrochemical Industry, Wiley-VCH GmbH, 2023, pp. 1–22.
17. M. Quraishi, D.S. Chauhan, Recent Trends in the Development of Corrosion Inhibitors, in: U.K. Mudali, T.S. Rao, S. Ningshen, P.R. G, R.P. George, T.M. Sridhar (Eds.) A Treatise on Corrosion Science, Engineering and Technology, Springer, 2022, pp. 783–799.
18. M. Quraishi, D.S. Chauhan, Environmentally Sustainable Corrosion Inhibitors in Oil and Gas Industry, in: C.B. Verma, C.M. Hussain, E.E. Ebenso (Eds.) Organic Corrosion Inhibitors: Synthesis, Characterization, Mechanism, and Applications, John Wiley & Sons, Inc., 2022, pp. 221–240.
19. D.S. Chauhan, F. El-Hajjaji, M. Quraishi, Heterocyclic Ionic Liquids as Environmentally Benign Corrosion Inhibitors: Recent Advances and Future Perspectives, in: Ionic Liquid-Based Technologies for Environmental Sustainability, Elsevier, 2022, pp. 279–294.
20. M. Quraishi, D.S. Chauhan, Drugs as Environmentally Sustainable Corrosion Inhibitors, in: C.M. Hussain, C.B. Verma (Eds.) Sustainable Corrosion Inhibitors II: Synthesis, Design, and Practical Applications, ACS Publications, 2021, pp. 1–17.
21. F. Ansari, D.S. Chauhan, M. Quraishi, Oleochemicals as Corrosion Inhibitors, in: C. Verma, C.M. Hussain (Eds.) Organic Corrosion Inhibitors: Synthesis, Characterization, Mechanism, and Applications, John Wiley & Sons, Inc., 2021, p. 343.
22. A. Singh, K.R. Ansari, D.S. Chauhan, M.A. Quraishi, S. Kaya, H. Yu, Y. Lin, Corrosion Mitigation by Planar Benzimidazole Derivatives, in: Corrosion, IntechOpen, 2020.

23. K.R. Ansari, D.S. Chauhan, A. Singh, V.S. Saji, M.A. Quraishi, Corrosion Inhibitors for Acidizing Process in Oil and Gas Sectors, in: V.S. Saji, S.A. Umoren (Eds.) Corrosion Inhibitors in the Oil and Gas Industry, Wiley-VCH Verlag GmbH & Co. KGaA, 2020, pp. 153–176. ISBN 978-3-527-34618-9.

24. M. Maanonen, Helsinki Metropolia University of Applied Sciences, 2014, pp. 1–32. https://www.theseus.fi/bitstream/handle/10024/70713/Steel_Pickling_in_Challenging_Conditions_2014_Thesis_Mika_Maanonen.pdf?sequence=1&isAllowed=y

25. A.I.H. Committee, ASM Handbook, ASM International, 1990.

26. V.S. Sastri, Corrosion Inhibitors: Principles and Applications, John Wiley & Sons, Inc., 1998.

27. C. Smith, F. Dollarhide, N.J. Byth, Journal of Petroleum Technology, 30 (1978) 737–746.

28. L. Kalfayan, Production Enhancement with Acid Stimulation, Pennwell Books, 2008.

29. M. Finšgar, J. Jackson, Corrosion Science, 86 (2014) 17–41.

30. R.S. Schechter, Oil well stimulation. United States; Web. (1992).

31. G. Schmitt, British Corrosion Journal, 19 (1984) 165–176.

32. M.A. Quraishi, K.R. Ansari, D.S. Chauhan, Google Patents, Saudi Arabia, 2022, pp. 1–30.

33. D.K. Yadav, M.A. Quraishi, Industrial & Engineering Chemistry Research, 51 (2012) 8194–8210.

34. P. Mourya, S. Banerjee, R.B. Rastogi, M.M. Singh, Industrial & Engineering Chemistry Research, 52 (2013) 12733–12747.

35. M. Behpour, S. Ghoreishi, N. Soltani, M. Salavati-Niasari, M. Hamadanian, A, Gandomi, Corrosion Science, 50 (2008) 2172–2181.

36. P. Dohare, D.S. Chauhan, A.A. Sorour, M.A. Quraishi, Materials Discovery, 9 (2017) 30–41.

37. M. Yadav, L. Gope, N. Kumari, P. Yadav, Journal of Molecular Liquids, 216 (2016) 78–86.

38. I. Obot, N. Obi-Egbedi, A. Eseola, Industrial & Engineering Chemistry Research, 50 (2011) 2098–2110.

39. P. Okafor, M.E. Ikpi, I. Uwah, E. Ebenso, U. Ekpe, S. Umoren, Corrosion Science, 50 (2008) 2310–2317.

40. K.R. Ansari, M.A. Quraishi, A. Singh, Corrosion Science, 79 (2014) 5–15.

41. I.B. Onyeachu, D.S. Chauhan, K.R. Ansari, I. Obot, M.A. Quraishi, A.H. Alamri, New Journal of Chemistry, 43 (2019) 7282–7293.

42. I. Obot, I.B. Onyeachu, Journal of Molecular Liquids, 249 (2018) 83–96.

43. M. Bedair, M. El-Sabbah, A. Fouda, H. Elaryian, Corrosion Science, 128 (2017) 54–72.

44. M.A. Quraishi, D.S. Chauhan, V.S. Saji, Heterocyclic Organic Corrosion Inhibitors: Principles and Applications, Elsevier Inc., Amsterdam, 2020.

45. S.K. Shukla, M.A. Quraishi, Corrosion Science, 51 (2009) 1007–1011.

46. S.K. Shukla, M.A. Quraishi, Materials Chemistry and Physics, 120 (2010) 142–147.

47. S.K. Shukla, M.A. Quraishi, Journal of Applied Electrochemistry, 39 (2009) 1517–1523.

48. I.B. Onyeachu, I.B. Obot, A.H. Alamri, C.A. Eziukwu, The European Physical Journal Plus, 135 (2020) 129.

49. I. Ahamad, M. Quraishi, Corrosion Science, 51 (2009) 2006–2013.

50. A.K. Singh, M. Quraishi, Corrosion Science, 52 (2010) 1529–1535.

51. A.K. Singh, S. Khan, A. Singh, S. Quraishi, M.A. Quraishi, E.E. Ebenso, Research on Chemical Intermediates, 39 (2013) 1191–1208.

52. C. Verma, M.A. Quraishi, D.S. Chauhan, Green Corrosion Inhibition: Fundamentals, Design, Synthesis and Applications, Royal Society of Chemistry, 2022.

53. C. Verma, D.S. Chauhan, M.A. Quraishi, Journal of Materials and Environmental Science, 8 (2017) 4040–4051.
54. D.S. Chauhan, A.A. Sorour, M.A. Quraishi, International Journal of Chemistry and Pharmaceutical Sciences, 4 (2016) 680–691.
55. P. Singh, D.S. Chauhan, S.S. Chauhan, G. Singh, M.A. Quraishi, Journal of Molecular Liquids, 286 (2019) 110903.
56. P. Singh, D.S. Chauhan, K. Srivastava, V. Srivastava, M.A. Quraishi, International Journal of Industrial Chemistry, 8 (2017) 363–372.
57. P. Dohare, D.S. Chauhan, B. Hammouti, M.A. Quraishi, Analytical and Bioanalytical Electrochemistry, 9 (2017) 762.
58. P. Dohare, D.S. Chauhan, M.A. Quraishi, International Journal of Corrosion and Scale Inhibition, 7 (2018) 25–37.
59. L. Moiseeva, Protection of Metals, 41 (2005) 76–83.
60. A. Singh, K. Ansari, X. Xu, Z. Sun, A. Kumar, Y. Lin, Scientific Reports, 7 (2017) 14904.
61. A. Singh, Y. Lin, I. Obot, E.E. Ebenso, K. Ansari, M.A. Quraishi, Applied Surface Science, 356 (2015) 341–347.
62. D.S. Chauhan, K.R. Ansari, A.A. Sorour, M.A. Quraishi, H. Lgaz, R. Salghi, International Journal of Biological Macromolecules, 107 (2018) 1747–1757.
63. J. Haque, V. Srivastava, S. Chauhan, H. Lgaz, M.A. Quraishi, ACS Omega, 3 (2018) 5654–5668.
64. V. Srivastava, D.S. Chauhan, P.G. Joshi, V. Maruthapandian, A.A. Sorour, M.A. Quraishi, ChemistrySelect, 3 (2018) 1990–1998.
65. D.K. Yadav, D.S. Chauhan, I. Ahamad, M.A. Quraishi, RSC Advances, 3 (2013) 632–646.
66. N. Baig, D.S. Chauhan, T.A. Saleh, M.A. Quraishi, New Journal of Chemistry, 43 (2019) 2328–2337.
67. B.J. Usman, S.A. Ali, Arabian Journal for Science and Engineering, 43 (2018) 1–22.
68. M.A. Mazumder, H.A. Al-Muallem, S.A. Ali, Corrosion Science, 90 (2015) 54–68.
69. M.A. Mazumder, H.A. Al-Muallem, M. Faiz, S.A. Ali, Corrosion Science, 87 (2014) 187–198.
70. S. Nešić, Corrosion Science, 49 (2007) 4308–4338.
71. D.S. Chauhan, M. Quraishi, A. Qurashi, Journal of Molecular Liquids, 326 (2021) 115117.
72. M. Askari, M. Aliofkhazraei, S. Ghaffari, A. Hajizadeh, Journal of Natural Gas Science and Engineering, 58 (2018) 92–114.
73. A. Edwards, C. Osborne, S. Webster, D. Klenerman, M. Joseph, P. Ostovar, M. Doyle, Corrosion Science, 36 (1994) 315–325.
74. X. Guan, Y. Hu, Expert Opinion on Therapeutic Patents, 22 (2012) 1353–1365.
75. R.A. Jaal, M.C. Ismail, B. Ariwahjoedi, MATEC Web of Conferences, EDP Sciences, 2014, p. 05012.
76. X. Liu, P. Okafor, Y. Zheng, Corrosion Science, 51 (2009) 744–751.
77. V. Jovancicevic, S. Ramachandran, P. Prince, Corrosion, 55 (1999) 449–455.
78. X. Liu, S. Chen, H. Ma, G. Liu, L. Shen, Applied Surface Science, 253 (2006) 814–820.
79. A. Singh, K.R. Ansari, Y. Lin, M.A. Quraishi, H. Lgaz, I.-M. Chung, Journal of the Taiwan Institute of Chemical Engineers, 95 (2019) 341–356.
80. A. Singh, K. Ansari, A. Kumar, W. Liu, C. Songsong, Y. Lin, Journal of Alloys and Compounds, 712 (2017) 121–133.
81. D.S. Chauhan, C. Verma, M. Quraishi, Journal of Molecular Structure, 1227 (2021) 129374.
82. C. Verma, L. Olasunkanmi, E.E. Ebenso, M.A. Quraishi, Journal of Molecular Liquids, 251 (2018) 100–118.

83. A. Singh, Y. Lin, K. Ansari, M.A. Quraishi, E.E. Ebenso, S. Chen, W. Liu, Applied Surface Science, 359 (2015) 331–339.
84. A. Singh, M. Talha, X. Xu, Z. Sun, Y. Lin, ACS Omega, 2 (2017) 8177–8186.
85. I.B. Onyeachu, D.S. Chauhan, M. Quraishi, I. Obot, Corrosion Engineering, Science and Technology, 56 (2020) 154–161.
86. M. Finšgar, I. Milošev, Corrosion Science, 52 (2010) 2737–2749.
87. G. Kear, B. Barker, F. Walsh, Corrosion Science, 46 (2004) 109–135.
88. F. King, C. Litke, M. Quinn, D. LeNeveu, Corrosion Science, 37 (1995) 833–851.
89. S. Li, M.T. Teague, G.L. Doll, E.J. Schindelholz, H. Cong, Corrosion Science, 141 (2018) 243–254.
90. G. Bengough, R. Jones, R. Pirret, Journal of the Institute of Metals, 23 (1920) 65–158.
91. C. Verma, E.E. Ebenso, M.A. Quraishi, Journal of Molecular Liquids, 248 (2017) 927–942.
92. V.S. Sastri, Green Corrosion Inhibitors: Theory and Practice, John Wiley & Sons, Inc., 2012.
93. M. Scendo, Corrosion Science, 49 (2007) 3953–3968.
94. N. Huynh, S. Bottle, T. Notoya, D. Schweinsberg, Corrosion Science, 42 (2000) 259–274.
95. R. Vera, F. Bastidas, M. Villarroel, A. Oliva, A. Molinari, D. Ramírez, R. del Río, Corrosion Science, 50 (2008) 729–736.
96. M. Finšgar, Corrosion Science, 72 (2013) 90–98.
97. R.G. Sundaram, G. Vengatesh, M. Sundaravadivelu, Surfaces and Interfaces, 22 (2021) 100841.
98. Y. Qiang, S. Fu, S. Zhang, S. Chen, X. Zou, Corrosion Science, 140 (2018) 111–121.
99. L. Lakhrissi, B. Lakhrissi, R. Touir, M.E. Touhami, M. Massoui, E.M. Essassi, Arabian Journal of Chemistry, 10 (2017) S3142–S3149.
100. H.O. Curkovic, E. Stupnisek-Lisac, H. Takenouti, Corrosion Science, 52 (2010) 398–405.
101. M.B.P. Mihajlović, M.B. Radovanović, Ž.Z. Tasić, M.M. Antonijević, Journal of Molecular Liquids, 225 (2017) 127–136.
102. E.-S.M. Sherif, A. El Shamy, M.M. Ramla, A.O. El Nazhawy, Materials Chemistry and Physics, 102 (2007) 231–239.
103. K.R. Ansari, M.A. Quraishi, Analytical and Bioanalytical Electrochemistry, 8 (2016) 136–144.
104. R. Padash, E. Jamalizadeh, A.H. Jafari, Anti-Corrosion Methods and Materials, 64 (2017) 550–554.
105. M. Abdallah, O. Hazazi, A. Fawzy, S. El-Shafei, A. Fouda, Protection of Metals and Physical Chemistry of Surfaces, 50 (2014) 659–666.
106. N. Chaubey, V.K. Singh, M.A. Quraishi, International Journal of Industrial Chemistry, 8 (2017) 75–82.
107. J. Dobryszycki, S. Biallozor, Corrosion Science, 43 (2001) 1309–1319.
108. R. Subramanian, V. Lakshminarayanan, Corrosion Science, 44 (2002) 535–554.
109. M. Ameer, A. Fekry, International Journal of Hydrogen Energy, 35 (2010) 11387–11396.
110. M. Saremi, E. Mahallati, Cement and Concrete Research, 32 (2002) 1915–1921.
111. A. Saraby-Reintjes, Electrochimica Acta, 30 (1985) 387–401.
112. A. Saraby-Reintjes, Electrochimica Acta, 30 (1985) 403–417.
113. J.-D. Kim, S.-I. Pyun, Corrosion Science, 38 (1996) 1093–1102.
114. M. Bhardwaj, R. Balasubramaniam, International Journal of Hydrogen Energy, 33 (2008) 2178–2188.
115. A. Fekry, M. Ameer, International Journal of Hydrogen Energy, 35 (2010) 7641–7651.
116. F. Altaf, R. Qureshi, S. Ahmed, Journal of Electroanalytical Chemistry, 659 (2011) 134–142.

117. M. Petrović, A. Simonović, M. Radovanović, S. Milić, M. Antonijević, Chemical Papers, 66 (2012) 664–676.
118. P. Kumari, J. Nayak, A.N. Shetty, Portugaliae Electrochimica Acta, 29 (2011) 445–462.
119. P.R. Kumari, J. Nayak, A.N. Shetty, Journal of Coatings Technology and Research, 8 (2011) 685.
120. R. Kalaivani, P.T. Arasu, S. Rajendran, Chemical Science Transactions, 2 (2013) 1352–1357.
121. I. Zaafarany, Portugaliae Electrochimica Acta, 30 (2012) 419–426.
122. R.A. Muzzarelli, Natural Chelating Polymers; Alginic Acid, Chitin and Chitosan, in: Natural Chelating Polymers; Alginic Acid, Chitin and Chitosan, Pergamon Press, 1973.

3 Quercetin- and Carbazole-based Corrosion Inhibitors

Ichraq Bouhouche[1], Khalid Bouiti[1], Nabil Lahrache[1], Najoua Labjar[1], Ghita Amine Benabdallah[2], and Souad El Hajjaji[2]
[1]Laboratory of Molecular Spectroscopy Modelling, Materials, Nanomaterials, Water and Environment, CERNE2D, ENSAM, Mohammed V University in Rabat, Morocco
[2]Laboratory of Molecular Spectroscopy Modelling, Materials, Nanomaterials, Water and Environment, CERNE2D, Faculty of Sciences, Mohammed V University in Rabat, Morocco

3.1 INTRODUCTION

Corrosion is a multidisciplinary issue, and it affects different industries—chemical, petroleum, and energetic—causing production stoppages, replacement of components, accidents, and harmful effects on the environment [1, 2]. It evolves according to the corroding medium such as moisture, O_2, organic or inorganic acids, higher temperature and pressure, and the characteristics and composition of the materials [3], and it is a naturally occurring event tending to induce the deterioration of the materials [2]. By adopting appropriate corrosion prevention strategies, much of that damage could be avoidable. From the diverse approaches to corrosion containment, using inhibitors is the most straightforward, economical, and cost-effective strategy commonly used in industries [4].

The employment of inhibitors is crucial to delay the deterioration trend by adding a chemical compound at a low concentration [5, 6]; it is a more economical, more efficient, and more useful implementation approach [5]. Heterocyclic organic compounds contain a unique electron pair in the heteroatoms that are easily accessible for donation to the targeted metal, resulting in an efficient adsorption process of the inhibitor molecule [4, 6]. The quercetin molecule is found in various fruits such as strawberries, apples, and even red onions [7, 8]. It is the finest flavonoid molecule [9]. It is a functional molecule that has enormous potential in the therapeutic field owing to its antioxidant and anticancer properties [10, 11]. Carbazole is commonly utilized in solar cells [12]. It is also utilized as a corrosion inhibitor. The usage of carbazole and its variants has expanded due of its specific properties: cheap cost, low reactive ecological toxicity, thermal stability, and improved solubility in monomers [13–15].

DOI: 10.1201/9781003377016-3

3.2   CORROSION INHIBITION

3.2.1   Inhibitors Molecules

The inhibitors are chemical agents that minimize corrosion when introduced into a medium at low concentrations [16]. It must not only be stable in the presence of other characteristics of the environment but also not impair the stability of the species contained in the environment [17]. It must also decrease the corrosion process of the metal while keeping the physicochemical features of the latter. The choice of corrosion inhibitors for practical applications is based on the understanding of their mode of action [18]. There are various inhibitor classes depending on formulation, electrochemical mechanism of activity, and surface principle of action. Organic compounds have some development potential as corrosion inhibitors: their usage is currently recommended over inorganic inhibitors, largely owing to environmental concerns [19].

Organic inhibitors are frequently formed as petroleum industry by-products [20]. They comprise at least one active center capable of exchanging electrons with a metal such as nitrogen, oxygen, phosphorus, or sulfur. Mineral substances are more typically employed in alkaline conditions, and less often in acidic ones. The products dissolve in solution, and it is the results of their dissociation that ensure the inhibitory phenomenon [21]. Oxo anions of type XO_4^{n-}, such as chromates, molybdates, phosphates, and silicates, are the principal inhibitory anions [22]. Ca^{2+} and Zn^{2+} are the most frequent cations, as are those that form insoluble salts with anions such as hydroxyl OH^-. Various molecules in use today are decreasing [23], mainly due to the fact that most effective products are hazardous to the environment.

3.2.2   Adsorption of Corrosion Inhibitors

The inhibitor of corrosion generates an interlayer on the surface, altering electrochemical processes by blocking anodic or cathodic sites through adsorbing the molecules of inhibitor on the metal interface [24–26].

3.2.2.1   Chemisorption and Physisorption

The phenomenon of adsorption is one that occurs when the chemical interactions between the atoms that make up a surface are incomplete. This surface has a propensity to bridge this gap by ensnaring atoms and molecules from their surroundings [27]. The physisorption retains the identity of the adsorbed molecules and is based on three kinds of interacting forces: hydrogen bonds, polar force, and van der Waals force. The force of electrostatic adsorption is proportional to the difference between the charges carried by the inhibitor and the metal's surface, which is proportional to the difference between the metal's corrosion potential and its potential for zero charges in the corrosive medium under consideration [24, 28] as well as the physically adsorbed substances, which condense rapidly on the metal but are easily degraded by desorption when the temperature increases.

Chemisorption, on the other hand, is based on the pooling of electrons between the polar part of the molecule and the metallic surface [29], which leads to the formation of more stable chemical bonds based on higher bond energies, the electrons come mainly from the nonbonding doublets of the inhibiting molecules such as the heteroatoms. The adsorption is accompanied by a deep modification of the electronic charge distribution of the adsorbed molecules [21, 29, 30]. The important parameter is the density of electrons around the effective center, leading to the strengthening of covalent bonds between the donor atom and the metal atom. The same reason applies to cyclic amines, which are better inhibitors than aliphatic amines. Unsaturated organic compounds are electron carriers capable of creating bonds with metal atoms. The presence of an unsaturated bond can be favorable to the inhibitory efficiency of an organic molecule in an acidic environment since it can adsorb in the same way on either a positively or negatively charged surface [20, 31, 32].

3.2.2.2 Adsorption Isotherm

The behavior of adsorbed inhibitors may be interpreted by fitting the experimental data to an adsorption isotherm. Langmuir, Freundlich, Temkin, and Frumkin investigate the molecular interaction and heterogeneity components of the adsorption layer. Dhar–Flory–Huggins, Flory–Huggins, or Bockris–Swindels, on the other hand, investigate the mechanism of water flow and develop an approach to calculate the rate of water molecules passing through each inhibitor molecule [26, 33]. The most widely used model is the Langmuir isothermal adsorption model. A monolayer of adsorbate is expected to bind to a particular number of localized adsorption sites. There must be no lateral contacts or steric hindrances between the molecules. Furthermore, each adsorbent site was anticipated to have the same affinity for the adsorbate as well as the same activation energy and enthalpy of sorption [34, 35].

The following equations describe the isotherm of Langmuir adsorption:

$$\frac{C}{\theta} = \frac{1}{K_{ads}} + C$$

where C is the inhibitor concentration, K_{ads} is the adsorption equilibrium constant, and θ is the recovery rate.

The slope of the graph must equal unity for the adsorption of a given inhibitor to adequately match the Langmuir model and the assumptions upon which it is based. The corrosion scientist must determine if the produced adsorbed film is a monolayer or a multilayer, and whether interactions take place in the adsorbed layer. The closer the slope is to unity, within the limits of systematic or experimental error, the more probable the inhibitor obeys the Langmuir model. The measured slope sometimes deviates significantly from unity, indicating that the adsorption is not monolayer and that the Langmuir model cannot be utilized to examine inhibitor adsorption. The Langmuir–Freundlich isotherm [36–38] is an isotherm hybrid that may be used to solve problems. The Temkin concept is an isotherm that is commonly utilized in the discussion of corrosion inhibitor mechanism of action. Unlike the other models examined so far, it sheds light on the nature of the interactions that occur in the adsorbed layer. It is assumed that adsorption takes place at the most energetically advantageous sites [39, 40].

The adsorption free energy of the adsorbate is a linear function of the coverage rate in this model, and the chemical rate constants are functions of θ. On the surface, there is attraction or repulsion between adsorbed species.

$$\theta = \frac{1}{f} * \ln K_{ads} * C$$

where C is the concentration of the inhibitor, θ is the degree of surface coverage, K_{ads} is a constant, and f is the energy parameter.

The Frumkin isotherm states that the adsorbed molecules interact and affect the subsequent adsorption by repulsion or attraction of the molecules [41]. It is usually expressed in a linear form as follows:

$$\left(\frac{\theta}{1-\theta}\right)e^{-2a\theta} = K_{ads} * C$$

where a denotes the molecular interactions in the adsorbed adsorbate layer and the heterogeneous character of the surface and may take positive or negative values. Positive values of a indicate that the interactions between molecules are favorable in nature, resulting in an increase in adsorption energy [39, 42] with higher surface coverage θ, where K_{ads} is the adsorption process's equilibrium constant and C is the inhibitor concentration. Based on the Freundlich isotherm, the adsorption mechanism on a heterogeneous adsorbent surface is multilayered, and the adsorption sites have different degrees of adsorbate attraction [43]. The Freundlich isotherm model's linear form is as follows:

$$\log\theta = \log K + \frac{\log C}{n}$$

where K is the constant associate to adsorption ability of adsorbent and n is the adsorption intensities related to the heterogeneity of adsorbent surface.

A plot of $\log\theta$ against $\log C$ gives a straight line with slope $1/n$ and an intercept equal to $\log K$. Favorable adsorption is generally indicated by a Freundlich constant (n) between 1 and 10. $n = 1$ represents a correlated adsorption mechanism with uniform energy over the entire surface of the adsorbent, so the binding strength increases as the adsorbate becomes fixed [39, 43]. The corrosion inhibitor's active mechanism is based on adsorption to the surface to establish a protective layer; this process is not purely physical or chemical. It is controlled by the chemical composition of the inhibitor, the charge and character of the metal surface, the charge distribution of the molecules, and the medium [44, 45].

Heteroatoms are necessary for corrosion inhibitors, notably those possessing lone electron pairs and π electrons in the case of aromatic cycles, as well as various bonds that may interact with the metal's free orbital and induce adsorption. The adsorption free energy informs the type of adsorption according to the value until −20 kJ/mol relate to electrostatic interaction or physisorption [46], the value lower than −40 kJ/mol corresponding to chemical adsorption, the negativity indicates the spontaneity of this process [1, 2, 47]. Also, the enthalpy informs about the inhibition mechanism, the positive value of this parameter corresponds to endothermic chemisorption and the negative one is attributed to an exothermic physisorption process [3, 47].

3.3 CORROSION MEASUREMENT

3.3.1 Gasometric and Gravimetric

Due to the evident relationship between the rate of H_2 gas evolution and the rate of corrosion, monitoring the quantity of hydrogen gas released at cathodic sites may offer valuable information about the corrosion process [48]: The quantity of H_2 gas is released during metallic corrosion in an aggressive solution with and without inhibitors [49]. The losing weight analysis is the most straightforward, economical, and widely used technique for determining the corrosion inhibition effectiveness. Typically, the initial stage involves cutting the metal specimen to be examined in tiny sections of predetermined size, then degrease, clean, and polish the samples, which are then submerged in the corrosive medium and cleaned according to conventional ASTM procedures [50–52] after a predetermined time. The average weight loss is estimated by subtracting the mass before and after immersion.

3.3.2 Electro-chemical Impedance Spectroscopy (EIS)

EIS is a nondestructive method that provides information on an electrochemical system's frequency behavior, such as storage and dispersion of energy. The primary benefits of this method are its suitability for low-conductivity systems, the availability of mechanistic information, and the ability to assess solution resistance [53–55]. The collected electrochemical data are fitted to a suitable equivalent circuit using the nonlinear least squares approach to explain the complicated response [56]. Nyquist, Bode, and phase angle graphs are used to illustrate the data. The Nyquist plot is the most often used graphical representation data, and it gives a simple comprehension of the corrosive system's electrochemical behavior. It also facilitates the prediction of analogous circuit elements [55, 57].

The reaction of an electrochemical system to a modest amplitude of the disturbance frequency reveals information about the corrosion system's internal dynamics. The analogous circuit over the corroding metal in an aqueous medium is often a combination of resistance and capacity representing the corrosion surface [55, 58]. The charge transfer resistance R_{ct} is the parallel resistance that controls the rate of corrosion. The capacitance at the metal–electrolyte barrier is often approximated as double-layer capacitance [59–61]. A high value of R_{ct} in corrosion inhibition studies indicates the establishment of a barrier that must be removed to permit the transfer of charge on corrosion, signifying a decline in the rate of corrosion [62]. R_{ct} represents the disparity between the highest and lowest frequencies of the resistance's actual component.

3.3.3 POLARIZATION MEASUREMENTS

PDP is a disruptive process that alters the surface of the electrode. The E_{ocp} is retained to indicate the potential without a cumulative current, as indicated by measurements fitted to the potential [63, 64]. In an ideal circumstance, the E_{ocp} and E_{corr} values would be the same. Modifications of the surface of the electrode while scanning the potential may account for the apparent disparity between the two results. If the potential of a working electrode is shifted sufficiently further in the direction positive to E_{ocp}, the current from cathodic reduction becomes insignificant and the measured current response comprises the anode contribution [63]. However, with significant negative potentials, the cathode current will predominate the net current responding.

In corrosion inhibition experiments, it is essential to determine the optimal sweep rate, since this varies depending on the kind of metal environment and the type of corrosion inhibitor used. When the scan speed is too fast, the polarization curves vary, which may lead to a misinterpretation of the polarized electrode process due to charge disturbance and insufficient time to reach the steady state [65]. If the measurement is performed too slowly, the interfacial structure of the polarized electrode may change.

3.3.4 SURFACE ANALYSIS

SEM is the most frequently used technique for microscopic topography [66]. This technique exploits the notion of electron–matter interactions to produce high-resolution images of the researched material's surface [67]. As a consequence, a material-interacting electron beam examines the surface of the sample. This electron–matter interaction leads to the emission of particles and radiation. The surface's topography, microstructure, and chemical composition may be investigated using detectors that allow the gathering of the various transmitted signals [68].

Likewise, the AFM provides data on the geometry of the surface of the metal for comparison and topographic imaging [69–71]. Photoelectron X-ray spectroscopy is often employed to determine oxidizing conditions, electronic and coequilibrium states [72–75]. Additional characterizations of corrosion inhibitors are commonly undertaken using FTIR spectroscopy to provide functional group and vibration modeling information [76]. Functional groups, electrical transitions, and bandgaps may also be identified via UV-visible spectroscopy: bandgaps in electrical and optical systems [77].

3.3.5 THEORETICAL CALCULATION METHOD

Current computer technology enables researchers to use computational analyses, by simulating molecules, to generate or forecast the performance of new inhibitors against corrosion [78]. Molecular dynamics approaches have previously been shown to be very useful in determining the molecular structure of strong inhibitors as well as understanding their electrical structure and reactivity. Quantum chemistry simulations may reduce research expenditures in addition to precisely predicting corrosion inhibition at the molecular level [79, 80]. This technique often employs the combination of molecular dynamics and the theory of density function to examine the mechanism of inhibition at the molecular level [81]. The lowest occluded and higher vacant orbitals act as a crucial factor in establishing the effectiveness of inhibiting corrosion in the DFT modeling [82]. Gaussian can determine the HOMO and LUMO energy, energy gap, chemical absolute density, chemical smoothness, and altitude of a transmitted electron [83].

3.4 CARBAZOLE AND ITS DERIVATIVES FOR CORROSION INHIBITION

Sulfur and nitrogen heterocyclic compounds have caught the imagination of researchers. Many natural and manmade drugs, such as papaverine, theobromine, emetine, theophylline, atropine, codeine, reserpine, morphine, diazepam, chlorpromazine, barbiturates, and antipyrine, have a heterocyclic structure [84–89]. Heterocyclic aromatic compounds have increased solubility and water polarity as a single carbon component has been substituted with nitrogen, sulfur, or oxygen [90]. When compared to analogous polycyclic aromatic hydrocarbons, the chemical characteristics result in enhanced bioavailability and mobility, which has a range of environmental consequences on these molecules [91, 92].

Due to electron delocalization in the ring, planar, cyclic, and conjugated compounds with aromaticity behave as nonsaturated compounds and are undergoing substituting processes instead of adding reactions. It may be considered as an instance of cyclic delocalization and resonance [93]. The fundamental unsaturated ring structure is regarded to be especially stable based on the predisposition to favor replacement over addition. Conjugation is important for aromatization. This is related to the fact that flatness and overlapping p-orbitals are essential for aromatic compounds so that electrons may be delocalized for improved quality. Electron delocalization also creates resonance, which appears in numerous ways depending on the organization and structure of a molecule [91].

When this heteroatom is hybridized to a sp^2 state, a pair of electrons is put in a p-orbital that is parallel to the p-orbitals of the carbon, resulting in the development of the ring system (plane ring). There are six electrons in the π-system. The four electrons come from the heteroatom and two double bonds, for a total of four [91, 94]. The aromatic features of pyrrole are owing to the six-electron planar ring produced by these five sp^2 hybridized atoms. The chemical manufacture of carbazoles has remained of interest, especially if they display odd patterns of substitutions and/or merged with other cyclic structures, potentially requiring the creation of an original synthetic approach to access them [95]. The chemical structure of carbazole is given in Figure 3.1. Since the chemistry of carbazoles permits access to widespread structures whose optical and chemical properties are easily tuned through structural modifications, carbazole-containing fragments are a sought-after constituent of improved materials designed for thermal and optical applications, such as light-emitting liquid crystals, light-emitting dyes, and conducting polymers [96–98].

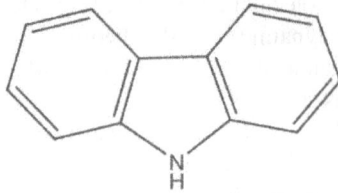

FIGURE 3.1   Chemical structure of carbazole.

As a result of its structure, carbazole has a high corrosion-inhibiting ability. As corrosion inhibitors, carbazole and its derivatives have been the focus of several studies. Çakmakcı et al. synthesized a poly(carbazole-*co*-pyrrole) copolymer via cyclic voltammetry in an acetonitrile solution containing tetrabutylammonium perchlorate and TiO_2-coated 304 stainless steel. The corrosion resistance performance was analyzed by OCP curves and PDP and EIS techniques. Corrosion experiments demonstrated that the composite layer increased the resistance of stainless steel to corrosion for a period of immersion of 50 days in 0.1 M HCl medium [99].

Ates and Zylmaz electrochemically and chemically produced polycarbazole/nano clay and Zn nanocomposites over a SS electrode. The electrodes underwent evaluation via electrochemical techniques, infrared Fourier transform spectroscopy, mass spectrometry, reduced transmit reflectance, SEM, and XRD. The corrosion resistance of the modified coatings on SS304 was examined using OCP monitoring, PDP, and EIS analysis. The films have higher corrosion resistance than both chemical and electrochemical films [100].

Nwankwo et al. explored the inhibiting effects of five carbazole derivatives against *Desulfovibrio vulgaris* in 1.0 M HCl environment by gravimetric method, PDP, and EIS methods. Carbazole compounds proved as mix inhibitors, with mild steel exhibiting mostly cathodic inhibition in 1.0 M HCl. According to the interface-morphed data, in an aqueous acidic medium and a culture of sulfate-reducing bacteria, the chemicals created an adsorbed coating on the surface of the mild steel. Quantum chemistry computations have been done to develop the molecular interpretations of the inhibitory properties of the substances. The molecules' interactions with the surface of the mild steel were modeled via a molecular dynamic modeling approximation, with the Fe(110) crystalline layer acting as a typical metal interface [101].

In addition, electrochemistry and gravimetry were used to determine the inhibitory corrosion capacities of 3(9*H*-carbazol-9-yl)-1,2-propanediol and 3,6-dibromo-9-phenylcarbazole for a 1.0 M HCl concentration and microbially affected mild steel corrosion. As catholically active mixture-type inhibitors, the inhibitors used prevent corrosion of mild steel in 1 M HCl medium by potentiodynamic polarization. The quantum chemical simulations reveal that the carbazole ring is the most crucial chemical component of the studied molecules that exchange acceptor–donor atoms with the Fe atom in mild steel. According to MC simulations, the rings of carbazole in inhibiting molecules approach Fe(110) in a nearly planar attitude, with estimated −114.24 kcal/mol for DBPCZ and −119.85 kcal/mol for CZPD adsorption energies [14].

The effects of three carbazole derivatives as anticorrosive additives for abiotic mild steel, as well as gravimetric method and electrochemical analysis in 1.0 M HCl, the components prevented corroding of the stainless steel. PDP data showed that the carbazoles exhibit cathodic and inhibitory characteristics. Scanning electron microscopy demonstrated that carbazoles produced a protective layer on stainless steel in 1 M HCl and SRB medium.

The compounds' aromatic–electronic moieties and heteroatoms reacted chemically with mild steel, based on FTIR spectra. The strongest interactions between O and N atoms and mild steel were discovered by simulations of quantum chemistry.

There was good agreement between experimental and estimated inhibition effectiveness. MC simulations show that molecules of carbazole bind aggressively to steel and remove water from metal surfaces [102]. The study by Duran et al. includes electropolymerizing and depositing poly(N-methylcarbazole) within tetrabutylammonium perchlorate bearing acetonitrile on 304 stainless sheets of steel. Using open-circuit potential, EIS, and PDP, the evaluation of the corrosion behavior of the steel electrodes coated with the polymer in 1.0 M solution of H_2SO_4 was examined. The coating shields the substrate anodically and considerably reduces the acidic steel corrosion rate [103].

3.5 QUERCETIN AND ITS DERIVATIVES FOR CORROSION INHIBITION

In nature, quercetin (Figure 3.2) is present in a diverse array of plants, including apples, berries, brassica vegetables, capers, grapes, onions, tea, and tomatoes. It is also present in a vast range of seeds, nuts, flowers, barks, and leaves [104, 105]. Its chemical structural formula is shown in Figure 3.2 and its molecular formula is $C_{15}H_{10}O_7$. It is an organic polar auxin transport inhibitor. The molecule of quercetin contains a ketocarbonyl group, and the first carbon's oxygen atom is based on and capable of generating salts with potent acids. Quercetin includes four effective groups in its molecular structure, and the presence of double bonds and a phenolic hydroxyl group boosts its antioxidant activity [106, 107]. It also exhibits antibacterial properties and efficacy in reducing the development of biofilms by blocking the growth of associated genes, antitumor properties, and antiangiogenic effects. Moreover, it has a major impact on the reduction of mycotoxins, protecting cells from damage [108].

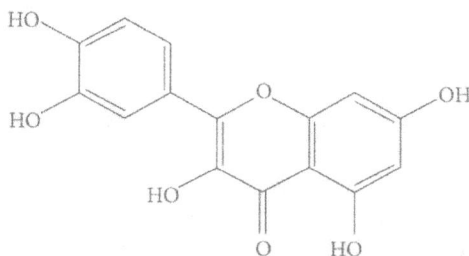

FIGURE 3.2 Chemical structure of quercetin.

Quercetin is mainly present as a combination with alcohols, phenolic acids, sugars, etc. Quercetin derivatives undergo hydrolysis within the gastrointestinal system after intake and are then incorporated and metabolized. Consequently, the quantity and formulation of each of its derivatives in food are important in terms of its bioavailability as a glycone [109, 110]. The major classes of derivatives of quercetin are the ethers and glycosides, and the phenyl and sulfate constituents, which are the least common [111]. *O*-Glycosides are derived from quercetin and contain a glycosidic linkage which is widespread in the phytochemical plant kingdom, all plants contain compounds of these groups, such as onions contain large amounts of these substances in a diversity of forms [109, 112].

The most common glycosylation site for quercetin is the OH group on the C-3 carbon. The 3-*O*-glycosides of quercetin exist as a monosaccharide with galactose, glucose, xylose, or rhamnose [112, 113]. Ether bonding can be performed for each quercetin molecule hydroxyl function with an alcoholic molecule, mainly methanol [114]. Quercetin can hold up to different combinations of ether functions. Figure 3.3 represents the chemical structures of some common quercetin derivatives.

3.5.1 QUERCETIN DERIVATIVES FOR CORROSION INHIBITION

Several works have been carried out to confirm the interest in using quercetin or synthesizing inhibitors based on it to develop and achieve the maximum possible inhibitory capacity toward metal corrosion. Ulaeto et al. suggested active anticorrosive film development. Active anticorrosion coatings were made by loading the phytochemical site quercetin into silica nanocontainers as a natural organic inhibitor. Active corrosion protection was activated by an engineered flaw in the epoxy nanocomposite covering and corrosion responses; 1.0 wt%-filled coatings performed best. The unstriped coatings outperformed the epoxy coating in corrosion resistance. Quercetin's pH-induced autoxidation and polymerization activities protected against corrosion. This stimulates aluminum alloy corrosion prevention research using bio-based coatings [115].

Sukul et al. synthesized mono- and di-4-(2-hydroxyéthyl)piperazin-1-yl) methyl derivatives of quercetin to test their corrosion-inhibiting efficacy in 1.0 M HCl using electrochemical techniques and weight loss. Comparing corrosion rates in uninhibited and inhibited medium versus temperature, the kinetic and thermodynamic parameters of the adsorption were evaluated. The functional density theory explains the corrosion inhibition propension between mono- and di-quercetin derivatives. Dynamic molecule simulations are used to calculate the interaction energy between inhibitor molecules and metal surfaces [116]. Baildya et al. used density functional theory to investigate the inhibition efficacy and reactive sites of quercetin and its derivatives as possible corrosion inhibitors. Quantum chemical variables included energies of molecular boundary orbitals, distribution of charges, electronic affinity, potential for ionization, momentum, softness, hardness, electrophilicity, electronegativity, electrophilicity, and quercetin-surface transfer of charge. Chemisorptive binding was shown by the adsorption energies of quercetin derivatives on iron. Thus, well-designed derivatives of quercetin may inhibit iron corrosion [117].

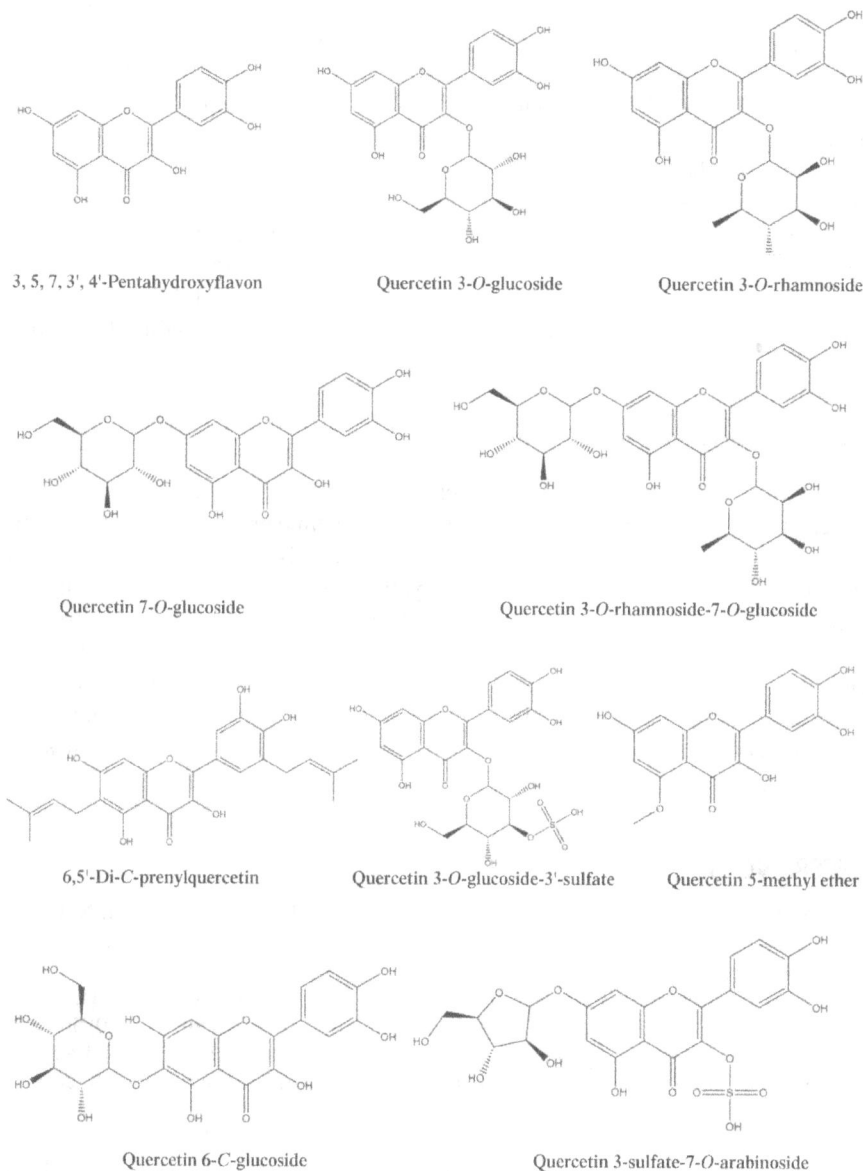

3, 5, 7, 3', 4'-Pentahydroxyflavon   Quercetin 3-*O*-glucoside   Quercetin 3-*O*-rhamnoside

Quercetin 7-*O*-glucoside   Quercetin 3-*O*-rhamnoside-7-*O*-glucoside

6,5'-Di-*C*-prenylquercetin   Quercetin 3-*O*-glucoside-3'-sulfate   Quercetin 5-methyl ether

Quercetin 6-*C*-glucoside   Quercetin 3-sulfate-7-*O*-arabinoside

FIGURE 3.3 Quercetin and related compounds.

Quercetin (3,3',4'5,7-pentahydroxyflavone) with Eu ions was added to 3.5% sodium chloride by Dehghani et al. to inhibit corrosion of mild steel. Tafel demonstrated that quercetin:Eu components inhibit steel specimen corrosion through a multifaceted process. After 48 hours of exposure to sodium chloride solution with 200:600 ppm quercetin:Eu, the effectiveness of EIS and Tafel was 98% and 90%, respectively. After injecting the combination, micrographs revealed that a dense, almost smooth layer of inhibition covered the surface [118]. Hijazi et al. tested the inhibitory effect of sumac, *Rhus coriaria*, a Lebanese plant containing quercetin, the tests were focused on handling mild steel in 0.5 M H_2SO_4 and 0.5 M HCl medium via EIS and PDP methods, AFM, and FTIR. *R. coriaria* and its phytochemicals behave as a mixture inhibitor in either environment, according to the potentiodynamic polarization curves. As indicated by the measurements, dissolution was controlled by activation, and inhibition of corrosion occurred by inhibitor adsorption on the metal surface. The extract inhibits more in 0.5 M HCl solutions compared to H_2SO_4. According to Flory–Huggins' model, *R. coriaria* and quercetin spontaneously adsorb on metal surfaces, inhibiting them. The creation of the protective coating on soft iron's surface was studied by surface analysis [119].

3.6 CONCLUSION

The best-known, most efficient, and most helpful corrosion inhibitors are organic molecules, and all research has proven that these chemicals, particularly those containing sulfur, nitrogen, and oxygen and numerous bonds in an aliphatic or aromatic system, have significant inhibitory efficiency. Electron delocalization of the ring systems offers the different aromatic compounds with tremendous inhibitory strength, particularly when they include heteroatoms such as oxygen, sulfur, and nitrogen. Quercetin and carbazole and its derivatives are used for corrosion inhibition to protect materials from corrosion damage due to their structures. Quercetin and carbazole corrosion inhibitors are most effective when lengthening immersion times and increasing the inhibitor concentration on the medium.

REFERENCES

1. K. Hossam, F. Bouhlal, L. Hermouche, I. Merimi, H. Labjar, A. Chaouiki, N. Labjar, S.-I. Malika, A. Dahrouch, M. Chellouli, B. Hammouti, S. El Hajjaji, Journal of Failure Analysis and Prevention, 21 (2021) 213–227.
2. Y. El Hamdouni, F. Bouhlal, H. Kouri, M. Chellouli, M. Benmessaoud, A. Dahrouch, N. Labjar, S. El Hajjaji, Journal of Failure Analysis and Prevention, 20 (2020) 563–571.
3. K. Bouiti, H. Aldeen Al-Sharabi, M. Bensemlali, F. Bouhlal, B. Abidi, N. Labjar, S. Laasri, S. El Hajjaji, The European Physical Journal Applied Physics, 97 (2022) 67.
4. M.A. Quraishi, D.S. Chauhan, V.S. Saji, Heterocyclic Organic Corrosion Inhibitors: Principles and Applications, Elsevier, Amsterdam: Cambridge, MA, 2020.
5. H. Al-sharabi, K. Bouiti, F. Bouhlal, N. Labjar, A. Dahrouch, M. El Mahi, E.M. Lotfi, B. El Otmani, G.A. Benabdellah, S. El Hajjaji, International Journal of Corrosion and Scale Inhibition, 11 (2022) 956–984.
6. T. Eicher, S. Hauptmann, A. Speicher, The Chemistry of Heterocycles: Structures, Reactions, Synthesis, and Applications, John Wiley & Sons, Inc., 2013.
7. E.Y. Jin, S. Lim, o Kim, Y.-S. Park, J.K. Jang, M.-S. Chung, H. Park, K.-S. Shim, Y.J. Choi, Food Science and Biotechnology, 20 (2011) 1727–1733.
8. J. Zotaj, K. Tare, J. Kokalari, A. Lame, Albanian Journal of Agricultural Science, 459–462 (2017) 5.
9. U. De Marchi, L. Biasutto, S. Garbisa, A. Toninello, M. Zoratti, Biochimica et Biophysica Acta (BBA): Bioenergetics, 1787 (2009) 1425–1432.

10. M. Massaro, G. Cavallaro, C.G. Colletti, G. Lazzara, S. Milioto, R. Noto, S. Riela, Journal of Materials Chemistry B, 6 (2018) 3415–3433.
11. E.U. Graefe, J. Wittig, S. Mueller, A.-K. Riethling, B. Uehleke, B. Drewelow, H. Pforte, G. Jacobasch, H. Derendorf, M. Veit, The Journal of Clinical Pharmacology, 41 (2001) 492–499.
12. Z. Chen, H. Li, X. Zheng, Q. Zhang, Z. Li, Y. Hao, G. Fang, ChemSusChem, 10 (2017) 3111–3117.
13. M. Ates, N. Uludag, Journal of Solid State Electrochemistry, 20 (2016) 2599–2612.
14. H.U. Nwankwo, E.D. Akpan, L.O. Olasunkanmi, C. Verma, A.M. Al-Mohaimeed, D.A.A. Farraj, E.E. Ebenso, Journal of Molecular Structure, 1223 (2021) 129328.
15. F. Dumur, European Polymer Journal, 125 (2020) 109503.
16. U. Eskişehir Osmangazi, İB. Topçu, A. Uzunömeroğlu, JSEAM, 3 (2020) 93–109.
17. S.A. Umoren, M.M. Solomon, Journal of Environmental Chemical Engineering, 5 (2017) 246–273.
18. P.B. Raja, M. Ismail, S. Ghoreishiamiri, J. Mirza, M.C. Ismail, S. Kakooei, A.A. Rahim, Chemical Engineering Communications, 203 (2016) 1145–1156.
19. D.K. Verma, R. Aslam, J. Aslam, M.A. Quraishi, E.E. Ebenso, C. Verma, Journal of Molecular Structure, 1236 (2021) 130294.
20. S. Marzorati, L. Verotta, S. Trasatti, Molecules, 24 (2018) 48.
21. I. Hamdani, O. Mokhtari, L. Lamri, S. Zaoui, D. Bouknana, A. Aouniti, M. Berrabah, A. Bouyanzer, B. Hammouti, Arabian Journal of Chemical and Environmental Researches, 5 (2018) 101–123.
22. M. Badji, D. Gassama, M. Bodian, R. Sylla-Gueye, K. Cissé, M. Fall, Materials Science (MEDŽIAGOTYRA), 9 (2022) 195–200.
23. S. Hooshmand Zaferani, M. Sharifi, D. Zaarei, M.R. Shishesaz, Journal of Environmental Chemical Engineering, 1 (2013) 652–657.
24. R.T. Loto, C.A. Loto, A.P.I. Popoola, Journal of Materials and Environmental Science 3 (2012) 885–894.
25. N.S. Patel, S. Jauhariand, G.N. Mehta, S.S. Al-Deyab, I. Warad, B. Hammouti, International Journal of Electrochemical Science, 8 (2013) 2635–2655.
26. T.J. Harvey, F.C. Walsh, A.H. Nahlé, Journal of Molecular Liquids, 266 (2018) 160–175.
27. A. Dabrowski, Advances in Colloid and Interface Science, 93 (2001) 135–224.
28. N.O. Obi-Egbedi, I.B. Obot, Corrosion Science, 53 (2011) 263–275.
29. O.D. Agboola, N.U. Benson, Frontiers in Environmental Science, 9 (2021) 678574.
30. S.R. Kachel, B.P. Klein, J.M. Morbec, M. Schöniger, M. Hutter, M. Schmid, P. Kratzer, B. Meyer, R. Tonner, J.M. Gottfried, Journal of Physical Chemistry C, 124 (2020) 8257–8268.
31. A.S. Fouda, M.A. Elmorsi, B.S. Abou-Elmagd, Polish Journal of Chemical Technology, 19 (2017) 95–103.
32. A. Khanra, M. Srivastava, M.P. Rai, R. Prakash, ACS Omega, 3 (2018) 12369–12382.
33. M. Králik, Chemical Papers, 68 (2014) 1–14.
34. M. Finšgar, Corrosion Science, 72 (2013) 82–89.
35. M.M.A. El-Sukkary, I. Aiad, A. Deeb, M.Y. El-Awady, H.M. Ahmed, S.M. Shaban, Petroleum Science and Technology, 28 (2010) 1158–1169.
36. H. Tian, Y.F. Cheng, W. Li, B. Hou, Corrosion Science, 100 (2015) 341–352.
37. M.A. Migahed, M.M. Attya, S.M. Rashwan, M. Abd El-Raouf, A.M. Al-Sabagh, Egyptian Journal of Petroleum, 22 (2013) 149–160.
38. S.M. Shaban, I. Aiad, M.M. El-Sukkary, E.A. Soliman, M.Y. El-Awady, Journal of Industrial and Engineering Chemistry, 21 (2015) 1029–1038.
39. E. Ituen, O. Akaranta, A. James, Chemical Science International Journal, 18 (2017) 1–34.
40. M.A. Al-Ghouti, D.A. Da'ana, Journal of Hazardous Materials, 393 (2020) 122383.

41. A. Popova, M. Christov, A. Zwetanova, Corrosion Science, 49 (2007) 2131–2143.
42. R. Saadi, Z. Saadi, R. Fazaeli, N.E. Fard, Korean Journal of Chemical Engineering, 32 (2015) 787–799.
43. N. Ayawei, A.N. Ebelegi, D. Wankasi, Journal of Chemistry, 2017 (2017) 1–11.
44. M. Boudalia, R.M. Fernández-Domene, M. Tabyaoui, A. Bellaouchou, A. Guenbour, J. García-Antón, Journal of Materials Research and Technology, 8 (2019) 5763–5773.
45. S. Saker, N. Aliouane, H. Hammache, S. Chafaa, G. Bouet, Ionics, 21 (2015) 2079–2090.
46. K. Bouiti, H. Aldeen Al-sharabi, F. Bouhlal, N. Labjar, A. Dahrouch, M.E. Mahi, E.M. Lotfi, B. El Otmani, G.A. Benabdellah, S. El Hajjaji, International Journal of Corrosion and Scale Inhibition, 11 (2022) 382–401.
47. O.M. Myina, R.A. Wuana, I.S. Eneji, R. Sha'Ato, Nigerian Journal of Chemical Research, 25 (2020) 15.
48. J. Ahmed E S, G.M. Ganesh, Buildings, 12 (2022) 1682.
49. S. El Wanees, M. Alahmdi, M. Alsharif, Y. Atef, Egyptian Journal of Chemistry, 62 (2019) 811–825.
50. M. Finšgar, J. Jackson, Corrosion Science, 86 (2014) 17–41.
51. A.Y. Kina, J.A.C. Ponciano, International Journal of Electrochemical Science, 8 (2013) 12600–12612.
52. S.A. Wade, Y. Lizama, Microscopy and Microanalysis, 12 (2015) 170–177.
53. X. Zhao, H. Zhuang, S.-C. Yoon, Y. Dong, W. Wang, W. Zhao, Journal of Food Quality, 2017 (2017) 1–16.
54. S. Feliu, Metals, 10 (2020) 775.
55. N. Meddings, M. Heinrich, F. Overney, J.-S. Lee, V. Ruiz, E. Napolitano, S. Seitz, G. Hinds, R. Raccichini, M. Gaberšček, J. Park, Journal of Power Sources, 480 (2020) 228742.
56. B. Boukamp, Solid State Ionics, 169 (2004) 65–73.
57. J. Huang, Z. Li, B.Y. Liaw, J. Zhang, Journal of Power Sources, 309 (2016) 82–98.
58. S.S. Abdel-Rehim, K.F. Khaled, N.S. Abd-Elshafi, Electrochimica Acta, 51 (2006) 3269–3277.
59. J.-M. Hu, J.-T. Zhang, J.-Q. Zhang, C.-N. Cao, Corrosion Science, 47 (2005) 2607–2618.
60. D.I. Njoku, M.A. Chidiebere, K.L. Oguzie, C.E. Ogukwe, E.E. Oguzie, Advances in Materials and Corrosion, 1 (2013) 54–61.
61. D.R. Cantrell, S. Inayat, A. Taflove, R.S. Ruoff, J.B. Troy, Journal of Neural Engineering, 5 (2008) 54–67.
62. D.S. Chauhan, K.R. Ansari, A.A. Sorour, M.A. Quraishi, H. Lgaz, R. Salghi, International Journal of Biological Macromolecules, 107 (2018) 1747–1757.
63. N. Al-Qahtani, J. Qi, A.M. Abdullah, N.J. Laycock, M.P. Ryan, International Journal of Science and Engineering Investigations, 10 (2021) 1–18.
64. I.B. Obot, I.B. Onyeachu, A. Zeino, S.A. Umoren, Journal of Adhesion Science and Technology, 33 (2019) 1453–1496.
65. J.R. Scully, Corrosion, 56 (2000) 199–218.
66. E. Ponz, J.L. Ladaga, R.D. Bonetto, Microscopy and Microanalysis, 12 (2006) 170–177.
67. B.J. Inkson, Scanning Electron Microscopy (SEM) and Transmission Electron Microscopy (TEM) for Materials Characterization, in: Materials Characterization Using Nondestructive Evaluation (NDE) Methods, Elsevier, 2016, pp. 17–43.
68. R. Brydson, Electron Energy Loss Spectroscopy, Oxford: Taylor & Francis, 2001.
69. M. Finšgar, Corrosion Science, 169 (2020) 108632.
70. R. Haldhar, D. Prasad, A. Saxena, Journal of Environmental Chemical Engineering, 6 (2018) 2290–2301.
71. X. Li, S. Deng, H. Fu, Materials Chemistry and Physics, 115 (2009) 815–824.
72. H. Zarrok, A. Zarrouk, B. Hammouti, R. Salghi, C. Jama, F. Bentiss, Corrosion Science, 64 (2012) 243–252.

73. M. Corrales Luna, T. Le Manh, R. Cabrera Sierra, J.V. Medina Flores, L. Lartundo Rojas, E.M. Arce Estrada, Journal of Molecular Liquids, 289 (2019) 111106.
74. M. Bouanis, M. Tourabi, A. Nyassi, A. Zarrouk, C. Jama, F. Bentiss, Applied Surface Science, 389 (2016) 952–966.
75. A. Singh, K.R. Ansari, D.S. Chauhan, M.A. Quraishi, S. Kaya, Sustainable Chemistry and Pharmacy, 16 (2020) 100257.
76. M. Kasaeian, E. Ghasemi, B. Ramezanzadeh, M. Mahdavian, G. Bahlakeh, Applied Surface Science, 462 (2018) 963–979.
77. H.F. Liang, C.T.G. Smith, C.A. Mills, S.R.P. Silva, Journal of Materials Chemistry C, 3 (2015) 12484–12491.
78. I. Lukovits, E. Kálmán, F. Zucchi, Corrosion, 57 (2001) 3–8.
79. L. Guo, R. Zhang, B. Tan, W. Li, H. Liu, S. Wu, Journal of Molecular Liquids, 310 (2020) 113239.
80. S.Z. Salleh, A.H. Yusoff, S.K. Zakaria, M.A.A. Taib, A. Abu Seman, M.N. Masri, M. Mohamad, S. Mamat, S. Ahmad Sobri, A. Ali, P.T. Teo, Journal of Cleaner Production, 304 (2021) 127030.
81. J. Tan, L. Guo, H. Yang, F. Zhang, Y. El Bakri, RSC Advances, 10 (2020) 15163–15170.
82. A. Pal, C. Das, Industrial Crops and Products, 151 (2020) 112468.
83. R. Hsissou, F. Benhiba, M. Khudhair, M. Berradi, A. Mahsoune, H. Oudda, A. El Harfi, I.B. Obot, A. Zarrouk, Journal of King Saud University – Science, 32 (2020) 667–676.
84. G.A. Cordell, M.L. Quinn-Beattie, N.R. Farnsworth, Phytotherapy Research, 15 (2001) 183–205.
85. F.E. Koehn, G.T. Carter, Nature Reviews. Drug Discovery, 4 (2005) 206–220.
86. A. Mital, Scientia Pharmaceutica, 77 (2009) 497–520.
87. Y.-W. Chin, M.J. Balunas, H.B. Chai, A.D. Kinghorn, The AAPS Journal, 8 (2006) 239–259.
88. S. Bhambhani, K.R. Kondhare, A.P. Giri, Molecules, 26 (2021) 3374.
89. D.D. Nekrasov, Chemistry of Heterocyclic Compounds, 37 (2001) 263–275.
90. S. Meyer, H. Steinhart, Chemosphere, 40 (2000) 359–367.
91. L.B. Salam, M.O. Ilori, O.O. Amund, 3 Biotech, 7 (2017) 111.
92. S. Peddinghaus, M. Brinkmann, K. Bluhm, A. Sagner, G. Hinger, T. Braunbeck, A. Eisenträger, A. Tiehm, H. Hollert, S.H. Keiter, Reproductive Toxicology, 33 (2012) 224–232.
93. A.T. Balaban, D.C. Oniciu, A.R. Katritzky, Chemical Reviews, 104 (2004) 2777–2812.
94. Q. Weng, X. Wang, X. Wang, Y. Bando, D. Golberg, Chemical Society Reviews, 45 (2016) 3989–4012.
95. K. Schofield, Hetero-Aromatic Nitrogen Compounds, Boston, MA: Springer, 1967.
96. K. Karon, M. Lapkowski, Journal of Solid State Electrochemistry, 19 (2015) 2601–2610.
97. H. Jiang, J. Sun, J. Zhang, Current Organic Chemistry, 16 (2012) 2014–2025.
98. M. Manickam, P. Iqbal, M. Belloni, S. Kumar, J.A. Preece, Israel Journal of Chemistry, 52 (2012) 917–934.
99. İ Çakmakcı Ünver, G. Bereket, B. Duran, Polymer-Plastics Technology and Engineering, 57 (2018) 242–250.
100. M. Ates, A.T. Özyılmaz, Progress in Organic Coatings, 84 (2015) 50–58.
101. H.U. Nwankwo, L.O. Olasunkanmi, E.E. Ebenso, Scientific Reports, 7 (2017) 2436.
102. H.U. Nwankwo, L.O. Olasunkanmi, E.E. Ebenso, Journal of Bio- and Tribo-Corrosion, 4 (2018) 13.
103. B. Duran, İ Çakmakcı Ünver, G. Bereket, Journal of Adhesion Science and Technology, 31 (2017) 1467–1479.
104. Y. Li, J. Yao, C. Han, J. Yang, M. Chaudhry, S. Wang, H. Liu, Y. Yin, Nutrients, 8 (2016) 167.
105. P.M. Shah, Journal of Pharmaceutical Sciences and Research, 8 (2016) 878–880.

106. A. Altomare, M. Fiore, G. D'Ercole, E. Imperia, R.M. Nicolosi, S. Della Posta, G. Pasqua, M. Cicala, L. De Gara, S. Ramella, M.P.L. Guarino, Nutrients, 14 (2022) 5374.
107. D. Yang, T. Wang, M. Long, P. Li, Oxidative Medicine and Cellular Longevity, 2020 (2020) 1–13.
108. W. Qi, W. Qi, D. Xiong, M. Long, Molecules 27 (2022) 6545.
109. M. Materska, Polish Journal of Food and Nutrition Sciences, 58 (2008) 407–413.
110. A.-S. Michala, A. Pritsa, Diseases, 10 (2022) 37.
111. N.A. Sagar, S. Pareek, N. Benkeblia, J. Xiao, Food Frontiers, 3 (2022) 380–412.
112. M.F. Nisar, C. Wan, Z. Manzoor, Y. Waqas, K. Niaz, M.M. Ayaz, Glycosidic Derivatives of Flavonoids, in: Recent Advances in Natural Products Analysis, Elsevier, 2020, pp. 57–84.
113. W.M. Dabeek, M.V. Marra, Nutrients, 11 (2019) 2288.
114. F. Khan, K. Niaz, F. Maqbool, F. Ismail Hassan, M. Abdollahi, K. Nagulapalli Venkata, S. Nabavi, A. Bishayee, Nutrients, 8 (2016) 529.
115. S.B. Ulaeto, A.V. Nair, J.K. Pancrecious, A.S. Karun, G.M. Mathew, T.P.D. Rajan, B.C. Pai, Progress in Organic Coatings, 136 (2019) 105276.
116. D. Sukul, A. Pal, S.K. Saha, S. Satpati, U. Adhikari, P. Banerjee, Physical Chemistry Chemical Physics, 9 (2018) 14.
117. N. Baildya, N.N. Ghosh, A.P. Chattopadhyay, SN Applied Sciences, 1 (2019) 735.
118. A. Dehghani, A.H. Mostafatabar, G. Bahlakeh, B. Ramezanzadeh, Journal of Molecular Liquids, 316 (2020) 113914.
119. K.M. Hijazi, A.M. Abdel-Gaber, G.O. Younes, R. Habchi, Portugaliae Electrochimica Acta, 39 (2021) 237–252.

4 Imidazole- and Imidazoline-based Corrosion Inhibitors

Khasan Berdimuradov[1], Elyor Berdimurodov[2,3],
Ilyos Eliboev[1], Lazizbek Azimov[1], Yusufboy
Rajabov[1], Jaykhun Mamatov[1], Bakhtiyor
Borikhonov[4], Abduvali Kholikov[1], Oybek
Mikhliev[5], and Khamdam Akbarov[1]
[1]Faculty of Industrial Viticulture and Food Production
Technology, Shahrisabz Branch of Tashkent Institute
of Chemical Technology, Shahrisabz, Uzbekistan
[2]Faculty of Chemistry, National University
of Uzbekistan, Tashkent, Uzbekistan
[3]Department of Chemistry, Akfa University,
Tashkent, Uzbekistan
[4]Faculty of Chemistry–Biology, Karshi State
University, Karshi, Uzbekistan
[5]Department of Chemistry, Karshi Engineering
Economics Institute, Karshi City, Uzbekistan

4.1 INTRODUCTION

4.1.1 IMIDAZOLE- AND IMIDAZOLINE-BASED ORGANIC COMPOUNDS AS EFFECTIVE INHIBITORS FOR METALS

Imidazole- and imidazoline-based organic compounds are used as effective inhibitors for metals. These compounds can be synthesized by various methods such as intermolecular nucleophilic substitution, ring opening of aziridines, and cyclization of oximes. The imidazole- and imidazoline-based organic compounds are a class of molecules that have been found to be effective inhibitors for metals. Imidazole is an aromatic heterocyclic compound, which means it contains carbon atoms bonded together in rings. The most common types of imidazoles are $1H$-imidazoles, $2H$-imidazoles, $3H$-imidazoles, and $4H$-imidazoles; however, there are many other variations that have been studied as well. The imidazole- and imidazoline-based organic compounds are interesting heterocyclic compounds in the corrosion inhibitors for the metal [1, 2]. The reason is that the imidazole- and imidazoline-based organic compounds are cost-effective, easily synthesized, and have good properties [3, 4]. There are several types of imidazole- and imidazoline-based organic

DOI: 10.1201/9781003377016-4

compounds that have been identified as effective inhibitors for metals. The charac-teristics of each type are as follows:

Imidazole-based Compounds: Imidazoles have a five-membered ring with one nitrogen atom bonded to two alkyl groups or aromatic rings (e.g., benzene). They are commonly used in pharmaceuticals and agrochemicals because they exhibit high stability at high temperatures, low toxicity, and good water solubility. The imidazole- and imidazoline-based organic compounds have nitrogen-based hetero rings, which promote inhibitor adsorption on the metal surface [4]. The following good characteristics are mainly responsible for the high inhibition effi-ciency [5–9]:

   i. The five-membered heterocyclic promotes the π-electron transformation between the corrosion inhibitor and metal surface [4, 10].
  ii. The imidazole- and imidazoline-based organic compounds are high polar molecules. This indicator supports the water solubility of these organic compounds [11, 12].
 iii. These compounds have a high chelation performance, which promotes donor–acceptor performance [13].
  iv. The hydrophobicity of the metal surface was increased after the adsorption of the imidazole- and imidazoline-based organic compounds [14].

Imidazole- and imidazoline-based organic compounds are the most effective inhibitors for metals. They can be used in a wide range of applications, from auto-motive lubricants to industrial coatings. The following are some of the advantages that make these compounds so popular:

Cost-effectiveness: Imidazoles have been shown to be more cost-effective than other types of inhibitors such as phosphates or amines. This is because they offer high performance at low concentrations while still being environment-friendly.

Imidazole- and imidazoline-based organic compounds are known to be effective inhibitors for metals. They can be used in the treatment of diseases caused by metal toxicity, such as Wilson's disease and copper poisoning. They also have applications in biochemistry research, where they can be used to study protein structure–function relationships at the molecular level.

The imidazole- and imidazoline-based organic compounds have shown to be effective inhibitors for metals (Figure 4.1). These compounds can be used as alter-natives to traditional chelating agents and are more efficient in removing heavy metals from water. They also have better stability in acidic conditions and are less expensive than other chelating agents such as EDTA or DTPA. This is an impor-tant finding since many countries around the world face water pollution caused by heavy metals which affect human health negatively when consumed over long periods of time (WHO). The future outlook for these types of research is promis-ing since there are still many unanswered questions regarding how these molecules interact with different types of metal ions at different pH levels and temperatures; however, it seems likely that they will continue being explored further due to their unique properties compared with other types of chelating agents currently being used today.

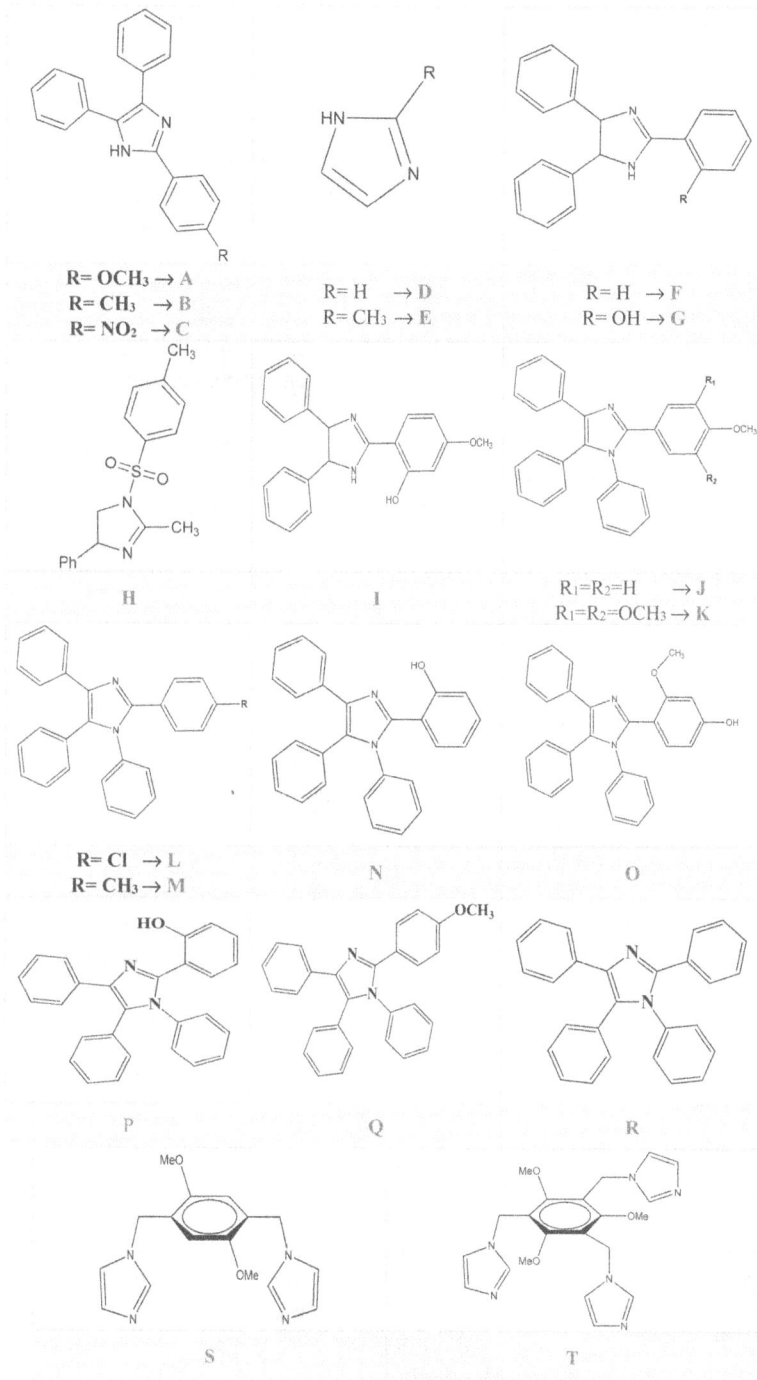

FIGURE 4.1 Imidazole- and imidazoline-based organic compounds as effective inhibitors for metals [5].

4.2   IMIDAZOLE-BASED CORROSION INHIBITORS

4.2.1   IMIDAZOLE-BASED CORROSION INHIBITORS FOR STEEL

In contemporary times, the advancement of efficient and eco-friendly corrosion inhibitors has emerged as a vital domain of study within the realm of materials science and engineering. One of the most promising classes of compounds in this regard is imidazole-based corrosion inhibitors, which have demonstrated exceptional performance in protecting steel structures from the deleterious effects of corrosion. This chapter aims to provide an in-depth overview of the chemistry, mechanisms, and applications of imidazole-based corrosion inhibitors, highlighting their potential to revolutionize the way we safeguard our critical steel infrastructure. The steel-based metallic materials are easily corroded in carbon dioxide, sulfide, amine, and other acidic gas-based aquatic solutions. For example, crude oil and gas contain carbon dioxide, sulfide, amine, and other acidic gas at a high concentration [15–17]. These types of gas are more acidic and easily react with metal materials. Steel, an alloy primarily composed of iron and carbon, is a widely used material in various industries such as construction, automotive, and aerospace, owing to its remarkable strength, flexibility, and cost-effectiveness. However, steel is highly susceptible to corrosion, a chemical or electrochemical process that leads to the gradual deterioration of the material. This presents considerable obstacles to the resilience, architectural stability, and lifespan of steel-reliant constructions, frequently requiring expensive upkeep and substitutions. Imidazole and its related compounds have surfaced as a powerful category of corrosion preventatives, due to their distinct molecular composition and capacity to generate safeguarding layers on the metallic exterior. The existence of both electron-contributing and electron-receiving groups within the imidazole circle enables the combination with metallic particles, leading to the creation of enduring and highly impervious coatings. This effectively reduces the rate of corrosion by hindering the electrochemical reactions responsible for material degradation. Numerous studies have reported the successful application of imidazole-based inhibitors in various settings, ranging from oil and gas pipelines to marine environments, demonstrating their versatility and adaptability. Additionally, these inhibitors are generally found to be biodegradable and nontoxic, making them an attractive choice from an environmental perspective. The steel-based metallic materials were used as the transportation metal pipe for crude oil gas in the chemical industry. Imidazole-based corrosion inhibitors were importantly investigated as good corrosion inhibitors for steel because of their high corrosion efficiency [18–20].

Singh et al. [21] researched the anti-polarizing efficiency of the following corrosion inhibitors: M-3, M-2, and M-1. Their inhibition performance was explored. The obtained results confirmed that the studied imidazole-based corrosion inhibitors were more efficient for the J55 steel in the carbon dioxide–saturated solutions. Among these corrosion inhibitors, the M-1 is more effective than others. The reason for this is that the M-1 contained the methoxy functional groups, which is attributed to the electron affinity of amino functional groups. As a result, the selected inhibitor is more effective than others. Figure 4.2 shows the anticorrosion

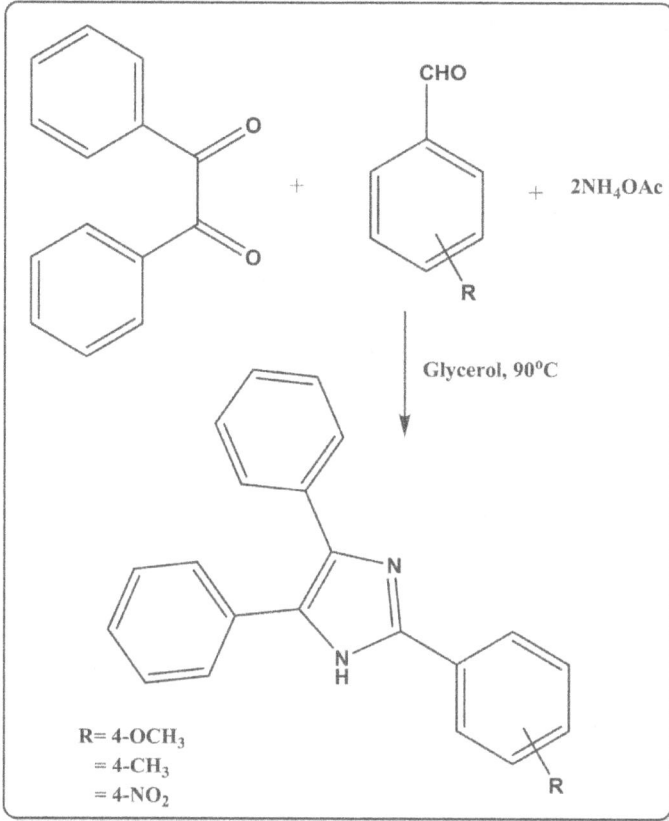

(a)

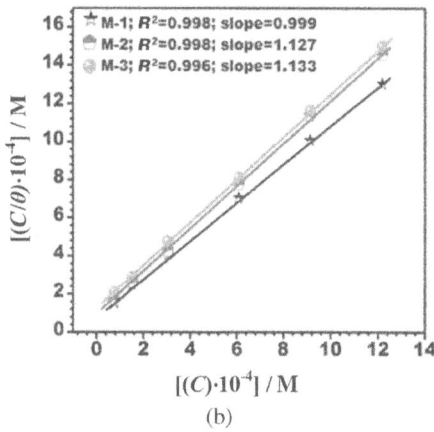

(b)

FIGURE 4.2 Anticorrosion efficiency of M-3, M-2, and M-1. (a) Synthesis procedures and their chemical structures. (b) Langmuir plots. (c) Contact angle properties. (d) Corrosion inhibition mechanism [21].

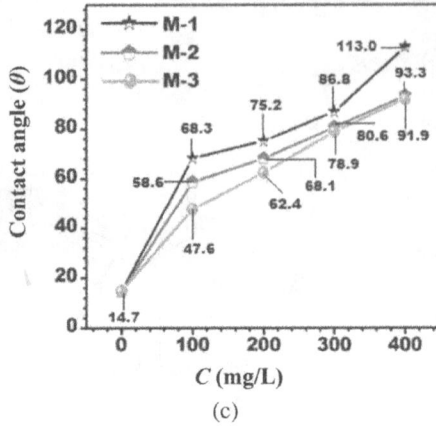

(c)

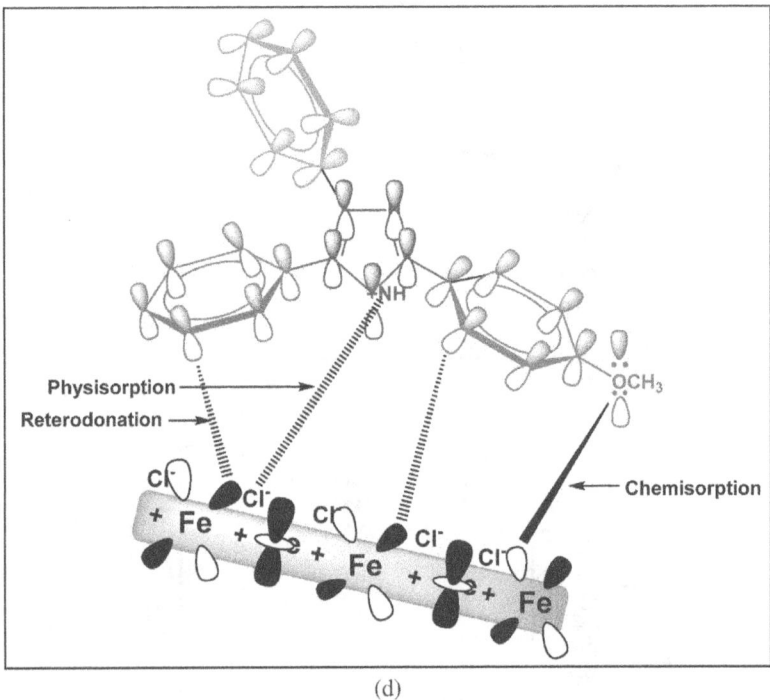

(d)

FIGURE 4.2 *(Continued)*

efficiency of M-3, M-2, and M-1: (a) synthesis procedures and their chemical structures, (b) polarization plots, (c) Langmuir plots, (d) contact angle properties, and
(e) corrosion inhibition mechanism. It was found that (i) the neutral and protonated
forms of selected corrosion inhibitors are more effective for steel corrosion; (ii) the
water-repellent nature of the mixture was improved by the addition of corrosion
deterrents; consequently, the metallic exterior was efficiently shielded from the
water-based medium.

Ouakki et al. [22] studied the corrosion inhibition of MTIPh-1 and TTIPh-1. The following was found:

i. The acidic attacks of the chloride ions were depleted by the presence of MTIPh-1 and TTIPh-1.
ii. The aromatic rings contained more nitrogen atoms, which promote the chelating efficiency of organic compounds.
iii. The proposed organic compound blend implies that the corrosion prevention process relies on both chemical and physical adsorption mechanisms.
iv. It is demonstrated that the corrosion inhibition of MTIPh-1 and TTIPh-1 occurs. The rationale behind this is that the methoxy functional groups facilitate the corrosion prevention process by contributing electrons to the metallic surface.
v. Theoretical computations, grounded in density functional concepts, verify that corrosion inhibition escalates with the increase of protonated centers. MTIPh-1 and TTIPh-1 readily become protonated in an acidic setting via nitrogen atoms. The protonated variant of the corrosion inhibitor exhibits a stronger inclination toward the anti-polarization agent.
vi. The theoretical estimations of global reactivity parameters ascertain that MTIPh-1 and TTIPh-1 organic substances function as more potent nucleophilic agents.

Zeng et al. [23] prepared the green corrosion inhibitors based on the following imidazole derivatives: [BMImB]Br$_2$, [APMIm]Br, and [PrMIm]Br. Their anticorrosion performances were explored by the electrochemical and theoretical methods. Among these green ionic liquids, the [BMImB]Br$_2$ was better than [APMIm]Br and [PrMIm]Br. This is due to the two bromide atoms and the molecular structures. The electron-rich regions were promoted by the presence of bromide atoms. The maximal corrosion inhibition efficiency was reached at the 303 K temperature. With the increase in temperature, the corrosion protection was increased effectively.

4.2.2 Imidazole-based Corrosion Inhibitors for Copper

In this study, we investigate the performance and underlying mechanisms of imidazole-based corrosion inhibitors for the protection of copper, a widely used metal in electrical wiring, plumbing, and electronics, which is highly susceptible to corrosion. The goal of this research is to explore the effectiveness of imidazole derivatives in preventing copper corrosion and to elucidate the factors contributing to their exceptional inhibitive properties. Through a series of electrochemical tests and surface analysis techniques, we evaluate the efficiency of various imidazole-based inhibitors in mitigating the corrosion of copper in different corrosive environments. Our results demonstrate that these inhibitors exhibit remarkable performance in reducing the corrosion rate, primarily due to their unique chemical structure, which enables the formation of stable, protective films on the copper surface. We further examine the influence of the inhibitor concentration, temperature, and other environmental factors on the inhibitive properties of imidazole derivatives, providing valuable insights for their practical implementation. Moreover, we discuss the environment-friendly nature and biodegradability of these inhibitors, making them a sustainable choice for corrosion protection.

The copper-based metallic materials are basic alloys in the automobile and electronic industries. The reason for this is that these materials are thermal conductivity and electrical conductivity. Nonetheless, copper-centered metal substances are prone to corrosion in acidic mixtures containing nitrate, sulfate, and chloride ions. Organic compounds predominantly serve as corrosion preventatives for copper-infused metallic materials in acid-induced deterioration. Within this group of inhibitors, imidazole-derived corrosion deterrents take precedence in safeguarding copper from the hostile onslaught of nitrate, sulfate, and chloride ions [24–27].

The copper corrosion in the saline solution is a large problem. Protecting copper-based metallic materials from salt accumulations on the metal surface is an urgent task in materials science. Hou et al. [24] studied the corrosion inhibition of the following three inhibitors: PMI, PAI, and PDI. These compounds were first prepared in this research work, and their chemical structures were confirmed by the FTIR spectra, ^{13}C NMR, ^{1}H NMR, and LC-MS. The corrosion inhibition of PMI, PAI, and PDI for copper in the aggressive saline medium contained sodium chloride (3.5%) was investigated. Figure 4.3 shows the (a) chemical structures, (b) solubility, and (c) Tafel plots of PMI, PAI, and PDI [4]. The following main comments were found:

   i. The salt and metal oxides or hydroxide accumulation on the metal surface was effectively depleted by the formation of the electrostatic bonds-based film on the metal surface.
  ii. Among the selected corrosion inhibitors were the PMI, PAI, and PDI. The reason for this is that the π-electrons of heterocycle are mainly attributed to the increased corrosion inhibition.
 iii. The metal–imidazole defensive layer was established on the metallic exterior via adsorption. The interplay between the metal and organic elements on the metal's surface was defined by the outcomes of molecular dynamics simulations. Additionally, this film exhibits an impressive water-repellent capability.

In an acidic solution, the copper was easily corroded. The imidazole-based organic compounds were dominant in the corrosion protection of copper-based materials from the acidic solution. For example, Costa et al. investigated the corrosion protection properties of the following imidazole-based corrosion inhibitors for copper: 4-(1H-imidazol-1-yl)phenol, (4-(1H-imidazol-1-yl)phenyl)methanol, 4-(1H-imidazol-1-yl)aniline, and 4-(1H-imidazol-1-yl)benzaldehyde. These types of organic compounds can protect copper from the aggressive sulfur acid solution. The inhibition performance of these inhibitors for copper in the acidic sulfur solution was researched and it was confirmed that these inhibitors adsorbed effectively on the metal surface by the electrostatic interactions [28]. El-Katori et al. introduced dithioglycouril (tetrahydroimidazo[4,5-d]imidazole-2,5(1H, 3H)-dithione)]

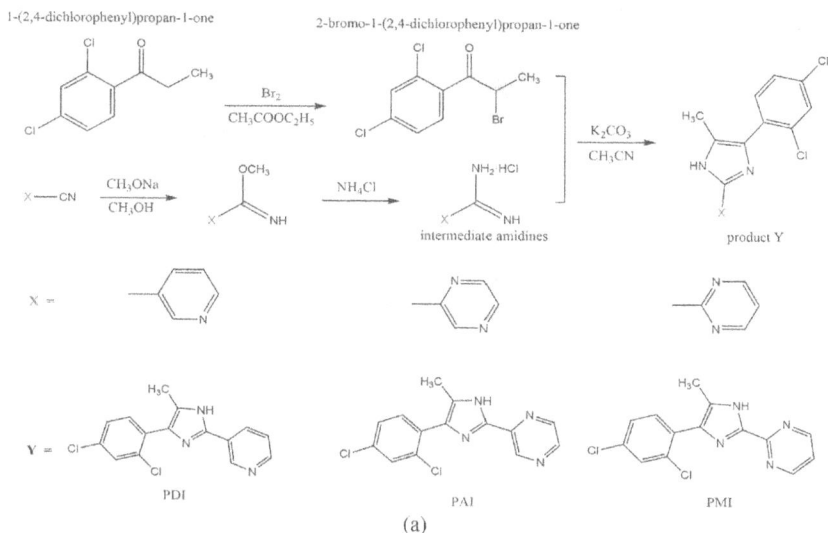

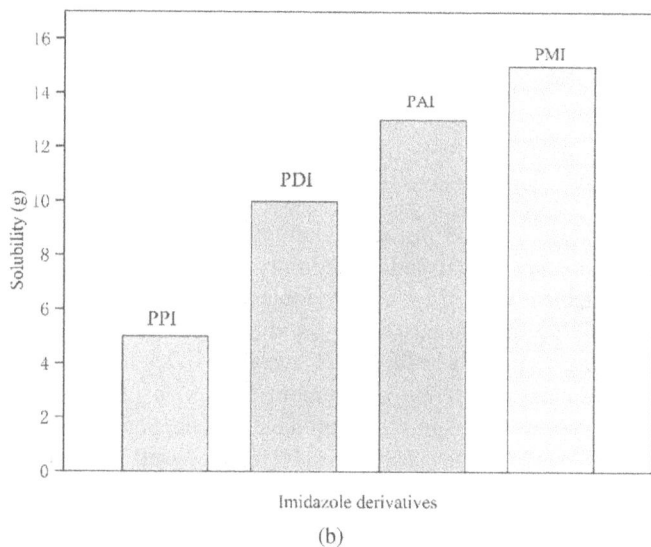

FIGURE 4.3 (a) Chemical structures. (b) Solubility of PMI, PAI, and PDI [24].

and glycouril (tetrahydroimidazo[4,5-*d*]imidazole-2,5(1*H*, 3*H*)-dione) (Figure 4.4) as the good corrosion inhibitors for the copper in the aggressive nitrate solution. It was found that their protection efficiency was over 90% at low concentrations [29].

Curkovic et al. [30] suggested that the 4-methyl-1-phenyl imidazole is a more efficient corrosion inhibitor for the copper-based metallic materials in the saline corrosion. The salt deposition on the copper surface was effectively depleted by the formation of defender layer. The alterations in electrochemical activity on the metallic exterior during the inhibition processes were examined through electrochemical quartz crystal microbalance and electrochemical impedance spectroscopy evaluations. Key electrochemical kinetic factors, such as corrosion pace and corrosion

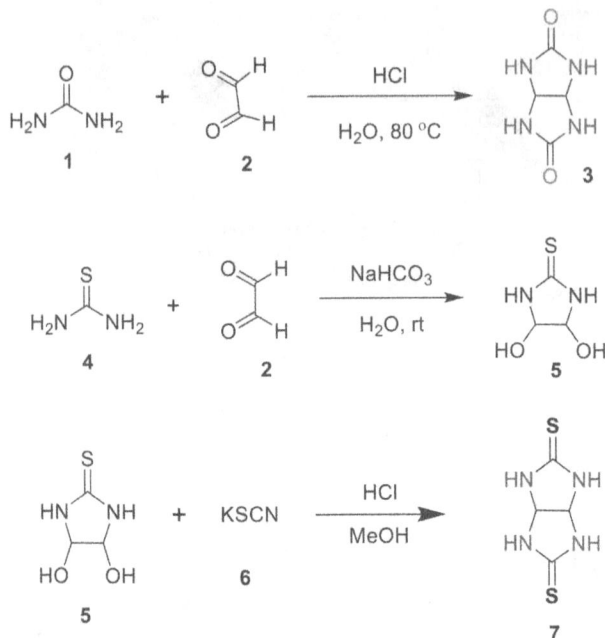

FIGURE 4.4 Synthesis of dithioglycouril (tetrahydroimidazo[4,5-*d*]imidazole-2,5(1*H*, 3*H*)-dithione)] and glycouril (tetrahydroimidazo[4,5-*d*]imidazole-2,5(1*H*, 3*H*)-dione) [29].

current density, were elevated in the noninhibited saltwater solution, while these parameters significantly diminished in the inhibited solution. The corrosion prevention of this imidazole is contingent upon the tolyl constituents.

The acidic raining is a serious factor for the copper corrosion. In near industrial regions, the acid raining was formed. The copper-based metallic materials were easily destructed under acidic raining solution. The imidazole-based organic compounds were suggested for the corrosion protection of copper. The reason for this is that the imidazole-based organic compounds interacted well with the metal surface by the interaction between the nitrogen atoms and d-orbitals on metal surface.

Simonović et al. [31] investigated the corrosion inhibition mechanism of 1,2-dimethyl-imidazole, 2-mercapto-1-methylimidazole, and 1,1′-sulfonylimidazole for the copper corrosion in the acid raining solutions. The surface analysis results confirmed the following:

i. The damage and abrasions on the copper exterior transpired due to the acidic rain–induced corrosion.

ii. A defensive coating formed on the metallic surface, consequently insulating the contact between the metal exterior and the acidic rain solution.

iii. The buildup of sodium chloride and copper chloride on the copper surface was halted by the presence of specific compounds.

iv. The adsorption of these inhibitors on the copper exterior occurs through spontaneous processes.

v. The methoxy, amino, phenyl, and alkaline functional groups primarily contribute to the anticorrosion actions of the aforementioned corrosion inhibitors.

vi. The quantum chemical calculations confirmed that the HOMO regions promote the nucleophilic performance of corrosion inhibitor.
vii. The LUMO regions support the antibonding interactions between the metal and organic compounds.

4.3 IMIDAZOLINE-BASED CORROSION INHIBITORS

4.3.1 IMIDAZOLINE-BASED CORROSION INHIBITORS FOR STEEL

In contemporary periods, imidazoline-derived organic substances represent the latest advancements in corrosion defense due to their distinct architecture and versatile characteristics. It has been noted that the imidazoline-derived organic substances consist of three isomer variations: 4-imidazoline, 3-imidazoline, and 2-imidazoline. These isomers are the basic core of some pharmaceutical drugs and natural compounds because their heterocycle is mainly responsible for the various biological–pharmacological functions. For example, they were used as an antidepressant, antihypercholesterolemic, antihypertensive, anti-inflammatory, and antihyperglycemic agents [32, 33].

The imidazoline-based organic compounds are in limited use in corrosion protection. In the present times, the core of the imidazoline-based organic compounds was modified with various functional groups. As a result, the corrosion inhibition increased importantly. This is due to the adsorption and electron negativity of the heterocyclic ring of imidazoline being enhanced [12, 34].

Zhang confirmed that halogen-modified imidazoline-based corrosion inhibitors are more efficient than others [35]. For example, the 2-(2-trichloromethyl-4,5-dihydro-imidazol-1-yl)-ethylamine (2-IM) and 2-(2-trifluoromethyl-4,5-dihydro-imidazol-1-yl)-ethylamine (1-IM) were investigated for the steel corrosion in the concentrated hydrochloric acid. It is noted that chloride is a better additive in the corrosion inhibition of imidazoline than other halogens [35].

Xiong et al. [36] introduced the SMID and SMIF imidazoline-based corrosion inhibitors for steel-based metallic materials. Figure 4.5 shows the (a) synthesis procedure and (b) molecular dynamic simulation of SMID and SMIF imidazoline-based corrosion inhibitors. The following was found:

i. The SMID and SMIF imidazoline-based corrosion inhibitors can protect the steel metal coupons from the saline and concrete solutions.
ii. The effectiveness of inhibition relies on alterations in the inhibitor's concentration, molecular configuration, placement within the aromatic ring, and the characteristics of connected functional groups. These factors play a crucial role in determining the overall performance of the inhibitor in preventing corrosion and ensuring the longevity of the materials being protected.
iii. The protonation procedures involving amino groups contribute to the increased solubility of imidazoline-derived organic substances. This increased solubility plays a vital role in the effectiveness of these compounds in corrosion protection, as it allows for better dispersion and interaction with the material surfaces. Consequently, this improved solubility contributes to the overall efficiency of imidazoline-derived substances in mitigating the damaging effects of corrosion on various materials and structures.

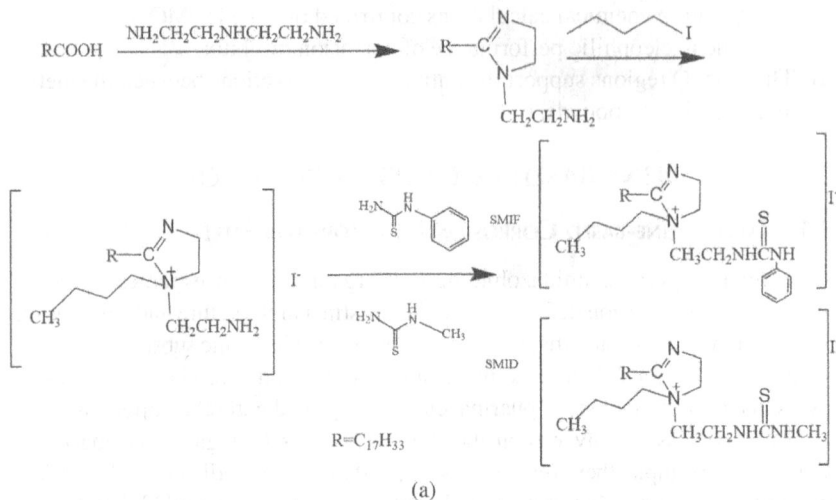

(a)

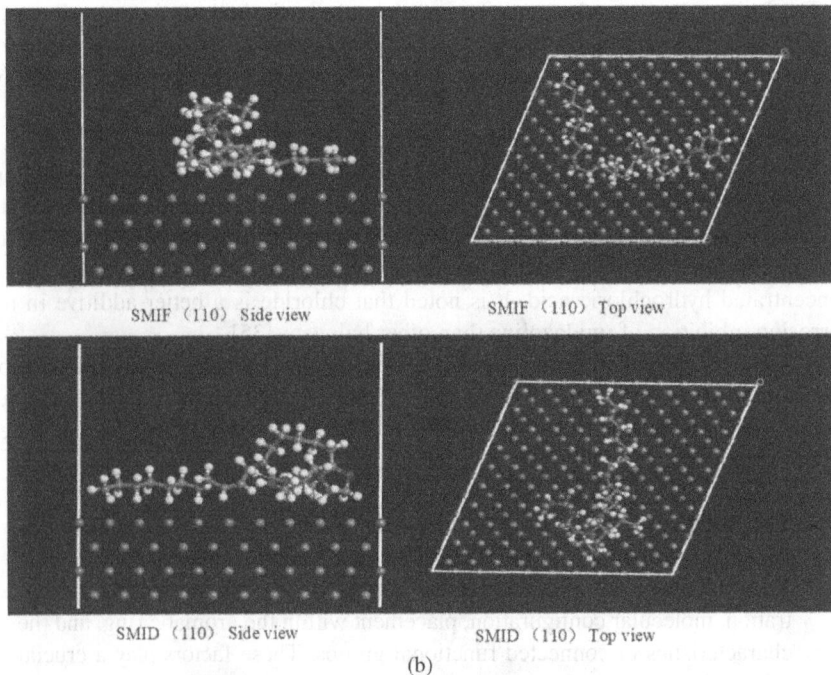

SMIF (110) Side view   SMIF (110) Top view

SMID (110) Side view   SMID (110) Top view

(b)

FIGURE 4.5 (a) Synthesis procedure and (b) molecular dynamic simulation of SMID and SMIF imidazoline-based corrosion inhibitors [36].

Steel corrosion in carbon dioxide–rich saline solutions is a primary concern in the chemical industry, as the aqueous phase of crude oil contains high levels of carbon dioxide and salts. Okafor et al. [37] studied the corrosion resistance of 2-undecyl-1-ethylamino imidazoline for steel-based materials in carbon dioxide–saturated saline solutions, employing scanning electron microscopy (SEM) analysis, electrochemical impedance spectroscopy techniques, and potentiodynamic polarization methods. Their findings included the following:

i. The enhancement of protection efficiency corresponded with the increase in inhibitor concentrations.
ii. The Temkin isotherm proved to be a more suitable model for describing inhibitor adsorption on the metal surface, suggesting that anionic and cationic activities on the metal surface were significantly impeded.

Zhang et al. [4] presented 2-methyl-4-phenyl-1-tosyl-4,5-dihydro-1H-imidazole as an effective corrosion inhibitor. Its anticorrosion properties were examined using electrochemical impedance spectroscopy, potentiodynamic polarization, and weight loss measurements. The chemical and electrostatic interactions between the multiple bonds of the chosen inhibitor and the metal surface promoted the adsorption behavior of the selected organic compound. The heteroatoms of this inhibitor also promote the adsorption behavior. The delocalization processes in the aromatic ring support the donor–acceptor performance of organic compound, because of the π-electrons of benzene ring. This corrosion inhibitor contained more heteroatoms, such as nitrogen and oxygen. The electron pairs of these atoms are also attributed to rise in the corrosion inhibition of this inhibitor.

The steel-based metallic materials also corroded in the underground systems. The aquatic solution in underground system is more alkaline and has higher salinity. To protect the metallic materials from the alkaline and salinity solution is important in chemical engineering. The imidazoline-based organic compounds were suggested in the protection of metallic materials from the alkaline–salinity systems.

The 1-[N,N'-bis(hydroxylethylether)-aminoethyl]-2-stearicimidazoline (HASI) [38] was introduced as the good corrosion inhibitor for steel-based metallic materials in the 5% NaCl saturated Ca(OH)$_2$ solution. It is indicated that the π-electrons in N=C−N region of selected compound promote the donor–acceptor mechanism. As can be seen from the obtained results, the studied corrosion inhibitor adsorbed on the metal surface by the parallel adsorption positions.

4.3.2 IMIDAZOLINE-BASED CORROSION INHIBITORS FOR COPPER

Copper corrosion poses a significant challenge in the electronics industry. To protect copper from aqueous solutions, imidazoline-based corrosion inhibitors have been utilized. This is due to the strong interaction between imidazoline and the metal surface through a donor–acceptor mechanism, where delocalized π-electrons are transferred from the imidazoline ring to the empty d-orbitals of copper [39, 40].

In some studies, imidazoline has been modified with different materials. For instance, Gemini imidazoline surfactants have been introduced as effective corrosion inhibitors for copper. It has been determined that the efficiency relies on fundamental factors such as surfactant concentrations, carbon chain length, and solution pH. Copper effectively resisted corrosion with the aid of Gemini imidazoline surfactants in neutral and alkaline solutions. The primary contributors to this anticorrosion efficiency are the carbon chains and fatty acid components [41].

Gonzalez-Rodriguez et al. explored the corrosion and inhibition processes on copper surfaces in the presence and absence of palm-oil-modified hydroxyethyl imidazoline, using electrochemical impedance spectroscopy (EIS), linear polarization resistance (LPR), and potentiodynamic polarization curves. The findings confirmed that imidazoline's heteroatoms are primarily responsible for the high corrosion inhibition. Langmuir adsorption isotherms suggested chemical and physical adsorption mechanisms employed by the palm-oil-modified hydroxyethyl imidazoline [42].

Wan et al. [43] suggested 2-phenyl imidazoline (2-PI) as the corrosion protection agent in the atmospheric environment. The corrosion of copper metals depends on the following basic parameters: NaCl loading density, temperature, and relative humidity. The effects of the above parameters were depleted by the presence of a corrosion inhibitor named 2-phenyl imidazoline (2-PI). Figure 4.6 presents the SEM micrographs of

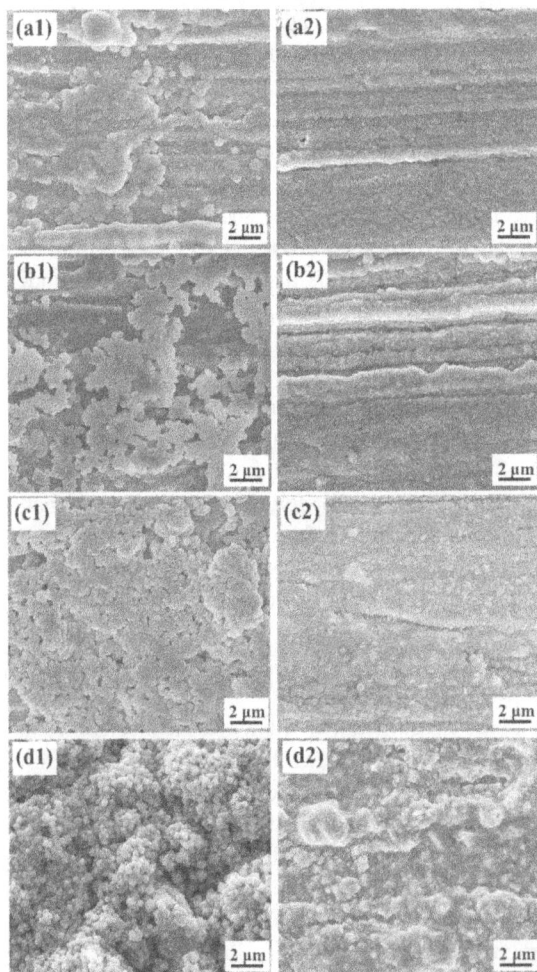

FIGURE 4.6 SEM images were taken of copper corrosion products under varying NaCl loading densities both with and without 2-PI. (2-phenyl imidazoline): 10 μg cm^{-2} (a1, a2), 20 μg cm^{-2} (b1, b2), 40 μg cm^{-2} (c1, c2), and 60 μg cm^{-2} (d1, d2) after 140 hours of exposure in 90% RH atmosphere [43].

Cu corrosion products under varying NaCl loading densities, in both the absence and presence of 2-PI (a corrosion inhibitor). The NaCl loading densities examined include 10 µg cm^{-2}, 20 µg cm^{-2}, 40 µg cm^{-2}, and 60 µg cm^{-2}, and the samples are exposed to a 90% relative humidity (RH) atmosphere for 140 hours. In the absence of 2-PI, the SEM micrographs reveal that as the NaCl loading density increases, corrosion products accumulate and pile up on the Cu surface. This observation is indicative of the gradual corrosion of Cu due to the increased presence of NaCl, a corrosive agent. Upon the addition of 2-PI, the corrosion inhibition performance becomes evident. For NaCl loading densities below 40 µg cm^{-2}, only a few nodular corrosion products are observed to form on the original scratches on the Cu surface. This suggests that 2-PI is effective in preventing the formation of a significant amount of corrosion products under these conditions. However, when the NaCl loading density reaches 60 µg cm^{-2}, the SEM micrographs show a substantial increase in corrosion products on the Cu surface. This observation suggests that the inhibitive performance of 2-PI has been significantly impaired at this higher NaCl loading density. Consequently, the copper surface is shielded from the detrimental impacts of factors such as NaCl loading density, temperature, and relative humidity. The d-orbitals of copper are linked to an increase in the corrosion inhibition processes, further enhancing the protection of the metal.

4.4 IMIDAZOLE AND IMIDAZOLINE AS EFFECTIVE INHIBITORS FOR ALUMINUM

The aluminum-based metallic alloys were widely performed in the automobile, machinery manufacturing, shipping, motor vehicle, welding, and airline industries. Imidazole and imidazoline were widely used as effective inhibitors for aluminum-based metallic materials. For example, He et al. [2] studied the corrosion inhibition processes of AA5052 Al alloys with the 2-phenyl-2-imidazoline by the SEM, contact angle measurements, and electrochemical and weight loss methods. As a result, the following main points were drawn:

  i. The high values in the activation energy of inhibited system show the formation of large activation energy. The corrosion processes were blocked with the high energetic barriers.
  ii. The adsorption of this compound on the aluminum surface is exothermic process, meaning the interaction between the corrosion inhibitor and metal surface releases extra energy into simulated system.
  iii. The chemisorption processes on the aluminum surface also release extra heat, because of the chemical interactions and formation of active chemical complex.

Quraishi et al. [44] synthesized the following imidazoline-based corrosion inhibitors for aluminum: HDI, NI, UDI, and PDI. Their chemical structures were confirmed by the spectroscopic methods. The following was found:

  i. The corrosion protection performance was changed during the rise in the inhibitor concentration and solution temperatures.

ii. The values of entropy of activation, enthalpy of activation, heat of adsorption, free energy of adsorption, and activation energy were estimated.

iii. UDI and NI were the most effective corrosion inhibitors among these imid-azoline-based compounds.

4.5   CONCLUSION

In this chapter, the imidazole- and imidazoline-based organic were reviewed and discussed as heterocyclic corrosion inhibitors for the metal. The reason is that the imidazole- and imidazoline-based organic compounds are cost-effective, easily synthesized, and have good properties. The imidazole- and imidazoline-based organic compounds have nitrogen-based hetero rings, which promote inhibitor adsorption on the metal surface. The five-membered heterocyclics promote the π-electron transformation between the corrosion inhibitor and the metal surface. The imidazole- and imidazoline-based organic compounds are high-polar molecules. This indicator supports the water solubility of these organic compounds. These compounds have a high chelation performance, which promotes donor–acceptor performance. The hydrophobicity of the metal surface was increased after the adsorption of the imidazole- and imidazoline-based organic compounds. The protection efficiency increased with the rise of inhibitor concentrations. The Temkin isotherm offers a more suitable description of inhibitor adsorption on the metal surface, signifying that the anion and cation activities were significantly inhibited. The corrosion protection effectiveness underwent alterations as the inhibitor concentration and solution temperatures increased. Measurements for the activation entropy, activation enthalpy, adsorption heat, adsorption free energy, and activation energy were calculated.

ABBREVIATIONS

[APMIm]Br	1-aminopropyl-3-methylimidazolium bromide
[BMImB]Br$_2$	1,4-bis(3-methylimidazolium-1-yl)butane dibromide
HDI	2-heptadecyl-1,3-imidazoline
M-3	2-(4-nitrophenyl)-4,5-diphenyl-imidazole
M-2	4,5-diphenyl-2-(*p*-tolyl)-imidazole
M-1	2-(4-methoxyphenyl)-4,5-diphenyl-imidazole
MTIPh-1	3-methoxy-4-(1,4,5-triphenyl-1*H*-imidazol-2-yl) phenol
NI	2-nonyl-1,3-imidazoline
PAI	2-(4-(2,4-dichlorophenyl)-5-methyl-1*H*-imidazol-2-yl)pyrazine
PDI	3-(4-(2,4-dichlorophenyl)-5-methyl-1*H*-imidazol-2-yl)pyridine
PMI	2-(4-(2,4-dichlorophenyl)-5-methyl-1*H*-imidazol-2-yl)pyrimidine
[PrMIm]Br	1-propyl-3-methylimidazolium bromide
TTIPh-1	2-(1,4,5-triphenyl-1*H*-imidazol-2-yl) phenol
UDI	2-undecyl-1,3-imidazoline

REFERENCES

1. He, Y., et al., Imidazoline derivative with four imidazole reaction centers as an efficient corrosion inhibitor for anti-CO_2 corrosion. Russian Journal of Applied Chemistry, 2015. **88**(7): 1192–1200.
2. He, X., et al., Inhibition properties and adsorption behavior of imidazole and 2-phenyl-2-imidazoline on AA5052 in 1.0 M HCl solution. Corrosion Science, 2014, **83**: 124–136.
3. Rbaa, M., et al., Synthesis of new halogenated compounds based on 8-hydroxyquinoline derivatives for the inhibition of acid corrosion: Theoretical and experimental investigations. Materials Today Communications, 2022, **33**: 104654.
4. Zhang, L., et al., A novel imidazoline derivative as corrosion inhibitor for P110 carbon steel in hydrochloric acid environment. Petroleum, 2015, **1**(3): 237–243.
5. Ouakki, M., M. Galai, and M. Cherkaoui, Imidazole derivatives as efficient and potential class of corrosion inhibitors for metals and alloys in aqueous electrolytes: A review. Journal of Molecular Liquids, 2022, **345**: 117815.
6. Mishra, A., et al., Imidazoles as highly effective heterocyclic corrosion inhibitors for metals and alloys in aqueous electrolytes: A review. Journal of the Taiwan Institute of Chemical Engineers, 2020, **114**: 341–358.
7. Berdimurodov, E., et al., Novel gossypol–indole modification as a green corrosion inhibitor for low-carbon steel in aggressive alkaline–saline solution. Colloids and Surfaces A: Physicochemical and Engineering Aspects, 2022, **637**: 128207.
8. Berdimurodov, E., et al., Green β-cyclodextrin-based corrosion inhibitors: Recent developments, innovations and future opportunities. Carbohydrate Polymers, 2022, **292**: 119719.
9. Berdimurodov, E., et al., Inhibition properties of 4,5-dihydroxy-4,5-di-*p*-tolylimidazolidine-2-thione for use on carbon steel in an aggressive alkaline medium with chloride ions: Thermodynamic, electrochemical, surface and theoretical analyses. Journal of Molecular Liquids, 2021, **327**: 114813.
10. Berdimurodov, E., et al., Experimental and theoretical assessment of new and eco-friendly thioglycoluril derivative as an effective corrosion inhibitor of St2 steel in the aggressive hydrochloric acid with sulfate ions. Journal of Molecular Liquids, 2021. **335**: 116168.
11. Rbaa, M., et al., Development process for eco-friendly corrosion inhibitors, in Eco-Friendly Corrosion Inhibitors, Elsevier, 2022, pp. 27–42.
12. Shahmoradi, A.R., et al., Theoretical and surface/electrochemical investigations of walnut fruit green husk extract as effective inhibitor for mild-steel corrosion in 1 M HCl electrolyte. Journal of Molecular Liquids, 2021, **338**: 116550.
13. Munis, A., et al., A newly synthesized green corrosion inhibitor imidazoline derivative for carbon steel in 7.5% NH_4Cl solution. Sustainable Chemistry and Pharmacy, 2020, **16**: 100258.
14. Berdimurodov, E., et al., Novel glycoluril pharmaceutically active compound as a green corrosion inhibitor for the oil and gas industry. Journal of Electroanalytical Chemistry, 2022, **907**: 116055.
15. Dewangan, Y., et al., *N*-Hydroxypyrazine-2-carboxamide as a new and green corrosion inhibitor for mild steel in acidic medium: Experimental, surface morphological and theoretical approach. Journal of Adhesion Science and Technology, 2022, **36**: 1–21.
16. Bereket, G., E. Hür, and C. Öğretir, Quantum chemical studies on some imidazole derivatives as corrosion inhibitors for iron in acidic medium. Journal of Molecular Structure: THEOCHEM, 2002, **578**(1–3): 79–88.
17. Mihajlović, M.B.P., et al., Imidazole based compounds as copper corrosion inhibitors in seawater. Journal of Molecular Liquids, 2017, **225**: 127–136.

18. Berdimurodov, E., et al., New and green corrosion inhibitor based on new imidazole derivate for carbon steel in 1 M HCL medium: Experimental and theoretical analyses. International Journal of Engineering Research in Africa, 2022, **58**: 11–44.

19. Curkovic, H.O., E. Stupnisek-Lisac, and H. Takenouti, The influence of pH value on the efficiency of imidazole based corrosion inhibitors of copper. Corrosion Science, 2010. **52**(2): 398–405.

20. Gašparac, R., C.R. Martin, and E. Stupnišek-Lisac, *In situ* studies of imidazole and its derivatives as copper corrosion inhibitors. I. Activation energies and thermodynamics of adsorption. Journal of the Electrochemical Society, 2000. **147**(2): 548.

21. Singh, A., et al., Electrochemical, surface and quantum chemical studies of novel imidazole derivatives as corrosion inhibitors for J55 steel in sweet corrosive environment. Journal of Alloys and Compounds, 2017. **712**: 121–133.

22. Ouakki, M., et al., Electrochemical, thermodynamic and theoretical studies of some imidazole derivatives compounds as acid corrosion inhibitors for mild steel. Journal of Molecular Liquids, 2020, **319**: 114063.

23. Zeng, X., et al., Three imidazole ionic liquids as green and eco-friendly corrosion inhibitors for mild steel in sulfuric acid medium. Journal of Molecular Liquids, 2021, **324**: 115063.

24. Hou, Y., et al., Synthesis of three imidazole derivatives and corrosion inhibition performance for copper. Journal of Molecular Liquids, 2022, **348**: 118432.

25. Berdimurodov, E., et al., Thioglycoluril derivative as a new and effective corrosion inhibitor for low carbon steel in a 1 M HCl medium: Experimental and theoretical investigation. Journal of Molecular Structure, 2021, **1234**: 130165.

26. Berdimurodov, E., et al., New anti-corrosion inhibitor (3*aR*, 6*aR*)-3*a*,6*a*-di-*p*-tolyltetrahydroimidazo[4,5-*d*]imidazole-2,5(1*H*, 3*H*)-dithione for carbon steel in 1 M HCl medium: Gravimetric, electrochemical, surface and quantum chemical analyses. Arabian Journal of Chemistry, 2020, **13**: 7504–7523.

27. Berdimurodov, E., et al., 8-Hydroxyquinoline is key to the development of corrosion inhibitors: An advanced review. Inorganic Chemistry Communications, 2022, **144**: 109839.

28. Costa, S.N., et al., Inhibition of copper corrosion in acid medium by imidazole-based compounds: Electrochemical and molecular approaches. Journal of the Brazilian Chemical Society, 2022, **34**: 309–324.

29. El-Katori, E.E., M. Ahmed, and H. Nady, Imidazole derivatives based on glycourils as efficient anti-corrosion inhibitors for copper in HNO_3 solution: Synthesis, electrochemical, surface, and theoretical approaches. Colloids and Surfaces A: Physicochemical and Engineering Aspects, 2022, **649**: 129391.

30. Otmacic Curkovic, H., E. Stupnisek-Lisac, and H. Takenouti, Electrochemical quartz crystal microbalance and electrochemical impedance spectroscopy study of copper corrosion inhibition by imidazoles. Corrosion Science, 2009, **51**(10): 2342–2348.

31. Simonović, A., et al., Inhibition of copper corrosion in acid rain solution using the imidazole derivatives. Russian Journal of Electrochemistry, 2021, **57**(5): 544–553.

32. Zhu, M., et al., Insights into the newly synthesized *n*-doped carbon dots for Q235 steel corrosion retardation in acidizing media: A detailed multidimensional study. Journal of Colloid and Interface Science, 2022, **608**: 2039–2049.

33. Verma, D.K., et al., *N*-Hydroxybenzothioamide derivatives as green and efficient corrosion inhibitors for mild steel: Experimental, DFT and MC simulation approach. Journal of Molecular Structure, 2021, **1241**: 130648.

34. Solomon, M.M., et al., Myristic acid based imidazoline derivative as effective corrosion inhibitor for steel in 15% HCl medium. Journal of Colloid and Interface Science, 2019, **551**: 47–60.

35. Zhang, K., et al., Halogen-substituted imidazoline derivatives as corrosion inhibitors for mild steel in hydrochloric acid solution. Corrosion Science, 2015, **90**: 284–295.

36. Xiong, L., et al., Corrosion behaviors of Q235 carbon steel under imidazoline derivatives as corrosion inhibitors: Experimental and computational investigations. Arabian Journal of Chemistry, 2021, **14**(2): 102952.
37. Okafor, P.C., X. Liu, and Y.G. Zheng, Corrosion inhibition of mild steel by ethylamino imidazoline derivative in CO_2-saturated solution. Corrosion Science, 2009, **51**(4): 761–768.
38. Feng, L., H. Yang, and F. Wang, Experimental and theoretical studies for corrosion inhibition of carbon steel by imidazoline derivative in 5% NaCl saturated $Ca(OH)_2$ solution. Electrochimica Acta, 2011, **58**: 427–436.
39. Kaur, J., et al., *Euphorbia prostrata* as an eco-friendly corrosion inhibitor for steel: Electrochemical and DFT studies. Chemical Papers, 2022, **77**: 957–976.
40. Haldhar, R., et al., Corrosion inhibitors: Industrial applications and commercialization, in: Sustainable Corrosion Inhibitors II: Synthesis, Design, and Practical Applications. American Chemical Society, 2021, pp. 10–219.
41. Wenchang, Z., et al., Imidazoline Gemini surfactants as corrosion inhibitor for copper in NaCl solution. International Journal of Electrochemical Science, 2020, **15**: 8786–8796.
42. Gonzalez-Rodriguez, J.G., et al., Use of a palm oil-based imidazoline as corrosion inhibitor for copper in 3.5% NaCl solution. International Journal of Electrochemical Science, 2016, **11**(10): 8132–8144.
43. Wan, S., Z. Dong, and X. Guo, Investigation on initial atmospheric corrosion of copper and inhibition performance of 2-phenyl imidazoline based on electrical resistance sensors. Materials Chemistry and Physics, 2021, **262**: 124321.
44. Quraishi, M.A., et al., Corrosion inhibition of aluminium in acid solutions by some imidazoline derivatives. Journal of Applied Electrochemistry, 2007, **37**(10): 1153–1162.

5 Indole- and Indoline-based Corrosion Inhibitors

Sheetal Ghangas[1], Ashish Kumar Singh[1,2],
Manjeet Singh[3], Balaram Pani[4], Sanjeeve Thakur[1]
[1]Department of Chemistry, Netaji Subhas
University of Technology, New Delhi, India
[2]Department of Chemistry, Hansraj College,
University of Delhi, New Delhi, India
[3]Department of Chemistry, School of Physical Sciences,
Mizoram University, Aizawl, Mizoram, India
[4]Department of Chemistry, Bhaskaracharya College of
Applied Sciences, University of Delhi, New Delhi, India

5.1 INTRODUCTION

The chemical and biological activities of heterocyclic compounds promote their use for many applications [1, 2]. Heterocyclic compounds, organic compounds of the cyclic framework, consist of heteroatoms in their structure, like nitrogen, oxygen, sulfur, etc. Among the number of heterocyclic compounds reported so far, a more eminent one is indole. The molecular formula of indole, a bicyclic heterocyclic moiety, is C_8H_7N. Here, a benzene ring and a pyrrole ring have been attached. Indole has been seen present in various natural components: serotonin, melatonin, and amino acids like tryptophan, etc. Additionally, certain synthetic drugs consist of indole, like fluvastatin, tadalafil, sumatriptan, etc., and a survey reported maximum sale of these drugs during 2010 [3].

Indole and its associated derivatives have been widely utilised in numerous biological and medicinal applications involving anti-cancer, anti-tumour, anti-fungal, etc. [4, 5]. It is also utilised in textile industry and agricultural pesticides and insecticides [6]. The availability of nonbonding electrons of nitrogen, π-electrons of double bonds and aromatic rings favour the corrosion inhibition potential of indole, indoline and their derivatives. Indoline is another heterocyclic compound with molecular formula C_8H_9N. Like indole, the chemical structure of indoline comprises benzene fused with five-membered rings containing nitrogen as the central atom. The structural aspect of indole is completely based on indole except for the double bond in the five-membered ring. These have been counted as significant constituents of various applications: biological, pharmaceutical, and industrial [7].

Literature evidence numerous reports on the synthesis and investigation of anti-corrosive behaviour of indole, indolines, and associated structures for metals and corresponding alloys. These have been seen to exhibit protection by getting adsorbed over metallic counterparts, and this chapter provides an overview of indole- and

DOI: 10.1201/9781003377016-5

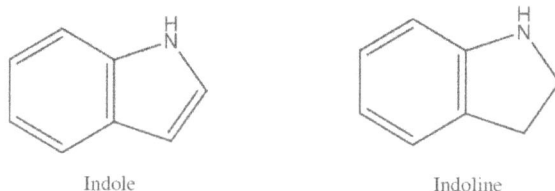

Indole   Indoline

FIGURE 5.1   Cheemical structures of indole and indoline.

indoline-based anti-corrosive moieties utilised for corrosion prevention [8, 9]. The chemical structures of indole and indoline are shown in Figure 5.1.

5.2   SYNTHESIS

Numerous techniques have been documented in the literature supporting the synthesis of indoles and indolines. The most reliable and conventional method reported so far is Fischer indole synthesis which involves the functionalisation of aromatic C—H. During the anti-corrosive studies of indole and its derivatives, substituted derivatives are preferred more owing to the presence of substituents. Some of the methods have been categorised and elaborated in Schemes 5.1–5.4:

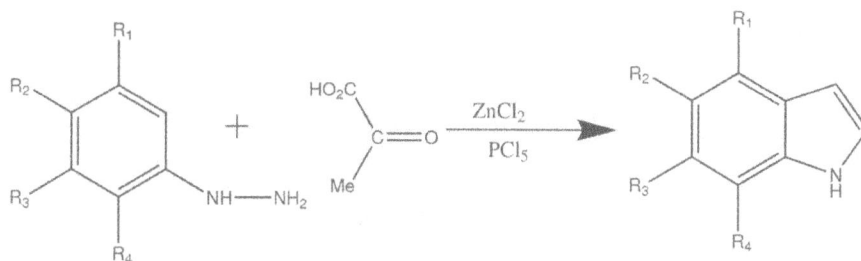

SCHEME 5.1   Reaction between phenylhydrazine and pyruvic acid using microwave irradiation [10].

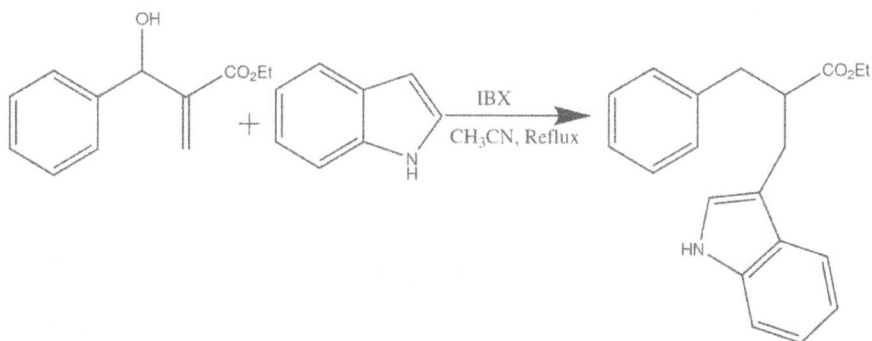

SCHEME 5.2   Michael reaction [11].

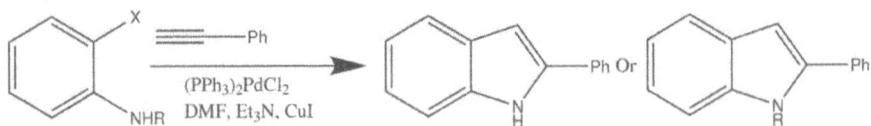

X is Br or I
R is Bn, H, Me, Ts, Ac

SCHEME 5.3   Synthesis of 2-substituted indoles by heteroannulation [12].

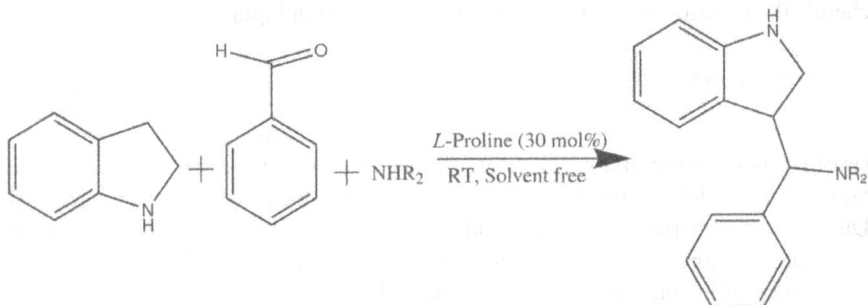

SCHEME 5.4   Mannich reaction involving solvent-free synthesis [13].

5.3   CORROSION AND CORROSION INHIBITORS

There is a variety of potential applications of metals and their derivatives; alloys find their use in the biomedical industry, building, etc. Practically, every industrial sector uses metals for a variety of functions. During industrial metal refinement, descaling, acid pickling, passivation, etc., metals are subject to pH, temperature, and humidity changes. Metals are somewhat sensitive to corrosion due to variations in these factors since a series of electrochemical processes occur on metals when they experience major environmental changes [14, 15].

Additionally, significant financial losses documented in literature occur due to the degradation of metals by various electrochemical reactions. It has been discovered that annual losses from corrosion covered 3.4% of the global GDP [16]. Various appraisals have been carried out to continually gauge the reported losses from direct corrosion. Additionally, in the present time, defending cultural heritage and economic losses has become tough. And for this, scientists working in the field of corrosion have made significant contributions by deploying corrosion inhibitors, nanoparticle coatings, cathodic protection, and galvanisation. The usage of corrosion inhibitors is the strategy that is currently being employed widely. These are the substances that, when included in corrosive solutions, slow down the mechanism of electrochemical reactions taking place over metallic and solution interfaces. These alter the rate of cathodic and anodic reactions taking place without giving rise to a change in electrolytic concentration [17].

Certain factors should be weighed while choosing a corrosion combater, like the solubility it manifests in the electrolyte, the molecular size of particles, and the electron-donating ability of inhibitor moiety. Literature is full of studies investigating the corrosion inhibition effect of numerous categories of heterocyclic compounds: pyrimidine, triazoles, pyridine, etc. [18, 19]. Formerly, investigatory studies focused primarily on increasing the corrosion inhibition effect of organic moieties. However, during the past few years, enhancing inhibitors' eco-friendliness has become a significant interest area. This has led to enormous work in the field of the development of environment-safe corrosion combaters. Indole- and indoline-derived corrosion inhibitors are ecologically safe organic moieties being researched extensively for their anti-corrosive properties [20].

5.4 INDOLES AND INDOLINES AS CORROSION INHIBITORS

Indoles and associated derivatives being heterocyclic moieties like pyridine, quinoline, etc., have been utilised extensively to combat corrosion [21]. Amongst the industrial sector, the petroleum industry has been reported to experience maximum utilisation of corrosion inhibitors as it involves several processes: pickling, acid cleaning, oil well acidising, etc. Alteration in operational conditions like temperature and pressure influences the corrosion rate.

Literature on indole supports its protonation in an acidic medium, and nitrogen atoms catalyse the protonation in the ring [22]. Later electrostatic interactions play a crucial role in the adsorption of protonated indole moiety over metallic counterparts [23]. Experimental studies on electrochemical and weight loss were performed to gain insight into indoles' corrosion mitigating properties. Various spectroscopic studies, like FTIR, XPS, UV-vis, etc., were involved to observe the metallic–inhibitor interactions. Further, surface studies like atomic force microscopy (AFM), scanning electron microscopy (SEM), and electron dispersive spectroscopy (EDS) have been used to evaluate the comparative roughness of metallic surfaces when they are exposed to inhibitor solutions containing indoles. Certain theoretical studies like DFT and molecular dynamics (MD) simulations are involved extensively to perceive the inhibition mechanism at the molecular level.

5.4.1 INDOLES AND INDOLINES AS CORROSION INHIBITORS FOR FERROUS ALLOYS

Ferrous alloys are metals and associated counterparts constituting iron, mild, medium, and high carbon steel. Due to its remarkable properties, low-carbon steel is preferred in almost every industrial sector: petrochemical, construction, etc. It has been reported that mild steel is a kind of recyclable material and is considered green [24]. Literature supports ample evidence favouring indole and derived compounds as corrosion inhibitors for mild steel, Q235 steel, and certain other iron alloys. Ojo et al. reported the anti-corrosive efficiency for two indole derivatives, 6-benzyl-2H-indol-2-one and 3-methylindole, over steel with less carbon. The anti-corrosive efficiency was reported to be maximum at low temperatures. Potentiodynamic studies confirmed that both synthesised inhibitors of mixed type slowed down both the anodic and cathodic reactions. SEM evidenced uniform coverage of low-carbon steel by organic inhibitors. Further, the Gibbs free energy values, i.e. −4.51–8.58 kJ/mol, support the inhibitor's adsorption with the low-carbon steel. Langmuir adsorption isotherm was seen to be followed primarily [25].

Sunil et al. studied two indole derivatives, 2-(1H-indol-3-yl)acetohydrazide (IAH) and 4-(1H-indol-3-yl) butane hydrazide (IBH), for their corrosion-combating potential over mild steel surface in 0.5 M HCl. Electrochemical impedance and gravimetric studies were utilised to evaluate the potential. Langmuir adsorption isotherm correlated with the adsorption data, and physical interactions were responsible for the adsorption of inhibitors over mild steel. The maximum corrosion-averting efficiency exhibited by both inhibitors was 80% for IAH and 94% for IBH, respectively. Surface analysis, SEM, and AFM were further utilised to get an insight into the blockage offered by indole derivatives over steel surfaces (Figure 5.2). IBH was proved comparatively more efficient as a long methylene chain enhanced the corrosion inhibition efficiency [26].

Berdimurodov et al. involved a gossypol-indole-modified compound (GIM) for low-carbon steel in a saline–alkaline solution. The modified moiety is observed to be water soluble owing to more electron-rich active centres in the form of heteroatoms. The corrosion retarding ability of gossypol-indole is also attributed to the presence of these heteroatoms. The potential of the same to retard corrosion was assessed by utilising EIS, PDP, and gravimetric in 1 M NaOH+ 1 M NaCl solution [27].

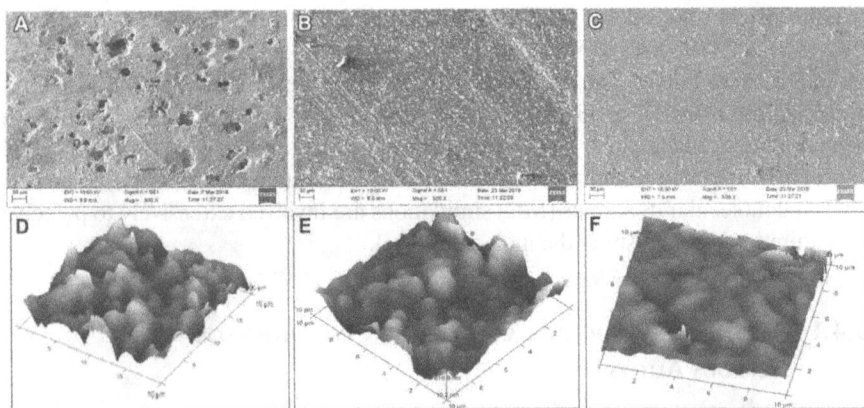

Sample	R_a (nm)	R_q (nm)
MS+HCl	318	238
MS+HCl+IAH	260	245
MS+HCl+IBH	165	126

FIGURE 5.2 SEM images of MS coupons exposed to (A) 0.5 M HCl, (B) IAH+0.5 M HCl, and (C) IBH+0.5 M HCl and AFM images of MS specimen immersed in (D) 0.5 M HCl, (E) IAH+0.5 M HCl, and (F) IBH+0.5 M HCl [26]. (Copyright 2021 Springer.)

Toukal et al. studied the inhibition efficiency of 1-(4-methoxybenzyl)-2-(4-methoxyphenyl)-1H-benzimidazole (MMBI) for XC48 type of iron alloy in 1 M HCl and 0.5 M H_2SO_4 solutions, respectively. Potentiodynamic studies along with electrochemical impedance spectroscopy were used to evaluate the corrosion-mitigating potential of the same. PDP studies signified the same as a mixed type of inhibitor in both electrolytic solutions. MMBI was seen to illustrate sound inhibition efficiency of 97% in HCl solution compared to H_2SO_4, i.e. 92% at a concentration of 10^{-4} M. Here Gibbs free energy values determine the adsorption of MMB over XC48 steel in chemisorption mode. The adsorption was seen to follow the Langmuir adsorption isotherm curve [28].

Liu et al. synthesised 3-(phenylsulfinyl) indoles; 3-(p-tolylsulfinyl)-1H-indole and 3-((4-fluorophenyl)sulfinyl)-1H-indole and surface analysis along with electrochemical impedance spectroscopy have been utilised as evaluating tools for the same in 1 M HCl solution over iron. PDP studies revealed the cathodic predominance of the self-assembled layer of 3-(phenylsulfinyl)indoles over a metallic surface. Inhibition efficiency in both the respective cases was found to be 91.5% for -(p-tolylsulfinyl)-1H-indole and 94.24% for 3-((4-fluorophenyl)sulfinyl)-1H-indole. Later surface studies, i.e. SEM, was utilised to gain insight into surface structure and monolayer self-assembled layers and XPS, on the other hand, revealed the chemical composition of adsorbed layers of indoline derivatives over the iron surface [29].

Taweel et al. investigated the ability of 3-((3-(2-hydroxyphenyl)-5-thioxo-1,2,4-triazol-4-yl)imino)indoline-2-one (HTTI) over mild steel surface in 0.5 M H_2SO_4 solution. The product was synthesised using equimolar concentrations of 2,3-dioxoindoline and 4-amino-5-thioxo-3-(2-hydroxyphenyl)-1,2,4-triazole dissolved in methanol. Certain studies, primarily including weight loss, were utilised as investigatory studies to evaluate the potential. HTTI involves a heterocyclic compound owing to the presence of certain aromatic rings along with lone pairs of heteroatoms incorporated in the structure, depicting an extraordinary range of inhibition efficiency [30].

Later Ashiari et al. investigated the anti-corrosive property of indole-3-carbaldehyde and 2-methylindole for mild steel in 1 M HCl solution. Polarisation and electrochemical impedance spectroscopy have been utilised to evaluate corrosion-preventing potential. The polarisation study revealed the anodic and cathodic nature of synthesised moiety, whereas the electrochemical study revealed an enhancement in the inhibition efficiency of inhibitor with concentration. The maximum inhibition efficiency reported in both cases was 95% and 94%, respectively, for a concentration of 1 mM. Analysis of contact angle measurements was also involved to gain insight into inhibitors' adsorption. Further, a positive charge over benzene enhanced the adsorption of the inhibitors over mild steel surfaces [31].

Fawzy et al. focused primarily on the investigation of three propane bis-oxoindoline indoline derivatives: 1,1'-(propane-1,3-diyl)bis(indoline-2,3-dione) (I), 2,2'-(propane-1,3-diylbis(2-oxoindoline-1-yl-3-ylidene))bis(hydrazine-1- carbothioamide (II), and 1,1'-propane-1,3-diylbis(1H-indole-2,3-dione)-3,3'-bis ({4-methyl-5-[4-methylphenyldiazenyl]-1,3-thiazol-2-yl}hydrazone) (III) as corrosion combaters for mild steel in 1 M H_2SO_4 solution via weight loss, electrochemical, and potentiodynamic studies. A rise in concentration of all three derivatives enhanced the inhibition efficiency, whereas a temperature rise retarded the protection ability of inhibitors. The predominance of the anodic character of inhibitors was seen.

Further, the adsorption was seen following the physical mode of interactions there [32]. Gupta et al. analysed the corrosion-combating potential of two new indole-derived compounds: 6′-amino-3′-methyl-2-oxo-1′H-spiro[indoline-3,4′-pyrano[2,3-c]pyrazole]-5′-carbonitrile (SIPP-1) and 6′-amino-3′-methyl-2-oxo-1′-phenyl-1′H-spiro[indoline-3,4′-pyrano[2,3-c]pyrazole]-5′-carbonitrile (SIPP-2) over mild steel surface in 1 M HCl solution. Synthesis of the utilised compounds was primarily based on green chemistry, i.e. multicomponent reactions. Weight loss studies evaluated the enhancement in corrosion inhibition efficiency, and the maximum inhibition efficiency reported in both cases were 96.65%% (SIPP-1) and 96.95% (SIPP-2) at an optimal concentration of 200 ppm. Adsorption data followed Langmuir adsorption isothermal curves and the potentiodynamic technique further supported the diverse nature of both inhibitors. SEM and AFM were used to gain insight into the surface component of mild steel [33]. A summary of significant works on indole and indoline as corrosion inhibitors for ferrous alloys is presented in Table 5.1.

5.4.2 INDOLES AND INDOLINES AS CORROSION INHIBITORS FOR NON-FERROUS ALLOYS

Literature has been explored all around, and more studies have been conducted on evaluating the anti-corrosive potential of indole and indoline derivatives over iron and associated alloys compared to non-ferrous alloys like aluminium, copper, nickel, zinc, etc. Non-ferrous alloys have several applications in the industrial field due to sound mechanical and thermal performance and less weight [34]. Though non-ferrous alloys possess corrosion resistance, exposure to electrolytic solutions of acids still makes them fall for corrosion and deterioration [35]. Available literature in this regard has been incorporated here in the text.

Damous et al. investigated the anti-corrosive potential of certain indole derivatives, namely indole, isatin, IND3-C2, IND3-C1, IND5- C1, IND2-C2, IND2-C1, and IND5-C2, for copper theoretically. Based on DFT, it has been concluded that the values of Fukui indices in the case of electrophilic and nucleophilic reactions support the adsorption of indole inhibitors over the copper surface by utilising the electrons of aromatic rings and electron pairs of heteroatoms, respectively. In their deprotonated state, it has been observed that the adsorption of indole inhibitors over copper takes place via the formation of chemical bonds, i.e. Cu—N bonds of indole. The inhibition efficiency trend was observed to be indole < IND3-C2 < IND3-C1 < IND5-C1 < IND2-C2 < IND2- C1 < IND5-C2 < isatin [36]. Feng et al. evaluated the corrosion potential of 3,3′-((4-(methylthio)phenyl)methylene)bis(1H-indole) (TPBI) over the copper surface in 0.5 M H_2SO_4 electrolytic solution. Electrochemical impedance spectroscopy has been utilised as an evaluating technique, confirming the synthesised inhibitor's diverse nature. Further, XPS and EDS affirmed the adsorption of TPBI over copper. Langmuir adsorption isotherm was seen to be followed, and chemisorption, as well as physisorption, was seen to prevail in the medium [37].

Scendo et al. investigated the corrosion-preventing potential of indole and 5-chloroindole for copper in acid chloride solution. Voltammetry on rotating disc electrodes was used as an investigating technique primarily. Both the inhibitors depicted potent

TABLE 5.1

Indole- and Indoline-based Corrosion Inhibitors for Ferrous Alloys

Chemical Structure with Abbreviation	Electrolytic Medium	Techniques Used	Inhibition Efficiency	Reference
6-Benzyl-2*H*-indol-2-one	HCl	PDP, SEM, WL	–	[25]
3-Methylindole	HCl	PDP, SEM, WL	–	[25]
IAH	0.5 M HCl	EIS, WL	80%	[26]
IBH	0.5 M HCl	EIS, WL, SEM, AFM	80%	[26]
GIM*	1 M NaOH + 1 M NaCl	EIS, PDP	96.01%	[27]

(*Continued*)

TABLE 5.1   (*Continued*)
Indole- and Indoline-based Corrosion Inhibitors for Ferrous Alloys

Chemical Structure with Abbreviation	Electrolytic Medium	Techniques Used	Inhibition Efficiency	Reference
MMBI	1 M HCl	MC, EIS	97%	[28]
MMBI	1 M H$_2$SO$_4$	MC, EIS	92%	[28]
3-(*p*-Tolylsulfinyl)-1*H*-indole	1 M HCl	EIS, PDP	91.5%	[29]
3-((4-Fluorophenyl)sulfinyl)-1*H*-indole	1 M HCl	EIS, PDP	94.24%	[29]

(*Continued*)

TABLE 5.1   (*Continued*)
Indole- and Indoline-based Corrosion Inhibitors for Ferrous Alloys

Chemical Structure with Abbreviation	Electrolytic Medium	Techniques Used	Inhibition Efficiency	Reference
HTTI	0.5 M HCl	WL	–	[30]
Indole-3-carbaldehyde	1 M HCl	EIS, PDP	95%	[31]
2-Methylindole	1 M HCl	EIS, PDP	94%	[31]
I	1 M H$_2$SO$_4$	EIS, WL, PDP	72%	[32]

(*Continued*)

TABLE 5.1   (Continued)
Indole- and Indoline-based Corrosion Inhibitors for Ferrous Alloys

Chemical Structure with Abbreviation	Electrolytic Medium	Techniques Used	Inhibition Efficiency	Reference
II	1 M H₂SO₄	EIS, PDP, WL	81%	[32]
III	1 M H₂SO₄	EIS, PDP, WL	77%	[32]
SIPP-1	1 M HCl	WL, SEM, AFM	96.65%	[33]
SIPP-2	1 M HCl	WL, SEM, AFM	96.95%	[33]

inhibition at an optimum concentration of 10^{-3} M. It was observed that a decrease in the pH of the electrolyte brought an enhancement in the inhibitory action of the two. The anodic nature of both the synthesised inhibitors was illustrated, and the necessary preventive action was inferred to form a coating over the copper surface [38]. Quartarone et al. investigated the anti-corrosive potential explained by various indole concentrations over copper surfaces in 0.5 M H_2SO_4. Weight loss technique along with potentiodynamic polarisation curves was used for investigation. Results revealed that indole depicted potent inhibition towards corrosion at a concentration of or greater than 2.5×10^{-3} M. In contrast, potentiodynamic polarisation proved indole as a mixed kind of inhibitor inhibiting both cathodic and anodic reactions, predominantly anodic ones. Frumkin adsorption isotherm was seen to be followed during the adsorption of indole over a copper surface. On the other hand, the values of Gibbs free energy revealed the involvement of physical and chemical interactions by indole during adsorption over the copper surface [39].

Later, Quartarone et al. performed the anti-corrosive analysis of indole-2-carboxylic acid for copper in 0.5 M H_2SO_4 solution. Anti-corrosive behaviour of indole-2-carboxylic acid was seen to be prevalent during the temperature range of 25–55°C. Polarisation studies illustrated the inhibitor's mixed inhibition behaviour, whereas chemisorption and physisorption were seen to be followed during inhibitor adsorption over metal surfaces [40]. Feng et al., in a recent study, focused on synthesising self-assembled monolayers by designing two indole-based moieties with O-heterocyclic and S-heterocyclic structures. The corrosion mitigation analysis was investigated for copper in 0.5 M H_2SO_4 solution. Electrochemical techniques and surface morphological studies were utilised to evaluate the same. It was concluded that compounds having sulfur (TYBI) as heterocyclic moieties exhibited better inhibition performance. Theoretical studies also revealed more electron donation from TYBI to vacant d-orbitals of copper than FYBI. Greater contact angle and film thickness supported the corrosion-inhibiting efficiency demarcated by TYBI compared to FYBI [41].

Feng et al. investigated a new compound, namely 4-(bis(5-bromo-1H-indol-3-yl) methyl)phenol (BMP), as a corrosion inhibitor for copper in 0.5 M H_2SO_4. Potentiodynamic studies revealed the mixed inhibition executed by BMP, and it has been observed that the effective inhibition efficiency was 99.06% at an optimal concentration of 10 mmol/L. Further, in the same manner, surface studies of copper were done using XPS, and the detailed insight into elements adsorbed over copper strip confirmed the coating of synthesised moiety over copper, respectively [42]. The FT-IR spectra of (a) BMP powder (b) and the protective film of BMP adsorbed on a copper surface are shown in Figure 5.3.

Amino acids like tryptophan consist of indole moiety as a prime constituent of its structure. Utilising this approach, Feng et al. investigated the anti-corrosive potential of tryptophan for aluminium in 1 M HCl+ 1 M H_2SO_4. A rise in inhibition efficiency was seen with temperature decrease and concentration increase. This corrosion inhibition efficiency is attributed to the involvement of sulfur and nitrogen in the structure of tryptophan. Anodic and cathodic reactions seemed to be affected due to tryptophan in the reaction media, thereby making it a mixed type of inhibitor. Langmuir adsorption isotherm was seen to be followed [43]. Shafei et al. studied three indole derivatives, indole (I), tryptamine (II), and tryptophan (III), as corrosion combaters

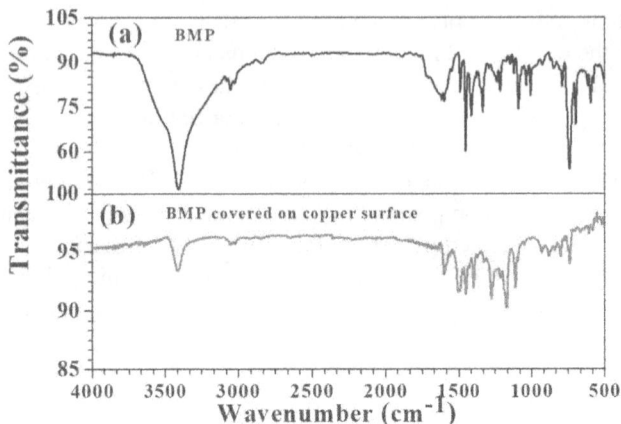

FIGURE 5.3 FT-IR spectra of (a) BMP powder (b) and the protective film of BMP adsorbed on the copper surface [42]. (Copyright 2020 Elsevier.)

for aluminium in NaCl solution. Electrochemical results depicted all three inhibiting corrosion when exposed to aggressive chloride ions. The order obtained for corrosion inhibition efficiency of all three was I < III < II. The least efficiency depicted by I is attributed to their formation of polymeric layers over the platinum surface during the cycle sweep [44]. A summary of significant works on indole and indoline as corrosion inhibitors for non-ferrous alloys is presented in Table 5.2.

5.5 MECHANISM OF CORROSION INHIBITION BY INDOLES AND INDOLINES

In general, one of the two basic types of interaction – physisorption (physical adsorption) or chemisorption (chemical adsorption) – can explain the adhesion of organic molecules to metal surfaces. Adsorption typically happens due to the combination of these two mechanisms [45, 46]. In physisorption, indoles adhere to the substrate's surface due to weak electrostatic force between the charged indoles and the electrically charged substrate surface. Conversely, chemisorption entails establishing a coordinating covalent connection through charge transfer or share of electrons between indoles and metal surfaces. Chemisorption has more considerable adsorption energy and irreversible nature, for which it has been proven to be more stable at higher temperatures. In contrast, physisorption is frequently linked with low energy and is regular only at low temperatures. Adsorption of indole and associated derivatives can be stated as a quasi-substitution process involving replacing water and other ions in the medium [47]. For instance,

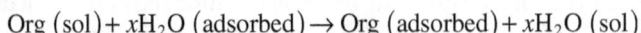

$$\text{Org (sol)} + x\text{H}_2\text{O (adsorbed)} \rightarrow \text{Org (adsorbed)} + x\text{H}_2\text{O (sol)}$$

Org (sol) and Org (adsorbed) refer to indoles in the solution and adsorbed layer, respectively, whereas x refers to the number of water molecules replaced with indoles in the medium. Similarly, H$_2$O (adsorbed) and H$_2$O (sol) refer to the water molecules

TABLE 5.2 (*Continued*)
Indole- and Indoline-based Corrosion Inhibitors for Non-Ferrous Alloys

Chemical Structure with Abbreviation	Electrolytic Medium	Techniques Used	Inhibition Efficiency	Reference
Indole	–	DFT	–	[36]
Isatin	–	DFT	–	[36]
IND-2	–	DFT	–	[36]
IND-3	–	DFT	–	[36]
IND-5	–	DFT	–	[36]

(*Continued*)

TABLE 5.2   (*Continued*)
Indole- and Indoline-based Corrosion Inhibitors for Non-Ferrous Alloys

Chemical Structure with Abbreviation	Electrolytic Medium	Techniques Used	Inhibition Efficiency	Reference
TBPI	0.5 M H₂SO₄	EIS, PDP	–	[37]
Indole	Acid chloride	Voltammetry	–	[38]
5-Chloroindole	Acid chloride	Voltammetry	–	[38]
Indole	0.5 M H₂SO₄	WL, PDP	–	[39]
Indole-2-carboxylic acid	0.5 M H₂SO₄	WL, PDP	–	[40]
TYBI	0.5 M H₂SO₄	EIS, PDP	98.58%	[41]

(Continued)

TABLE 5.2 *(Continued)*

Indole- and Indoline-based Corrosion Inhibitors for Non-Ferrous Alloys

Chemical Structure with Abbreviation	Electrolytic Medium	Techniques Used	Inhibition Efficiency	Reference
FYBI	0.5 M H$_2$SO$_4$	EIS, PDP	98.40%	[41]
BMP	0.5 M H$_2$SO$_4$	EIS, PDP	99.06%	[42]
Tryptophan	1 M H$_2$SO$_4$ + 1 M HCl	EIS, PDP	–	[43]
I	NaCl	EIS, PDP	–	[44]

(Continued)

TABLE 5.2   (Continued)

Indole- and Indoline-based Corrosion Inhibitors for Non-Ferrous Alloys

Chemical Structure with Abbreviation	Electrolytic Medium	Techniques Used	Inhibition Efficiency	Reference
  II	NaCl	EIS, PDP	–	[44]
  III	NaCl	EIS, PDP	–	[44]

in adsorbed form and solution. Additionally, adsorption isotherms like Langmuir, Temkin, and Frumkin are being utilised to explicate the adsorption of indoles over metallic surface [48]. From the above-mentioned studies, it has been derived that most of the inhibitors during adsorption over metallic counterparts follow Langmuir adsorption isothermal curves. Values of adsorption constant, i.e. K_{ads}, reveal the strength of adsorption adopted by indoles and metal counterparts.[48] Further, values of Gibbs free energy correspond to the nature of adsorption opted by indoles, i.e. obtained values of $G < 20$ kJ/mol corresponds to the prevalence of electrostatic interactions. In contrast, $G > 40$ kJ/mol correspond to chemical interactions [49]. There could be many factors responsible for organic corrosion molecules over metal surfaces (Figure 5.4) [50]:

1. Donation of electrons from the aromatic ring to vacant d-orbitals of the metal
2. Donation of unshared or lone pairs of electrons from heteroatoms to vacant d-orbitals of the metal
3. Interactions (electrostatic) of protonated organic inhibitors with negative charges like Cl⁻ adsorbed over metal strips

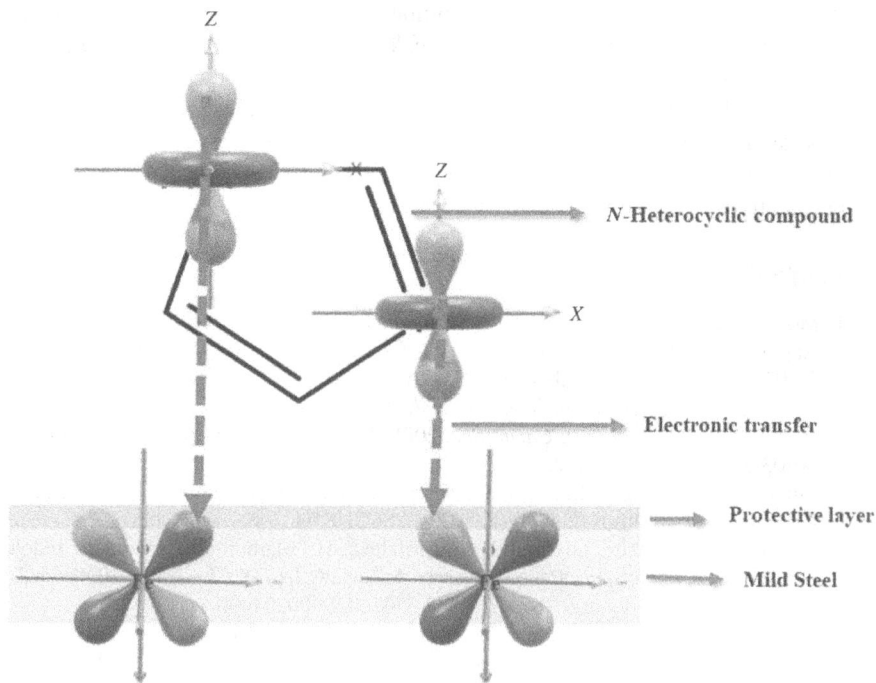

FIGURE 5.4 Proposed model preferred by nitrogen moiety (imidazole) for adsorption over mild steel [50]. (Copyright 2022 Taylor & Francis.)

5.6 CONCLUSION AND OUTLOOK

Corrosion scientists and engineers must thoroughly examine and capitalise on the inhibitory abilities of environment-safe organic moieties given the rising demands and the prediction that the market for inhibitors will boom shortly due to the acceleration in the economic growth of emerging economies. As per requirement, indoles and indolines have been seen as affordable and efficient corrosion inhibitors. The studies that investigated the inhibitory potentials of indoles in all researched mediums and for all studied metal kinds have been collected in this chapter. Testing approaches such as chemical, electrochemical, surface morphology, and theoretical modelling techniques have been used to explore indole derivatives. As a result, these techniques depict indoles' adsorption over metallic counterparts and, at this moment, offer a sense of protection. Primarily, the adsorption data in such cases have been reported to follow Langmuir adsorption isotherms, and a mixed type of inhibition has been seen.

Several theoretical studies like DFT, MD, QSAR, etc. have given an insight into molecular-level interactions observed at the metal–inhibitor interface. Substituents significantly affect the corrosion inhibition efficiency; electron-donating groups have been seen enhancing the inhibition efficiency, whereas electron-withdrawing groups have been seen retarding the corrosion inhibition efficiency provided. Indole compounds must be commercialised for large-scale industrial usage to provide a secure and environment-safe replacement to traditional chemicals that have now been employed in the surface protection of metal. All heterocyclic corrosion inhibitors incorporated so far in literature for corrosion retardation do not add to the environment, so there is a need to develop organic moieties that act as potent corrosion combaters and have environment-friendly nature.

REFERENCES

1. Mohammadi Ziarani, G.; Moradi, R.; Ahmadi, T.; Lashgari, N. Recent Advances in the Application of Indoles in Multicomponent Reactions. *RSC Adv.* 2018, *8* (22), 12069–12103. https://doi.org/10.1039/C7RA13321A.
2. Da Silva, J. F. M.; Garden, S. J.; Pinto, A. C. The Chemistry of Isatins: A Review from 1975 to 1999. *J. Braz. Chem. Soc.* 2001, *12* (3), 273–324. https://doi.org/10.1590/S0103-50532001000300002.
3. Inman, M.; Moody, C. J. Indole Synthesis: Something Old, Something New. *Chem. Sci.* 2012, *4* (1), 29–41. https://doi.org/10.1039/C2SC21185H.
4. Sachdeva, H.; Mathur, J.; Guleria, A.; Sachdeva, H.; Mathur, J.; Guleria, A. Indole Derivatives as Potential Anticancer Agents: A Review. *J. Chil. Chem. Soc.* 2020, *65* (3), 4900–4907. https://doi.org/10.4067/S0717-97072020000204900.
5. Singh, T. P.; Singh, O. M. Recent Progress in Biological Activities of Indole and Indole Alkaloids. *Mini-Reviews Med. Chem.* 2017, *18* (1). https://doi.org/10.2174/13895575176 66170807123201.
6. Tang, L.; Peng, T.; Wang, G.; Wen, X.; Sun, Y.; Zhang, S.; Liu, S.; Wang, L. Design, Synthesis and Preliminary Biological Evaluation of Novel Benzyl Sulfoxide 2-Indolinone Derivatives as Anticancer Agents. *Molecules.* 2017, *22* (11), 1979. https://doi.org/10.3390/MOLECULES22111979.
7. Barden, T. C. Indoles: Industrial, Agricultural and Over-the-Counter Uses. 2010, 31–46. https://doi.org/10.1007/7081_2010_48.
8. Costa, Â. C. F.; Cavalcanti, S. C. H.; Santana, A. S.; Lima, A. P. S.; Brito, T. B.; Oliveira, R. R. B.; Macêdo, N. A.; Cristaldo, P. F.; Araújo, A. P. A.; Bacci, L. Insecticidal Activity of Indole Derivatives against *Plutella xylostella* and Selectivity to Four Non-Target Organisms. *Ecotoxicology.* 2019, *28* (8), 973–982. https://doi.org/10.1007/S10646-019-02095-1/FIGURES/5.
9. Mudila, H.; Prasher, P.; Kumar, M.; Kumar, A.; Zaidi, M. G. H.; Kumar, A. Critical Analysis of Polyindole and Its Composites in Supercapacitor Application. *Mater. Renew. Sustain. Energy.* 2019, *8* (2), 1–19. https://doi.org/10.1007/S40243-019-0149-9/FIGURES/11.
10. Bratulescu, G. A New and Efficient One-Pot Synthesis of Indoles. *Tetrahedron Lett.* 2008, *49* (6), 984–986. https://doi.org/10.1016/J.TETLET.2007.12.015.
11. Yadav, J. S.; Reddy, B. V. S.; Singh, A. P.; Basak, A. K. The First One-Pot Oxidative Michael Reaction of Baylis–Hillman Adducts with Indoles Promoted by Iodoxybenzoic Acid. *Tetrahedron Lett.* 2007, *48* (24), 4169–4172. https://doi.org/10.1016/J.TETLET.2007.04.086.
12. Oskooie, H. A.; Heravi, M. M.; Behbahani, F. K. A Facile, Mild and Efficient One-Pot Synthesis of 2-Substituted Indole Derivatives Catalyzed by Pd(PPh3)2Cl2. *Molecules.* 2007, *12* (7), 1438–1446. https://doi.org/10.3390/12071438.
13. Kumar, A.; Gupta, M. K.; Kumar, M. L-Proline Catalysed Multicomponent Synthesis of 3-Amino Alkylated Indoles via a Mannich-type Reaction under Solvent-free Conditions. *Green Chem.* 2012, *14* (2), 290–295. https://doi.org/10.1039/C1GC16297G.

14. Active Protective Coatings. 2016, *233*. https://doi.org/10.1007/978-94-017-7540-3.

15. Olasunkanmi, L. O.; Obot, I. B.; Kabanda, M. M.; Ebenso, E. E. Some Quinoxalin-6-yl Derivatives as Corrosion Inhibitors for Mild Steel in Hydrochloric Acid: Experimental and Theoretical Studies. *J. Phys. Chem. C.* 2015, *119* (28), 16004–16019. https://doi.org/10.1021/ACS.JPCC.5B03285/ASSET/IMAGES/LARGE/JP-2015-03285A_0013.JPEG.

16. Biezma, M. V.; San Cristóbal, J. R. Methodology to Study Cost of Corrosion. *40* (4), 344–352. https://doi.org/10.1179/174327805X75821.

17. Palanisamy, G. Corrosion Inhibitors. *Corros. Inhib.* 2019. https://doi.org/10.5772/INTECHOPEN.80542.

18. Olasunkanmi, L. O.; Obot, I. B.; Ebenso, E. E. Adsorption and Corrosion Inhibition Properties of *N*-{*n*-[1-*R*-5-(Quinoxalin-6-yl)-4,5-Dihydropyrazol-3-yl]Phenyl}Methanesulfonamides on Mild Steel in 1 M HCl: Experimental and Theoretical Studies. *RSC Adv.* 2016, *6* (90), 86782–86797. https://doi.org/10.1039/C6RA11373G.

19. Verma, C.; Olasunkanmi, L. O.; Ebenso, E. E.; Quraishi, M. A. Substituents Effect on Corrosion Inhibition Performance of Organic Compounds in Aggressive Ionic Solutions: A Review. *J. Mol. Liq.* 2018, *251*, 100–118. https://doi.org/10.1016/J.MOLLIQ.2017.12.055.

20. Verma, C.; Quraishi, M. A.; Ebenso, E. E.; Obot, I. B.; El Assyry, A. 3-Amino Alkylated Indoles as Corrosion Inhibitors for Mild Steel in 1 M HCl: Experimental and Theoretical Studies. *J. Mol. Liq.* 2016, *219*, 647–660. https://doi.org/10.1016/J.MOLLIQ.2016.04.024.

21. Marinescu, M. Recent Advances in the Use of Benzimidazoles as Corrosion Inhibitors. *BMC Chem.* 2019, *13* (1), 1–21. https://doi.org/10.1186/S13065-019-0655-Y/TABLES/4.

22. Düdükcü, M.; Yazici, B.; Erbil, M. The Effect of Indole on the Corrosion Behaviour of Stainless Steel. *Mater. Chem. Phys.* 2004, *87* (1), 138–141. https://doi.org/10.1016/J.MATCHEMPHYS.2004.05.043.

23. Tussolini, M.; Viomar, A.; Gallina, A. L.; do Prado Banczek, E.; da Cunha Taras, M.; Pinto Rodrigues, P. R. Electrochemical Behavior of Indole for AISI 430 Stainless Steel in Changing the Media from 1 Mol L^{-1} H$_2$SO$_4$ to 1 Mol L^{-1} HCl. *Rem Rev. Esc. Minas.* 2013, *66* (2), 215–220. https://doi.org/10.1590/S0370-44672013000200012.

24. Broadbent, C. Steel's Recyclability: Demonstrating the Benefits of Recycling Steel to Achieve a Circular Economy. *Int. J. Life Cycle Assess.* 2016, *21* (11), 1658–1665. https://doi.org/10.1007/S11367-016-1081-1/FIGURES/8.

25. Ojo, F. K.; Adejoro, I. A.; Lori, J. A.; Oyeneyin, O. E.; Akpomie, K. G. Indole Derivatives as Organic Corrosion Inhibitors of Low Carbon Steel in HCl Medium: Experimental and Theoretical Approach. *Chem. Africa.* 2022, *5* (4), 943–956. https://doi.org/10.1007/S42250-022-00378-5.

26. Sunil, D.; Kumari, P.; Shetty, P.; Rao, S. A. Indole Hydrazide Derivatives as Potential Corrosion Inhibitors for Mild Steel in HCl Acid Medium: Experimental Study and Theoretical Calculations. *Trans. Indian Inst. Met.* 2022, *75* (1), 11–25. https://doi.org/10.1007/S12666-021-02382-8/FIGURES/9.

27. Berdimurodov, E.; Kholikov, A.; Akbarov, K.; Guo, L.; Kaya, S.; Katin, K. P.; Verma, D. K.; Rbaa, M.; Dagdag, O.; Haldhar, R. Novel Gossypol–Indole Modification as a Green Corrosion Inhibitor for Low-Carbon Steel in Aggressive Alkaline–Saline Solution. *Colloids Surfaces A: Physicochem. Eng. Asp.* 2022, *637*, 128207. https://doi.org/10.1016/J.COLSURFA.2021.128207.

28. Toukal, L.; Foudia, M.; Haffar, D.; Aliouane, N.; Al-Noaimi, M.; Bellal, Y.; Elmsellem, H.; Abdel-Rahman, I. Monte Carlo Simulation and Electrochemical Performance Corrosion Inhibition Whid Benzimidazole Derivative for XC48 Steel in 0.5 M H$_2$SO$_4$ and 1.0 M HCl Solutions. *J. Indian Chem. Soc.* 2022, *99* (9), 100634. https://doi.org/10.1016/J.JICS.2022.100634.

29. Liu, Y.; Wang, Z.; Zhang, Z.; Wang, Y.; Cheng, J.; Li, H. J.; Wu, Y. C. Investigation of 3-(Phenylsulfinyl)Indoles Self-assembled Monolayer for the Inhibition of Iron Corrosion in Acidic Media. *Mater. Corros.* 2022, *73* (9), 1490–1504. https://doi.org/10.1002/MACO.202213120.

30. Protection of Mild Steel in H$_2$SO$_4$ Solution with 3-((3-(2-Hydroxyphenyl)-5-Thioxo-1,2,4-Triazol-4-yl)Imino)Indolin-2-One. *Int. J. Corros. Scale Inhib.* doi: 10.17675/2305-6894-2020-9-3-14.

31. Ashhari, S.; Sarabi, A. A. Indole-3-carbaldehyde and 2-Methylindole as Corrosion Inhibitors of Mild Steel during Pickling. *Pigment Resin Technol.* 2015, *44* (5), 322–329. https://doi.org/10.1108/PRT-11-2014-0104/FULL/PDF.

32. Fawzy, A.; Farghaly, T. A.; Bahir, A. A. A.; Hameed, A. M.; Alharbi, A.; El-Ossaily, Y. A. Investigation of Three Synthesized Propane Bis-oxoindoline Derivatives as Inhibitors for the Corrosion of Mild Steel in Sulfuric Acid Solutions. *J. Mol. Struct.* 2021, *1223*, 129318. https://doi.org/10.1016/J.MOLSTRUC.2020.129318.

33. Gupta, N. K.; Haque, J.; Salghi, R.; Lgaz, H.; Mukherjee, A. K.; Quraishi, M. A. Spiro[indoline-3,4′-pyrano[2,3-*c*]pyrazole] Derivatives as Novel Class of Green Corrosion Inhibitors for Mild Steel in Hydrochloric Acid Medium: Theoretical and Experimental Approach. *J. Bio- Tribo-Corros.* 2018, *4* (2), 1–16. https://doi.org/10.1007/S40735-018-0132-5/FIGURES/14.

34. Khaled, K. F. Guanidine Derivative as a New Corrosion Inhibitor for Copper in 3% NaCl Solution. *Mater. Chem. Phys.* 2008, *112* (1), 104–111. https://doi.org/10.1016/J.MATCHEMPHYS.2008.05.052.

35. Xhanari, K.; Finšgar, M.; Hrnčič, M. K.; Maver, U.; Knez, Ž.; Seiti, B. Green Corrosion Inhibitors for Aluminium and Its Alloys: A Review. *RSC Adv.* 2017, *7* (44), 27299–27330. https://doi.org/10.1039/C7RA03944A.

36. Damous, M.; Allal, H.; Belhocine, Y.; Maza, S.; Merazig, H. Quantum Chemical Exploration on the Inhibition Performance of Indole and Some of Its Derivatives against Copper Corrosion. *J. Mol. Liq.* 2021, *340*, 117136. https://doi.org/10.1016/J.MOLLIQ.2021.117136.

37. Feng, Y.; Feng, L.; Sun, Y.; He, J. The Inhibition Mechanism of a New Synthesized Indole Derivative for Copper in Acidic Environment via Experimental and Theoretical Study. *J. Mater. Res. Technol.* 2020, *9* (1), 584–593. https://doi.org/10.1016/J.JMRT.2019.10.087.

38. Scendo, M.; Poddebniak, D.; Malyszko, J. Indole and 5-Chloroindole as Inhibitors of Anodic Dissolution and Cathodic Deposition of Copper in Acidic Chloride Solutions. *J. Appl. Electrochem.* 2003, *33* (3), 287–293. https://doi.org/10.1023/A:1024117230591.

39. Quartarone, G.; Moretti, G.; Bellomi, T.; Capobianco, G.; Zingales, A. Using Indole to Inhibit Copper Corrosion in Aerated 0.5 M Sulfuric Acid. *Corrosion.* 1998, *54* (8), 606–618. https://doi.org/10.5006/1.3287636.

40. Quartarone, G.; Zingales, A.; Bellomi, T.; Bonaldo, L.; Gajo, M.; Gajo, G.; Paolucci, G. Corrosion Inhibition of Copper in Aerated 0.5 M Sulfuric Acid by Indole-2-Carboxylic Acid. *Corrosion.* 2005, *61* (11), NACE-05111041. https://onepetro.org/corrosion/article-abstract/116683/Corrosion-Inhibition-of-Copper-in-Aerated-0-5-M.

41. Feng, L.; Ren, X.; Feng, Y.; Tan, B.; Zhang, S.; Li, W.; Liu, J. Self-Assembly of New O- and S-Heterocycle-based Protective Layers for Copper in Acid Solution. *Phys. Chem. Chem. Phys.* 2020, *22* (8), 4592–4601. https://doi.org/10.1039/C9CP06910K.

42. Feng, Y.; Feng, L.; Wang, Z.; Zhang, X. Surface Analysis of 4-(Bis(5-bromo-1*H*-indol-3-yl)methyl)phenol Adsorbed on Copper by Spectroscopic Experiments. *Spectrochim. Acta Part A Mol. Biomol. Spectrosc.* 2020, *228*, 117752. https://doi.org/10.1016/J.SAA.2019.117752.

43. Ashassi-Sorkhabi, H.; Ghasemi, Z.; Seifzadeh, D. The Inhibition Effect of Some Amino Acids towards the Corrosion of Aluminum in 1 M HCl + 1 M H$_2$SO$_4$ Solution. *Appl. Surf. Sci.* 2005, *249* (1–4), 408–418. https://doi.org/10.1016/J.APSUSC.2004.12.016.

44. El-Shafei, A. A.; El-Maksoud, S. A. A.; Fouda, A. S. The Role of Indole and Its Derivatives in the Pitting Corrosion of Al in Neutral Chloride Solution. *Corros. Sci.* 2004, *46* (3), 579–590. https://doi.org/10.1016/S0010-938X(03)00067-2.

45. Ansari, K. R.; Quraishi, M. A.; Singh, A. Isatin Derivatives as a Non-Toxic Corrosion Inhibitor for Mild Steel in 20% H_2SO_4. *Corros. Sci.* 2015, *95*, 62–70. https://doi.org/10.1016/J.CORSCI.2015.02.010.

46. Lebrini, M.; Robert, F.; Vezin, H.; Roos, C. Electrochemical and Quantum Chemical Studies of Some Indole Derivatives as Corrosion Inhibitors for C38 Steel in Molar Hydrochloric Acid. *Corros. Sci.* 2010, *52* (10), 3367–3376. https://doi.org/10.1016/J.CORSCI.2010.06.009.

47. Quartarone, G.; Ronchin, L.; Vavasori, A.; Tortato, C.; Bonaldo, L. Inhibitive Action of Gramine towards Corrosion of Mild Steel in Deaerated 1.0 M Hydrochloric Acid Solutions. *Corros. Sci.* 2012, *64*, 82–89. https://doi.org/10.1016/J.CORSCI.2012.07.008.

48. Ahamad, I.; Prasad, R.; Quraishi, M. A. Adsorption and Inhibitive Properties of Some New Mannich Bases of Isatin Derivatives on Corrosion of Mild Steel in Acidic Media. *Corros. Sci.* 2010, *52* (4), 1472–1481. https://doi.org/10.1016/J.CORSCI.2010.01.015.

49. Singh, A. K.; Pandey, A. K.; Banerjee, P.; Saha, S. K.; Chugh, B.; Thakur, S.; Pani, B.; Chaubey, P.; Singh, G. Eco-Friendly Disposal of Expired Anti-Tuberculosis Drug Isoniazid and Its Role in the Protection of Metal. *J. Environ. Chem. Eng.* 2019, *7* (2), 102971. https://doi.org/10.1016/J.JECE.2019.102971.

50. Sheetal; Batra, R.; Singh, A. K.; Singh, M.; Thakur, S.; Pani, B.; Kaya, S. Advancement of Corrosion Inhibitor System through N-Heterocyclic Compounds: A Review. 2022, 1–29. https://doi.org/10.1080/1478422X.2022.2137979.

6 Naphthyridine, Phthalocyanine, and Their Derivative-based Corrosion Inhibitors

Sandeep Yadav,[1,2] Anirudh Pratap Singh Raman,[1,2] Madhur Babu Singh,[1,2] Pallavi Jain,[2] Prashant Singh,[1] and Kamlesh Kumari[3]
[1]Department of Chemistry, Atma Ram Sanatan Dharma College, University of Delhi, Delhi, India
[2]Department of Chemistry, Faculty of Engineering and Technology, SRM Institute of Science and Technology, Uttar Pradesh, India
[3]Department of Zoology, University of Delhi, Delhi, India

6.1 INTRODUCTION

Exploring organic molecules as inhibitors is an effective and economical way to avoid or minimise corrosion. These compounds provide a hydrophobic layer on the metal surface, preventing corrosive species from attacking it. Polar functional groups and numerous electron-rich bonds serve as adsorption centres [1]. Nitrogen-containing heterocyclic compounds, including pyridine, imidazole, and quinolone, have performed better than other examined organic compounds and are more commonly explored [2, 3]. The reason is that the presence of lone pairs of electrons on nitrogen improves the adsorption of these molecules on the surface of metal [4, 5].

Heterocyclic compounds interact chemically and electrostatically with metal surfaces using their electron-rich sites to bind to the metal surface. Even ionic or electrostatic interactions between metallic surfaces and inhibitors contribute significantly to physisorption or electrostatic adsorption [6]. Chemisorption occurs during the sharing of the charges with one another. The standard Gibb's free energy (G) value can be used to evaluate the nature of the adsorption of inhibitors. Physisorption is related to the G value of -20 kJ mol^{-1} or higher, and chemisorption is related to the G value of -40 kJ mol^{-1} or higher. According to a review of the literature, majority of times, organic corrosion inhibitors adhere to metallic surfaces through the physiochemisorption process, for which the value of G ranges from -20 kJ mol^{-1} to -40 kJ mol^{-1}[7, 8].

DOI: 10.1201/9781003377016-6

6.1.1 STRUCTURES AND BASIC PROPERTIES OF NAPHTHYRIDINE AND PHTHALOCYANINE

6.1.1.1 Naphthyridine: Fundamental

Naphthyridines (NTDs) are heterocyclic derivatives of naphthalenes that contain nitrogen, and, therefore, NTDs are referred to as diazanaphthalenes in the nomenclature replacement system. Similarly, it is known as pyridopyridines in the nomenclature for the fused heterocyclic system [9]. "Naphthyridines" refer to the family of six diazanaphthalenes (Figure 6.1) with a nitrogen atom in each ring [10, 11].

6.1.1.2 Synthesis of Naphthyridine Derivatives

From Nicotinaldehydes

Using nicotinaldehydes as an initial material, it is possible to create various derivatives of NTD, including groups like carboxylic acid, aryl, and pyrazolo. By employing the Friedlander reaction with 4-aminonicotinaldehyde or its derivative, the sodium salt of 2-oxopropanoic acid, and NaOH as a catalyst, Chan et al. (1999) produced the carboxylic acid derivative (1,6-naphthyridine-2-carboxylic acid) in substantial amounts (Scheme 6.1). In the Friedlander reaction, a 2-amino-substituted carbonyl molecule (aromatic) is combined with a suitable carbonyl-substituted derivative that contains a reactive methylene group, followed by cyclodehydration in the acidic or basic catalyst [12].

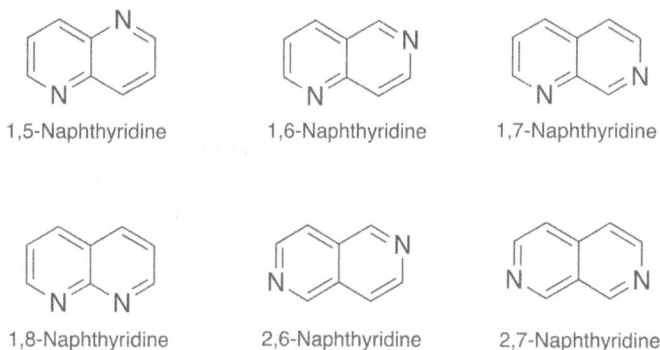

1,5-Naphthyridine   1,6-Naphthyridine   1,7-Naphthyridine

1,8-Naphthyridine   2,6-Naphthyridine   2,7-Naphthyridine

FIGURE 6.1 Some familiar members of the naphthyridine family.

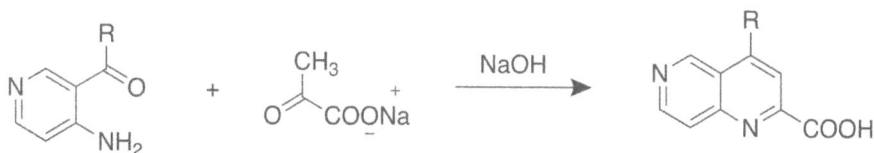

R = H, CH₃

SCHEME 6.1 Synthesis of 1,6-naphthyridine-2-carboxylic acid derivative of naphthyridine [12].

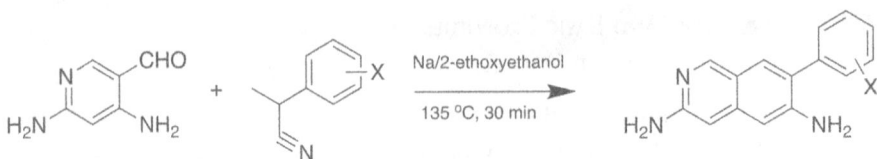

X = H / 2,6-dichloro / 3,5-dimethoxy

SCHEME 6.2 Synthesis of 1,6-naphthyridine derivative [13].

Thompson et al. (2000) used a similar method (Friedlander condensation) the following year to create 1,6-naphthyridine derivatives (Scheme 6.2). In this, sodium pyruvate and NaOH were switched out for sodium alkoxide and phenyl acetonitrile to obtain 1,6-naphthyridines in quantifiable yields [13].

From Nicotinamides

From 2-substituted nicotinamide, Ikekawa (1958) created 1,6-naphthyridin-5(6H)-one derivative by oxidising it with chromium trioxide. The ketone compounds, produced by oxidation, spontaneously cycled to produce 1,6-naphthyridines. Both substituted and unsubstituted products had average yields. The aminolysis of 7-methyl-5H-pyrano[4,3-b]pyridin-5-one and 5H-pyrano[4,3-b]pyridin-5-one produced the starting materials: 2-(2-hydroxyethyl)nicotinamide and 2-(2-hydroxypropyl)nicotinamide, respectively [14].

6.1.2 PHTHALOCYANINE

Phthalocyanine (Pc) is planar and has multiple aromatic molecules that are planar and highly conjugated. These molecules have four iminoisoindoline units with 18 π-delocalising electrons [15]. Its four fused benzene rings resemble the porphine's structure (Figure 6.2). Stable dye complexes with most of the metal atoms are reported. Braun and Tcherniac first described the free molecule in 1907 [16], although they did not determine its structure. When scientists attempted to manufacture phthalonitrile in 1927, they accidentally created the Cu–Pc complex. The two leading absorption bands for phthalocyanines are the Q band, typically in the range of 600–700 nm, and the Soret band at 300–400 nm. Due to electron–hole

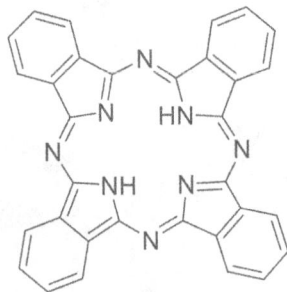

FIGURE 6.2 Chemical structure of phthalocyanine.

interactions between overlapping molecules, these bands typically break into at least two components in the solid state. Varied polymorphic forms can have highly different absorption spectra because this splitting is strongly influenced by the arrangement and distance of neighbouring molecules [17].

Pc molecules are being studied extensively due to their chemical and thermal stability. Various derivatives of Pc are prepared by attaching groups with the benzene rings and incorporating metal atoms in the centre. Due to the possibilities of preparing a large number of derivatives, phthalocyanine is being studied for its application in various fields like catalysis [18], anti-cancerous therapy [19], energy storage [20], medicinal photosensitisers [21], supercapacitors [22], hydrogen evolution reactions [23], and corrosion inhibition [24]. In this work, authors have focused on the corrosion-inhibiting properties of Pc molecules for various metals.

6.2 NAPHTHYRIDINE, PHTHALOCYANINE, AND THEIR DERIVATIVES AS CORROSION INHIBITORS

6.2.1 NAPHTHYRIDINE

The production of a protective and inhibitive coating due to the adsorption of organic compounds typically delays the dissolution of metals by preventing direct contact between metals and hostile media. NTD has natural properties such as anti-hypertensive, antibacterial, anti-platelet, anti-inflammatory, anti-arrhythmic, and herbicide safener actions; because of this, NTD derivatives have recently received much attention. NTD derivatives are excellent anti-corrosive possibilities because of the high amount of functionality and solubility in polar solutions. Employing experimental techniques, the impact of NTDs on steel taken in an acidic medium was assessed. The concentration of NTD molecules improves their capacity for protection. NTD-3 was reported to have a maximum inhibitory efficiency of 98.69% at a concentration of 4.11×10^{-5} mol L^{-1}. NTD-1 had the highest inhibitory efficiency (96.1%), followed by NTD-2 (97.4%) and then NTD-3 (98.7%). Polarisation research revealed that NTDs performed as mixed-type inhibitors and efficiently blocked the corrosion sites studied by Verma et al. [25]. Ansari et al. synthesised the three NTD derivatives and studied their corrosion inhibition ability on N-80 steel (15% HCl) [26].

Scheme 6.3 depicts the synthetic pathway. From ethanol, the chemicals were recrystallised. The corrosion inhibition impact of NTDs was studied by employing

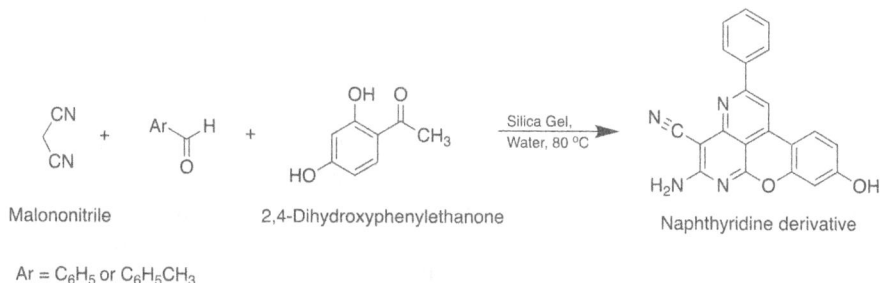

Ar = C$_6$H$_5$ or C$_6$H$_5$CH$_3$

SCHEME 6.3 Synthetic route for the preparation of naphthyridines [27].

EIS, weight loss, and other electrochemical techniques. Also, the steric and electronic effects by neighbouring groups on mild steel in 1 M HCl were calculated. AFM and SEM were used for surface investigations. To look into the relationships between experimental findings and reactivity indices produced from quantum chemistry, quantum chemical simulations were also performed [27].

The results of the study demonstrate the potential of the synthesised pyrazolo[1,5-a][1, 8]naphthyridine, pyrazolo[1,5-a]pyridine, and pyrazolo[1,5-a]pyrimidine and their derivatives as effective corrosion inhibitors, with a best inhibitory activity of 81% and an ideal concentration of 500 ppm for MS surface corrosion in 1.0 M H_2SO_4. By the results of the experiments, theoretical calculations revealed a strong link between the chemical parameters of the compounds under study and their inhibition efficiency for the corrosion process [28].

Salman et al. synthesised three chromeno naphthyridine derivatives (CNs) using one-pot multicomponent reaction malononitrile, aromatic aldehydes, and 2,4-hydroxyacetophenone in a single water reaction vessel with silica gel acting as a catalyst. The synthesised derivative of chromeno naphthyridine (CN-1) exhibits maximum inhibitory efficiency of 98.3%. EIS measurement demonstrates that charge transfer management is what prevents corrosion from occurring. The results of PDP point to the heterogeneous character of CNs adsorption. Adsorption Langmuir model exhibits perfect fitting. FTIR and UV-vis measurements supported the adhesion of CNs onto the surface of the metal. According to computational methods with DFT data, neutral and protonated forms' adsorption on steel surfaces prevents corrosion [29].

With the help of corrosion weight loss testing, PP, EIS, and solution analysis techniques, 1,8-naphthyridine compounds have been created, and their ability to inhibit the corrosion of mild steel in 1 M HCl has been examined. The studied substances function as mixed inhibitors, according to polarisation measurements, and the inhibition's effectiveness rises with the inhibitor's concentration. Variations in the impedance characteristics indicate that NTD compounds have been adsorbing to the metal surface and creating a protective coating. Atomic absorption spectroscopy was used to analyse the solution, and the results demonstrate that iron dissolves less slowly in the presence of inhibitors [30].

6.2.2 PHTHALOCYANINE

Amide-based organic chemicals are widely used in the metal industry as corrosion inhibitors. The available data demonstrate that organic macrocyclic molecules functioning as efficient inhibitors via adsorption on a metal surface have many electrons in a conjugated system containing non-bonded pairs of electrons in O, N, S, and P. Samal et al. investigated various M-Pc systems (M = Ni, Mn, Zn, Co, Fe, or Cu) for their effectiveness in preventing the corrosion of copper surfaces. According to the study, free-base phthalocyanine coatings have 88% corrosion inhibition efficacy compared to metal equivalents, making them suitable corrosion inhibitors for Cu in HCl. The free N-centres, which attach to the Cu atoms of the surface, are responsible for the free base phthalocyanine's maximum adsorption energy. N-atom's ability to donate electrons improves the electron transfer properties, causing a powerful chemical connection with the copper atoms. The same pattern can be seen in the average bond lengths between Cu and N in adsorbed complexes (Cu−N = 1.97 Å Pc and 3.32 Å in Mn-Pc). As the N atoms in M-pc molecules are bound to the metal centres, these molecules show low adsorption energy [31].

Nnaji et al. synthesised Pc-based dyes cobalt(II)2,9,16-tris(4-(*tert*-butyl)phenoxy)-23-(pyridin-4-yloxy)phthalocyanine (D1) and cobalt(II)2,9,16,24-tetrakis(4-(*tert*-butyl)phenoxy)phthalocyanine (D2) for inhibition of aluminium corrosion in 1 M hydrochloric acid. Figure 6.3 shows the synthesis route for the preparation of D1 and D2. Anodic Tafel slopes are less positive, and cathodic Tafel slopes are less negative when the inhibitors are present. Calculated inhibitory efficacy values (IE%) are higher for D1 than D2, demonstrating stronger asymmetric derivative corrosion inhibition [32].

Khan et al. created an amide-substituted zinc metal–centred macrocyclic phthalocyanine (TAZnPc), a macrocyclic metal complex. They used spectrochemical methods to analyse it to establish the substance's actions as a mild steel corrosion inhibitor in HCl solutions. Through the use of EIS and PDP techniques, the inhibition efficiency of TAZnPc was examined, and it was discovered that the corrosion inhibition efficiency of TAZnPc in HCl increased with a higher inhibitor's quantity and dropped with a rise in temperature from 303 K to 323 K. The study's results also showed that TAZnPc was adsorbed on mild steel by physisorption and chemisorption, both of which adhered to the Langmuir adsorption model. With an inhibition efficiency of 84.62%, TAZnPc behaved like a mixed inhibitor for mild steel's

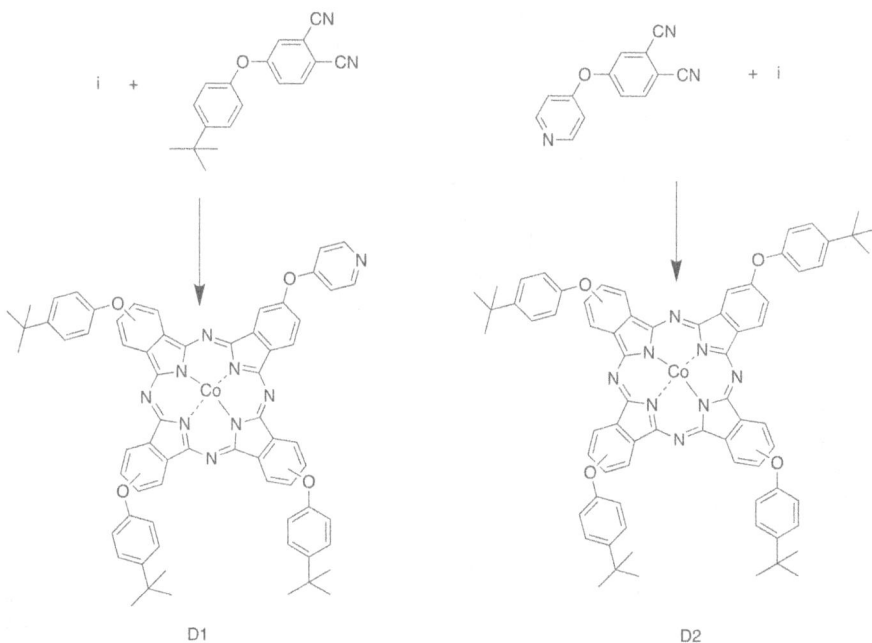

D1   D2

FIGURE 6.3 Synthesis route for the preparation of D1 and D2 (i = *n*-pentanol, DBU, Co(OAc)$_2$) [32].

Co(OAc)₂)[32]

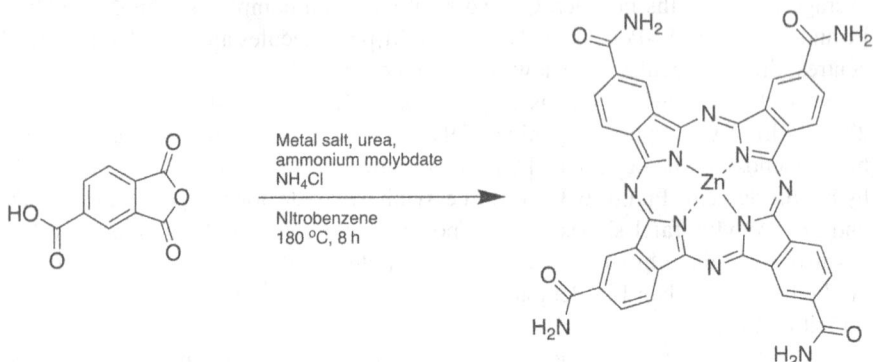

FIGURE 6.4 Synthesis route for the preparation of TAZnPc [33].

corrosion inhibition in 0.25 M HCl. Figure 6.4 shows the synthesis route for the preparation of TAZnPc [33].

Dibetsoe et al. synthesised four phthalocyanines and three naphthalocyanines and investigated them for aluminium corrosion inhibition in 1 M HCl. It was found that all the derivatives showed decent corrosion inhibition efficiency. They also studied the synergetic effect of KI in corrosion inhibition. It was observed that the addition of 0.1% KI increased corrosion inhibition efficiency (Figure 6.5). Temperature-dependent studies revealed that corrosion-inhibiting property decreases with increasing temperature. Adhesion of Pcs on metals is found to be immediate and involves chemisorption and physisorption. The results also correlated with the computational studies performed with B3LYP/6-31G(*d,p*) parameters [34].

Liyanage et al. prepared polyphthalocyanine polymer coatings for corrosion prevention, which were applied to additive-manufactured (AM) 316L steel samples. Using a drop casting technique, coatings were applied to the metallic surface, and later the layer was thermally polymerised. Open-circuit potential (OCP) study (15 days) of coated and non-coated samples revealed that Pc films protect AM steel surface from corrosion for an extended period (in 3.5 wt% NaCl). As the coating is hydrophobic and an oxide layer is present on the metal surface, corrosive ions are prevented from attacking the metallic surface; thus, Pc-coated samples demonstrate higher corrosion resistance. According to potentiodynamic and EIS analysis, the property of corrosion protection of the coating is slightly reduced after a 15-day exposure to the destructive, corrosive ions. But it still offers superior protection than bare metal against corrosion [35].

The abilities of benzothiazole-derived phthalonitrile (BTThio) and phthalocyanine (ClGaBTThioPc) as corrosion inhibitors were investigated electrochemically in 1.0 M HCl. ClGaBTThioPc has a higher corrosion efficiency than BTThio due to its higher π-conjugation. Adsorption experiments showed that BTThio and ClGaBTThioPc adhered to the Freundlich, Temkin, and El-Awady isotherms, and the adsorption mechanism is quite complex. EIS showed that by using an adsorption mechanism, ClGaBTThioPc and BTThio limit the corrosion of aluminium. It

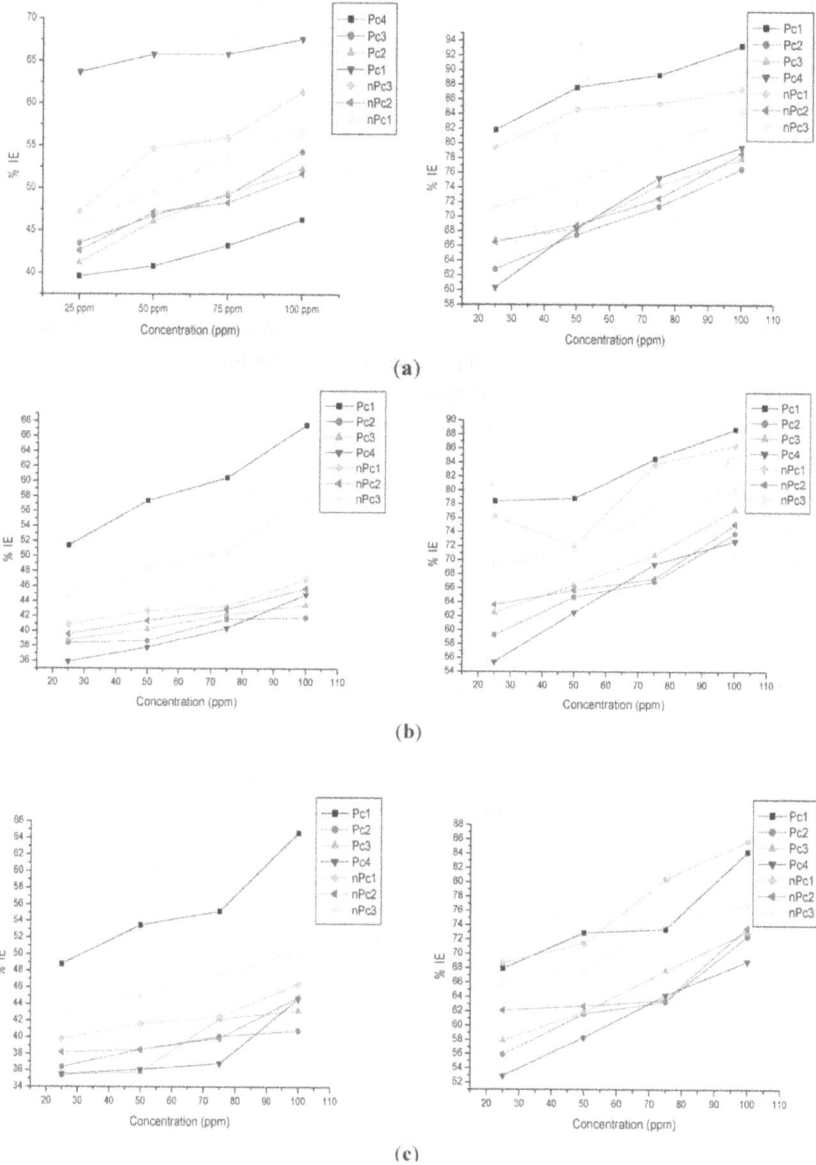

FIGURE 6.5 The corrosion inhibition efficiency of various Pcs and nPcs without KI (left-hand side) and with KI (right-hand side) at (a) 303 K, (b) 313 K, and (c) 323 K [34].

is speculated that inhibitors preserve the metal from corrosion by defending it from Cl⁻ ions, which cause corrosion. According to theoretical calculations, BTThio and ClGaBTThioPc can suppress corrosion by adhering to aluminium. The experimental findings show that ClGaBTThioP has a better ability to inhibit Al-corrosion than BTThio, indicating that inhibitors with higher unsaturation perform better than corrosion inhibitors [36].

Nnaji et al. studied ball-type phthalocyanines, which were complex with metals. These complexes performed well as Al-corrosion inhibitors in an acidic medium (1 M HCl); it was observed that the corrosion rate was significantly reduced when the concentration of metallated phthalocyanines was increased. Inhibitors affected the corrosion kinetics and processes by adsorbing on the Al-surface; this resulted in retardation of the corrosion rate. Inhibitor adsorption followed Temkin, Langmuir, Frundlich, and El-Awady isotherms. In the presence of reduced graphene oxide, ball-type M-Pc complexes showed synergetic effects, enhancing the corrosion inhibitor efficiency values [37].

Akin et al. used SEM-EDS analysis and electrochemical, gravimetric, and quantum chemical calculations to evaluate the impact of various water-soluble metallophthalo-cyanines (CoPc, CuPc, and ZnPc) on copper metal. The highest corrosion inhibitor efficiency from the results of electrochemical tests and gravimetric analyses was found with CoPc at 0.01 M concentration. Quantum chemical calculations determined the following corrosion inhibition efficacy order: CoPc > CuPc > ZnPc [38].

Metal-free (M), Co (y), and ClGa (x) tetrakis(4-acetamidophenoxy) phthalocy-anines were tested for inhibition of corrosion and their adsorption characteristics on Al in an acidic medium (HCl). The study included electrochemical techniques and FTIR, SEM, and X-ray diffraction observations. At 301 K, at the maximum tested inhibitor concentration (10 M), the potentiodynamic polarisation approach yielded inhibitory efficiency values of 93.3% (M), 69.7% (x), and 87.7% (y). As mixed-type corrosion inhibitors, these compounds demonstrated excellent corrosion inhibition effectiveness [39].

Nnaji and colleagues effectively electrodeposited tetrakis[(benzothiazol-2-yl-thio) phthalocyaninato] gallium(III)chloride (a) and tetrakis[(benzothiazol-2ylphenoxy) phthalocyaninato] gallium(III)chloride (b) on Al surface for testing their corrosion inhibition efficiency (acidic medium). It was compared to how well different electro-deposited metallated phthalocyanines resisted corrosion. With inhibitor efficiency values from EIS measurements: 82% for complex "b" and 86% for complex "a" in 1.0 M HCl, it was evident from EIS and polarisation techniques that complex "a" functioned better than complex "b."

Using Tafel polarisation and EIS at temperatures between 303 K and 323 K, tetra-nitro cobalt phthalocyanine (TNCoPc) was studied to check its properties to constrain corrosion in mild steel (acidic medium). The inhibitor was used at a concentration of 1.25–5 mM. It was discovered that as the temperature dropped and the inhibitor con-centration increased, the inhibition's effectiveness increased. According to polarisation experiments, TNCoPc functions as a mixed-type inhibitor. At its ideal concentration of 5 mM, TNCoPc had a maximum inhibitory effect of 86.48%. According to adsorption experiments, this inhibitor followed Langmuir's adsorption isotherm on the surface of the metal and experienced both physisorption and chemisorption [40].

Tetranitro cobalt phthalocyanine (TNCoPc) was tested in an acidic medium (0.25 M HCl) at varied temperatures (303–323 K) as a potential mild steel corrosion inhibitor. The inhibitor was used at varied concentrations (1.25–5mM). It was discovered that as the temperature dropped and the inhibitor concentration increased, the effectiveness of the inhibition increased. According to polarisation experiments, TNCoPc behaved similarly to a mixed-type inhibitor at all concentrations. At its ideal concentration of 5 mM, TNCoPc had a maximum inhibitory effect of 86.48%. According to adsorption experiments, this inhibitor followed Langmuir's adsorption isotherm on the surface of the metal and experienced both chemisorption and physisorption [40].

On the surface of iron, self-assembled monolayers (SAMs) of phthalocyanine-$H_2[Pc(OBNP)_2]$ (OBNP = binaphthylpthalocyanine) were created, and their property to suppress iron corrosion in 0.5 M H_2SO_4 was investigated. The findings demonstrated that $H_2[Pc(OBNP)_2]$ molecules could create SAMs on the iron surface to prevent corrosion successfully, and the inhibition efficiency varied with immersion time. The inhibition efficiency increased as immersion times increased from 1 to 16 hours, with 16 hours producing the maximum efficiency. As the immersion time exceeded 16 hours, the effectiveness of the inhibition was slightly reduced, which may have been caused by the desorption of some $H_2[Pc(OBNP)_2]$ molecules from the iron surface. The adsorption arrangement of $H_2[Pc(OBNP)_2]$ molecules on the surface of iron was also determined using molecular modelling studies [41]. There are many other reported systems of phthalocyanine derivatives complexed with various metal ions to date; some of them are mentioned in Table 6.1.

TABLE 6.1
List of Phthalocyanine Derivatives Reported as Corrosion Inhibitors

S/N	Name	Metal	Electrolyte	Inhibition Efficiency	Reference
1.	ZnPcS	Al	1 M HCl	96.3%	[42]
2.	H_2Pc.Sx	Al alloy	1 M HCl	91.96%	[43]
3.	CoPc, CuPc, and ZnPc	Cu	0.1 M HCl	76.88%, 67.06%, and 64.04%	[38]
4.	CuPc and TNCuPc	Al	1 M HCl	98.8% and 97.9%	[44]
5.	CuPc	1018 Carbon steel	0.5 M H_2SO_4	89.5%	[45]
6.	Cu-phcy	ASTM A606-4 steel	16% HCl	94.0%	[46]
7.	H_2Pc CuPc, CuPc·S_4·Na_4	Mild steel	1 M HCl	87%.1, 88.6%, and 94.6%	[47]
8.	TNCoPc	Mild steel	0.25 M HCl	86.48%	[40]
9.	ClGaBTPc	Al	1 M HCl	82.7%	[48]
10.	1,(4)-tetrakis[(2-mercapto) pyridine]phthalocyanine	Al	1 M HCl	83%	[49]
11.	ClGaBTThioPc	Al	1 M HCl	82.7%	[36]
12.	H_2PPc.(SO$_3$Na)$_x$	Carbon steel/ iron	1 M HCl	97.9%	[24]
13.	TAPcNi	Carbon steel	1 M HCl	–	[50]
14	Tetrakis[(4-benzothiazol-2ylphenoxy)phthalocyaninato] gallium(III)chloride	Al	1 M HCl	87.05%	[51]
15.	BTSPA filled with Cu-Ph	Carbon steel	0.1M HCl	–	[52]
16.	Co(II)-TCPC	Mild steel	1% NaCl	68.5%	[53]
17.	Fe(III)-TCPC	Mild steel	1% NaCl	82.4%	[54]
18.	Alkyd@HoPc$_2$	Carbon steel	0.1M HCl	87.4%	[55]

6.3 CONCLUSION

As corrosion inhibitors, naphthalocyanine, phthalocyanine, and their derivatives are widely utilised and are being explored. It is known that phthalocyanine spontaneously reacts with the metallic surface and produces chelating complexes because they contain 8 nitrogen atoms and 18 π-electrons. These molecules are insoluble in polar electrolytes due to their macrocyclic structure. Still, they exhibit excellent solubility in acidic solutions due to nitrogens bridging the pyrrole rings. These substances adhere to metallic surfaces by the Langmuir adsorption isotherm model. A potentiodynamic polarisation investigation shows that these compounds primarily function as mixed-type corrosion inhibitors. Although research indicates that these compounds can reasonably suppress corrosion, their use as anti-corrosive materials has not advanced much. Although there is relatively little research on the use of naphthalocyanine, phthalocyanine, and their derivatives as corrosion inhibitors in industries, these compounds are expected to demonstrate relatively high inhibitory efficacy due to their macrocyclic structure. In light of this, the usage of these compounds for industrial applications, such as high-temperature corrosion inhibitors, should be investigated.

REFERENCES

1. C. Verma, M. H. Abdellattif, A. Alfantazi, and M. A. Quraishi, "N-heterocycle compounds as aqueous phase corrosion inhibitors: A robust, effective and economic substitute," *J Mol Liq*, vol. 340, p. 117211, 2021. doi: 10.1016/j.molliq.2021.117211
2. A. Singh, K. R. Ansari, M. A. Quraishi, S. Kaya, and P. Banerjee, "The effect of an N-heterocyclic compound on corrosion inhibition of J55 steel in sweet corrosive medium," *New J Chem*, vol. 43, no. 16, pp. 6303–6313, 2019. doi: 10.1039/C9NJ00356H
3. I. B. Obot, N. O. Obi-Egbedi, and N. W. Odozi, "Acenaphtho [1,2-*b*] quinoxaline as a novel corrosion inhibitor for mild steel in 0.5 M H2SO4," *Corros Sci*, vol. 52, no. 3, pp. 923–926, 2010. doi: 10.1016/J.CORSCI.2009.11.013
4. C. Verma, L. O. Olasunkanmi, E. E. Ebenso, and M. A. Quraishi, "Substituents effect on corrosion inhibition performance of organic compounds in aggressive ionic solutions: A review," *J Mol Liq*, vol. 251, pp. 100–118, 2018. doi: 10.1016/J.MOLLIQ.2017.12.055
5. C. Verma, E. E. Ebenso, and M. A. Quraishi, "Molecular structural aspects of organic corrosion inhibitors: Influence of –CN and –NO2 substituents on designing of potential corrosion inhibitors for aqueous media," *J Mol Liq*, vol. 316, p. 113874, 2020. doi: 10.1016/J.MOLLIQ.2020.113874
6. N. Kovačević and A. Kokalj, "Chemistry of the interaction between azole type corrosion inhibitor molecules and metal surfaces," *Mater Chem Phys*, vol. 137, no. 1, pp. 331–339, 2012. doi: 10.1016/J.MATCHEMPHYS.2012.09.030
7. C. Verma, E. E. Ebenso, M. A. Quraishi, and C. M. Hussain, "Recent developments in sustainable corrosion inhibitors: Design, performance and industrial scale applications," *Mater Adv*, vol. 2, no. 12, pp. 3806–3850, 2021. doi: 10.1039/d0ma00681e
8. M. Rbaa *et al.*, "8-Hydroxyquinoline based chitosan derived carbohydrate polymer as biodegradable and sustainable acid corrosion inhibitor for mild steel: Experimental and computational analyses," *Int J Biol Macromol*, vol. 155, pp. 645–655, 2020. doi: 10.1016/J.IJBIOMAC.2020.03.200
9. A. R. Katritzky and C. W. Rees, *Comprehensive Heterocyclic Chemistry*, Pergamon Press, 1984.

10. W. W. Paudler and R. M. Sheets, "Recent developments in naphthyridine chemistry," *Adv Heterocycl Chem*, vol. 33, pp. 147–184, 1983. doi: 10.1016/S0065-2725(08)60053-7

11. T. Devadoss, V. Sowmya, and R. Bastati, "Synthesis of 1,6-naphthyridine and its derivatives: A systematic review," *ChemistrySelect*, vol. 6, no. 15, pp. 3610–3641, 2021. doi: 10.1002/SLCT.202004462

12. L. Chan *et al.*, "Discovery of 1,6-naphthyridines as a novel class of potent and selective human cytomegalovirus inhibitors [4]," *J Med Chem*, vol. 42, no. 16, pp. 3023–3025, 1999.

13. A. M. Thompson, H. D. H. Showalter, and W. A. Denny, "Synthesis of 7-substituted 3-aryl-1,6-naphthyridin-2-amines and 7-substituted 3-aryl-1,6-naphthyridin-2(1*H*)-ones via diazotization of 3-aryl-1,6-naphthyridine-2,7-diamines," *J Chem Soc Perkin 1*, no. 12, pp. 1843–1852, 2000. doi: 10.1039/B002599M

14. N. Ikekawa, "Studies on naphthyridines. I. Synthesis of 1, 6-naphthyridine," *Chem Pharm Bull (Tokyo)*, vol. 6, no. 3, pp. 263–269, 1958. doi: 10.1248/CPB.6.263

15. M. Yahya, Y. Nural, and Z. Seferoğlu, "Recent advances in the nonlinear optical (NLO) properties of phthalocyanines: A review," *Dyes Pigm*, vol. 198, no. July 2021, 2022. doi: 10.1016/j.dyepig.2021.109960

16. A. Braun and J. Tcherniac, "Über die Produkte der Einwirkung von Acetanhydrid auf Phthalamid," *Berichte der deutschen chemischen Gesellschaft*, 1907. doi: 10.1002/CBER.190704002202

17. J. D. Wright, "Phthalocyanines," in *Encyclopedia of Materials: Science and Technology*, pp. 6987–6991, 2001. doi: 10.1016/B0-08-043152-6/01238-9

18. S. Yuan *et al.*, "Tuning the catalytic activity of Fe-phthalocyanine-based catalysts for the oxygen reduction reaction by ligand functionalization," *ACS Catal*, vol. 12, no. 12, pp. 7278–7287, 2022.

19. C. C. Rennie and R. M. Edkins, "Targeted cancer phototherapy using phthalocyanine–anticancer drug conjugates," *Dalton Trans*, vol. 51, no. 35, pp. 13157–13175, 2022. doi: 10.1039/D2DT02040H

20. J. Fan *et al.*, "Fully conjugated poly(phthalocyanine) scaffolds derived from a mechanochemical approach towards enhanced energy storage," *Angew Chem Int Ed*, vol. 61, no. 38, p. e202207607, Sep. 2022. doi: 10.1002/ANIE.202207607

21. K. Huang, H. Zhang, M. Yan, J. Xue, and J. Chen, "A novel zinc phthalocyanine-indometacin photosensitizer with 'Three-in-one' cyclooxygenase-2-driven dual targeting and aggregation inhibition for high-efficient anticancer therapy," *Dyes Pig*, vol. 198, p. 109997, 2022. doi: 10.1016/J.DYEPIG.2021.109997

22. M. B. Arvas, S. Yazar, and Y. Sahin, "An ultra-high power density supercapacitor: Cu(II) phthalocyanine tetrasulfonic acid tetrasodium salt doped polyaniline," *J Alloys Compd*, vol. 919, p. 165689, 2022. doi: 10.1016/J.JALLCOM.2022.165689

23. S. Aralekallu, L. K. Sannegowda, and V. Singh, "Developments in electrocatalysts for electrocatalytic hydrogen evolution reaction with reference to bio-inspired phthalocyanines," *Int J Hydrogen Energy*, vol. 48, pp. 16569–16592, 2023. doi: 10.1016/J.IJHYDENE.2023.01.169

24. P. Zhao, L. Niu, L. Huang, and F. Zhang, "Electrochemical and XPS investigation of phthalocyanine oligomer sulfonate as a corrosion inhibitor for iron in hydrochloric acid," *J Electrochem Soc*, vol. 155, no. 10, p. C515, 2008.

25. C. Verma, A. A. Sorour, E. E. Ebenso, and M. A. Quraishi, "Inhibition performance of three naphthyridine derivatives for mild steel corrosion in 1 M HCl: Computation and experimental analyses," *Results Phys*, vol. 10, pp. 504–511, 2018. doi: 10.1016/J.RINP.2018.06.054

26. K. R. Ansari and M. A. Quraishi, "Experimental and computational studies of naphthyridine derivatives as corrosion inhibitor for N80 steel in 15% hydrochloric acid," *Physica E Low Dimens Syst Nanostruct*, vol. 69, pp. 322–331, 2015. doi: 10.1016/J.PHYSE.2015.01.017

27. P. Singh, E. E. Ebenso, L. O. Olasunkanmi, I. B. Obot, and M. A. Quraishi, "Electrochemical, theoretical, and surface morphological studies of corrosion inhibition effect of green naphthyridine derivatives on mild steel in hydrochloric acid," *J Phys Chem C*, vol. 120, no. 6, pp. 3408–3419, 2016.

28. S. A. Mousa, "Synthesis, characterization and corrosion inhibition of pyrazolo[1,5-*a*][1,8]naphthyridine, pyrazolo[1,5-*a*]pyridine and pyrazolo[1,5-*a*]pyrimidine derivatives, quantum chemical approach," *Int J Chem Stud*, vol. 5, no. 3, pp. 290–296, 2017.

29. M. Salman, K. R. Ansari, V. Srivastava, D. S. Chauhan, J. Haque, and M. A. Quraishi, "Chromeno naphthyridines based heterocyclic compounds as novel acidizing corrosion inhibitors: Experimental, surface and computational study," *J Mol Liq*, vol. 322, p. 114825, 2021. doi: 10.1016/J.MOLLIQ.2020.114825

30. K. Kalaiselvi, V. Nijarubini, and J. M.-J. Chem, "Investigation of the inhibitive effect of 1,8-naphthyridine derivatives on corrosion of mild steel in acidic media," *RASĀYAN J Chem*, vol. 6, pp. 52–64, 2013.

31. P. P. Samal, A. Dekshinamoorthy, S. Arunachalam, S. Vijayaraghavan, and S. Krishnamurty, "Free base phthalocyanine coating as a superior corrosion inhibitor for copper surfaces: A combined experimental and theoretical study," *Colloids Surf A Physicochem Eng Asp*, vol. 648, p. 129138, 2022. doi: 10.1016/J.COLSURFA.2022.129138

32. N. Nnaji, P. Sen, and T. Nyokong, "Symmetry effect of cobalt phthalocyanines on the aluminium corrosion inhibition in hydrochloric acid," *Mater Lett*, vol. 306, p. 130892, 2022. doi: 10.1016/J.MATLET.2021.130892

33. F. Khan, S. M. Sudhakara, Y. M. Puttaigowda, and Pushpanjali, "Amide substituted zinc centered macrocyclic phthalocyanines for corrosion inhibition of mild steel in hydrochloric acid medium," *Surf Eng Appl Electrochem*, vol. 58, no. 6, pp. 613–624, 2023. doi: 10.3103/S1068375523010076

34. M. Dibetsoe *et al.*, "Some phthalocyanine and naphthalocyanine derivatives as corrosion inhibitors for aluminium in acidic medium: Experimental, quantum chemical calculations, QSAR studies and synergistic effect of iodide ions," *Molecules*, vol. 20, no. 9, pp. 15701–15734, 2015. doi: 10.3390/MOLECULES200915701

35. A. Liyanage, D. J. Karunarathne, S. Nasrazadani, F. D'Souza, H. R. Siller, and T. D. Golden, "Polyphthalocyanine coatings for corrosion protection on additive manufactured steel materials," *Prog Org Coat*, vol. 170, p. 106990, 2022. doi: 10.1016/J.PORGCOAT.2022.106990

36. N. Nnaji, N. Nwaji, J. Mack, and T. Nyokong, "Corrosion resistance of aluminum against acid activation: Impact of benzothiazole-substituted gallium phthalocyanine," *Molecules*, vol. 24, no. 1, p. 207, 2019. doi: 10.3390/MOLECULES24010207

37. N. Nnaji, N. Nwaji, J. Mack, and T. Nyokong, "Ball-type phthalocyanines and reduced graphene oxide nanoparticles as separate and combined corrosion inhibitors of aluminium in HCl," *J Mol Struct*, vol. 1236, p. 130279, 2021. doi: 10.1016/J.MOLSTRUC.2021.130279

38. M. Akin *et al.*, "The water-soluble peripheral substituted phthalocyanines as corrosion inhibitors for copper in 0.1 N HCl: Gravimetric, electrochemical, SEM-EDS, and quantum chemical calculations," *Prot Met Phys Chem Surf*, vol. 56, no. 3, pp. 609–618, 2020. doi: 10.1134/S207020512003003X/TABLES/5

39. N. Nnaji, P. Sen, and T. Nyokong, "Aluminum corrosion retardation properties of acetamidophenoxy phthalocyanines: Effect of central metal," *J Mol Struct*, vol. 1242, p. 130806, 2021. doi: 10.1016/J.MOLSTRUC.2021.130806

40. M. S. Sarvajith, Pushpanjali, and Fasiulla, "Electrochemical investigation of tetranitro cobalt phthalocyanine on corrosion control of mild steel in hydrochloric acid medium," *Surf Eng Appl Electrochem*, vol. 55, no. 3, pp. 324–334, 2019. doi: 10.3103/S1068375519030153/TABLES/6

41. S. Li *et al.*, "Inhibition studies of phthalocyanine self-assembled monolayers for iron corrosion in 0.5 M H_2SO_4 solution," *Nanosci Nanotechnol Lett*, vol. 11, no. 6, pp. 855–860, 2019. doi: 10.1166/NNL.2019.2955

42. A. O. Ogunsipe, E. C. Ogoko, and O. K. Abiola, "Corrosion inhibition of aluminium in 1.0 M HCl by zinc phtalocyanine sulfonate," *J Chem Soc Niger*, vol. 45, no. 3, pp. 533–539, 2020.

43. P. Zhao, X. Han, W. Wang, M. Zhao, and F. Zhang, "Exfoliation inhibition on aluminum alloy 7075 by water-soluble phthalocyanine derivates in 1 mol/L HCl," *J Adhes Sci Technol*, vol. 34, no. 12, pp. 1331–1347, 2020. doi: 10.1080/01694243.2019.1707562

44. T. Pesha *et al.*, "Inhibition effect of phthalocyaninatocopper(II) and 4-tetranitro(phthalocyaninato)copper(II) inhibitors for protection of aluminium in acidic media," *Int J Electrochem Sci*, vol. 14, pp. 137–149, 2019. doi: 10.20964/2019.01.17

45. J. C. Valle-Quitana, G. F. Dominguez-Patiño, and J. G. Gonzalez-Rodriguez, "Corrosion inhibition of carbon steel in 0.5 M H_2SO_4 by phtalocyanine blue," *ISRN Corrosion*, p. 945645, 2014. doi: 10.1155/2014/945645

46. I. Aoki, I. C. Guedes, and S. L. A. Maranhão, "Copper phthalocyanine as corrosion inhibitor for ASTM A606-4 steel in 16% hydrochloric acid," *J Appl Electrochem*, vol. 32, no. 8, pp. 915–919, 2002. doi: 10.1023/A:1020506432003/METRICS

47. P. Zhao, Q. Liang, and Y. Li, "Electrochemical, SEM/EDS and quantum chemical study of phthalocyanines as corrosion inhibitors for mild steel in 1 mol/l HCl," *Appl Surf Sci*, vol. 252, no. 5, pp. 1596–1607, 2005. doi: 10.1016/J.APSUSC.2005.02.121

48. N. Nnaji, N. Nwaji, G. Fomo, J. Mack, and T. Nyokong, "Inhibition of aluminium corrosion using benzothiazole and its phthalocyanine derivative," *Electrocatalysis*, vol. 10, no. 4, pp. 445–458, 2019. doi: 10.1007/S12678-019-00538-1/FIGURES/12

49. O. K. Özdemir, A. Aytaç, D. Atilla, and M. Durmuş, "Corrosion inhibition of aluminum by novel phthalocyanines in hydrochloric acid solution," *J Mater Sci*, vol. 46, no. 3, pp. 752–758, 2011. doi: 10.1007/S10853-010-4808-6/FIGURES/10

50. S. L. A. Maranhão, I. C. Guedes, F. J. Anaissi, H. E. Toma, and I. Aoki, "Electrochemical and corrosion studies of poly(nickel-tetraaminophthalocyanine) on carbon steel," *Electrochim Acta*, vol. 52, no. 2, pp. 519–526, 2006. doi: 10.1016/J.ELECTACTA.2006.05.033

51. N. Nnaji, N. Nwaji, and T. Nyokong, "Electrodeposited benzothiazole phthalocyanines for corrosion inhibition of aluminium in acidic medium," *Int J Electrochem*, vol. 2020, pp. 1–11, 2020. doi: 10.1155/2020/8892559

52. P. H. Suegama and I. Aoki, "Electrochemical behavior of carbon steel pre-treated with an organo functional bis-silane filled with copper phthalocyanine," *J Braz Chem Soc*, vol. 19, no. 4, pp. 744–754, 2008. doi: 10.1590/S0103-50532008000400019

53. S. Hettiarachchi, Y. W. Chan, R. B. Wilson, and V. S. Agarwala, "Macrocyclic corrosion inhibitors for steel in acid chloride environments," *Corrosion*, vol. 45, no. 1, pp. 30–34, 1989. doi: 10.5006/1.3577884

54. S. Hettiarachchi, Y. W. Chan, R. B. Wilson, and V. S. Agarwala, "Phthalocyanine and polyphthalocyanine coatings for corrosion protection of metals," *Online Proc Library*, vol. 125, p. 321, 1988. doi: 10.1557/PROC-125-321

55. M. A. Deyab, R. Słota, E. Bloise, and G. Mele, "Exploring corrosion protection properties of alkyd@lanthanide bis-phthalocyanine nanocomposite coatings," *RSC Adv*, vol. 8, no. 4, pp. 1909–1916, 2018. doi: 10.1039/C7RA09804A

7 Purine-, Pyran-, Pyrazole-, and Pyrazine-based Corrosion Inhibitors

Pragnesh Dave[1,2]
[1] Department of Chemistry, Sardar Patel University, Anand, Gujarat, India
[2] Government Engineering College, Bhuj, Gujarat, India

7.1 INTRODUCTION

Metals are found to have critical applications in various fields. Corrosion of metals caused numerous economic devastation. Corrosion of metals is the natural process which transfers refined metal into a chemically more stable form. It is the slow decomposition of metals by chemical or electrochemical reactions. Preserving such valuable metals using appropriate technology is primarily essential [1]. The various corrosion protection methods, which embrace barrier coating, hot-dip galvanization, metal alloying, cathodic protection, and corrosion inhibitors, are demonstrated [2]. The applications of inhibitors are widely shown because of their easy availability, and they effectively bind with metals [3]. The corrosion inhibitor is the chemical material that suppresses metal corrosion at a suitable concentration by decreasing the corrosion rate without fluctuating the concentration of the corrosion agent.

Chromate, molybdate, and nitrate are the furthermost generally used corrosion inhibitors. Organic inhibitors [4] showed good corrosion inhibition efficiency as they possessed essential features to control metal corrosion. The heterocycles have N, S, and O, phenyl rings, π-bonds, and various functional groups that deliver substantial metal surface coverage and permit corrosion protection. The corrosion inhibitor can be classified as cathodic or anodic, or mixed type depending on influencing cathodic or anodic reaction or both [5]. This chapter highlights purine (PU), pyran, pyrazole, and pyrazine as corrosion inhibitors for mitigating metal corrosion.

7.2 PURINE-, PYRAN-, PYRAZOLE-, AND PYRAZINE-BASED CORROSION INHIBITORS

7.2.1 PURINE AS A CORROSION INHIBITOR

PU is a heterocyclic aromatic organic compound with two rings of pyrimidine and imidazole combined. PU exhibited remarkable corrosion protection efficiency for 304 austenitic stainless steel (SS) in an HCl solution [6]. The augmented concentration

DOI: 10.1201/9781003377016-7

of PU led to enhancing the corrosion inhibition efficiency. The highest corrosion inhibition efficiency was found to be 85% at 10 mM concentration as determined by PDP. The influence of temperature suggested that an increased temperature led to a decrease in inhibition efficiency. Tafel polarization showed that PU emerged as a cathodic–anodic mixed type of corrosion inhibitor. The increased influence of the PU concentration for Cu was studied in 1 M NaCl [7]. The concentration of PU raises the inhibition efficiency (IE). The Langmuir adsorption isotherm confirms the adsorption of PU on the surface of copper. The adsorption of PU on the copper surface followed the chemical adsorption mechanism.

The two compounds of PU derivatives, 6-furfurylaminopurine (FAP) and N-benzoylaminopurine (N-BAP), have been demonstrated as corrosion inhibitors for mild steel (MS) in 1 M HCl [8]. Both inhibitors displayed high-corrosion activity. N-BAP exhibited 97% corrosion inhibition, increasing to 98.6% after 24 hours. The heteroatoms N and O of the corrosion inhibitors contribute to high corrosion inhibition performance. The heteroatoms N and O provided robust bonding to Fe metal. The corrosion inhibitors adenine (AD) and 2,6-diamino PU showed corrosion-inhibitive properties for Cu in the simulated body fluid. The corrosion of Cu was suppressed by around 90% by both corrosion inhibitors, and further corrosion protection of Cu was enhanced in the presence of potassium sorbate [9]. PUs and their analogues, such as guanine, AD, 2,6-diamino PU, 6-thioguanine, and 2,6-thiopurine, have been examined as corrosion inhibitors for MS in 1 M HCl solution [10]. All undertaking compounds contributed to increasing corrosion inhibition efficiency by enhancing the corrosion inhibitor concentration.

The corrosion inhibition efficiency of inhibitors followed the efficiency order as 2,6-dithiopurine > 6-thioguanine > 2,6-diamino purine > AD > guanine. The 88% is the highest corrosion inhibition efficiency observed at 10^{-3} M 2,6-thiopurine. PU and AD were studied as corrosion inhibitors for copper in 0.5 M Na_2SO_4 solutions at pH 6.8 [11]. The employed techniques (EQCM and PDP) revealed that increased concentration of inhibitors influenced corrosion protection. Corrosion inhibition efficiency enhanced as the concentration of inhibitors increased. The increase in the concentration of PU and AD led to a decrease in the corrosion of Cu in 0.5 M $NaNO_3$ [12]. Corrosion protection of copper was found through the adsorption of inhibitors over the surface of copper, and adsorption phenomena followed the Langmuir adsorption isotherm. The evaluated values of standard free energies (thermodynamic parameter) of adsorption depicted that adsorption of PU and AD occurred through chemical adsorption on the copper surface.

Various electrochemical techniques, scanning electron microscopy (SEM), X-ray spectroscopy (XPS), quantum chemical calculation, and molecular dynamics simulation [13] were used to study the corrosion inhibition mechanism and adsorption process of three PU compounds such as guanine (G), adenine (A), and hypoxanthine (I) for copper in alkaline artificial seawater. The obtained results indicated that the three inhibitors protected the corrosion of copper by forming a protective film on the surface of copper. The three inhibitors followed corrosion inhibition efficiency: $I > G > A$. Three inhibitors obeyed the Langmuir adsorption model, and these three inhibitors adsorbed on the surface of copper through physical and chemical adsorption.

PU was demonstrated as a corrosion inhibitor for Al in 1.0 M deaerated stirred H_3PO_4 solution at 25 °C [14]. PU itself showed a poor corrosion inhibition effect, but its corrosion inhibitive property improved in the presence of I^- ions. The enhancement of corrosion activity is due to the synergistic effect of PU and I^- ions. The I^- ions assisted in the adsorption of PU on the Al surface. PDP method showed that PU itself alone and its mixture with I^- ions emerged as mixed-type corrosion inhibitors for the mitigation of the corrosion of Al in 1.0 M H_3PO_4 solution. The impedance diagram showed three-time constants or semicircles. The impedance diagram size relies on corrosion inhibitor concentration and dipping time. Guanine inhibited corrosion of copper corrosion in 0.1 M HCl. WL and EIS methods proved that good corrosion inhibition efficiency was achieved by guanine [15]. Around 87% corrosion protection efficiency has been displayed by guanine, and there is no change in the corrosion inhibition activity up to 333 K.

7.2.2 PYRAN AS A CORROSION INHIBITOR

Pyran, or oxine, is a six-membered non-aromatic heterocyclic with five carbon atoms and one oxygen atom containing two double bonds. The pyran derivatives, 2-amino-5-oxo-4-(p-tolyl)-4H,5H-pyrano [3,2-c]chromene-3-carbonitrile (PY-CH$_3$) and 2-amino-5-oxo-4-phenyl-4H,5H-pyrano [3,2-c]chromene-3-carbonitrile (PY-H) exhibited anti-corrosion activity in MS in HCl [16]. PY-H showed 90% corrosion protection activity, while PY-CH$_3$ attained 80% corrosion inhibition efficiency. The Langmuir adsorption isotherm consisted of two corrosion inhibitors adsorbed on the metal surface. 4-Hydroxy-6-methyl-3-(3-quinolin-8-yl-acryloyl)-pyran-2-one (HMQP) and 3-[3-(4-dimethylamino-phenyl)-acryloyl]-4-hydroxy-6-methyl-pyran-2-one (DMPHP) were found to be potential corrosion inhibitors for MS 1 M HCl [17]. The compound was studied in varied concentrations (0.001–1 mM) and four different temperatures (298, 308, 318, and 328 K) to find the optimal concentration and temperature for estimating top corrosion-inhibiting conditions. The maximum inhibition efficiency has been achieved at 90% and 85.4% at 298 K in the presence of 1 mM concentration of DMPHP and HMQP, respectively. The appearance of experimental adsorption results showed that corrosion inhibitors obeyed the Langmuir isotherm model. The PDP parameters indicated that DMPHP and HMQP behaved as mixed corrosion inhibitors.

The corrosion inhibition performance of pyran derivatives known as AP was evaluated as corrosion inhibitors for N80 steel in 15% HCl [18]. The corrosion inhibition action of various pyran compounds such as 2-amino-4-(4-methoxyphenyl)-7,7-dimethyl-5-oxo-5,6,7,8-tetrahydro-4H-chromene-3- carbonitrile (AP-1), 2-amino-7,7-dimethyl-5-oxo-4-phenyl-5,6,7,8-tetrahydro-4H-chromene-3-carbonitrile (AP-2), and 2-amino-7,7-dimethyl-4-(4-nitrophenyl)-5-oxo-5,6,7,8-tetrahydro-4H-chromene-3-carbonitrile (AP-3) was assessed by various methods like WL, EIS, and PDP. They showed that AP-1 bore the highest corrosion inhibitive efficiency, followed by AP-2 and AP-3. It indicated that the electron releasing of AP-1 is a responsible factor for the increased corrosion activity of the inhibitor, while having an electron-withdrawing group on the AP-3 suppressed the corrosion performance of the inhibitor. The SEM images also supported the corrosion protection of three inhibitors. The adsorption of APs on the surface of N80 steel conformed Langmuir isotherm. The PDP method revealed that inhibitors are mixed corrosion inhibitors with a cathodic predominance effect.

Two novel pyran-2-one derivatives, (E)-4-hydroxy-6-methyl-3-(1-(phenylimino) ethyl)-2H-pyran-2-one (Py-1) and (E)-3-(1-((2-chlorophenyl)imino)ethyl)-4-hydroxy-6-methyl-2H-pyran-2-one (Py-2), emerged as suitable corrosion protectors for MS in 1 M HCl [19]. The corrosion performance of both inhibitors has been investigated by EIS and PDP methods, respectively. The corrosion inhibition efficiency has been enhanced by elevating of concentration of corrosion inhibitors, and corrosion control activity was achieved to 90% at 10^{-3} M concentration of inhibitor. Pyran-2-one derivatives (Py-1 and Py-2) showed mixed adsorption behaviour, as revealed by the PDP method. The increase in temperature contributed to the decline in the corrosion inhibition efficacy of both compounds and the thermodynamic parameters showed that the corrosion inhibitors followed chemical adsorption (chemisorption) rather than physical adsorption.

The three pyran derivatives, ethyl 2-amino-4-(4-hydroxyphenyl)-6-(p-tolyl)-4H-pyran-3-carboxylate (HP), ethyl 2-amino-4-(4-methoxyphenyl)-6-(p-tolyl)-4H-pyran-3-carboxylate (MP), and ethyl 2-amino-4-(4-hydroxy-3,5-dimethoxyphenyl)-6-(p-tolyl)-4H-pyran-3-carboxylate (HDMP) exhibited excellent corrosion inhibitive activity for MS in 1 M H_2SO_4 [20]. The increased concentration of inhibitor led to enhanced corrosion inhibitive efficiency. Three inhibitors, HP, MP, and HDMP, showed maximum corrosion inhibition efficiency of 91.6%, 92.5%, and 95%, respectively, at 2 mM concentrations followed by Langmuir isotherm that well behaved as a mixed type of corrosion inhibitors as evaluated by the PDP method. Schiff base (SB) resulting from 3-acetyl-4-hydroxy-6-methyl-(2H)-pyran-2-one and 2,2'-(ethylenedioxy)diethylamine was demonstrated for the control corrosion of MS in 1 M HCl using PDP and EIS methods [21]. SB was found to be a potential corrosion inhibitor as corrosion inhibition efficiency is enhanced by enhancing corrosion inhibitor concentration. The highest corrosion inhibition efficiency was 80.61% and 77.21% at 400 ppm inhibitor concentration. Quantum chemical calculations revealed the incidence of atomic sites bearing potential nucleophilic and electrophilic features of corrosion inhibitors that interacted with charged metal surfaces.

7.2.3 PYRAZOLE AND PYRAZINE AS CORROSION INHIBITORS

Pyrazole is the five-membered heterocycle containing two nitrogen atoms that are ortho-substituted. It is reported that pyrazole is a suitable corrosion inhibitor for the nullifying corrosion of various metals like iron, aluminium, copper, steel, etc. The compound 3,5-dimethyl-1H-pyrazol-1-yl m(4-((4-hydroxybenzylidene) amino)phenyl) methanone (DPHM) reported as corrosion inhibitor for low-carbon steel in 1 M HCl [22]. The WL method suggested that the corrosion inhibitor concentration amplified from 100 ppm to 400 ppm enhanced corrosion inhibition efficiency. The effect of temperature showed that temperature increased from 40 °C to 60 °C also contributed to the increase in corrosion inhibition efficiency. The highest corrosion inhibition was observed at 400 ppm and 60 °C. At this condition, the corrosion inhibitor exhibited 89.5% corrosion inhibition efficiency. The electrochemical methods EIS and PDP also revealed good corrosion performance of the inhibitor.

Pyrazole derivatives, ethyl 1-amino-3-(2-chlorophenyl)-5,10-dioxo-5,10-dihydro-1*H*-pyrazolo[1,2-*b*] phthalazine-2 carboxylate and ethyl 1-amino-5,10-dioxo-3-(*p*-tolyl)-5,10-dihydro-1*H*-pyrazolo[1,2-*b*]phthalazine-2-carboxylate[23] abbreviated as Py-1 and Py-2 exhibited excellent corrosion inhibition efficiency for SS in 2.0 M H_2SO_4. The PDP method indicated that both compounds Py-1 and Py-2 respectively displayed 96.8% and 91.2% corrosion inhibition efficiency at 10^{-2} M concentration of inhibitors, while the EIS method showed that Py-1 and Py-2 respectively showed 96.7% and 90.9% corrosion inhibition efficiency at same corrosion inhibitors concentration. The adsorption of both corrosion inhibitors followed the Langmuir adsorption isotherm model and acted as an anodic type of corrosion inhibitor. The SEM/EDX method also maintained the corrosion protection of both inhibitors.

Four new pyrazole acylhydrazone Schiff bases have been prepared and hired as corrosion inhibitors to control the corrosion of AZ91D, a magnesium alloy in 0.5 wt% NaCl [24]. The undertaking corrosion inhibitors possessed worthy corrosion inhibition activity for Mg alloy. The obtained outcomes proved that the aryl ring (Ar1) and aryl heterocyclic ring (Ar2) bore by corrosion inhibitor molecules could be responsible for the inflexibility of the molecular structure and enabled easy adsorption of corrosion inhibitor molecules on the surface of magnesium alloy. The surface analysis techniques SEM and AFM showed good corrosion performance achieved by the inhibitors. The inhibitors formed the protective layer and enhanced corrosion protection efficiency (92.2%). Corrosion inhibitive efficiencies of pyrazole–perimidine (PYR–PER) hybrids consisting of several attaching groups were evaluated for the 304 SS in 1.0 M hydrochloric acid [25]. PYR–PER hybrids successfully declined the corrosion current density, augmented corrosion resistance, and repressed the corrosion of the 304 SS. The various attaching groups (methyl, fluoride, chloride, and bromide) over pyrazole derivatives assisted in developing the corrosion inhibition efficiency of the PYR–PER hybrids. The compound PYR–PER3 having the methyl and bromide groups presented maximum corrosion inhibition productivity (97.45%).

Corrosion inhibition performance of two pyrazole products, called *N*-((1*H*-pyrazol-1-yl)methyl)pyrimidin-2-amine (PPA) and 2-(((1*H*-pyrazol-1-yl)methyl) amino)benzoic acid (PMB), have been assessed as corrosion inhibitors for carbon steel (CS) in 1 M HCl [26] by various techniques (WL, EIS, PDP, UV-visible spectroscopy, and SEM/EDX). The gained results suggested that the corrosion inhibition efficacy increased by the concentration of inhibitor but decreased correspondingly by rising temperature. The highest corrosion inhibition efficiencies of 94% and 92% for PPA PMB at 10^{-3} M and 303 K have been observed, respectively, while corrosion inhibition efficiency decreased to 76.9% and 72.3%, respectively, at 10^{-3} M and 333 K. PDP method showed the anodic behaviour of PPA and the mixed-type behaviour of PMB. The EIS indicated that charge transfer resistance increased and double-layer capacitance decreased by increasing the concentration of both corrosion inhibitors.

Bi-pyrazole-carbohydrazide compounds such as 1,1'-(propane-1,3-diyl)bis(5-methyl-1*H*-pyrazole-3-carbohydrazide) (P2PZ) and 1,1'-(oxy-bis(ethane-2,1-diyl))bis (5-methyl-1*H*-pyrazole-3-carbohydrazide) (O2PZ) have performed outstanding corrosion protection property for MS in 1.0 M HCl [27]. Corrosion-inhibitive action of

P2PZ and O2PZ was evaluated by WL, PDP, and EIS methods. The EIS method showed that corrosion inhibition efficiency was 95% and 84% at 308 K and 10^{-3} mol L^{-1} of P2PZ and O2PZ, respectively. Corrosion protection efficiency is enhanced by the augmentation of concentration and decreased by increased temperature. Polarization curves showed that the P2PZ and O2PZ were found to be mixed-type inhibitors and lined with the Langmuir adsorption isotherm. The free energy of adsorption ($\Delta G°_{ads}$) values lay within the range of −39.94 to −37.36 kJ mol^{-1}, indicating a physicochemical adsorption process. SEM images showed that both inhibitors reduced the corrosion by forming a protective layer on the surface of the metal. An ionic liquid 2-benzyl-1-ethyl-1H-pyrazolium bis(trifluoromethylsulfonyl) amides ([EBPz][NTf2]) outperformed corrosion prevention action for AZ91D Mg alloy in 0.05 wt% NaCl [28]. The corrosion inhibition efficiency of [EBPz][NTf2] attained 91.4% at 200 ppm concentration.

The compounds named N,N-bis (3-carbohydrazide-5-methylpyrazol-1-yl) methylene (M2PyAz) and 1, 4-bis (3-carbohydrazide-5-methylpyrazol-1-yl)butane (B2PyAz) were evaluated as corrosion inhibitors for MS in 1 M HCl [29]. The corrosion inhibition efficiency of M2PyAz and B2PyAzwas 98.6% and 87.8%, respectively, at 10^{-3} M corrosion inhibitors concentration. EIS method showed that the charge transfer resistance was enhanced by increasing concentration for both inhibitors. PDP curves showed that the M2PyAz and B2PyAz represented mixed-type corrosion inhibitors. Free energy of adsorption calculated from the Langmuir isotherm model for both M2PyAz and B2PyAz signified that both compounds prevented acid entrance through chemisorption. WL measurements showed that in the 308–348 K temperature range, corrosion inhibition efficiency was almost 98.5% and 89% for M2PyAz and B2PyAz, respectively, at 6 hours soaking time. SEM images indicated the protective layer made on the metal surface in the appearance of both corrosion inhibitors.

The corrosion inhibitory effect of 1,5-dimethyl-1H-pyrazole-3-carbohydrazide (PyHz) on MS in 1 M HCl has been conducted by EIS, PDP, WL, and SEM/EDX methods [30]. The results indicated that PyHz possessed a noble corrosion mitigation property, and corrosion inhibition efficiency went to 96% at 10^{-3} M inhibitor concentration. The PDP method confirmed that the PyHz proceeded as a mixed-type corrosion inhibitor. The adsorption process observed the Langmuir isotherm. Four pyrazole derivatives (PYRs) are diverse by nature as substituents. The compounds were synthesized by multicomponent reaction (MCR) [31]. The corrosion inhibition actions of corrosion inhibitors were studied for MS in acidic conditions. The corrosion inhibition competency of PYRs was substituent- and concentration-reliant. The corrosion inhibition efficiency increased in the presence of electron-withdrawing and electron-releasing substituents. The highest corrosion inhibition efficiency was accomplished in the presence of electron-releasing (−OH, −OHOCH3) groups. The corrosion inhibition adeptness of the PYRs followed the corrosion inhibition order as PYR−OHCH$_3$ (94.88%) > PYR–OH (91.47%) > PYR–NO$_2$ (90.90%) > PYR–H (89.77%). The PDP curve recommended that PYRs were mixed-type corrosion inhibitors, and their adsorption process confirmed the Langmuir isotherm.

The corrosion inhibitor is soluble in water, known as bis azo pyrazolin-5-one (ABP), has been constructed resourcefully by the regioselective reaction between hydrazine and coumarin hydrazone (CMH) [32]. Numerous electrochemical and surface morphology methods measured the anti-corrosion inhibition efficiency of ABP in an acidic medium (1.0 M HCl). The novel bis pyrazole-based azo dye ABP showed 93.3% corrosion protection capacity at 16×10^{-6} M corrosion inhibitor concentration. Tafel curves proposed that ABP was a mixed-type corrosion inhibitor. The adsorption of ABP inhibitor line up with the Langmuir isotherm model. SEM/EDX, AFM, and XPS surface analysis also established the enhancement of an adsorption film which protected the CS surface from acid corrosion.

The six-membered cyclic compounds have two nitrogen atoms at 1- and 4-locations known as pyrazine. Various types of pyrazine products were deliberated for the prevention of corrosion. The inhibitive effect of pyrazine appendages, 2,3-pyrazine dicarboxylic acid (pyrazine C), pyrazine carboxamide (pyrazine E), and 2-methoxy-3-(1-methylpropyl) pyrazine (pyrazine H) were assessed on the X-60 steel in 15% HCl [33]. The WL measurement was conducted at 60 °C and 90 °C for 6 hours, and PDP, EFM, and EIS were completed at 25 °C. The corrosion inhibition efficiency of pyrazines C and E have not been altered by augmentation of their concentrations at 60 °C. Still, the corrosion inhibition efficiency of pyrazine E enlarged by concentration at 90 °C, while the corrosion inhibition efficiency of pyrazine C did not change. The corrosion inhibition performance of pyrazine H increased by concentration at both temperatures. The corrosion inhibition effectiveness of pyrazines C and E enhanced in the presence of sodium iodide and glutathione (Glu), but has not showed pronounced influence by pyrazine H. Potentiodynamic test revealed that the three pyrazine derivatives were evaluated as mixed-type corrosion inhibitors. Pyrazines C and H bore largely cathodic and pyrazine E showed more anodic effects.

The three compounds such as 2-methylpyrazine (MP), 2-aminopyrazine (AP), and 2-amino-5-bromopyrazine (ABP) were investigated as corrosion inhibitors on the corrosion of cold rolled steel (CRS) in 1.0 M H_2SO_4 [34]. All three pyrazine compounds were found to be potential corrosion inhibitors, and corrosion inhibition proficiency followed the ABP > AP > MP order. The adsorption of three inhibitors on the surface of CRS was followed by Langmuir adsorption isotherm. 2,3-Pyrazine dicarboxylic acid (PDA) was evaluated as a corrosion inhibitor for 90 Cu–10 Ni and 70 Cu–30 Ni alloys in 2% HCl [35]. Corrosion protection efficiency derived from WL, EIS, and PDP measurements indicated that corrosion inhibitors exhibited 39% and 55% corrosion inhibition efficiency for 90 Cu–10 Ni and 70 Cu–30 Ni, respectively, decreasing slightly after 72-hour immersion time. Adding 5 mM KI enhanced the corrosion inhibition to 73% and 92% for 90 Cu–10 Ni and 70 Cu–30 Ni, respectively, at room temperature. The 51% and 63% corrosion inhibition activity were found at 60 °C for 90 Cu–10 Ni and 70 Cu–30 Ni, respectively. PDP test indicated that corrosion inhibitors worked as mixed-type with somewhat more anodic effect on 70 Cu–30 Ni but the cathodic outcome on 90 Cu–10 Ni.

The corrosion inhibition activity of MS in 1 M H_2SO_4 has been conducted by employing acenaphtho[1,2-b]quinoxaline and acenaphtho[1,2-b]pyrazine corrosion inhibitors at 303–333 K [36]. PDP measurement reflected that corrosion inhibitors were evaluated to be good as mixed-type. EIS method reflected that the charge transfer resistance of MS increased in the presence of corrosion inhibitors. The effect of diethyl pyrazine-2,3-dicarboxylate (P1) was studied by WL, PDP, LRP (linear polarization resistance), and EIS methods for steel in 0.5 M H_2SO_4 solution [37]. The inhibiting action increased by elevating the concentration of the pyrazine compound, and it attained 82% at 10^{-2} M. PDP method indicates that pyrazine acted as a cathodic inhibitor. The cathodic curves suggest that the lessening of proton at the steel surface occurred with an activating mechanism. The yellow deposit on the steel surface was examined by infrared method. P1 adsorbed on the steel surface conferring to Langmuir adsorption model. EIS was used to determine the corrosion inhibition process of steel in 0.5 M H_2SO_4 solution at the open-circuit potential (OCP) [38]. Diethyl pyrazine-2,3-dicarboxylate (Prz) as a non-ionic surfactant (NS) corrosion inhibitor has been inspected. The Nyquist diagrams entailed a capacitive semicircle at high frequencies and a distinct inductive loop at low frequency values. The impedance measurements were understood according to suitable equivalent circuits. The results obtained showed that Prz is a decent corrosion inhibitor.

The corrosion inhibition efficiency increased by increased surfactant concentration, and the highest corrosion inhibition efficiency attained 80% at 5×10^{-3} M. Prz is adsorbed on the steel surface bestowing to a Langmuir isotherm adsorption model. Pyrido[2, 3-b]pyrazine derivative (P1) on MS in acidic media were studied [39]. The effect of corrosion inhibition of P1 was explored for MS in 1 M HCl. The corrosion inhibitor P1 showed a mixed inhibition mechanism without altering the hydrogen reduction mechanism, as revealed by the PDP method. The corrosion inhibition efficiency was augmented by uplifting the concentration of (P1) and extending a maximum value to 93% at 10^{-3} M concentration. The upsurging of temperature cuts the corrosion inhibition efficiency of (P1). It has been set up that (P1) adsorbed on the surface of metal steel, allowing the Langmuir isotherm model. Quantum chemical studies were also agreed to explain the data.

The study defined the preparation and characterization of thiophene/pyrazine Schiff base known as (E)-N-{(thiophene-2-yl) methylene}pyrazine-2-carboxamide (HL). The prepared compound was synthesized and scrutinized by melting point and micro (CHN) analysis, vibrational, nuclear magnetic resonance (^{1}H- and ^{13}C-NMR), and electronic (UV-vis) spectroscopies [40]. HL was also taken for adsorption and corrosion properties. The spectral data suggested tridentate bonding nature and enolimine tautomeric arrangement within the ligand grouping giving integrity to its creation. Density functional theory (DFT) calculations associated with the examination data, while its percentage corrosion inhibition efficiency (IE%), became bigger using intensification in concentration and deteriorated by way of an increase in temperature (T). The decreased corrosion prevention by increased temperature showed that adsorption phenomena are mixed-type, physisorption governing the corrosion inhibition route. Similarly, the adsorption of HL on the MS inclined Langmuir adsorption isotherm. However, the negative sign of thermodynamics parameters showed that HL adsorption on the MS surface in 1.0 M HCl was instantaneous and exothermic.

7.3   CONCLUSION AND FUTURE PROSPECTIVE

Corrosive media lead to the corrosion of metals. Corrosion led to a decreased quality of metals; hence protection of metals became a significant aspect. The protection of metals using corrosion inhibitors improved the corrosion resistance of the metal.

Organic compounds, specially heterocycles, mitigated metal corrosion to a large extent. Heterocycle compounds PU, pyran, pyrazole, and pyrazine are a prominent class of corrosion inhibitors for preventing metal corrosion due to heteroatoms in the ring, which facilitates the adsorption of molecules over the metal surface. Corrosion protection of PU, pyran, pyrazole, and pyrazine can be further enhanced in combination with nanomaterials as surface properties of nanomaterial can assist in the more adsorption of these heterocycles over the metal surface.

REFERENCES

1. T. Alemayehu, M. Birahane, Corrosion and its protection, Int J Acad Sci Res 2(2014)1–7.
2. S. Harsimran, K. Santosh, K. Rakesh, Overview of corrosion and its control: A critical review, Proc Eng Sci 3(2021)13–24.
3. I. A. Wonnie Ma, S. Ammar, S. S. A. Kumar, K. Ramesh, S. Ramesh, A concise review on corrosion inhibitors: Types, mechanisms and electrochemical evaluation studies, J Coat Technol Res 19(2021)241–268.
4. K. Xhanari, M. Finšgar, Organic corrosion inhibitors for aluminum and its alloys in chloride and alkaline solutions: A review, Arab J Chem 12(2019)4646–4663. https://doi.org/10.1016/j.arabjc.2016.08.009
5. N. A. Ahmed, D. S. Ahmed, H. Mohammed, M. Al-mashhadani, Inhibition of corrosion: Mechanisms and classifications an overview, Al-Qadisiyah J Pure Sci 25(2020)1–9. http://qu.edu.iq/journalsc/index.php/jops
6. M. Scendo, J. Trela, N. Radek, Purine as an effective corrosion inhibitor for stainless steel in chloride acid solutions, Corros Rev 30(2012)33–45. https://doi.org/10.1515/corrrev-2011-0039
7. M. Scendo, The effect of purine on the corrosion of copper in chloride solutions, Corros Sci 49(2007)373–390. https://doi.org/10.1016/j.corsci.2006.06.022
8. Z. Jiang, Y. Li, Q. Zhang, B. Hou, W. Xiong, H. Liu, G. Zhang, Purine derivatives as high efficient eco-friendly inhibitors for the corrosion of mild steel in acidic medium: Experimental and theoretical calculations, J Mol Liquids 323(2021)114809. https://doi.org/10.1016/j.molliq.2020.114809
9. M. B. P. Mihajlović, M. B. Radovanović, A. T. Simonović, Z. Z. Tasić, M. M. Antonijević, Evaluation of purine based compounds as the inhibitors of copper corrosion in simulated body fluid, Results Phys 14(2019)102357. https://doi.org/10.1016/j.rinp.2019.102357
10. Y. Yan, W. Li, L. Cai, B. Hou, Electrochemical and quantum chemical study of purines as corrosion inhibitors for mild steel in 1 M HCl solution, Electrochim Acta 53(2008)5953–5960. https://doi.org/10.1016/j.electacta.2008.03.065
11. M. Scendo, Corrosion inhibition of copper by purine or adenine in sulphate solutions, Corros Sci 49(2007)3953–3968. https://doi.org/10.1016/j.corsci.2007.03.037
12. M. Scendo, Inhibition of copper corrosion in sodium nitrate solutions with nontoxic inhibitors, Corros Sci 50(2008)1584–1592. https://doi.org/10.1016/j.corsci.2008.02.015
13. X. Guo, H. Huang, D. Liu, The inhibition mechanism and adsorption behavior of three purine derivatives on the corrosion of copper in alkaline artificial seawater: Structure and performance, Colloid Surf A: Physiochem Eng Aspects 622(2021)126644. https://doi.org/10.1016/j.colsurfa.2021.126644.
14. M. A. Amin, Q. Mohsen, O. A. Hazzazi, Synergistic effect of I⁻ ions on the corrosion inhibition of Al in 1.0 M phosphoric acid solutions by purine, Mater Chem Phys 114(2009)908–914. https://doi.org/10.1016/j.matchemphys.2008.10.057

15. E. F. Silva, J. S. Wysard, C. E. Bandeira, O. R. Mattos, Electrochemical and surface enhanced Raman spectroscopy study of Guanine as inhibitor for copper corrosion, Corros Sci 191(2021)109714. https://doi.org/10.1016/j.corsci.2021.109714

16. M. Ouakki, M. Galai, Z. Aribou, Z. Benzekri, E. H. El Assiri, K. Dahmani, E. Ech-chihbi, A. S. Abousalem, S. Boukhris, M. Cherkaoui, Detailed experimental and computational explorations of pyran derivatives as corrosion inhibitors for mild steel in 1.0 M HCl: Electrochemical/surface studies, DFT modeling, and MC simulation, J Mol Struct 1261(2022)132784. https://doi.org/10.1016/j.molstruc.2022.132784

17. M. Khattabi, F. Benhiba, S. Tabti, A. Djedouani, A. El Assyry, R. Touzani, I. Warad, H. Oudda, A. Zarrouk, Performance and computational studies of two soluble pyran derivatives as corrosion inhibitors for mild steel in HCl, J Mol Struct 1196(2019)231–234. https://doi.org/10.1016/j.molstruc.2019.06.070

18. A. Singh, K. R. Ansari, M. A. Quraishi, H. Lgaz, Y. Lin, Synthesis and investigation of pyran derivatives as acidizing corrosion inhibitors for N80 steel in hydrochloric acid: Theoretical and experimental approaches, J Alloy Compd 762(2018)347–362. https://doi.org/10.1016/j.jallcom.2018.05.236

19. A. El Hattak, S. Izzaouihda, Z. Rouifi, F. Benhiba, S. Tabti, A. Djedouani, N. Komiha, H. Abou El Makarim, R. Touzani, H. Oudda, I. Warad, A. Zarrouk, Anti-corrosion performance of pyran-2-one derivatives for mild steel in acidic medium: Electrochemical and theoretical study, Chem Data Collect 32(2021)100655. https://doi.org/10.1016/j.cdc.2021.100655

20. J. Saranya, F. Benhiba, N. Anusuya, R. Subbiah, A. Zarrouk, S. Chitra, Experimental and computational approaches on the pyran derivatives for acid corrosion, Colloid Surf A: Physiochem Eng Asp 603(2020)125231. https://doi.org/10.1016/j.colsurfa.2020.125231

21. J. N. Asegbeloyin, M. Ejikeme, L. O. Olasunkanmi, A. S. Adekunle, E. E. Ebenso, A novel Schiff base of 3-acetyl-4-hydroxy-6-methyl-(2H)pyran2-one and 2,2′-(ethylene-dioxy)diethylamine as potential corrosion inhibitor for mild steel in acidic medium, Materials 8(2015)2918–2934. https://doi.org/10.3390/ma8062918

22. A. S. Jasim, K. H. Rashid, K. F. AL-Azawi, A. A. Khadom, Synthesis of a novel pyrazole heterocyclic derivative as corrosion inhibitor for low-carbon steel in 1 M HCl: Characterization, gravimetrical, electrochemical, mathematical, and quantum chemical investigations, Results Eng 15(2022)100573. https://doi.org/10.1016/j.rineng.2022.100573

23. H. Chahmout, M. Ouakki, S. Sibous, M. Galai, N. Arrousse, E. Ech-chihbi, Z. Benzekri, S. Boukhris, A. Souizi, M. Cherkaoui, New pyrazole compounds as a corrosion inhibitor of stainless steel in 2.0 M H_2SO_4 medium: Electrochemical and theoretical insights, Inorg Chem Commun 147(2023)110150. https://doi.org/10.1016/j.inoche.2022.110150

24. D. Ma, J. Zhao, Q. Huang, G. Li, J. Liu, T. Ren, Pyrazole acylhydrazone Schiff bases as magnesium alloy corrosion inhibitor: Synthesis, properties and mechanism investigation, J Mol Struct 1281(2023)135056. https://doi.org/10.1016/j.molstruc.2023.135056

25. Ö Uğuz, M. Gümüş, Y. Sert, I. Koca, A. Koca, Utilization of pyrazole-perimidine hybrids bearing different substituents as corrosion inhibitors for 304 stainless steel in acidic media, J Mol Struct 1262 (2022)133025. https://doi.org/10.1016/j.molstruc.2022.133025

26. G. Laadam, M. El Faydy, F. Benhiba, A. Titi, H. Amegroud, A. S. Al-Gorair, H. Hawsawi, R. Touzani, I. Warad, A. Bellaouchou, A. Guenbour, M. Abdallah, A. Zarrouk, Outstanding anti-corrosion performance of two pyrazole derivatives on carbon steel in acidic medium: Experimental and quantum-chemical examinations, J Mol Liquids 375(2023)121268. https://doi.org/10.1016/j.molliq.2023.121268

27. K. Cherrak, O. M. A. Khamaysa, H. Bidi, M. El Massaoudi, I. A. Ali, S. Radi, Y. El Ouadi, F. El-Hajjaji, A. Zarrouk, A. Dafali, Performance evaluation of newly synthetized bi-pyrazole derivatives as corrosion inhibitors for mild steel in acid environment, J Mol Struct 1261 (2022)132925. https://doi.org/10.1016/j.molstruc.2022.132925

28. X. Gao, D. Ma, Q. Huang, T. Ren, G. Li, L. Guo, Pyrazole ionic liquid corrosion inhibitor for magnesium alloy: Synthesis, performances and theoretical explore, J Mol Liquids 353(2022)118769. https://doi.org/10.1016/j.molliq.2022.118769

29. C. Karima, M. El Massaoudi, H. Outada, M. Taleb, Electrochemical and theoretical performance of new synthetized pyrazole derivatives as promising corrosion inhibitors for mild steel in acid environment: Molecular structure effect on efficiency, J Mol Liquids 342(2021)117507. https://doi.org/10.1016/j.molliq.2021.117507

30. K. Cherrak, M. E. Belghiti, A. Berrissoul, M. El Massaoudi, M. El Faydy, M. Taleb, S. Radi, A. Zarrouk, A. Dafali, Pyrazole carbohydrazide as corrosion inhibitor for mild steel in HCl medium: Experimental and theoretical investigations, Surf Interf 20(2020)100578. https://doi.org/10.1016/j.surfin.2020.100578

31. C. Verma, V. S. Saji, M. A. Quraishi, E. E. Ebenso, Pyrazole derivatives as environmental benign acid corrosion inhibitors for mild steel: Experimental and computational studies, J Mol Liquids 298(2020)111943. https://doi.org/10.1016/j.molliq.2019.111943

32. E. E. El-Katori, A. El-Saeed, M. M. Abdou, Anti-corrosion and anti-microbial evaluation of novel water-soluble bis azo pyrazole derivative for carbon steel pipelines in petroleum industries by experimental and theoretical studies, Arabian J Chem 15(2022)104373. https://doi.org/10.1016/j.arabjc.2022.104373

33. I. B. Obot, S. A. Umoren, N. K. Ankah, Pyrazine derivatives as green oil field corrosion inhibitors for steel, J Mol Liquids 277(2019)749–761. https://doi.org/10.1016/j.molliq.2018.12.108

34. S. K. Saha, S. Deng, H. Fu, Three pyrazine derivatives as corrosion inhibitors for steel in 1.0 M H_2SO_4 solution, Corros Sci 53(2011)3241–3247. https://doi.org/10.1016/j.corsci.2011.05.068

35. I. C. Ukaga, P. C. Okafor, I. B. Onyeachu, A. I. Ikeuba, D. I. Njoku, The inhibitive performance of 2,3-pyrazine dicarboxylic acid and synergistic impact of KI during acid corrosion of 70/30 and 90/10 copper–nickel alloys, Mater Chem Phys 296(2023)127313. https://doi.org/10.1016/j.matchemphys.2023.127313

36. J. Saranya, P. Sounthari, K. Parameswari, S. Chitra, Acenaphtho[1,2-*b*]quinoxaline and acenaphtho[1,2-*b*]pyrazine as corrosion inhibitors for mild steel in acid medium, Measurement 77(2016)175–186. https://doi.org/10.1016/j.measurement.2015.09.008

37. M. Bouklah, A. Attayibat, S. Kertit, A. Ramdani, B. Hammouti, A pyrazine derivative as corrosion inhibitor for steel in sulphuric acid solution, Appl Surf Sci 242(2005)399–406. https://doi.org/10.1016/j.apsusc.2004.09.005

38. M. Kissi, M. Bouklah, B. Hammouti, M. Benkaddour, Establishment of equivalent circuits from electrochemical impedance spectroscopy study of corrosion inhibition of steel by pyrazine in sulphuric acidic solution, Appl Surf Sci 252(2006)4190–4197. https://doi.org/10.1016/j.apsusc.2005.06.035

39. M. Y. Hjouji, M. Djedid, H. Elmsellem, Y. K. Rodi, Y. Ouzidan, F. O. Chahdi, N. K. Sebbar, E. M. Essassi, I. Abdel-Rahman, B. Hammouti, Corrosion inhibition of mild steel in hydrochloric acid solution by pyrido[2,3-*b*]pyrazine derivative: Electrochemical and theoretical evaluation, J Mater Environ Sci 7(2016)1425–1435.

40. F. Chioma, O. W. Nnenna, O. Moses, Preparation, spectral characterization and corrosion inhibition studies of (*E*)-*N*-{thiphene-2-yl)methylene}pyrazine-2-carboxamide Schiff base ligand, Protect Metals Phys Chem Surf 56(2020)651–662.

8 Pyridines and Piperidines as Corrosion Inhibitors

*Taiwo W. Quadri[1], Nnadozie C. Ebenezer[2],
Lukman O. Olasunkanmi[3,4], and Eno E. Ebenso[1]*
[1]Centre for Material Science, College of Science,
Engineering and Technology, University of
South Africa, Johannesburg, South Africa
[2]David Umahi Federal University of Health
Sciences, Uburu, Ebonyi State, Nigeria
[3]Department of Chemistry, Faculty of Science,
Obafemi Awolowo University, Ile-Ife, Nigeria
[4]Department of Chemical Sciences, University of
Johannesburg, Johannesburg, South Africa

8.1 INTRODUCTION

Corrosion is a well-researched aspect of material science because of the damage it causes to metallic materials, which are indispensable in homes, industries, and the larger society. The substantial annual record of economic losses from corrosion amongst metal-based sectors such as petrochemical, transport, manufacturing, and packaging industries is staggering [1–3]. It is reported that 1 tonne of steel, the most versatile industrial metallic material, could degrade to rust every 90 seconds. In addition, 50% of every tonne of steel alloy is used as a replacement for degraded steel [4, 5]. Apart from steel, non-ferrous metals and alloys such as zinc, copper, aluminium, brass, and tin have experienced depletion of their essential properties due to corrosive ions in the surroundings. Therefore, it becomes imperative for scientists and engineers to proffer innovative solutions to tackle the corrosion phenomena in metals. Amongst the proposed solutions to retard the corrosion phenomena, the use of chemical additives, popularly referred to as corrosion inhibitors, has gained prominent attention because of its numerous advantages [6]. In particular, organic inhibitors have shown preferential advantages such as facile preparation, cost-effectiveness, excellent protection abilities, eco-friendliness, and ease of application [3, 7].

Based on literature outcomes from numerous corrosion studies conducted over the years, heterocyclic compounds containing aromatic rings, multiple bonds, heteroatoms, and polar substituents display outstanding adsorption behaviour on metal surfaces in corrosive solutions [8, 9]. These compounds adhere to the metal surfaces, creating an impenetrable blockade against chemical attacks. Adsorption often takes the form of physisorption or chemical adsorption or both. The effectiveness of heterocyclic compounds as additives for corrosion mitigation depends on the metal's nature, the substituents' degree of corrosiveness, operating temperature, molecular size, and the electronic structure of the chemical inhibitors [10].

DOI: 10.1201/9781003377016-8

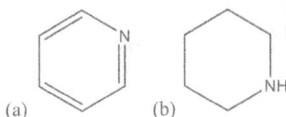

FIGURE 8.1 Chemical structure of (a) pyridine (b) piperidine.

The versatility of pyridine and piperidine derivatives has ensured their varied applications in several fields of scientific research, such as dyes, chemical synthesis, agrochemicals, pharmaceuticals, etc. They are widely considered as fundaments in the formulation of several pharmaceutical drugs and belong to the commonest *N*-heterocycles present in FDA-approved drugs [11–13]. Numerous reports have documented these heterocyclic compounds' production, analyses, and applications as antiviral, anti-cancer, anti-malarial, anti-hypertension, anti-microbial, analgesic, anti-inflammatory, and anti-psychotic agents [14–17]. The pyridine molecule (C_5H_5N) is a benzene compound with an N-atom in its structure (Figure 8.1a). On the other hand, piperidine, which is also a six-membered *N*-heterocycle, is associated with the chemical formula $(CH_2)_5NH$ in which one of the aromatic carbons is substituted with an amine (—NH) group (Figure 8.1b). In terms of their corrosion inhibition potentials, a study by Hackerman and Makrides in 1954 [18] revealed that piperidine, a stronger saturated base than pyridine, has greater inhibition potency than pyridine on account of the functional N-atom and not the delocalized π-electrons of the pyridine structure. A few other authors have shared this opinion based on their research findings [19–22].

8.2 INHIBITION OF METALLIC CORROSION BY PYRIDINE DERIVATIVES

From the literature, several early reports documented the inhibition capacities of pyridine-based inhibitors to restrain the corrosion rate in diverse media [18, 23–27]. The utilization of pyridine derivatives to assuage the corrosion process in different media has gained wide attention, evident from the extensive literature on the subject [7, 28–30]. Based on available literature reports, it is unarguable that the inhibitive potential of non-substituted pyridine molecules is limited. Interestingly, upon introducing substituents, especially at the second and sixth positions, the pyridine compound becomes effective by establishing an insoluble stable film on the studied metal surfaces [7]. Besides the role of substitution in ensuring excellent adsorption behaviour of pyridines, the molecular size, presence of heteroatoms, aromatic rings, and type/concentration of electrolyte medium are also vital parameters to be considered. This section reviews reported pyridines as inhibitors of metallic deterioration from 2020 till date.

8.2.1 EXPERIMENTAL AND COMPUTATIONAL ASSESSMENT OF SUBSTITUTED PYRIDINES AS CORROSION INHIBITORS

Generally, the abilities of organic inhibitors, including pyridines, to restrain the deterioration of metallic materials such as carbon steel (CS), X-65 steel, mild steel (MS), and other non-ferrous metals/alloys are often determined using experimental and computational methods. Weight loss (WL), called gravimetric analysis, is an orthodox and popular procedure employed to measure the speed of metallic corrosion and the inhibition efficiencies (IE) of pyridine-based inhibitors. Electrochemical techniques such as potentiodynamic polarization (PDP) and electrochemical impedance spectroscopy (EIS) are widely reported to measure the IE of pyridines. Corrosion current densities and charge transfer resistances are important parameters from these techniques, which are useful in understanding the adsorption behaviour of pyridine compounds. Surface characterization techniques, especially scanning electron microscopy (SEM), attenuated total refraction infrared (ATR-IR) spectroscopy, X-ray photoelectron spectroscopy (XPS), energy dispersive X-ray spectroscopy (EDS), and atomic force microscopy (AFM), have been utilized to probe the interaction at the metal–solution interface. Valuable quantum chemical variables such as E_{HUMO} (energy of the highest unoccupied molecular orbital), E_{LUMO} (energy of the lowest unoccupied molecular orbital), energy bandgap (ΔE), dipole moment (μ), etc. that give information on the reactivity of pyridines and the donor–acceptor interaction are calculated using quantum chemical methods such as semi-empirical methods, density functional theory (DFT), and Hartree Fock (HF). Molecular dynamic simulation (MDS) and Monte Carlo simulation (MCS) are modern computational techniques that provide useful insight into the orientation of pyridines on metal surfaces and the nature of adsorption. Furthermore, some studies have attempted to correlate molecular parameters of pyridines with their corrosion protection abilities using quantitative structure activity–property relationship (QSAR) [31–36].

Fouda et al. [37] conducted an extensive electrochemical, morphological, and computational assessment of PdC-Me, PdC-OH, and PdC-H for MS degradation in 1 M HCl to investigate the inhibition properties of three pyridine-based compounds. Adsorption via chemisorption on the metal surface was accredited to the donor–acceptor interaction between the vacant d-orbital and the conjugated π-charge of the inhibitor molecules. Also, the electron-donating substituent groups, —OH and —CH₃ with Hammett sigma constant (σ) of −0.17 and −0.37, respectively, demonstrated higher IE than —H, ($\sigma = 0.0$), due to increased electron density. Surface studies using AFM, XPS, and ATR-IR revealed an interaction between PdC-Me, PdC—OH, and PdC—H and the steel specimen that afforded metallic surface protection. Another research evaluated the effect of —Cl, —H, and —OCH₃ substituted pyridines on the protection capacities of X-65 steel degradation in acid using different established approaches [38]. The order of IE is as follows: PYR—OCH₃ > PYR—H > PYR—Cl suggesting the positive influence of electron-donating groups on IE. A study by Salim et al. demonstrated that 6-chloro-2-(4-chlorophenyl)imidazo[1,2-*a*] pyridine-3-carbal-dehyde (IPCl-2) lowered the corrosion current density of MS in HCl better than 6-chloro-2-(4-chlorophenyl)imidazo[1,2-*a*]pyridine (IPCl-1). It was observed that both chemisorbed onto the MS according to the Langmuir adsorption isotherm (LAI); while IPCl-1 had a mixed-inhibition effect, IPCl-2 was found to display a cathodic inhibition predominance. Several researchers have also documented the influence of substituents on the protection abilities of pyridine-based inhibitors [39–46]. The contribution of aromatic rings and N-atom to the protection performance of organic compounds was established in a study conducted by Hau and Huong using 1,10-phenathroline, quinoline, and pyridine as test compounds. Pyridine with a singular aromatic ring and N-atom yielded the least IE [47]. Also, the effect of steric hindrance on the protection abilities of two novel pyridine-based inhibitors has been reported [48]. Table 8.1 presents a summary of

TABLE 8.1

An Array of Investigated Pyridines for the Corrosion Protection of Metals in Different Media

S/N	Pyridine Compound	Metal/ Electrolyte	Nature of Inhibitor	Conc/Max IE	Reference
1.	BBTP	MS/1 M HCl	Mixed-type/ LAI	1 mM/ 95%	[58]
2.	NVAIP	MS/0.5 M HCl	Mixed-type/ LAI	500 ppm/ 92.98%	[59]
3.	4HT	MS/1 M HCl	LAI	0.005 M/ 96.20%	[60]
4.	I	API X-65 steel/6 M H_2SO_4	Mixed type/ LAI	2.66×10^{-3} M/ 92.70%	[61]
5.	III	API X-65 steel/6 M H_2SO_4	Mixed-type/ LAI	2.66×10^{-3} M/ 94.70%	[61]
6.	PD	Al/1 M HCl	Cathodic	10^{-3} M/ 80.20%	[62]

(Continued)

TABLE 8.1 (*Continued*)
An Array of Investigated Pyridines for the Corrosion Protection of Metals in Different Media

S/N	Pyridine Compound	Metal/ Electrolyte	Nature of Inhibitor	Conc/Max IE	Reference
7.	NIF	API 5L X-52 steel/2 M HCl	Mixed-type/ LAI	500 ppm/ 94%	[63]
8.	DMPIP	MS/0.5 M HCl	Mixed-type/ LAI	500 ppm/ 96.70%	[64]
9.	4BY	MS/1 M HCl	LAI	0.005 M/ 93.80%	[65]
10.	CTDP	MS/15% HCl	Mixed-type/ LAI	80 ppm/ 89.53%	[66]
11.	P2T	Brass/0.5 M H_2SO_4	Mixed-type/ LAI	0.5 M/ 84%	[67]
12.	2,6DP	Al/1 M NaCl	Mixed-type/ LAI	3 mM/ 66.92%	[68]

(Continued)

TABLE 8.1 *(Continued)*

An Array of Investigated Pyridines for the Corrosion Protection of Metals in Different Media

S/N	Pyridine Compound	Metal/ Electrolyte	Nature of Inhibitor	Conc/Max IE	Reference
13.	2,6DP	Al/1 M NaOH	Mixed-type/ LAI	0.4 M/ 55.80%	[68]
14.	2,6DP	Al/1 M HCl	Mixed-type/ Freundlich	0.4 M/ 51.73%	[68]
15.	THP	CS/1 M HCl	Mixed-type/ LAI	10^{-3} M/ 95.22%	[69]
16.	HBPA	X70 steel/2 M HCl	Mixed-type/ LAI	1.00 mM/ 96%	[70]
17.	B-HBPA	X70 steel/2 M HCl	Mixed-type/ LAI	1.00 mM/ 92%	[70]
18.	HL	MS/1 M HCl	Mixed-type/ LAI	100 ppm/ 93.78%	[71]
19.	MAEP	CS/1 M HCl	Mixed-type/ LAI	25 mM/ 91%	[72]
20.	MPPO	N80 steel/1 M HCl`	Mixed-type/ LAI	300 ppm 91.4%	[73]

(Continued)

TABLE 8.1 *(Continued)*

An Array of Investigated Pyridines for the Corrosion Protection of Metals in Different Media

S/N	Pyridine Compound	Metal/ Electrolyte	Nature of Inhibitor	Conc/Max IE	Reference
21.	 CPP	CS/1 M HCl	LAI	1 mM/ 92.8%	[74]
22.	 CPYA	MS/1 M HCl	Mixed-type/ LAI	4.59 mM/ 96%	[75]
23.	 PyTA	Cu/1 M HCl	Mixed-type/ LAI	1 mM/ 94%	[76]
24.	 BPDA	MS/0.5 M HCl	Mixed-type/ LAI	400 ppm/ 81.60%	[77]
25.	 TPP	MS/1 M HCl	Mixed-type/ LAI	0.075/ 90%	[78]
26.	 BAPN	CS/0.5 M H$_2$SO$_4$	Mixed-type/ LAI	5×10^{-4} M/ 97.45%	[79]
27.	 PAPN	CS/0.5 M H$_2$SO$_4$	Mixed-type/ LAI	5×10^{-4} M/ 95.43%	[79]

(Continued)

TABLE 8.1   (*Continued*)
An Array of Investigated Pyridines for the Corrosion Protection of Metals in Different Media

S/N Pyridine Compound	Metal/ Electrolyte	Nature of Inhibitor	Conc/Max IE	Reference
28. MAPN	CS/0.5 M H$_2$SO$_4$	Mixed-type/ LAI	5×10^{-4} M/ 91.73%	[79]
29. PIP	MS/1 M HCl	Mixed-type/ LAI	10^{-4} M/ 92%	[80]
30. 2PB	API X60 steel/ synthetic brine	Mixed-type	2.56 mM/ 88.30%	[81]
31. HMAP	MS/ 1 M HCl	LAI	0.5 g/L/ 90.80%	[82]

some recently conducted studies on the inhibition of metallic deterioration using pyridine-based compounds.

It is essential to highlight the efforts of several authors to generate beneficial knowledge on the protection mechanism of pyridine using modern computational methods. Before synthesis, Zhu et al. [49] carried out a thorough computational analysis (DFT and MDS) on the adsorption potential of six pyridine molecules labelled A–F on an iron surface. The study concluded that F – (3,4-bis(benzimidazole-20-yl)pyridine) – demonstrated the best adsorption behaviour and is suitable for corrosion protection. After correlating analytical and computational studies, a study by Mohamed et al. [50] established that 4-amino-3-choloro-2,5,6-trifluoro-pyridine (ACTFP) would be an efficient inhibitor of metallic corrosion because the obtained high E_{HOMO} indicates its strong potential for electron donation. Mollamin and Monjjemi conducted an extensive analytical and computational investigation to establish the protection abilities of pyridine compounds for aluminium alloy [51]. The potential inhibition action of four pyridine dicarboxylic acids was also assessed via the DFT method at 6-311G (d, p) basis set and the study concluded that 2,3-pyridinedicarboxylic acid would act as a high-performing anti-corrosive agent [52].

The anti-corrosion potentials of three pyridine-based Schiff bases in gas and aqueous phases were probed using three different DFT basis sets. According to the study, methyl and chloride substituents increased μ, ΔE, and high total energy (TE) which boosted the adsorption properties of the studied pyridines more than the basic structure. The study achieved an alignment between the experiment and theoretical outcomes. As a result, a new Schiff base with a different moiety was designed and its inhibition performance was proposed to be better in the aqueous phase [53]. Ser et al. performed a QSAR study on 41 selected pyridine and quinoline molecules used as corrosion inhibitors for MS. The authors obtained several DFT and physicochemical parameters in developing models via genetic algorithm-based techniques [54]. In another study, using a genetic algorithm approximation method, the QSPR model was developed for 28 pyridine and quinoline compounds. Three variables yielded a highly accurate, reliable model for the studied N-heterocycles [55].

Oyeneyin et al. [56] reported the adsorption potentials of six aminopyridine Schiff bases using computational techniques. Using DFT 6-31G(d), relevant quantum chemical indices were obtained that are employed in model development. MCS was adopted to evaluate the interaction energies of the stable configurations of the pyridine inhibitors. Among the tested compounds, N-(5-amino-2-hydroxybenzylidene) pyridine-4-amine demonstrated the expected trends affirming it as a high-performing inhibitor. Moreover, MCS revealed strong metal–inhibitor bonding and stable configuration of the inhibiting molecules. Recently, deeper insights into the bonding mechanism of 2-pyridylaldoxime (2POH) and 3-pyridylaldoxime (3POH) were provided by Lgaz et al. [57] using DFT, MDS, and self-consistent charge density functional tight-binding (SCC-DFTB) simulations. The study concluded that the studied molecules chemisorbed on the iron metal and that the observed chelating ability was due to the hard and soft acid and bases (HSAB) principle and ΔE. Results obtained from MDS demonstrated that all neutral and protonated pyridines displayed a parallel adsorption mode on the iron metal.

8.3 INHIBITION OF METALLIC CORROSION BY PIPERIDINE DERIVATIVES

The assessment of polar organic inhibitors, including piperidine, as potential chemical inhibitors have been documented for over 70 years. One such early study reported by Hackerman and Makrides in 1954 probed the inhibitive effects of organic substances such as diphenylamine, cyclohexylamine, dicyclohexylamine, aniline, pyridine, and piperidine [18]. In a 1973 paper, Aramaki [83] reported the role of stearic hindrances on the protection capacities of piperidine derivatives for iron deterioration in 6.1 M HCl. The study observed an anodic inhibition performance of methyl and ethyl substituents of piperidines which correlated with the polar substituent constant, molecular coverage area, and steric substituent constant. Furthermore, a cathodic inhibition performance was correlated with the dissociation constant. In 1976, Ramakrishnaiah and Subramanyan demonstrated from polarization measurements that seven *N*-heterocycles, including pyridine- and piperidine-based compounds, are defective in protecting aluminium from deteriorating in 1 M NaOH. However, their inhibition abilities improved upon introducing calcium as a synergistic additive. On the other hand, the *N*-heterocycles were found to greatly deter corrosion in 1 M HCl without the need for synergism [20]. Since then, other related works have been documented in the literature [22, 84].

8.3.1 Inhibition of Metallic Degradation for Iron-Based Metals by Piperidines

The degradation of iron-based metals is popularly studied because of its unique electrical, mechanical, and physical properties. It has also grown to become the most widely utilized metallic material in the metallurgical and material industry. Unfortunately, it suffers from corrosion in the presence of corrosive species. In a bid to impede rusting of MS, Khaled et al. studied the effect of piperidine-based compounds in HCl medium. Piperidine alongside six other derivatives were studied using electrochemical and semi-empirical theoretical methods. All the tested inhibitors behaved with mixed actions and their adsorption showed a good Temkin isotherm fit. In this study, there was no correlation between the experimental and theoretical findings, which could be attributed to the complexity of the corrosion phenomenon, particularly the competition for adherence of inhibitors at the metal–electrolyte interface [85]. Three newly synthesized pyridine- and piperidine-based compounds were tested and proven to be efficient in mitigating MS corrosion in HCl [86]. From WL experiments, the attained optimum concentration of 3.4×10^{-4} M at 303 K yielded excellent IE as shown in the reported order: N-PMPPC (80.22%) < N-BPPC (93.40%) < 4-MPPM (95.71%). Electrochemical corrosion tests showed the mixed-type effect of the piperidine compounds with a noted preference towards anodic reduction and a single charge transfer mechanism. The electron density and spatial configuration of 4-MPPM, N-BPPC, and N-PMPPC significantly contributed to their adsorption capacities on MS.

As part of efforts to promote the utilization of eco-friendly and inexpensive materials to protect the dissolution of metals, several studies have examined the potential of piperine, a predominant alkaloid isolated from black pepper (*Piper nigrum*) to retard steel corrosion reaction in different electrolytic systems. Piperine extracted from black pepper [87] and piperanine, another extract from black pepper, was exploited for their anti-corrosive properties towards C38 steel in acid using experimental techniques [88]. Both were found to effectively impede steel dissolution with an IE exceeding 95% at low amounts. Other authors have explored the protection capacities of *Piper* species for the corrosion protection of metals with astounding outcomes [89–93]. Sequel to the experimental studies conducted by Dahmani and his team [87, 88], the group performed a DFT investigation on the theoretical assessment of the anti-corrosive effect of two piperidine derivatives using 6-31 G(d) basis set with obtained results showing remarkable consistency with the experimental outcomes [94]. In another study, the team performed an elaborate theoretical investigation into the anti-corrosive potential of three piperidines using the DFT protocol at the B3LYP/LANL2LDZ basis set and MCS. Information regarding the most reactive sites of the inhibiting molecules with the iron surface and the thermodynamic variables were gleaned from the DFT analysis. The calculated interaction energies obtained from MCS aligned with the experimental and surface examinations [95].

Senthikumar et al. [96] investigated the protective capacity of three substituted piperidin-4-one oxime compounds for MS degradation in HCl using several established protocols. Compound I with a methyl group at C-1 of piperidine molecule yielded high IE than Compound II and III having dual methyl groups and an isopropyl group, respectively. The depreciation of IE was related to the rise of steric effect at C-3 of the piperidine molecule. These compounds were reported as mixed-type as they interfered with both electrode reactions at the same time. In this case, the adsorption mode was physical adsorption and the mechanism followed Temkin's adsorption isotherm model. From AM1 semi-empirical calculations, the study observed that the phenyl rings and N-ring served as the adsorption centres for the inhibition phenomenon.

Another study had reported the protection abilities of N-heterocyclic compounds for iron degradation in perchloric acid. The study showed retardation of corrosion processes at the anode and cathode regions and the experimental data yielded a superlative fit with the Langmuir isotherm [97]. The protection capacities of three novel piperidine sulfonamide compounds, labelled as FMPPMBS, FMPPNBS, and FMPPDBS, towards MS deterioration in acid were examined using WL, EIS, PDP, and SEM. According to the obtained impedance outcomes, the IE of the studied piperidines followed the order: FMPPDBS (96.9%) > FMPPMBS (94.0%) > FMPPNBS (92.10%) and Tafel plots indicated that the piperidines had a mixed control on the corrosion process. The obtained IE was correlated to the presence of phenyl moieties and increased electron density in the chemical structure of the substituted piperidines. The adsorption study showed the obtained data accurately aligned with the Langmuir isotherm. The experimental and analytical study was complemented with a theoretical investigation carried out using different semi-empirical techniques [98]. Kaya et al. further carried out an insightful computational analysis of the aforementioned piperidine sulfonamides using MDS, HF, and DFT with different basis sets and delineated the mechanism of action of the studied pyridine derivatives [99]. Table 8.2 summarizes other cases of iron-based corrosion inhibition in various media using piperidines.

8.3.2   Inhibition of Metallic Degradation for Non-Ferrous Metals by Piperidines

Literature reports on the exploitation of piperidine-based plant materials as chemical additives for effective corrosion control of non-ferrous metals and alloys in corrosive electrolytes are quite limited [100–102]. A prominent work in this regard was

TABLE 8.2
A List of Piperidines Utilized in the Corrosion Protection of Different Metals in Corrosive Solutions

S/N	Piperidine	Metal/ Electrolyte	Nature of Adsorption	Conc/Max IE	Reference
1.	NPD	Soft cast steel/1 M HCl	Freundlich/ mixed-type	100 ppm/ 93.85%	[108]
2.	MDPO	MS/1 N H_2SO_4	Temkin/ cathodic-type	20 mM/ 90.30%	[109]
3.	N-OPEP	MS/1 M HCl	Langmuir/ mixed-type	500 ppm/ 91.8%	[110]
4.	MFPP	MS/1 M HCl	Langmuir/ mixed-type	1000 ppm/ 96.4%	[111]
5.	NPB	MS/1 M HCl	Langmuir/ mixed-type	0.5 mM/ 96.80%	[112]
6.	PMBP	N80 steel/15% HCl	Langmuir/ mixed-type	64.6×10^{-5} M/ 91.70%	[113]
7.	F1	MS/0.5 M HCl	Langmuir/ mixed-type	0.4 mM/ 96.10%	[114]

(Continued)

TABLE 8.2 (*Continued*)
A List of Piperidines Utilized in the Corrosion Protection of Different Metals in Corrosive Solutions

S/N	Piperidine	Metal/ Electrolyte	Nature of Adsorption	Conc/Max IE	Reference
8.	F2	MS/0.5 M HCl	Using a genetic algorithm approximation method, the QSPR model was developed for 28 pyridine and quinoline compounds	Using a genetic algorithm approximation method, the QSPR model was developed for 28 pyridine and quinoline compounds	[114]
9.	F3	MS/0.5 M HCl	Using a genetic algorithm approximation method, the QSPR model was developed for 28 pyridine and quinoline compounds	Using a genetic algorithm approximation method, the QSPR model was developed for 28 pyridine and quinoline compounds	[114]
10.	CYH	MS/0.5 M H_2SO_4	Using a genetic algorithm approximation method, the QSPR model was developed for 28 pyridine and quinoline compounds	Using a genetic algorithm approximation method, the QSPR model was developed for 28 pyridine and quinoline compounds	[115]
11.	AAI-3	MS/1 M HCl	Langmuir/ mixed-type	0.689 mM/ 97.78%	[116]

reported by Cai et al. [103] where amide alkaloids, including a piperidine compound, piperine, pipernonatine, and piperanine, were extracted from *Piper longum* L., a local Chinese medicine peppers, and examined for their corrosion protection ability of copper in HCl. Piperine showed the best adsorption property on copper than the other isolated compounds owing to its relatively shorter carbon length and additional double bond.

In another study, Khaled studied nickel corrosion in the presence of four piperidine compounds using theoretical and electrochemical test methods [104]. The studied piperidine inhibitors repressed nickel corrosion in 1 M HNO_3 at both electrode areas, thus acting as mixed-type compounds. Increase in the amount of piperidines in HNO_3 correspondingly amplified the thickness of the electric double layer and the inhibition potencies of the pyridines. Both electrostatic adsorption and chemical adsorption were found to be responsible for the effective adherence of the tested compounds on the metal. As per the theoretical explanation, 4-methyl pyridine (4mp) which had the highest E_{HOMO} and ΔN values was found to be the best inhibitor candidate. The study successfully established that inhibition potency depends not only on molecular size/structure, but also on the nature of the metal under study. Other studies have documented the protection performance of piperidine-based inhibitors on Cu in 1.5% NaCl medium [105], Cu in brass in natural water [106], and Zn–Al–Cu alloys in HCl [107].

8.4   SUMMARY AND CONCLUSIONS

This chapter has successfully established the potentials of pyridine and piperidine compounds to inhibit metallic degradation in various metals/alloys and electrolytic solutions. The influence of polar substituents either as electron-withdrawing or electron-donating has been shown to display high inhibition potency than non-substituted pyridine and piperidine. These N-heterocyclic compounds have demonstrated abilities to adsorb on metal surfaces and displace water molecules to proffer corrosion protection to the metal under study. Electrochemical studies demonstrated that the inhibitors often possess mixed inhibitive action on the anodic and cathodic regions as well as simple charge transfer mechanism. In many cases, their mode of adsorption was found to comply with the famous Langmuir adsorption isotherm model and the mechanism of action often involve both physisorption and chemisorption. Interesting highlights on the donor–metal interaction of the inhibitor molecules have become clarified using computational techniques. As per the literature review, there is need for further investigations and exploitations on both compounds to have richer understanding of their inhibition mechanism.

REFERENCES

1. T.W. Quadri, E.D. Akpan, L.O. Olasunkanmi, O.E. Fayemi, E. Ebenso, Fundamentals of Corrosion Chemistry, in: Environmentally Sustainable Corrosion Inhibitors, Elsevier, 2022, pp. 25–45.
2. T.W. Quadri, L.O. Olasunkanmi, O.E. Fayemi, E.E. Ebenso, Utilization of ZnO-based materials as anticorrosive agents: A review, Inorganic Anticorrosive Materials, (2022) 161–182.
3. T.W. Quadri, L.O. Olasunkanmi, O.E. Fayemi, E.E. Ebenso, Nanomaterials and Nanocomposites as Corrosion Inhibitors, in: Sustainable Corrosion Inhibitors II: Synthesis, Design, and Practical Applications, ACS Publications, 2021, pp. 187–217.

4. R. Javaherdashti, How corrosion affects industry and life, Anti-Corrosion Methods and Materials, 47 (2000) 30–34.

5. R. Javaherdashti, Corrosion management: A guide for industry managers, Corrosion Reviews, 21 (2003) 311–326.

6. C. Verma, T.W. Quadri, E.E. Ebenso, M. Quraishi, Polymer nanocomposites as industrially useful corrosion inhibitors: Recent developments, Handbook of Polymer Nanocomposites for Industrial Applications, Elsevier Science Publishing Co Inc., 2021, pp. 419–435.

7. C. Verma, M.H. Abdellattif, A. Alfantazi, M. Quraishi, N-Heterocycle compounds as aqueous phase corrosion inhibitors: A robust, effective and economic substitute, Journal of Molecular Liquids, 340 (2021) 117211.

8. R. Aslam, G. Serdaroglu, S. Zehra, D.K. Verma, J. Aslam, L. Guo, C. Verma, E.E. Ebenso, M. Quraishi, Corrosion inhibition of steel using different families of organic compounds: Past and present progress, Journal of Molecular Liquids, 348 (2021) 118373.

9. L.T. Popoola, Organic green corrosion inhibitors (OGCIs): A critical review, Corrosion Reviews, 37 (2019) 71–102.

10. L. Chen, D. Lu, Y. Zhang, Organic compounds as corrosion inhibitors for carbon steel in HCl solution: A comprehensive review, Materials, 15 (2022) 2023.

11. E. Vitaku, D.T. Smith, J.T. Njardarson, Analysis of the structural diversity, substitution patterns, and frequency of nitrogen heterocycles among US FDA approved pharmaceuticals: Mini perspective, Journal of Medicinal Chemistry, 57 (2014) 10257–10274.

12. P. Martins, J. Jesus, S. Santos, L.R. Raposo, C. Roma-Rodrigues, P.V. Baptista, A.R. Fernandes, Heterocyclic anticancer compounds: Recent advances and the paradigm shift towards the use of nanomedicine's tool box, Molecules, 20 (2015) 16852–16891.

13. M.M. Heravi, V. Zadsirjan, Prescribed drugs containing nitrogen heterocycles: An overview, RSC Advances, 10 (2020) 44247–44311.

14. Y. Ling, Z.-Y. Hao, D. Liang, C.-L. Zhang, Y.-F. Liu, Y. Wang, The expanding role of pyridine and dihydropyridine scaffolds in drug design, Drug Design, Development and Therapy, 15 (2021) 4289–4338.

15. A.A. Altaf, A. Shahzad, Z. Gul, N. Rasool, A. Badshah, B. Lal, E. Khan, A review on the medicinal importance of pyridine derivatives, Journal of Drug Design and Medicinal Chemistry, 1 (2015) 1–11.

16. N.A. Frolov, A.N. Vereshchagin, Piperidine derivatives: Recent advances in synthesis and pharmacological applications, International Journal of Molecular Sciences, 24 (2023) 2937.

17. C.R. Quijia, V.H. Araujo, M. Chorilli, Piperine: Chemical, biological and nanotechnological applications, Acta Pharmaceutica, 71 (2021) 185–213.

18. N. Hackerman, A. Makrides, Action of polar organic inhibitors in acid dissolution of metals, Industrial & Engineering Chemistry, 46 (1954) 523–527.

19. R. Chaudhary, T. Namboodhiri, I. Singh, A. Kumar, Effect of pyridine and its derivatives on corrosion of 0040, 2826 MB, and 2605–8–2 metallic glasses in sulphuric acid solution at 25° C, British Corrosion Journal, 24 (1989) 273–278.

20. K. Ramakrishnaiah, N. Subramanyan, Effect of some nitrogen containing organic compounds on the corrosion and polarization behaviour of aluminium in 1 M solutions of sodium hydroxide and hydrochloric acid with and without calcium, Corrosion Science, 16 (1976) 307–316.

21. R. Barradas, P. Hamilton, Chemical structure and adsorption of organic molecules at the polarized mercury–electrolyte interface, Canadian Journal of Chemistry, 43 (1965) 2468–2485.

22. S. Sampat, J. Vora, Corrosion inhibition of 3s aluminium in trichloroacetic acid by methyl pyridines, Corrosion Science, 14 (1974) 591–595.

23. F.M. Donahue, K. Nobe, Theory of organic corrosion inhibitors adsorption and linear free energy relationships, Journal of the Electrochemical Society, 112 (1966) 886–891.

24. J. Vosta, J. Eliasek, Study on corrosion inhibition from aspect of quantum chemistry, Corrosion Science, 11 (1971) 223–229.

25. A. Afanas' ev, E. Chankova, A. Burmistrova, V. Lunichenko, Effect of technical corrosion inhibitors on the variation in ductility of low-carbon steel during acid pickling, Soviet Materials Science: A Transl. of Fiziko-Khimicheskaya Mekhanika materialov/ Academy of Sciences of the Ukrainian SSR, 4 (1969) 111–113.

26. R.C. Ayers, N. Hackerman, Corrosion in HCl using methyl pyridines, Journal of the Electrochemical Society, 110 (1963) 507.

27. K. Kishi, S. Ikeda, Ultraviolet study for the adsorption of pyridine and 2,2′-bipyridyl on evaporated metal films, The Journal of Physical Chemistry, 73 (1969) 2559–2564.

28. C. Verma, M. Quraishi, C.M. Hussain, Pyridine and its derivatives as corrosion inhibitors, in: Organic Corrosion Inhibitors: Synthesis, Characterization, Mechanism, and Applications, John Wiley & Sons, Inc., 2021, pp. 123–148.

29. C. Verma, K.Y. Rhee, M. Quraishi, E.E. Ebenso, Pyridine based *N*-heterocyclic compounds as aqueous phase corrosion inhibitors: A review, Journal of the Taiwan Institute of Chemical Engineers, 117 (2020) 265–277.

30. M.A. Quraishi, D.S. Chauhan, V.S. Saji, Heterocyclic biomolecules as green corrosion inhibitors, Journal of Molecular Liquids, 341 (2021) 117265.

31. C. Verma, M.A. Quraishi, Adsorption behavior of 8,9-bis(4 (dimethyl amino)phenyl) benzo[4,5]imidazo[1,2-*a*]pyridine-6,7-dicarbonitrile on mild steel surface in 1 M HCl, Journal of the Association of Arab Universities for Basic and Applied Sciences, 22 (2017) 55–61.

32. M.A. Quraishi, 2-amino-3,5-dicarbonitrile-6-thio-pyridines: New and effective corrosion inhibitors for mild steel in 1 M HCl, Industrial & Engineering Chemistry Research, 53 (2014) 2851–2859.

33. Y. Ji, B. Xu, W. Gong, X. Zhang, X. Jin, W. Ning, Y. Meng, W. Yang, Y. Chen, Corrosion inhibition of a new Schiff base derivative with two pyridine rings on Q235 mild steel in 1.0 M HCl, Journal of the Taiwan Institute of Chemical Engineers, 66 (2016) 301–312.

34. K.R. Ansari, M.A. Quraishi, A. Singh, Corrosion inhibition of mild steel in hydrochloric acid by some pyridine derivatives: An experimental and quantum chemical study, Journal of Industrial and Engineering Chemistry, 25 (2015) 89–98.

35. C. Verma, L.O. Olasunkanmi, T.W. Quadri, E.-S.M. Sherif, E.E. Ebenso, Gravimetric, electrochemical, surface morphology, DFT, and Monte Carlo simulation studies on three N-substituted 2-aminopyridine derivatives as corrosion inhibitors of mild steel in acidic medium, The Journal of Physical Chemistry C, 122 (2018) 11870–11882.

36. T.W. Quadri, L.O. Olasunkanmi, O.E. Fayemi, E.D. Akpan, C. Verma, E.-S.M. Sherif, K.F. Khaled, E.E. Ebenso, Quantitative structure activity relationship and artificial neural network as vital tools in predicting coordination capabilities of organic compounds with metal surface: A review, Coordination Chemistry Reviews, 446 (2021) 214101.

37. A.E.-A.S. Fouda, A.H. El-Askalany, A.F. Molouk, N.S. Elsheikh, A.S. Abousalem, Experimental and computational chemical studies on the corrosion inhibitive properties of carbonitrile compounds for carbon steel in aqueous solutions, Scientific Reports, 11 (2021) 21672.

38. N. El Basiony, E.H. Tawfik, M.A. El-raouf, A.A. Fadda, M.M. Waly, Synthesis, characterization, theoretical calculations (DFT and MC), and experimental of different substituted pyridine derivatives as corrosion mitigation for X-65 steel corrosion in 1 M HCl, Journal of Molecular Structure, 1231 (2021) 129999.

39. S. Satpati, S.K. Saha, A. Suhasaria, P. Banerjee, D. Sukul, Adsorption and anti-corrosion characteristics of vanillin Schiff bases on mild steel in 1 M HCl: Experimental and theoretical study, RSC Advances, 10 (2020) 9258–9273.

40. A. Berisha, Experimental, Monte Carlo and molecular dynamic study on corrosion inhibition of mild steel by pyridine derivatives in aqueous perchloric acid, Electrochem, 1 (2020) 188–199.

41. W. Daoudi, L. Guo, M. Azzouzi, T. Pooventhiran, A.E. Boutaybi, S. Lamghafri, A. Oussaid, A. El Aatiaoui, Evaluation of the corrosion inhibition of mild steel by newly synthesized imidazo [1, 2-*a*] pyridine derivatives: Experimental and theoretical investigation, Journal of Adhesion Science and Technology, (2023) 1–24.

42. W. Daoudi, M. Azzouzi, O. Dagdag, A. El Boutaybi, A. Berisha, E.E. Ebenso, A. Oussaid, A. El Aatiaoui, Synthesis, characterization, and corrosion inhibition activity of new imidazo [1.2-*a*] pyridine chalcones, Materials Science and Engineering: B, 290 (2023) 116287.

43. B.A. Al Jahdaly, Preparation and evaluation of new pyridone derivatives and their investigation corrosion depletion property for copper corrosion in HCl acid solution, Biomass Conversion and Biorefinery, (2023) 1–16.

44. A. Saady, Z. Rais, F. Benhiba, R. Salim, K.I. Alaoui, N. Arrousse, F. Elhajjaji, M. Taleb, K. Jarmoni, Y.K. Rodi, Chemical, electrochemical, quantum, and surface analysis evaluation on the inhibition performance of novel imidazo [4,5-*b*] pyridine derivatives against mild steel corrosion, Corrosion Science, 189 (2021) 109621.

45. M. Rezaeivala, S. Karimi, B. Tuzun, K. Sayin, Anti-corrosion behavior of 2-((3-(2-morpholino ethylamino)-N_3-((pyridine-2-yl) methyl) propylimino) methyl) pyridine and its reduced form on carbon steel in hydrochloric acid solution: Experimental and theoretical studies, Thin Solid Films, 741 (2022) 139036.

46. H.M. Abd El-Lateef, M.M. Khalaf, K. Shalabi, A.A. Abdelhamid, Efficient synthesis of 6,7-dihydro-5*H*-cyclopenta[*b*]pyridine-3-carbonitrile compounds and their applicability as inhibitor films for steel alloy corrosion: Collective computational and practical approaches, ACS Omega, 7 (2022) 24727–24745.

47. N.N. Hau, D.Q. Huong, Effect of aromatic rings on mild steel corrosion inhibition ability of nitrogen heteroatom-containing compounds: Experimental and theoretical investigation, Journal of Molecular Structure, 1277 (2022) 134884.

48. A.A. Farag, E.A. Mohamed, G.H. Sayed, K.E. Anwer, Experimental/computational assessments of API steel in 6 M H_2SO_4 medium containing novel pyridine derivatives as corrosion inhibitors, Journal of Molecular Liquids, 330 (2021) 115705.

49. J. Zhu, G. Zhou, F. Niu, Y. Shi, Z. Du, G. Lu, Z. Liu, Understanding the inhibition performance of novel dibenzimidazole derivatives on Fe(110) surface: DFT and MD simulation insights, Journal of Materials Research and Technology, 17 (2022) 211–222.

50. T.A. Mohamed, U.A. Soliman, I.A. Shaaban, W.M. Zoghaib, L.D. Wilson, Raman, DRIFT and ATR-IR spectra, corrosion inhibition, DFT and solid-state calculations of 4-amino-3-choloro-2,5,6-trifluoropyridine, Journal of Molecular Structure, 1207 (2020) 127837.

51. F. Mollaamin, M. Monajjemi, Corrosion inhibiting by some organic heterocyclic inhibitors through Langmuir adsorption mechanism on the Al-X (X = Mg/Ga/Si) alloy surface: A study of quantum three-layer method of CAM-DFT/ONIOM, Journal of Bio-and Tribo-Corrosion, 9 (2023) 33.

52. A. Hassan, R. Hussein, M. Abou-krisha, M.I. Attia, Density functional theory investigation of some pyridine dicarboxylic acids derivatives as corrosion inhibitors, International Journal of Electrochemical Science, 15 (2020) 4274–4286.

53. M. Kaur, K. Kaur, H. Kaur, Quest of Schiff bases as corrosion inhibitors: A first principle approach, Journal of Physical Organic Chemistry, 34 (2021) e4260.

54. C.T. Ser, P. Žuvela, M.W. Wong, Prediction of corrosion inhibition efficiency of pyridines and quinolines on an iron surface using machine learning-powered quantitative structure–property relationships, Applied Surface Science, 512 (2020) 145612.

55. R.L. Camacho-Mendoza, L. Feria, L. Zárate-Hernández, J.G. Alvarado-Rodríguez, J. Cruz-Borbolla, New QSPR model for prediction of corrosion inhibition using conceptual density functional theory, Journal of Molecular Modeling, 28 (2022) 238.

56. O.E. Oyeneyin, N.D. Ojo, N. Ipinloju, A.C. James, E.B. Agbaffa, Investigation of corrosion inhibition potentials of some aminopyridine Schiff bases using density functional theory and Monte Carlo simulation, Chemistry Africa, 5 (2022) 319–332.

57. H. Lgaz, H.-s Lee, S. Kaya, R. Salghi, S.M. Ibrahim, M. Chafiq, L. Bazzi, Y.G. Ko, Unraveling bonding mechanisms and electronic structure of pyridine oximes on Fe(110) surface: Deeper insights from DFT, molecular dynamics and SCC–DFT tight binding simulations, Molecules, 28 (2023) 3545.

58. A. Suhasaria, M. Murmu, S. Satpati, P. Banerjee, D. Sukul, Bis-benzothiazoles as efficient corrosion inhibitors for mild steel in aqueous HCl: Molecular structure–reactivity correlation study, Journal of Molecular Liquids, 313 (2020) 113537.

59. K.S. Vranda, P.D.R. Kumari, Corrosion inhibition of mild steel by 6-Bromo-(4, 5-dimethoxy-2-nitrophenyl) methylidene] imidazo [1, 2-a] pyridine-2-carbohydrazide in 0.5 M hydrochloric acid solution, in: AIP Conference Proceedings, AIP Publishing LLC, 2020, p. 040006.

60. A. Al-Amiery, L. Shaker, Corrosion inhibition of mild steel using novel pyridine derivative in 1 M hydrochloric acid, Koroze a ochrana materiálu, 64 (2020) 59–64.

61. K.E. Anwer, A.A. Farag, E.A. Mohamed, E.M. Azmy, G.H. Sayed, Corrosion inhibition performance and computational studies of pyridine and pyran derivatives for API X-65 steel in 6 M H_2SO_4, Journal of Industrial and Engineering Chemistry, 97 (2021) 523–538.

62. N. Abdelshafi, M. Sadik, M.A. Shoeib, S.A. Halim, Corrosion inhibition of aluminum in 1 M HCl by novel pyrimidine derivatives, EFM measurements, DFT calculations and MD simulation, Arabian Journal of Chemistry, 15 (2022) 103459.

63. M.E. Ikpi, F.E. Abeng, Electrochemical and quantum chemical investigation on adsorption of nifedipine as corrosion inhibitor at API 5L X-52 steel/HCl acid interface, Archives of Metallurgy and Materials, 65 (2020), 1, 125–131.

64. K.V. Shenoy, P.P. Venugopal, P.R. Kumari, D. Chakraborty, Effective inhibition of mild steel corrosion by 6-bromo-(2,4-dimethoxyphenyl)methylidene]imidazo [1,2-a] pyridine-2-carbohydrazide in 0.5 M HCl: Insights from experimental and computational study, Journal of Molecular Structure, 1232 (2021) 130074.

65. A. Resen, M. Hanoon, R. Salim, A. Al-Amiery, L. Shaker, A. Kadhum, Gravimetrical, theoretical, and surface morphological investigations of corrosion inhibition effect of 4-(benzoimidazole-2-yl) pyridine on mild steel in hydrochloric acid, KOM–Corrosion and Material Protection Journal, 64 (2020) 122–130.

66. V. Saraswat, M. Yadav, I. Obot, Investigations on eco-friendly corrosion inhibitors for mild steel in acid environment: Electrochemical, DFT and Monte Carlo simulation approach, Colloids and Surfaces A: Physicochemical and Engineering Aspects, 599 (2020) 124881.

67. D.J. Karunarathne, A. Aminifazl, T.E. Abel, K.L. Quepons, T.D. Golden, Corrosion inhibition effect of pyridine-2-thiol for brass in an acidic environment, Molecules, 27 (2022) 6550.

68. R. Padash, G.S. Sajadi, A.H. Jafari, E. Jamalizadeh, A.S. Rad, Corrosion control of aluminum in the solutions of NaCl, HCl and NaOH using 2,6-dimethylpyridine inhibitor: Experimental and DFT insights, Materials Chemistry and Physics, 244 (2020) 122681.

69. F. Benhiba, Z. Benzekri, Y. Kerroum, N. Timoudan, R. Hsissou, A. Guenbour, M. Belfaquir, S. Boukhris, A. Bellaouchou, H. Oudda, Assessment of inhibitory behavior of ethyl 5-cyano-4-(furan-2-yl)-2-methyl-6-oxo-1, 4,5,6-tetrahydropyridine-3-carboxylate as a corrosion inhibitor for carbon steel in molar HCl: Theoretical approaches and experimental investigation, Journal of the Indian Chemical Society, 100 (2023) 100916.

70. N.B. Iroha, C.U. Dueke-Eze, T.M. Fasina, V.C. Anadebe, L. Guo, Anticorrosion activity of two new pyridine derivatives In protecting X70 pipeline steel in oil well acidizing fluid: Experimental and quantum chemical studies, Journal of the Iranian Chemical Society, 19 (2022) 2331–2346.

71. M. Bozorg, M. Rezaeivala, S. Borghei, M. Darroudi, Anti-corrosion behavior of 2-(((4-((2-morpholinoethyl)(pyridin-2-ylmethyl)amino)butyl)imino)methyl) naphthalen-1-ol on mild steel in hydrochloric acid solution: Experimental and theoretical studies, Thin Solid Films, 762 (2022) 139558.

72. A. Alrebh, M.B. Rammal, S. Omanovic, A pyridine derivative 2-(2-methylaminoethyl) pyridine (MAEP) as a 'green' corrosion inhibitor for low-carbon steel in hydrochloric acid media, Journal of Molecular Structure, 1238 (2021) 130333.

73. Y. Chen, Z. Chen, Y. Zhuo, Newly synthesized morpholinyl Mannich bases as corrosion inhibitors for N80 steel in acid environment, Materials, 15 (2022) 4218.

74. M. Dawood, Z. Alasady, M. Abdulazeez, D. Ahmed, G. Sulaiman, A. Kadhum, L. Shaker, A. Alamiery, The corrosion inhibition effect of a pyridine derivative for low carbon steel in 1 M HCl medium: Complemented with antibacterial studies, The International Journal of Corrosion and Scale Inhibition, 10 (2021) 1766.

75. C.M. Fernandes, M.V.P. de Mello, N.E. dos Santos, A.M.T. de Souza, M. Lanznaster, E.A. Ponzio, Theoretical and experimental studies of a new aniline derivative corrosion inhibitor for mild steel in acid medium, Materials and Corrosion, 71 (2020) 280–291.

76. R. Farahati, H. Behzadi, S.M. Mousavi-Khoshdel, A. Ghaffarinejad, Evaluation of corrosion inhibition of 4-(pyridin-3-yl) thiazol-2-amine for copper in HCl by experimental and theoretical studies, Journal of Molecular Structure, 1205 (2020) 127658.

77. X. Liu, Y. Sun, M. Lu, X. Pan, Z. Wang, Electrochemical and surface analytical studies of transition metal bipyridine dicarboxylic acid complexes as corrosion inhibitors for a mild steel in HCl solution, Journal of Adhesion Science and Technology, 36 (2022) 567–583.

78. L. Tabti, R.M. Khelladi, N. Chafai, A. Lecointre, A.M. Nonat, L.J. Charbonnière, E. Bentouhami, Corrosion protection of mild steel by a new phosphonated pyridines inhibitor system in HCl solution, Advanced Engineering Forum, 36 (2020) 59–75.

79. T. Attar, F. Nouali, Z. Kibou, A. Benchadli, B. Messaoudi, E. Choukchou-Braham, N. Choukchou-Braham, Corrosion inhibition, adsorption and thermodynamic properties of 2-aminopyridine derivatives on the corrosion of carbon steel in sulfuric acid solution, Journal of Chemical Sciences, 133 (2021) 1–10.

80. A. Saady, E. Ech-Chihbi, F. El-Hajjaji, F. Benhiba, A. Zarrouk, Y.K. Rodi, M. Taleb, A. El Biache, Z. Rais, Molecular dynamics, DFT and electrochemical to study the interfacial adsorption behavior of new imidazo[4,5-*b*] pyridine derivative as corrosion inhibitor in acid medium, Journal of Applied Electrochemistry, 51 (2021) 245–265.

81. I.B. Onyeachu, I.B. Obot, A.Y. Adesina, Green corrosion inhibitor for oilfield application II: The time–evolution effect on the sweet corrosion of API X60 steel in synthetic brine and the inhibition performance of 2-(2-pyridyl) benzimidazole under turbulent hydrodynamics, Corrosion Science, 168 (2020) 108589.

82. A. Al-Amiery, T.A. Salman, K.F. Alazawi, L.M. Shaker, A.A.H. Kadhum, M.S. Takriff, Quantum chemical elucidation on corrosion inhibition efficiency of Schiff base: DFT investigations supported by weight loss and SEM techniques, International Journal of Low-Carbon Technologies, 15 (2020) 202–209.

83. K. Aramaki, Effect of steric hindrances on corrosion Inhibition effectiveness of piperidine derivatives, Denki Kagaku Oyobi Kogyo Butsuri Kagaku, 41 (1973) 321–326.

84. S. Sankarapapavinasam, F. Pushpanaden, M. Ahmed, Piperidine, piperidones and tetrahydrothiopyrones as inhibitors for the corrosion of copper in H_2SO_4, Corrosion Science, 32 (1991) 193–203.

85. K. Khaled, K. Babi-Samardžija, N. Hackerman, Piperidines as corrosion inhibitors for iron in hydrochloric acid, Journal of Applied Electrochemistry, 34 (2004) 697–704.

86. A. Ramachandran, P. Anitha, S. Gnanavel, Structural and electronic impacts on corrosion inhibition activity of novel heterocyclic carboxamides derivatives on mild steel in 1 M HCl environment: Experimental and theoretical approaches, Journal of Molecular Liquids, 359 (2022) 119218.

87. M. Dahmani, A. Et-Touhami, S. Al-Deyab, B. Hammouti, A. Bouyanzer, Corrosion inhibition of C38 steel in 1 M HCl: A comparative study of black pepper extract and its isolated piperine, International Journal of Electrochemical Science, 5 (2010) 1060–1069.

88. M. Dahmani, S. Al-Deyab, A. Et-Touhami, B. Hammouti, A. Bouyanzer, R. Salghi, A. El Mejdoubi, Investigation of piperanine as HCl ecofriendly corrosion inhibitors for C38 steel, International Journal of Electrochemical Science, 7 (2012) 2513–2522.

89. N. Mohd, A.S. Ishak, Thermodynamic study of corrosion inhibition of mild steel in corrosive medium by *Piper nigrum* extract, Indian Journal of Science and Technology, 8 (2015) 1.

90. M. Quraishi, D. Kumar Yadav, I. Ahamad, Green approach to corrosion inhibition by black pepper extract in hydrochloric acid solution, The Open Corrosion Journal, 2 (2009) 56–60.

91. A. Singh, V.K. Singh, M. Quraishi, Inhibition of mild steel corrosion in HCl solution using Pipali (*Piper longum*) fruit extract, Arabian Journal for Science and Engineering, 38 (2013) 85–97.

92. M. Idham, S. Nasip, N. Hazirah, B. Abdullah, S. Alias, *Piper nigrum* (green corrosion inhibitor) as a modified quenchant in heat treatment of ductile iron, IOP Conference Series: Materials Science and Engineering, 1176 (2021) 012027.

93. E. Ebenso, N. Eddy, A. Odiongenyi, Corrosion inhibitive properties and adsorption behaviour of ethanol extract of *Piper guinensis* as a green corrosion inhibitor for mild steel in H_2SO_4, African Journal of Pure and Applied Chemistry, 2 (2008) 107–115.

94. Y. Karzazi, M.E.A. Belghiti, A. Dafali, B. Hammouti, A theoretical investigation on the corrosion inhibition of mild steel by piperidine derivatives in hydrochloric acid solution, Journal of Chemical and Pharmaceutical Research, 6 (2014) 689–696.

95. M. Belghiti, S. Echihi, A. Mahsoune, Y. Karzazi, A. Aboulmouhajir, A. Dafali, I. Bahadur, Piperine derivatives as green corrosion inhibitors on iron surface; DFT, Monte Carlo dynamics study and complexation modes, Journal of Molecular Liquids, 261 (2018) 62–75.

96. A. Senthilkumar, K. Tharini, M. Sethuraman, Studies on a few substituted piperidin-4-one oximes as corrosion inhibitor for mild steel in HCl, Journal of Materials Engineering and Performance, 20 (2011) 969–977.
97. K. Babić-Samardžija, K. Khaled, N. Hackerman, N-Heterocyclic amines and derivatives as corrosion inhibitors for iron in perchloric acid, Anti-Corrosion Methods and Materials, 52 (2005) 11–21.
98. C.P. Kumar, K. Mohana, H. Muralidhara, Electrochemical and thermodynamic studies to evaluate the inhibition effect of synthesized piperidine derivatives on the corrosion of mild steel in acidic medium, Ionics, 21 (2015) 263–281.
99. S. Kaya, L. Guo, C. Kaya, B. Tüzün, I. Obot, R. Touir, N. Islam, Quantum chemical and molecular dynamic simulation studies for the prediction of inhibition efficiencies of some piperidine derivatives on the corrosion of iron, Journal of the Taiwan Institute of Chemical Engineers, 65 (2016) 522–529.
100. A. Singh, M. Quraishi, Pipali (*Piper longum*) and Brahmi (*Bacopa monnieri*) extracts as green corrosion inhibitor for aluminum in NaOH solution, Journal of Chemical and Pharmaceutical Research, 4 (2012) 322.
101. P. Mourya, N. Chaubey, S. Kumar, V. Singh, M. Singh, *Strychnos nuxvomica, Piper longum* and *Mucuna pruriens* seed extracts as eco-friendly corrosion inhibitors for copper in nitric acid, RSC Advances, 6 (2016) 95644–95655.
102. A. Singh, I. Ahamad, M.A. Quraishi, *Piper longum* extract as green corrosion inhibitor for aluminium in NaOH solution, Arabian Journal of Chemistry, 9 (2016) S1584–S1589.
103. L. Cai, Q. Fu, R. Shi, Y. Tang, Y.-T. Long, X.-P. He, Y. Jin, G. Liu, G.-R. Chen, K. Chen, 'Pungent' copper surface resists acid corrosion in strong HCl solutions, Industrial & Engineering Chemistry Research, 53 (2014) 64–69.
104. K. Khaled, M.A. Amin, Computational and electrochemical investigation for corrosion inhibition of nickel in molar nitric acid by piperidines, Journal of Applied Electrochemistry, 38 (2008) 1609–1621.
105. M. Singh, R. Rastogi, B. Upadhyay, Inhibition of copper corrosion in aqueous sodium chloride solution by various forms of the piperidine moiety, Corrosion, 50 (1994) 620–625.
106. X.J. Raj, N. Rajendran, Inhibition effect of newly synthesised piperidine derivatives on the corrosion of brass in natural seawater, Protection of Metals and Physical Chemistry of Surfaces, 49 (2013) 763–775.
107. S. Mahmoud, Hydrochloric acid corrosion inhibition of Zn–Al–Cu alloy by methyl-substituted piperidines, Portugaliae Electrochimica Acta, 26 (2008) 245–256.
108. S. Rajendraprasad, S. Ali, B. Prasanna, Electrochemical behavior of N1-(3-methylphenyl) piperidine-1,4-dicarboxamide as a corrosion inhibitor for soft-cast steel carbon steel in 1 M HCl, Journal of Failure Analysis and Prevention, 20 (2020) 235–241.
109. J.R. Xavier, R. Nallaiyan, Corrosion inhibitive properties and electrochemical adsorption behaviour of some piperidine derivatives on brass in natural sea water, Journal of Solid State Electrochemistry, 16 (2012) 391–402.
110. A. Alamiery, Study of corrosion behavior of N'-(2-(2-oxomethylpyrrol-1-yl) ethyl) piperidine for mild steel in the acid environment, Biointerface Research in Applied Chemistry, 12 (2022) 3638–3646.
111. K. Ravichandran, N.M. Kumar, K. Subash, T.S. Narayanan, Mannich base derivatives: A novel class of corrosion inhibitors for cooling water systems, Corrosion Reviews, 19 (2001) 29–42.
112. M. Jeeva, G. Venkatesa Prabhu, C. Rajesh, Inhibition effect of nicotinamide and its Mannich base derivatives on mild steel corrosion in HCl, Journal of Materials Science, 52 (2017) 12861–12888.

113. M. Yadav, S. Kumar, T. Purkait, L. Olasunkanmi, I. Bahadur, E. Ebenso, Electrochemical, thermodynamic and quantum chemical studies of synthesized benzimidazole derivatives as corrosion inhibitors for N80 steel in hydrochloric acid, Journal of Molecular Liquids, 213 (2016) 122–138.

114. C.P. Kumar, K. Mohana, M. Raghu, M. Jagadeesha, M. Prashanth, N. Lokanath, Fluorine substituted thiomethyl pyrimidine derivatives as efficient inhibitors for mild steel corrosion in hydrochloric acid solution: Thermodynamic, electrochemical and DFT studies, Journal of Molecular Liquids, 311 (2020) 113311.

115. V.M. Rangaswamy, J. Keshavayya, Anticorrosive ability of cycloheximide on mild steel corrosion in 0.5 M H_2SO_4 solution, Chemical Data Collections, 37 (2022) 100795.

116. C. Verma, M. Quraishi, E. Ebenso, I. Obot, L. El Assyry, 3-Amino alkylated indoles as corrosion inhibitors for mild steel in 1 M HCl: Experimental and theoretical studies, Journal of Molecular Liquids, 219 (2016) 647–660.

9 N-Heterocycles Having One and Two Nitrogen Atoms as Corrosion Inhibitors

Mehmet Gümüş[1], İrfan Çapan[2], and İrfan Koca[3]
[1]Akdagmadeni Health College, Yozgat Bozok University, Yozgat, Turkey
[2]Department of Material and Material Processing Technologies, Technical Sciences Vocational College, Gazi University, Ankara, Turkey
[3]Department of Chemistry, Faculty of Art & Sciences, Yozgat Bozok University, Yozgat, Turkey

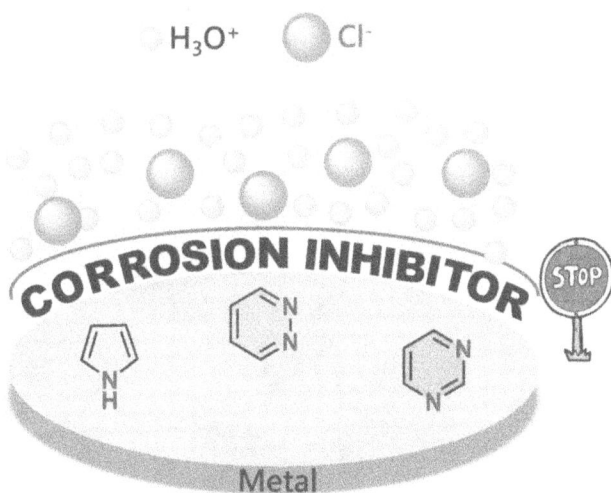

9.1 PYRROLE-, PYRROLIDINE- BASED CORROSION INHIBITORS

Pyrrole and pyrrolidine compounds are important heterocyclic structures containing nitrogen atoms. The pyrrole molecule is a heterocyclic aromatic compound with the formula C_4H_4NH. It is a volatile liquid purified before use, usually darkening immediately when left in the open air. The pyrrolidine molecule is a heterocyclic

DOI: 10.1201/9781003377016-9

secondary amine derivative with the formula $(CH_2)_4NH$. This molecule, called tetrahydropyrrole, is a colorless liquid that forms a homogeneous mixture of water and many organic solvents. Because of their biological and medicinal significance, these quintuple-ring heterostructures have been the topic of several investigations [1, 2]. Chlorophyll, hemoglobin, myoglobin, and cytochrome molecules are natural pyrrole derivatives. These compounds, which have vital functions in living systems, are metal complexes of pyrrole. Pyrrole compounds attract attention with their roles in photosynthesis, oxygen transport, and redox cycle reactions [3–5]. Nicotine, scalusamide, bgugaine, D-ribitol, and aegyleptolidine compounds are biologically active natural pyrrolidine analogues. Pyrrole and pyrrolidine analogues are included in the structure of many medicinal drugs as active ingredients. They are popular compounds in therapeutic applications with their antitumor, anti-tuberculosis, antiviral, antibiotic, anti-inflammatory, and cholesterol-lowering effects (Figure 9.1) [6].

Another area where pyrrole and pyrrolidine compounds are widely used is corrosion, which causes high costs in the industry. An organic corrosion inhibitor's efficiency is governed by its functional groups, molecular geometry, steric considerations, and the planarity of the aromatic structure, and these inhibitors typically work by surface adsorption. The temperature and acid concentration of the corrosive environment, as well as the composition of the metal, all influence the course of the corrosion event. One of the important variables used to control corrosion parameters is the concentration of organic compounds used as corrosion inhibitors. Inhibitors against corrosion must have rings of varying sizes, such as macrocyclic compounds, or exhibit superior surface coating properties with their π-bonds and heteroatoms [7]. These compounds have been chosen as corrosion inhibitors because they contain

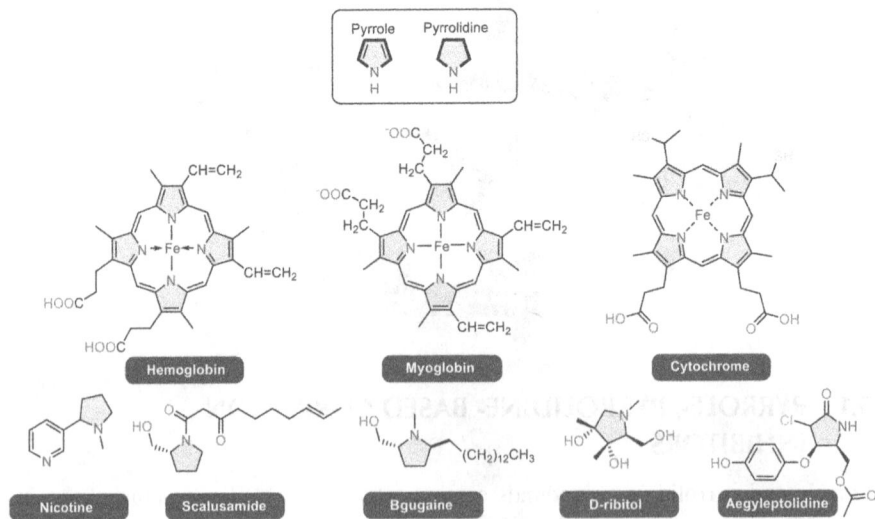

FIGURE 9.1   Some representative examples of pyrrole and pyrrolidine derivatives.

numerous π-electrons and heteroatoms, which produce greater adsorption on the steel surface than other organic inhibitors [8, 9].

Copper metal is utilized in heat conductors, pipe manufacturing, alloy materials, heat exchangers, and a variety of other industrial uses. The easy processing of copper, the advantages of its electrical and thermal properties, its relatively low cost, and its resistance to many corrosive chemicals are among the most important reasons why it is preferred for industrial use. It is particularly resistant in corrosive environments such as sodium carbonate and diluted mineral acid solutions. It is, nevertheless, sensitive in corrosive settings with various types of abrasives. To strengthen the resilience of copper metal in corrosive settings, heterocyclic compounds are favored as corrosion inhibitors. It is also possible to use these inhibitors together with surfactants. Surfactants are functional structures containing non-polar hydrophobic groups and polar hydrophilic groups. These compounds provide wide surface coverage with the long hydrocarbon chain they have and offer significant advantages for the prevention of corrosion. The multiple active centers contained in the surfactant facilitate the surface interaction in the adsorption event. Gopi et al. investigated the corrosion inhibitory properties of 1-(2-pyrrole carbonyl)benzotriazole (**Pr1**) and 1-(2-thienyl carbonyl)-benzotriazole (TBTA) compounds by using them together with sodium dodecyl sulfate (SDS) surfactant and molybdate (Figure 9.2). Studies on copper metal corrosion at various temperatures have been conducted in groundwater. When combined with SDS and molybdate, the synthetic TBTA and **Pr1** compounds were discovered to be effective corrosion inhibitors for copper in groundwater [10]. Gopi et al. investigated compound **Pr1** as a mild steel (MS) corrosion inhibitor in groundwater in another study. The efficiency of Zn^{2+}, 3-phosphonopropionic acid (3-PPA), and **Pr1** alone and in combination to prevent corrosion was evaluated. Electrochemical impedance spectroscopy (EIS) and potentiodynamic polarization measurements were performed. Studies on the surface of MS immersed in the **Pr1** + Zn^2 + 3-PPA system revealed that the benzotriazole derivative and Fe(II)–phosphonate complex were adsorbed [11].

Verma et al. examined how MS corrosion in 1 M HCl was affected by pyrrole-4-carbonitrile compounds (**Pr2–Pr4**). Acetophenone derivatives (1 equiv.), thiophenol (2 equiv.), and malononitrile (2 equiv.) were reacted in a one-step multicomponent coupling reaction with water as the solvent and triethyl amine as the catalyst to create

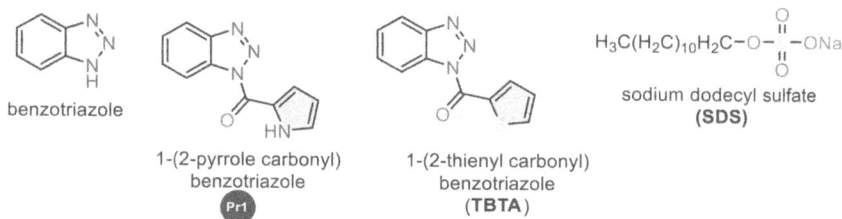

benzotriazole

1-(2-pyrrole carbonyl)
benzotriazole
(Pr1)

1-(2-thienyl carbonyl)
benzotriazole
(TBTA)

$H_3C(H_2C)_{10}H_2C-O-\overset{O}{\underset{O}{\overset{\|}{\underset{\|}{S}}}}-ONa$

sodium dodecyl sulfate
(SDS)

FIGURE 9.2 Chemical structure of benzotriazole, 1-(2-pyrrole carbonyl) benzotriazole (**Pr1**), 1-(2-thienyl carbonyl)-benzotriazole (TBTA), and sodium dodecyl sulfate (SDS) surfactant.

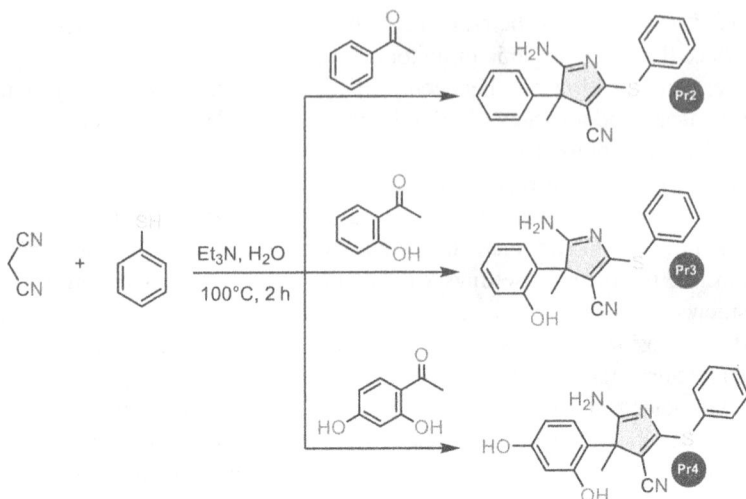

SCHEME 9.1   Synthesis of amino pyrrole derivatives.

the **Pr2–Pr4** used in this study (Scheme 9.1). The effect of the obtained compounds on corrosion was investigated experimentally and theoretically and it was found that the concentration increases the effectiveness of inhibition. Surface morphology investigations included atomic force microscopy (AFM) and scanning electron microscopy (SEM). It was discovered that **Pr2–Pr4** adsorbed on the metal surface and stopped MS corrosion in 1 M HCl, with the optimum inhibition efficacy occurring at a concentration of 50 mg L^{-1}. The heteroatoms in the inhibitor's structure can be neutral or protonated (cationic); they can be easily protonated in a hydrochloric acid solution. **Pr2–Pr4** were also discovered to be anodic-type inhibitors, with inhibition efficacy increasing with the amount of electron-donating —OH groups [12].

Multicomponent reactions (MCR) can be used to create pyrrole derivatives by combining three or more starting components. For this purpose, pyrrole derivatives can be obtained whenever 1,3-dicarbonyl compounds, amines, aldehydes, and nitroalkanes are used. Louroubi et al. used this method to synthesize pyrrole derivatives (**Pr5–Pr7**; Figure 9.3) and evaluated them as corrosion inhibitors. According

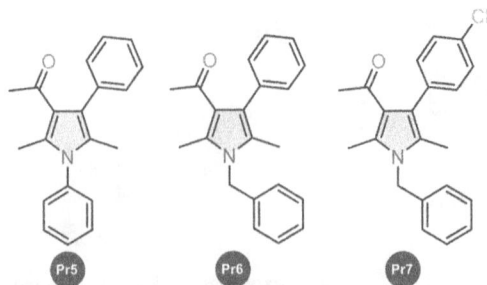

FIGURE 9.3   Chemical structures of (**Pr5–Pr7**) pyrrole derivatives.

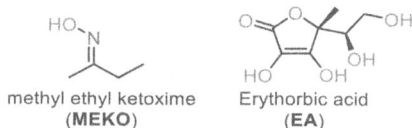

methyl ethyl ketoxime   Erythorbic acid
(MEKO)   (EA)

FIGURE 9.4   Chemical structures of methyl ethyl ketoxime (**MEKO**) and erythorbic acid (**EA**).

to the findings of the experiments, pyrrole derivatives are adsorbed on the steel surface and prevent corrosion. The steel surface's unoccupied d-orbital and the inhibitor's heteroatom's lone-pair electron interact as donors and acceptors throughout this adsorption phase. In 1 M HCl solution at 10^{-3} M, the inhibition efficiency of potential corrosion inhibitor compounds for S300 steel ranged from 82% to 96%. The correlation between the density functional theory (DFT) and electrochemical results was more than 99%. It has been established that the pyrrole chemical investigated efficiently blocks the active areas on the surface of steel S300. Furthermore, Hirshfeld surface analysis studies together with electrostatic potential surfaces (EPS) studies have provided important data in explaining the corrosion efficiency of the inhibitor's active sites [13, 14].

Kassim et al. employed the linear polarization resistance (LPR) technique to evaluate the inhibitory influence of pyrrole in the presence of methyl ethyl ketoxime (**MEKO**) and erythorbic acid (**EA**) (Figure 9.4) on carbon steel corrosion under static circumstances of 3.5 wt% NaCl solution. When different concentrations of MEKO and EA were added to the brine environment, the number of oxygen scavenger molecules in the solution was depleted and the corrosion rate decreased. This study demonstrated that combining pyrrole with oxygen scavengers (MEKO, EA) enhances inhibitory efficacy because of the multiple effects of effective corrosion inhibition [9].

Zarrouk et al. synthesized 1-phenyl-1*H*-pyrrole-2,5-dione (**Pr8**) and 1-(4-methylphenyl)-1*H*-pyrrole-2,5-dione (**Pr9**) (Figure 9.5) compounds and described these chemicals' corrosion inhibiting properties in 1 M HCl solution of carbon steel. These studies conducted experiments at 35 °C with weight loss, PDP, and EIS methods. It was concluded that **Pr8** and **Pr9** showed excellent inhibition for carbon steel at 1 M HCl at 308 K. While the potentiodynamic polarization curves show that inhibitors prevent metal dissolution, Langmuir adsorption isotherms proved that the adsorption mechanism on the carbon steel surface is due to chemical adsorption. Data obtained by

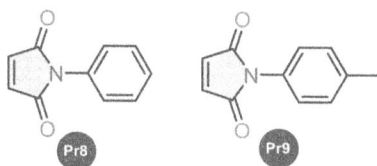

FIGURE   9.5   Chemical   structures   of   1-phenyl-1*H*-pyrrole-2,5-dione   (**Pr8**)   and 1-(4-methylphenyl)-1*H*-pyrrole-2,5-dione (**Pr9**).

X-ray photoelectron spectroscopy (XPS) support the adsorption of pyrrole-derived compounds on carbon steel surfaces [8].

Guo et al. investigated the effects of three heterocyclic compounds thiophene, pyrrole, and furan with the Fe(110) surface by DFT calculations. Theoretical calculations have shown that the inhibitory activities of these inhibitors containing heteroatoms increase in the order of O < N < S [15].

Charles et al. synthesized and characterized four pyrrole-derived Schiff base compounds and investigated their inhibitory activities against MS corrosion in 0.1 M HCl medium. The synthesized compounds are respectively N-((1H-pyrrol-2-yl) methylene)aniline (**Pr10**), N((1H-pyrrol-2-yl)methylene)-4-methylaniline (**Pr11**), N-((1H)-pyrrol-2-yl)methylene)-4 methoxyaniline (**Pr12**), and N-((1H-pyrrol-2-yl) methylene)-3,5-dimethylaniline (**Pr13**) (Figure 9.6). Among the Schiff base compounds tested, the molecule with the strongest inhibitory effect was the compound with a methoxy group on the aromatic ring (**Pr12**). In this case, the inhibitor has significant adsorption characteristics on the steel surface due to the methoxy group, which raises the electron density in the aromatic ring with its electron donor feature and activates the ring [16].

On metal surfaces, corrosion typically takes place in environments with oxygen and moisture. Electrochemical reactions take place in the anodic and cathodic regions while corrosion occurs. The reduction process occurs in the cathodic region, whereas the oxidation reaction occurs in the anodic region. Acidic corrosion media typically lead to hydrogen formation reactions. Corrosion inhibitors reduce or stop corrosion by affecting these reactions in anodic and cathodic regions. We can classify inhibitors into three groups according to their electrochemical behavior:

   i. Anodic inhibitors
  ii. Cathodic inhibitors
 iii. Mixed inhibitors

To briefly mention them, the first of these is anodic inhibitors, also called passivating inhibitors. These increase the anodic polarization, and consequently move

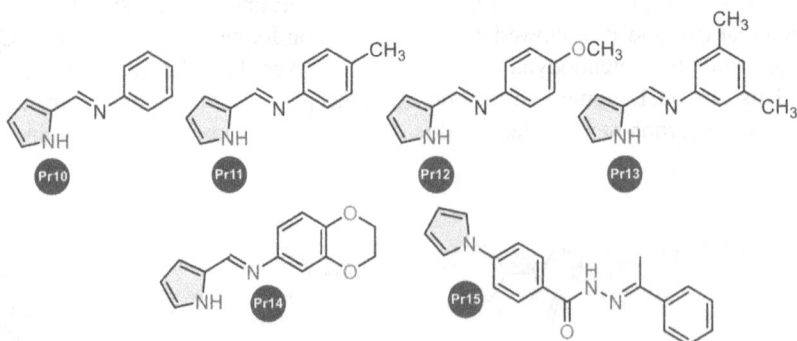

FIGURE 9.6 Chemical structures of N-((1H-pyrrol-2-yl)methylene)aniline (**Pr10**), N((1H-pyrrol-2-yl)methylene)-4-methylaniline (**Pr11**), N-((1H)-pyrrol-2-yl)methylene)-4 methoxyaniline (**Pr12**), and N-((1H-pyrrol-2-yl)methylene)-3,5-dimethylaniline (**Pr13**).

the corrosion potential in the cathodic direction. These often lower corrosion rates by producing a protective coating or barrier on the metal surface [17]. Cathodic inhibitors prevent corrosion by reducing the rate of reduction in the electrochemical corrosion cell. The slowdown in corrosion occurs when cations move toward the cathode surfaces where they precipitate chemically or electrochemically, and the cathode surfaces are blocked. Cathodic inhibitors shift the corrosion potential to the anodic direction [18]. Mixed inhibitors have both anodic and cathodic inhibitor characteristics and delay both anodic and cathodic processes involved in the corrosion process. These inhibitors are generally film-forming compounds that indirectly block both anodic and cathodic sites and cause the formation of precipitates on the surface. The benefits of such inhibitors include the ability to regulate both cathodic and anodic corrosion processes [19].

The compound N-(1H-pyrrole-2-ylmethylidene)-2,3-dihydro-1,4-benzodioxin-6-amine (**Pr14**) was synthesized by Pandimuthu et al. The inhibitory effect of this pyrrole derivative on MS corrosion in HCl (1.0 M) and H_2SO_4 (0.5 M) environments was investigated by impedance and polarization tests. It has been stated that increasing the concentration in acidic environments increases the inhibition efficiency: 600 ppm inhibitor concentration showed 81.6% inhibitory efficiency in 0.5 M H_2SO_4. In 1.0 M HCl medium, it was observed that it reached 84.2% inhibitory efficiency. It was stated that **Pr14** is a mixed inhibitor with Tafel polarization curves, and the morphology of the steel surface examined by SEM and AFM techniques also supports the chemical adsorption mechanism [20].

Alamiery et al. synthesized the molecule N'-(1-phenylethylidene)-4-(1H-pyrrol-1-yl)benzohydrazide (**Pr15**) using acetophenone and 4-(1H-pyrrol-1-yl)benzohydrazidine compounds. Its effectiveness as a corrosion inhibitor for carbon steel in a solution of 1 M hydrochloric acid was tested. Where the maximum inhibitory efficacy was achieved at 500 ppm concentration (94.5%), it was observed that a protective adsorption layer was formed on the low-carbon steel surface, which was consistent with the Langmuir adsorption isotherm. DFT calculations, gravimetric analyses, and SEM images provided evidence supporting the inhibitory property of the synthesized molecule [21].

As an organic coating for corrosion protection, polypyrrole has significant advantages. The good adhesion to metal surfaces, lower toxic effect compared to other paints, good corrosion resistance, synthesis ability, and high conductivity of polypyrrole provide significant advantages. Furthermore, conductive polymer films provide significant corrosion resistance by catalyzing the passivation of rusting metal. Corrosion studies involving electrochemical pyrrole coating on metal surfaces are becoming increasingly important. Asan et al. investigated the corrosion resistance of polypyrrole coating on MS in 0.1 M hydrochloric acid medium. It was investigated how the corrosion resistance of the coating was changed by adding MoS_2 to the coating solutions. It was determined by the Tafel polarization method that the corrosion resistance of polypyrrole coatings with MoS_2 additives increased and it was stated that the MoS_2 additive made the coating surface impermeable. It was explained that the corrosion potential of the steel shifted to positive and the oxidation current decreased, and the MoS_2 additive had a significant effect on the polypyrrole coating [22]. El Jaouhari et al. performed corrosion tests by coating polypyrrole films on

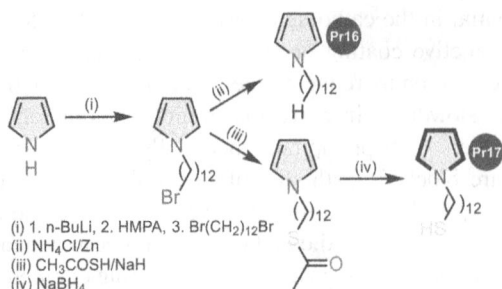

(i) 1. n-BuLi, 2. HMPA, 3. Br(CH₂)₁₂Br
(ii) NH₄Cl/Zn
(iii) CH₃COSH/NaH
(iv) NaBH₄

SCHEME 9.2   Synthesis of 1-dodecylpyrrole derivatives.

carbon steel in sodium saccharinate aqueous medium. The corrosion resistance of the coating was measured in 3% sodium chloride solution by electrochemical methods. X-ray photoelectron spectroscopy (XPS) and SEM analyses showed smooth and homogeneous coating on the carbon steel surface [23].

In order to prevent corrosion by polymer coating, Lallemand et al. electrochemically polymerized the compounds 1-dodecylpyrrole (**Pr16**) and 12-(pyrrole-1-yl) dodecane-1-thiol (**Pr17**) on nickel metal (Scheme 9.2). Various spectroscopic and microscopic techniques were used to analyze the chemical composition and topography of the surface. It has been observed that **Pr16** polymer coating formed on the nickel surface reduces the anodic dissolution rate of the nickel electrode negligibly with the barrier it creates against corrosion [24].

Lebrini et al. investigated the corrosion inhibitory properties of 2-amino-1-(4-aminophenyl)-1*H*-pyrrolo(2,3-*b*)quinoxaline-3-carbonitrile (**Pr18**) (Figure 9.7) on C38 steel. In corrosion tests in HCl medium, impedance data showed that there was adsorption at the metal–electrolyte interface; polarization studies showed that the pyrrole derivative compound works as a mixed inhibitor. The Langmuir adsorption isotherm and XPS data both have been used to demonstrate that physical adsorption and chemical absorption, respectively, occur on the surface of steel [25].

Gadow et al. synthesized pyrrole derivatives as corrosion inhibitors for carbon steel in a 1.0 M HCl solution (Figure 9.8). The results of potentiodynamic polarization, EIS, and electrochemical frequency modulation (EFM) in the corrosive medium proved that the inhibitors showed mixed inhibitory properties. **Pr20** showed the best corrosion activity with a value of 82.8% in 1×10^{-6} M. Electrochemical measurement results showed that inhibitor compounds were adsorbed on the carbon steel

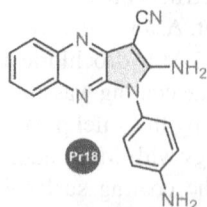

FIGURE 9.7   Chemical structures of 2-amino-1-(4-aminophenyl)-1*H*-pyrrolo(2,3-*b*)quinoxaline-3-carbonitrile (**Pr18**).

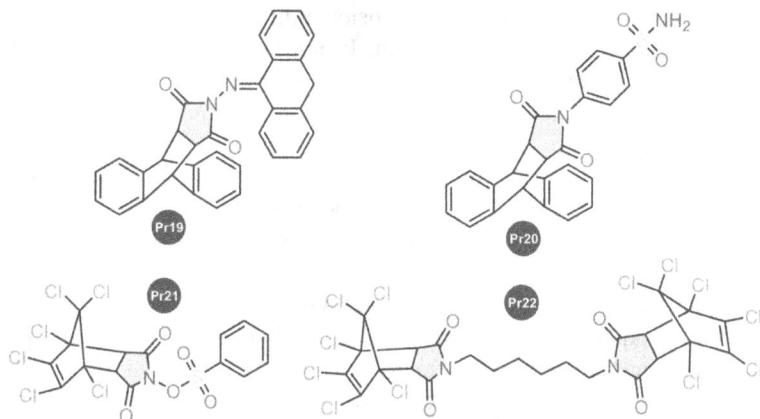

FIGURE 9.8 Chemical structures of pyrrole derivatives (**Pr19–Pr22**).

surface by chemisorption mechanism and their adsorption behavior was compatible with Langmuir adsorption isotherm [26].

Thin coatings made of conductive polymers form protective layers against corrosion, and the conductive polymer's potentials provide anodic protection effects. Adamczyk et al. coated stainless steel with poly(3,4-ethylenedioxythiophene)-based composite coatings in the presence of 4-(pyrrol-1-yl)benzoic acid (**Pr23**) (Figure 9.9) and phosphododecamolybdic acid (PMo$_{12}$). The production of the obtained composite coating resulted in very effective results against pitting corrosion in a strongly acidic environment containing chloride anions. According to reports, the pyrrole additive in conductive polymer coatings is important in protecting steel in extremely corrosive environments [27].

Dithiocarbamates, which are considered green corrosion inhibitors, may be readily adsorbed on the metal surface. These compounds have excellent chelation ability with metal ions due to their sulfur and nitrogen atoms. In this respect, compounds containing dithiocarbamate functional groups attract attention as important structures in the protection of alloys in corrosive environments. Bagherzadeh et al. studied the corrosion of Monel alloy in a 5% HCl solution. The efficiency of the pyrrolidine dithiocarboxylate (**Pr24**) (Figure 9.10) molecule by self-adsorption on the Monel surface was investigated, and electrochemical data revealed that the corrosion rate was reduced [28].

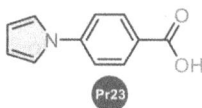

FIGURE 9.9 Chemical structures of 4-(pyrrol-1-yl)benzoic acid (**Pr23**).

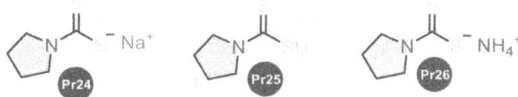

FIGURE 9.10 Chemical structures of pyrrolidine dithiocarboxylate (**Pr24–Pr26**).

Qafsaoui et al. investigated the corrosion inhibitory effect of 1-pyrrolidine dithiocarbamate (**Pr25**) on copper metal. It has been stated that this study is a preliminary study for the protection of Al alloys and was carried out in 0.5 M NaCl medium and three different inhibitor concentrations. Experimental data showed that **Pr25** has a protective effect against corrosion by reducing the anodic dissolution and oxygen reduction reaction. It has been said that corrosion inhibition occurs as a result of the development of stable compounds on the copper surface [29]. In another study, the effect of 1-pyrrolidine dithiocarbamate (**Pr25**) on the galvanic bonding resistance of AA 2024-T3 alloy in NaCl environment was investigated by Qafsaoui et al. Based on electrochemical, gravimetric, and surface analysis tests, **Pr25** forms an adsorbed film by forming the CuI–PDTC complex and protects against corrosion by reducing alloy reactivity [30].

Zhang et al. investigated the effect of ammonium pyrrolidine dithiocarbamate (**Pr26**) on the corrosion of copper surface in a 3% NaCl solution. Optimum conditions for self-assembling monolayers (SAMs) of **Pr26** have been determined. It was observed that the maximum inhibition efficiency reached 98.7% under the conditions of 8-hour immersion time of the corrosion inhibitor and 1 mmol L^{-1} inhibitor concentration. The adsorption of **Pr26** on the copper surface was supported by infrared spectrum and energy dispersive spectroscopy analyses. In addition, theoretical studies have shown that the sulfur atoms in the inhibitor are the most active adsorption sites. Electrochemical measurements indicate that **Pr26** SAMs act as a mixed-type inhibitor, conducive to the suppression of anodic oxidation and cathodic reduction reactions [30]. In addition, Al-Rawajfeh et al. studied **Pr26** corrosion inhibition in steel water pipes used in the Jordan National Water Supply Network (JNWSN) in various corrosive environments. Variables in the experimental studies included aggressive solution concentration, **Pr26** concentration, and immersion time. The inhibitory effectiveness decreases as the concentration of the corrosive solution increases, and the inhibitory effectiveness changes in direct proportion to the inhibitor concentration and immersion time [31].

Haque et al. synthesized quaternary ammonium compounds, including pyrrolidine and propargyl substituent and hydrophobic alkyl groups, and performed corrosion studies on MS at 1 M HCl. The corrosion studies of *N,N*-dipropargylpyrrolidium bromide (**Pr27**), *N*-dodecyl, *N*-propargylpyrrolidium bromide (**Pr28**), and *N*-hexadecyl, *N*-propargylpyrrolidium bromide (**Pr29**) (Figure 9.11) compounds were shown to exhibit very good inhibitory properties in yields ranging from 92.6% to 96.2%. Electrochemical studies have shown that **Pr27** and **Pr28** encoded compounds act as mixed-type inhibitors, while **Pr29** acts as a cathodic inhibitor. The findings from UV-vis, SEM, and XPS analyses to determine surface morphology support the adsorption of compounds on the surface of MS. Furthermore, theoretical studies (DFT analysis) on the molecular structure and corrosion inhibitor properties of the inhibitors were stated to support the experimental data [32].

Bouklah et al. looked into the corrosion-inhibiting capabilities of pyrrolidin-2-one derivatives (**Pr32–Pr33**) (Figure 9.12) for steel in a corrosive medium containing 0.5

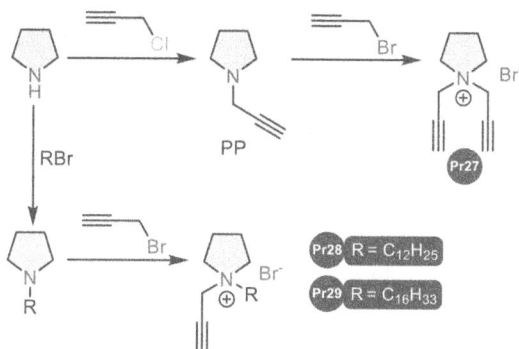

FIGURE 9.11 Chemical structures of N,N-dipropargylpyrrolidium bromide (**Pr27**), N-dodecyl, N-propargylpyrrolidium bromide (**Pr28**), and N-hexadecyl, N-propargylpyrrolidium bromide (**Pr29**).

M H_2SO_4. In addition, the inhibitory properties of the synthesized compounds were compared with the reagents 1-vinylpyrrolidin-2-one (**Pr30**), 2-mercaptoethanol, and mercaptoacetic acid compounds. According to the weight loss tests, R_p polarization, and impedance measurements, it was stated that **Pr31** and **Pr32** molecules showed cathodic inhibitor properties and were adsorbed on the steel surface. In particular, it was indicated that **Pr32** demonstrated inhibitory effectiveness, with a maximum value of 89% at 5×10^{-3} M concentration [33].

Verma et al. synthesized three 3-amino alkyl substituted indole derivatives. It was stated that the pyrrolidine derivative 3-(phenyl(pyrrolidin-1-yl)methyl)-1H-indole (**Pr33**), one of the synthesized compounds, inhibited MS corrosion with a 96.08% efficiency in a 1 M HCl environment, and the efficiency improved with increasing concentration. The inhibitor adsorbs on the surface of the steel and acts as a cathodic inhibitor. The SEM and AFM data obtained for the surface morphology also showed that the adsorbed film formed on the steel surface and the adsorption followed the Langmuir isotherm. In addition, theoretical data obtained from quantum chemical calculations and dynamic simulation studies support the inhibition activities of the molecule. It has also been theoretically investigated that the coating on the MS surface is related to the increase in molecular volume in the inhibition mechanism [34].

Jeeva et al. investigated the corrosion inhibitor properties of pyridine-tethered Mannich base compounds in 1.0 M hydrochloric acid (HCl) solution for MS. Among the compounds studied, the pyrrolidine derivative, and 1-(pyridin-4-yl (pyridin-1-yl)

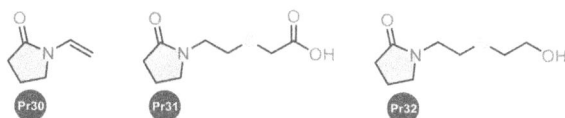

FIGURE 9.12 Chemical structures of pyrrolidin-2-one derivatives (**Pr32–Pr33**).

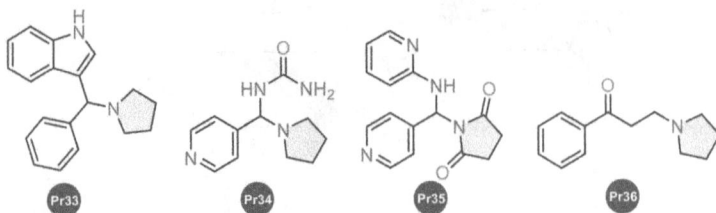

FIGURE 9.13 Chemical structures of 1-(pyridin-4-yl (pyridin-1-yl)methyl)urea (**Pr34**), 1-((pyridin-2-ylamino)(pyridin-4-yl)methyl)pyrrolidin-2,5-dione (**Pr35**), and 1-phenyl-3-(1-pyrrolidinyl)-propanone (**Pr36**).

methyl)urea (**Pr34**) (Figure 9.13), was found to be a mixed-type inhibitor with dominant control of the anodic reaction, preventing corrosion via physisorption and chemisorption mechanisms. SEM micrographs revealed that a good protective film was formed on the MS surface, while DFT calculations revealed that the inhibitor's heteroatoms and bonds aided in film formation [35].

Jeeva et al. also investigated the compound 1-((pyridin-2-ylamino)(pyridin-4-yl) methyl)pyrrolidin-2,5-dione (**Pr35**), which is a Mannich base synthesized using succinimide, 2-aminopyridine, and pyridine-4-carboxaldehyde reagents in equivalent stoichiometric ratios and characterized. The corrosion inhibitory properties of the **Pr35** compound were studied on MS in a 1.0 M HCl medium. The inhibitor is an excellent mixed-type inhibitor that controls the cathodic process and obeys the Langmuir adsorption isotherm, according to the experimental data. In addition, it has been suggested that adsorption is physical adsorption with FTIR, SEM-EDX, and AFM analyses and DFT calculations to get an idea about the mechanism of the corrosion prevention process [36].

In another study, Zhan and colleagues synthesized and characterized the Mannich base 1-phenyl-3-(1-pyrrolidinyl)-propanone (**Pr36**) using acetophenone, pyrrolidine, and formaldehyde reagents. Corrosion studies were carried out for N80 steel of this synthesized compound in 15% hydrochloric acid (HCl) environment. The results showed that **Pr36** had excellent corrosion inhibition in a 15% HCl environment. It was stated that Mannich base tested as a corrosion inhibitor showed 99.8% corrosion inhibition at 0.6% inhibitor concentration. According to the corrosion test results, it has been suggested that **Pr36** can be used as an effective inhibitor against corrosion in oil wells [37].

In an acidic medium, the corrosion inhibitor properties of the 2-pyrrolidin-1-yl-1,3-thiazol-5-carboxylic acid (**Pr37**) (Figure 9.14) compound were investigated. Corrosion studies included electrochemical tests, SEM analyses for surface

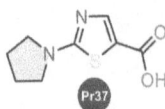

FIGURE 9.14 Chemical structures of 2-pyrrolidin-1-yl-1,3-thiazol-5-carboxylic acid (**Pr37**).

morphology, and theoretical calculations. According to experimental and theoretical investigations, **Pr37** suppresses both anodic and cathodic corrosion processes, and physical adsorption occurs on the MS surface. Furthermore, at 0.1 M HCl conditions, **Pr37** was shown to be an excellent mixed-type inhibitor for preserving MS against corrosion [38].

9.2 PYRIDAZINE-BASED CORROSION INHIBITORS

Pyridazine is pyridine-like aromatic heterocyclic organic compound, also known as diazine. It has two nitrogen atoms in the six-membered ring and there are three diazine isomers. If these atoms are in positions 1 and 2, they are named "pyridazine"; in positions 1 and 3, they are named "pyrimidine"; and in positions 1 and 4, they are named "pyrazine" (Figure 9.15).

Pyridazine is a planar molecule with an N—N bond in its ring structure. While pyridazines were first named by Knorr, the first samples were synthesized by Fischer. However, Tauber was the first to synthesize unsubstituted pyridazine [39]. Since the pyridazine molecule is basic and aromatic like pyridine, it is easily soluble in water compared to hydrocarbons. The chemical activity of adjacent nitrogen atoms in the pyridazine ring offers unique chemistry for this compound. This allows pyridazines to interact with suitable substrates with acidic functional groups to form supramolecular complexes [40]. They are also of great interest in the field of medicine and pharmacy such as antiviral, antibacterial, antibiotic, anticancer, antidiabetic, antidepressant, etc. [41]. It is also possible to see the structure of pyridazine in the structure of some compounds used as herbicides [42]. Pyridazines also have characteristic properties such as good resistance at room temperature, low density, as well as electrical conductivity properties [43]. They can also be used to produce thin coatings on the surface of metals. Because of the aforementioned electrical qualities, pyridazine compounds have recently found application as corrosion inhibitors. In this section, information about pyridazine compounds used as corrosion inhibitors will be given.

Pyridazine compounds used as corrosion inhibitors are generally seen to be derivatized in three different ways. These are the replacement of substituents in the ring, the attachment of a different heteroaromatic ring to the pyridazine ring, or the use of condensed ring systems. The selection of substituent and aromatic ring systems is selected by considering factors affecting corrosion such as heteroatom, conjugation, and π system.

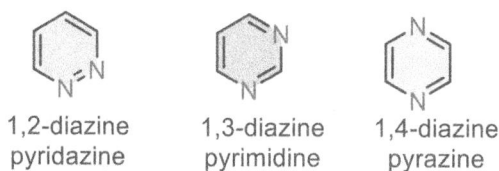

1,2-diazine   1,3-diazine   1,4-diazine
pyridazine   pyrimidine   pyrazine

FIGURE 9.15   Chemical structures of 1,2-, 1,3-, and 1,4-diazines.

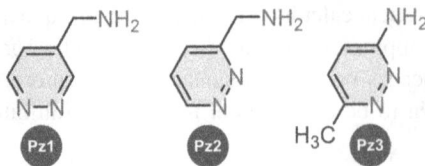

FIGURE 9.16 Chemical structures of pyridazin-4-ylmethylamine (**Pz1**), pyridazin-3-yl-methylamine (**Pz2**), and 3-amino-6-methylpyridazine (**Pz3**).

Pyridazin-4-ylmethylamine (**Pz1**), pyridazin-3-ylmethylamine (**Pz2**), and 3-amino-6-methylpyridazine (**Pz3**) (Figure 9.16) are the three isomers of the aminomethyl pyridazine compound. The anticorrosion properties of these three compounds were investigated by Mashuga et al. [44]. PDP, electrochemical EIS, and computational techniques were used to investigate the inhibitory effects of these chemicals on MS in a 1 M HCl environment. According to the results of the experiments, the inhibitor molecules are adsorbed on the steel surface via coordinating covalent bonds established between Fe and the atoms (C and N) in the molecule.

Molecules adsorbed on the steel surface both chemical and physical adsorption mechanisms in hydrochloric acid follow to Langmuir and Temkin adsorption isotherms. Increasing the amount of inhibitor caused a decrease in the amount of corrosion. Based on the experimental data and DFT calculations, the corrosion inhibition mechanism for the investigated **Pz** molecules is as follows: First, the neutral molecule is absorbed on the metal surface. This absorption is caused by the interaction of the p-orbitals of the nitrogen atoms or electrons of the pyridazine ring with the 4d-orbitals of the Fe. Later, the Fe surface usually has positive charges as a result of the oxidation of Fe to Fe^{2+}. The positively charged Fe surface can bind with Cl^- ions, the source of which is acid. By the way, the inhibitors are found in a protonated state. We can schematize this event as follows:

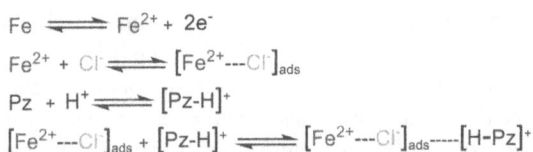

$$Fe \rightleftharpoons Fe^{2+} + 2e^-$$
$$Fe^{2+} + Cl^- \rightleftharpoons [Fe^{2+}\text{---}Cl^-]_{ads}$$
$$Pz + H^+ \rightleftharpoons [Pz\text{-}H]^+$$
$$[Fe^{2+}\text{---}Cl^-]_{ads} + [Pz\text{-}H]^+ \rightleftharpoons [Fe^{2+}\text{---}Cl^-]_{ads}\text{-----}[H\text{-}Pz]^+$$

As a result of the PDP experiments, **Pz1** and **Pz2** were shown to have more anodic effects, whereas **Pz3** was found to have more cathodic inhibitive effects.

Three 3-chloropyridazine derivatives (**Pz4–Pz6**) (Figure 9.17) were tested for their potential to the protection of the MS surface and inhibit MS corrosion in 1 M

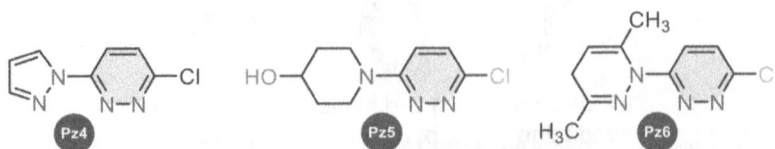

FIGURE 9.17 Chemical structures of 3-chloropyridazine derivatives (**Pz4–Pz6**).

HCl solution [45]. Corrosion tests were carried out using PDP and EIS techniques. FTIR spectroscopy was used to explore the interactions of pyridazine molecules with MS.

In general, organic inhibitors prevent corrosion in aqueous solution by adsorbing organic molecules onto the metallic surface. Water and organic molecules in the solution form a semi-equilibrium. Organic molecules likely to remove water molecules from the metal–electrolyte contact and adsorb to active sites on the metal surface. In order to determine the adsorption mode of the studied pyridazine molecules, the compatibility of the experimental surface coverage values with isotherms such as Langmuir, Temkin, Frumkin, and Freundlich was examined and it was determined that these values matched the Langmuir adsorption isotherm.

The linear form of the Langmuir adsorption isotherm can be expressed by the following equation:

$$\frac{C_{inh}}{\theta} = C_{inh} + \frac{1}{K_{ads}}$$

where C_{inh} is the concentration of the inhibitor, θ is the surface coverage ($\theta = \%IZ/100$), and K_{ads} is the adsorption equilibrium constant.

Gibb's adsorption free energy (ΔG_{ads}°) is calculated from the following equation:

$$\Delta G_{ads}^{\circ} = -RT \ln\left(1 \times 10^{6} K_{ads}\right)$$

where R is the universal gas constant, T is the absolute temperature, and K is the concentration of water in the bulk solution (in ppm).

ΔG_{ads}° values are often used to classify an adsorption as physisorption ($\Delta G_{ads}^{\circ} \leq -20$ kJ mol^{-1}) or chemisorption (-40 kJ mol$^{-1} \leq \Delta G_{ads}^{\circ}$) [46, 47]. The calculated ΔG_{ads}° values of the **Pz4–Pz6** are between -20 and -40 kJ mol^{-1}, suggesting that the adsorption of the studied **Pz4–Pz6** in 1 M HCl on the MS surface provides both physical and chemical adsorption mechanisms. According to both experimental and theoretical studies, the compounds' surface protection efficiency is **Pz6 > Pz5 > Pz4**.

The inhibitory properties of 2-((6-chloropyridazin-3-yl)thio)-N,N-diethyl acetamide (**Pz7**) (Figure 9.18) for Cu metal in 0.5 M H$_2$SO$_4$ were investigated using electrochemical/surface-analytical experiments and theoretical calculations [48].

Electrochemical tests revealed that **Pz7** operated as a cathodic inhibitor, and the pyridazine compound's absorption on the metal surface led to an improvement in charge transfer resistance as well as a decrease in Cdl. According to SEM and AFM measurements, the copper surface occurred a smoother surface after **Pz7** was absorbed there. XPS analyses were performed to determine how the **Pz7** absorption

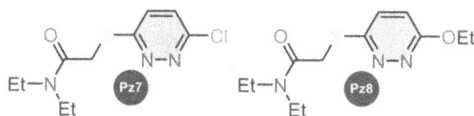

FIGURE 9.18 Chemical structures of 3-chloropyridazine derivatives (**Pz7–Pz8**).

mechanism occurs on the copper surface. These analyses show that **Pz7** is adsorbed on the Cu surface via the formation of coordination bonds through the N heteroatoms and copper atoms.

Compound **Pz8** was synthesized by replacing the Cl atom with the methoxy group in the **Pz7** compound and its protective effect on copper metal was also investigated [49]. Acting as a modest cathodic corrosion inhibitor for copper in 0.5 M H_2SO_4, **Pz8** had a maximum inhibition of 94.1% at 298 K, 93.2% at 308 K, and 91.3% at 318 K at 4 mM concentration. Its effectiveness was determined by EIS. XPS study demonstrated that **Pz8** molecules are securely adsorbed on the copper surface by the N—Cu bond, Cu—S bond, and other chemical interactions. When comparing **Pz7** and **Pz8** compounds, studies have shown that Cl and methoxy groups have similar activity in corrosion. In addition to experimental studies, it has been theoretically demonstrated by quantum chemical calculations and molecular dynamics simulation studies that **Pz7** and **Pz8** have an excellent corrosion inhibition capacity for copper in a strong acid environment.

Two different sulfanyl pyridazine compounds, including 3-substituted ester group with different alkyl chains (**Pz10**, **Pz11**), were investigated in 3.5% NaCl in terms of their protective mechanisms against iron corrosion and the results were compared with **Pz9** [50]. It was found that the investigated pyridazine derivatives shifted both the corrosion potential and the pitting potential of iron to higher values. As a result of the experiments, it was discovered that the greater inhibition efficiency for 400 ppm **Pz10** was 92%. The long alkyl chain at **Pz10** behaves more ideally capacitively than in the case of **Pz11** due to the structure of the Fe–electrolyte interface, and the adsorption of insulating barrier layers on the Fe–electrolyte interface. This is explained by the active bonding of nitrogen and sulfur atoms in the structure of the pyridazine compound, as well as the increase in charge density around the adsorption active sites as a result of the longer S-alkylated side chain.

Abdel Hameed et al. synthesized four different sulfanyl pyridazine compounds based on the **Pz11** compound [25] (Scheme 9.3). Together with the **Pz11** compound, these synthesized compounds were investigated as corrosion inhibitors against

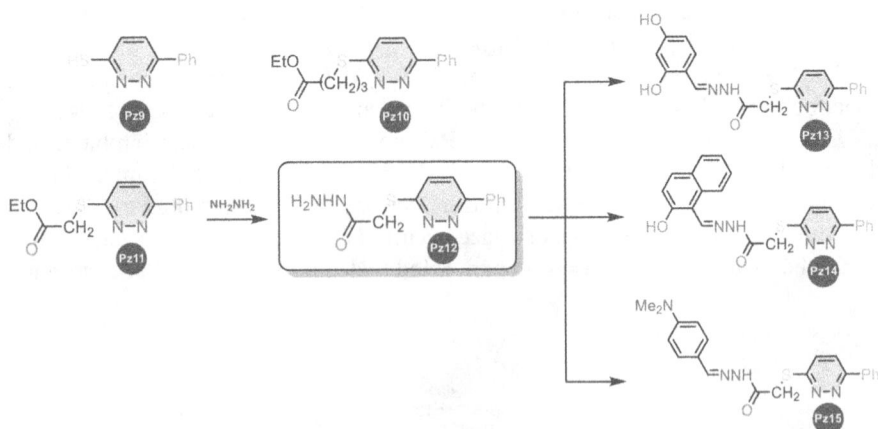

SCHEME 9.3   Synthesis of some sulfanyl pyridazine compounds.

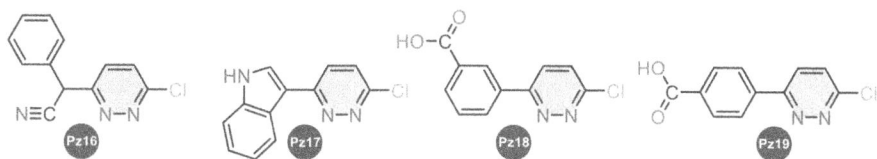

FIGURE 9.19 Chemical structures of pyridazine derivatives (**Pz16–Pz19**).

carbon steel in a 1 M HCl medium. As a result of the experiments, the most effective inhibitor with 93% inhibition efficiency was C3, and the order of activity was found as **Pz13 > Pz15 > Pz14 > Pz12 > Pz11**.

Using electrochemical, spectroscopic, and theoretical computational chemistry approaches, four compounds with a 6-chloro-3-pyridazine structure were examined for their inhibitory characteristics for MS in 1 M HCl [51]. The common finding for these four molecules is that the inhibitory activity increases with increasing inhibitor concentration and they are mixed-type inhibitors. **Pz16** and **Pz17** (Figure 9.19) showed the best protection performances. This is due to the excess of nitrogen atoms and π-bonds. The spectroscopic investigations indicated that the compounds chemically interact with MS in acidic media, and that the pyridazine ring is actively involved in molecule adsorption on the steel surface. Adsorption data, consistent with the Langmuir and Temkin isotherm models, showed that the metal–inhibitor interaction is both physisorption and chemisorption.

In general, the inhibitory efficacy of the *N*-heterocyclic organic inhibitor improves with the number of aromatic systems and the presence of electronegative atoms in the molecule [52]. In order to investigate this situation, pyridine-substituted pyridazine molecule (**Pz20**) was designed and a series of studies, including polarization and EIS, XPS analysis, and quantum chemical calculations were carried out on the corrosion inhibition of carbon steel in 1 M HCl solutions [53]. The findings of the experiments revealed that this organic compound inhibited the corrosion of carbon steel in 1 M HCl solutions at 30 °C, and the inhibition efficacy improved with increasing inhibitor concentration. It was determined that the adsorption of the compound, which was revealed as a mixed-type inhibitor as a result of potentiodynamic polarization experiments, in 1 M HCl solution on the carbon steel surface complies with the Temkin isotherm with a very high negative value for the standard Gibbs free adsorption energy ΔG°_{ads} (chemisorption).

In another study, different substituents were attached to the pyridazine ring in compound **Pz20** (Figure 9.20) and its corrosion inhibitory properties on MS in a

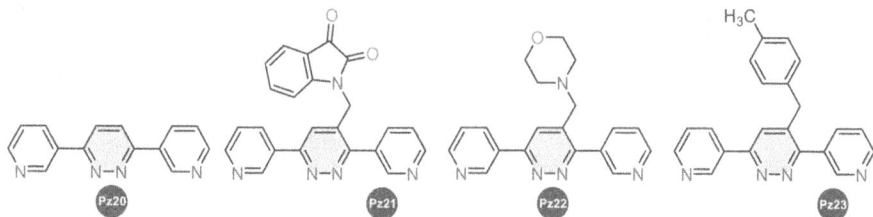

FIGURE 9.20 Chemical structures of pyridazine derivatives (**Pz20–Pz23**).

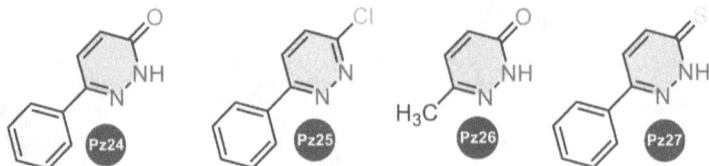

FIGURE 9.21 Chemical structures of pyridazine derivatives (**Pz24–Pz27**).

1 M HCl environment were investigated with similar methods. In the series in which the most active substituent was isatin, the order of activity was **Pz21 > Pz22 > Pz23**, and the maximum efficiency values were determined as 87%, 72%, and 70%, respectively, at a concentration of 1.0×10^{-3} M. The absorption isotherm of these compounds, which act as mixed inhibitors, conforms to the Langmuir isotherm model instead of the Temkin isotherm.

Corrosion inhibition performances of four molecules of 1,4-disubstituted pyridazine derivative for MS in 1 M HCl solution were measured in two separate studies [54, 55]. In the first study, the effects of carbonyl and chlorine atoms on the ring for **Pz24**, **Pz25** compounds, and the effects of alkyl/aryl groups and carbonyl/thiocarbonyl groups on the corrosion inhibition for **Pz24**, **Pz26**, and **Pz27** (Figure 9.21) compounds were investigated in the second study.

As a result of thermodynamic measurements, it has been revealed that the interactions of the investigated inhibitors with the metal surface are more chemical than physical. The adsorption of the studied molecules on the MS conforms to Langmuir adsorption isotherm. In the first study, SEM analyses showed that the compounds protected the surface of MS in the tested acid, and the protective performance of **Pz25** was better than **Pz24**, with 96% inhibition efficiency. In the second study, the inhibitory activities of the inhibitors were listed as **Pz27 > Pz24 > Pz26**. The efficacy of PPYS, which is the most effective, was measured as 98.5% at 0.5 mM. The compounds act as mixed-type inhibitors. In the light of these data, it was concluded that the chlorine atom is more effective than the carbonyl group; the aryl group is more effective than the alkyl group; and the thiocarbonyl group is more effective than the carbonyl group in corrosion inhibition.

In order to investigate the effect of C=O and C=S groups on inhibition, **Pz28** and **Pz29** molecules (Figure 9.22) were investigated by Ghazoui et al. for C38 steel in 1 M HCl [56]. In another study, **Pz30** and **Pz31** molecules were investigated for iron metal [57]. In both studies, the high inhibition efficiency of compounds containing C=S group is because of the presence of sulfur atom, which has more

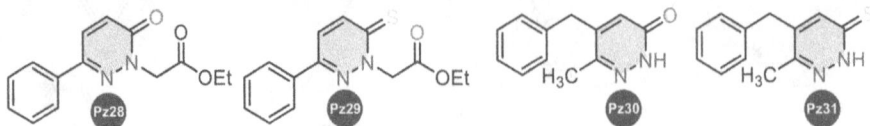

FIGURE 9.22 Chemical structures of pyridazine derivatives (**Pz28–Pz31**).

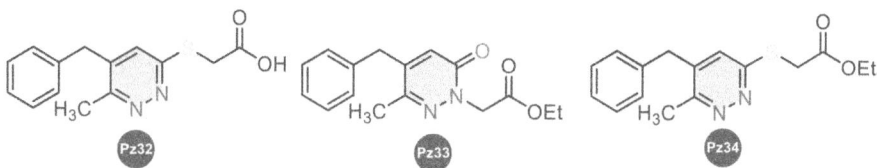

FIGURE 9.23 Chemical structures of pyridazine derivatives (**Pz32–Pz34**).

electron-releasing properties than the oxygen atom, in addition to the pyridazine ring, which provides strong bonding of this compound to the surface.

In another study proving that the sulfur atom is effective in inhibition, **Pz32–Pz34** compounds (Figure 9.23) were examined for the iron atom in 1 M HCl. **Pz32** and **Pz34** molecules were shown to have better inhibition efficacy than **Pz33** molecules [58].

Compounds **Pz35** and **Pz36** (Figure 9.24) were compared to investigate the effect of alkyl and aryl groups. Although the corrosion efficiency of aryl groups increased due to π-electrons, this increase was not as high as expected. While the inhibition efficiency was 83.1% when the methyl group was attached to the pyridazine ring, this value was 83.9% when the phenyl group was substituted [59].

The ester group of compound **Pz36** is changed into a hydrazide when it reacts with hydrazine, yielding compound **Pz37**. Nahle et al. investigated **Pz37** compound as corrosion inhibitor on C38 MS in 1.0 M HCl solution [60]. The inhibitory activity of **Pz37** acting as a mixed-type inhibitor increases with increasing concentration and reached 85% at 1.0×10^{-3} M concentration. The nitrogen atoms in the hydrazide group are seen to boost inhibition. Impedance studies revealed that the molecule is adsorbed on the surface of C38 MS, resulting in an increase in charge transfer resistance and a decrease in double-layer capacitance. Its adsorption complied with the Langmuir adsorption isotherm.

Two pyridazine molecules (**Pz38** and **Pz39**) (Figure 9.25) are employed as corrosion inhibitors for steel in 0.5 M H_2SO_4 solution [61]. The interactions of these compounds with the steel surface are realized by means of the pyridazine ring, phenyl

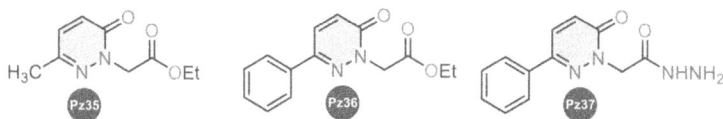

FIGURE 9.24 Chemical structures of pyridazine derivatives (**Pz35–Pz37**).

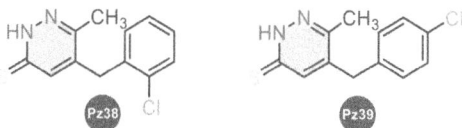

FIGURE 9.25 Chemical structures of pyridazine derivatives (**Pz38–Pz39**).

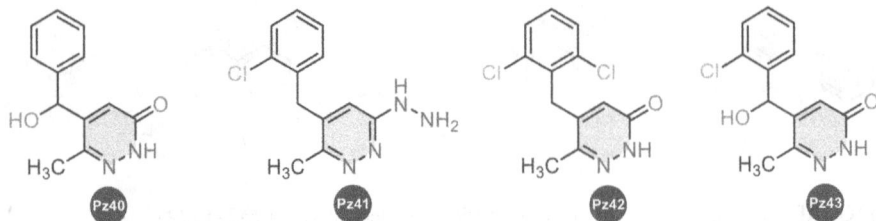

FIGURE 9.26 Chemical structures of pyridazine derivatives (**Pz40–Pz43**).

ring, sulfur, and chlorine atom. In this study, it was intended to investigate the effect of the positions of chlorine atoms on adsorption. Popova and others have conducted a study on this topic and tried to explain the adsorption of molecules with *ortho-*, *meta-*, and *para*-substituents using the Hammett equation [62]. They concluded that in cases where the substituent position is *ortho*, it is expected to inhibit chemisorption to a greater extent than physical adsorption. In this study, it was also observed that the presence of Cl in the *ortho*-position of the molecule strengthens the adsorption compared to the *para*-position, so the **Pz38** has a higher inhibitory activity than the **Pz39**.

Pyridazine molecules, the structures of which are given in Figure 9.26, were studied with copper metal in a 2NHNO$_3$ [63]. Among these molecules acting as mixed inhibitors, the **Pz42** and **Pz43** (Figure 9.26) molecules have the best inhibitory efficiency. **Pz42** and **Pz43** molecules showed 96% and 94% inhibition efficiency at 10^{-3} M, respectively.

Studies investigating the effect of pyridazine molecules and substituents on corrosion have been expressed so far. Several working groups have also investigated the potential of the condensed pyridazine systems to be corrosion inhibitors. One of them is the study of Bentiss et al. In this study, 1,4-dipyridine-substituted pyridazino[4,5-*b*]indole (**Pz44**) compound was investigated against MS in 1 M HCl medium [64]. Acting as an anodic inhibitor, **Pz44** prevents corrosion by forming a chemically absorbed film on the steel surface. It has been discovered that the adsorption of this inhibitor on the steel surface follows the Langmuir isotherm.

In another study using condensed pyridazine compounds as inhibitors for Al alloy in 0.1 M HCl and 0.5 M HCl solutions, imidazo-pyridazine and bromo/chloro-imidazo-pyridazine compounds (**Pz45–Pz47**) (Figure 9.27) are used [65]. The

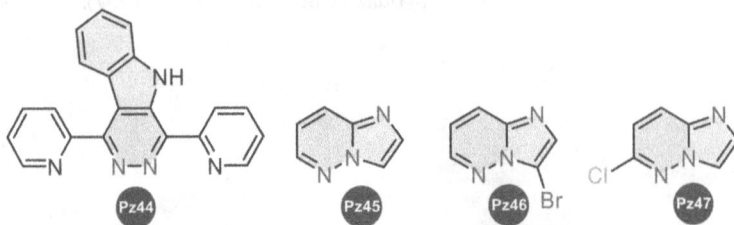

FIGURE 9.27 Chemical structures of pyridazine derivatives (**Pz44–Pz47**).

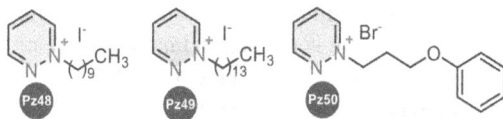

FIGURE 9.28 Chemical structures of pyridazine derivatives (**Pz48–Pz50**).

electrochemical tests and adsorption model analysis indicated that all three substances are mixed-type inhibitors, with **Pz45** exhibiting the best physicochemical anticorrosion performance. At the same inhibitor doses, **Pz45–Pz47** compounds demonstrate more potent anticorrosion efficacy for Al in 0.5 M HCl than 0.1 M HCl. These findings show that despite being caustic ions for Al electrodes, H^+ and Cl^- concentrations have an impact on the absorption of inhibitors. The reaction process in the HCl solution is considered to proceed as follows:

1. Cl^- ions adsorb on the surface of the Al electrode and produce $AlCl_{ads}$.
2. Inhibitors bind to H^+ via the production of N-onium ions.
3. The imidazo-pyridazine compounds adsorb on the metal surface due to the electrostatic interaction between $AlCl_{ads}$ and N-onium ions, preventing additional assaults by chloride ions. Furthermore, protonated inhibitors are absorbed on cathodic sites in competition with hydrogen ions. The production of $AlCl_{ads}$ and N-onium ions is more likely in 0.5 M HCl than in 0.1 M HCl. As a result, the corrosion effect of the studied compounds is greater in 0.5 M HCl solution. Because $AlCl_{ads}$ are limited in 0.1 M HCl solution, inhibition efficacy is mostly determined by the adsorption capacity of the inhibitive molecule itself. In the event, we may conclude that **Pz45** has a more appealing adsorption ability than the other derivatives.

Some pyridazine derivative salts have also been evaluated as corrosion inhibitors [66, 67]. The impact of three pyridazin-1-ium salt derivatives on steel corrosion in HCl (1 M) solution was studied using EI and XPS spectroscopy. The obtained findings showed that **Pz48**, **Pz49**, and **Pz50** (Figure 9.28) reached a value of 86.7%, 88.6%, and 95.2% at 10^{-3} M, respectively. The adsorption of inhibitors on the steel surface in the investigated acidic solution follows the Langmuir isotherm, and the ΔG_{ads}° values show that all inhibitors undergo chemical adsorption, leading to the formation of a strong protective coating on the metal surface.

9.3 CORROSION INHIBITORS, INCLUDING PYRIMIDINE COMPOUNDS

Pyrimidine is an aromatic chemical substance with nitrogen atoms in positions 1 and 3 with less basic properties and a lower solubility profile than pyridine. As in the pyridine ring, the unshared electron pairs here also affect the basicity of the structure. It has a boiling point of 123.8 °C, a melting point of 22 °C, is easily soluble in water, and is a colorless liquid with an irritating odor.

Pyrimidines are widely found in living organisms in nature, especially in their condensed derivatives. Pyrimidine is the general name for the nitrogenous aromatic bases found in nucleic acids, which form the core of life, and in some coenzymes and vitamins. Of considerable biological interest, cytosine, uracil, and thymine, the nucleobases in nucleic acids, are also pyrimidine derivatives. The purine molecule containing a fused pyrimidine imidazole ring is also a compound of the nucleic acids found in cell structure.

The pyrimidine ring structure can also be found in vitamin B1, folic acid, riboflavin or azathioprine, an immunosuppressive drug, zidovudine, which was developed as an anticancer drug, sulfadiazine, which has an antibacterial effect, substituted and fused cyclic compounds and their derivatives, including the treatment of hypertension and heart rhythm disorders (Figure 9.29) [68].

The pyrimidine ring, which is present in the skeleton of many biologically active compounds, and pyrimidine derivatives can be characterized as environment-friendly compounds. Additionally, many of the characteristics necessary to be efficient corrosion inhibitors are present in pyrimidine derivatives. Some of them are the capacity to move electrons to the metal's surface's vacant d-orbital in order to generate coordinate covalent bonds as well as the capacity to draw unoccupied free electrons to the metal surface by using antibonding orbitals to create reverse bonds. Therefore, pyrimidine derivatives are predicted to be excellent corrosion inhibitors that can be used in the industrial field not only because of their inhibitory activity but also because of their non-toxic chemical structure [69, 70].

Acidic conditions are widely used in many industrial processes, often causing severe metallic corrosion. Carbon steel is used extensively in various industries despite having only a modest amount of corrosion resistance because of its excellent mechanical qualities and affordable price. Corrosion inhibitors are often used to control or prevent corrosion in metals or carbon steel. For this purpose, many different organic inhibitors have been tested, among them being organic compounds containing many heteroatoms, which contain many π-electrons, Through the provision

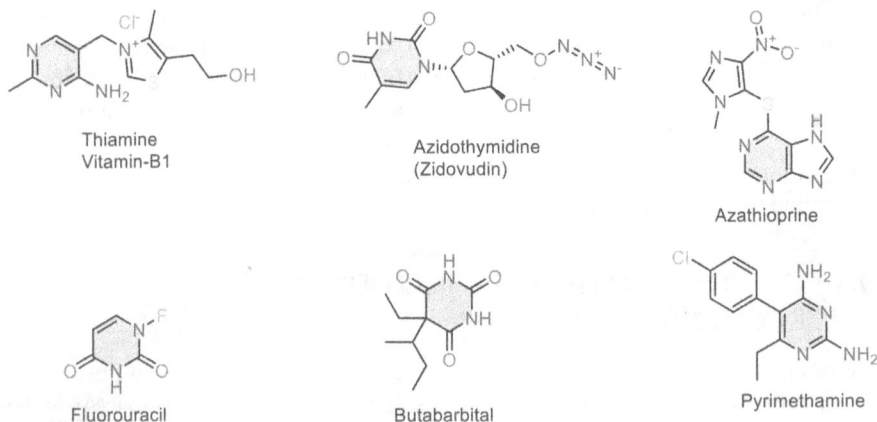

Thiamine
Vitamin-B1

Azidothymidine
(Zidovudin)

Azathioprine

Fluorouracil

Butabarbital

Pyrimethamine

FIGURE 9.29 Chemical structures of vitamin B$_1$, folic acid, riboflavin or azathioprine, zidovudine, and sulfadiazine.

of electrons that interact either directly or indirectly with the metal surface, they have been discovered to exhibit significant inhibitory properties. However, due to their toxicity and difficulty in biodegrading, the use of organic inhibitors with different heterocyclic rings has a negative impact on the environment. The ecologically beneficial characteristics of pyrimidine derivatives have come to light in this context, and their usage as corrosion inhibitors has garnered interest [71, 72].

To date, many pyrimidine derivatives have been synthesized, and their suitability as corrosion inhibition against various steel samples in acidic environments formed using different acids has been investigated.

As a result of the investigations, the adsorption of organic compounds containing pyrimidine ring on the metal surface is directly dependent on the physicochemical properties of the pyrimidine compounds, the electronic density of the donor atom, and the interactions of π-orbitals with the d-orbitals of the surface atom. The results obtained showed that the inhibition efficiency of the studied pyrimidine compounds depends on the concentration and functional structure of the pyrimidine compounds. The activity of the inhibitors increases with the increase in the concentration of the pyrimidine compounds; however, it decreases with the increase in the ambient temperature [73].

In the study carried out by H.S. Awad and S. Abdel Gawad, the following eight pyrimidine compounds were used in the experiments using carbon steel as the metal surface. The inhibition effect of the compounds (**Py1–Py8**) (Figure 9.30) containing different mono-substituent groups to the pyrimidine ring was investigated in the corrosion weight loss tests carried out in a 2 M HCl medium. The compounds containing pyrimidine ring (**Py2–Py4** with amino, hydroxyl, and mercapto functional groups) showed better inhibitory effects than the others. When the concentrations of these compounds were increased, it was observed that their inhibitory activity increased even more [74].

The compounds examined can be listed as follows according to their inhibition effects:

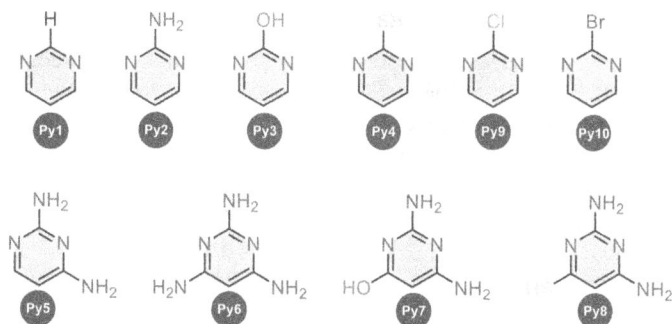

Py4 > Py3 > Py2 > Py1

Py4 > Py3 > Py2 > Py1

FIGURE 9.30   Chemical structures of **Py1–Py8**.

This shows that **Py4** with mercapto group is the compound with the best inhibition activity, followed by the compound with amino group and then the compound with hydroxyl group attached to pyrimidine ring. This is not really surprising because when we look at the electronegativity of the heteroatoms contained in these compounds, the order of oxygen, nitrogen, and sulfur atoms is as follows: S > N > O. According to these results, the substituent group in a single pyrimidine ring shows its effect on corrosion inhibition directly dependent on the properties of the functional group. As a result of cathodic (hydrogen evolution) and anodic (metal dissolution) reactions for compound **Py4**, it was observed that this compound behaves as a mixed-type corrosion inhibitor, with the anodic effect being more prominent, as it is affected by the presence of this compound in the corrosive environment.

In the next step of the study, **Py2** (2-aminopyrimidine) was chosen as the main compound, and the change in the corrosion inhibition effect was examined by adding different substituent groups attached to the 2-aminopyrimidine ring and then increasing their amount. The inhibition efficiency was found to be **Py8** > **Py7** > **Py6** > **Py5** > **Py2** for these compounds. Firstly, the inhibition increased when one more amino group was added to the meta-position of the 2-aminopyrimidine compound. The inhibition efficiency of **Py6**, which was obtained by adding one more amino group to **Py5**, also surpassed **Py5**. **Py6–Py8** have three substituents attached to the pyrimidine ring. It is seen that the compound with the most active inhibition property is **Py8** which contains thiol group. The functional groups of these three compounds (**Py6–Py8**) are mercapto, amino, and hydroxyl groups. When the functional groups are ordered according to the electronegativity of the heteroatom contained in them, mercapto < amino< hydroxyl, the inhibition ability of compounds that contain these functional groups should follow an order opposite to the order of electronegativity, as a result, the ranking is **Py8** > **Py7**> **Py6**. As a result, depending on their concentrations and molecular structures, the chemicals examined in the study displayed varying degrees of inhibition. The most efficient chemical was **Py8**, which demonstrated a strong inhibitory activity of 96% at 1×10^{-2} M. These pyrimidine-containing compounds are thought to adsorb to the metal surface through a chemisorption process that forms donor–acceptor interactions between the inhibitor molecules and the unoccupied d-orbital of iron surface atoms [74].

In the study by Li et al., the inhibition of corrosion on steel surface exposed to acidic medium formed by using 0.1 M HNO_3 solution of five different pyrimidine derivatives (**Py2–Py4**, **Py9**, **Py10**) mentioned above with their formulas was investigated. As a result of the study, the effect of substituent groups on inhibition values follows the following ranking: SH > Br > NH_2 > OH > Cl, which is in good agreement with the experiments carried out in acidic media prepared with HCl and H_2SO_4 [75].

Among these compounds, **Py4** was found to be a highly effective inhibitor to prevent the corrosion of steel in 0.1 M HNO_3. **Py4** gave >95% protection while the other compounds gave about 35% protection. As anticipated, an increase in the inhibition efficiency was observed as the concentration of **Py4** increased and a significant decrease in the inhibition efficiency was observed with increasing the temperature.

When the corrosion inhibition values of the above compounds (4-amino-*N*-(pyrimidin-2-yl)benzenesulfonamide)-(**Py11**) and 4-amino-*N*-(4,6-dimethylpyrimidin-2-yl)benzenesulfonamide-(**Py12**) (Figure 9.31) were examined on steel at room

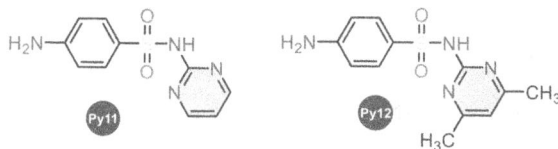

FIGURE 9.31 Chemical structures of 4-amino-*N*-(pyrimidin-2-yl)benzenesulfonamide)-(**Py11**) and 4-amino-*N*-(4,6-dimethylpyrimidin-2-yl)benzenesulfonamide-(**Py12**).

temperature with 1 M HCl and 0.5 M H$_2$SO$_4$ acids respectively, it was observed that **Py11** had more effective inhibition levels [76].

4-Phenylpyrimidine (**Py13**) is more effective than 5-phenylpyrimidine (**Py14**) (Figure 9.32) in the corrosion inhibition of steel in HCl solutions. The inhibition level decreases slightly when the temperature is increased [77].

When the corrosion inhibition values of vitamin B1, which contains a pyrimidine ring and is one of the important bioactive compounds, were examined, the inhibition efficiency on various steels in cold HCl solutions was investigated. In the experiment carried out in 0.5 M HCl, only vitamin B1 with a concentration of 10 mM had an inhibition value of 91.5% [78, 79].

In another study, following the literature data obtained from sulfur-containing pyrimidine derivatives such as 2-mercaptopyrimidine, more complex compounds containing thiol group were synthesized to inhibit the corrosion of steel in HNO$_3$ and H$_2$SO$_4$ solutions, and their inhibitory properties were investigated. The compound 6,6′-(1,4-phenylene)bis(4-(2,5-dimethylthiophen-3-yl)pyrimidine-2-thiol) (**Py15**) (Figure 9.33), a 2-mercaptopyrimidine derivative, inhibited the corrosion of metal very effectively [75].

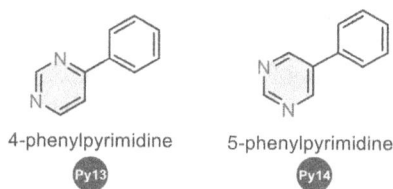

4-phenylpyrimidine **Py13**   5-phenylpyrimidine **Py14**

FIGURE 9.32 Chemical structures of phenylpyrimidine (**Py13**) is more effective than 5-phenylpyrimidine (**Py14**).

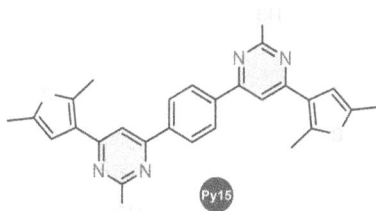

FIGURE 9.33 Chemical structures of 6,6′-(1,4-phenylene)bis(4-(2,5-dimethylthiophen-3-yl)pyrimidine-2-thiol) (**Py15**).

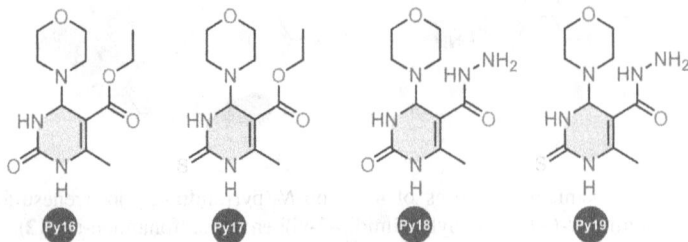

FIGURE 9.34   Chemical structures of pyrimidine derivatives **Py16–Py19**.

As examples of corrosion inhibitors with the hydrogenated form of the pyrimidine ring, the above compounds are mentioned. When the corrosion inhibition activities are ranked, they appear as **Py16 < Py17 < Py18 < Py19** (Figure 9.34). In this study, unlike the others, tetrahydropyrimidine derivatives further increased the effectiveness against corrosion of steel with increasing temperature. While compound 4 protected the steel by 89.4% in 0.5 M HCl solution at 30 °C, the protection rate was 92.4% when the temperature was increased to 60 °C. In general, the inhibition values of the four compounds were ranked as **Py16 < Py17 < Py18 < Py19**. As a result, the replacement of an oxygen atom with a sulfur atom increases the corrosion inhibitory activity of the molecule. Likewise, when the ester functional group which contains ethoxy group is replaced by hydrazine, the inhibitory activity of the molecule increases considerably [80].

9.4   CONCLUSIONS AND OUTLOOK

In particular, heterocyclic compounds containing heteroatoms such as nitrogen and sulfur are used as corrosion inhibitors on different metal surfaces, especially on MS surfaces, various steels, and non-ferrous metals (Al, Cu, Sn, Zn, and their alloys) in the presence of mineral acids. In general, electrochemical impedance, weight loss, and potentiodynamic polarization techniques have been applied in studies to demonstrate corrosion inhibition using acidic media. In corrosion studies, HCl is generally used as an acid to inhibit metal corrosion, while H_2SO_4, $HClO_4$, and H_3PO_4 are used as acids to inhibit metal corrosion relatively weakly. The inhibitors described in this chapter are likely to bind to metals by physical forces in addition to the mechanism that slows cathodic and anodic reactions. When investigating inhibitors, compounds containing sulfur atoms or bulky substituents have attracted particular attention in corrosion inhibition. These compounds adsorb more strongly on metals and prevent corrosion of metals with a stronger effect. Experiments have shown that the protective effect of inhibitors is good at cold ambient temperature, while their inhibition effect is significantly reduced at high temperatures. However, it has been observed that inhibitors generally inhibit the corrosion of metals in acidic media to a greater extent as the concentration of the inhibitor increases, that is, a higher effective inhibition property. Inhibition processes by forming synergistic corrosion inhibitors for metals exposed to different acid solutions have become more common in recent years.

In this chapter, concise information about the corrosion prevention of some molecules with pyrrole, pyrrolidine, pyridazine, and pyrimidine skeleton is given. The factors we have described above for these compounds are actually valid for all organic corrosion inhibitors. Therefore, scientists working in this field are advised to consider the above information when designing a new inhibitor.

REFERENCES

1. S.S. Gholap, European Journal of Medicinal Chemistry, 110 (2016) 13–31.
2. S. Kotha, D. Singh, Journal of Molecular Structure, 1275 (2023) 134600.
3. A. Domagala, T. Jarosz, M. Lapkowski, European Journal of Medicinal Chemistry, 100 (2015) 176–187.
4. M.-Y. Han, J.-Y. Jia, W. Wang, Tetrahedron Letters, 55 (2014) 784–794.
5. M. Pichon, B. Figadère, Tetrahedron: Asymmetry, 7 (1996) 927–964.
6. N. Jeelan Basha, S. Basavarajaiah, K. Shyamsunder, Molecular Diversity, 26 (2022) 2915–2937.
7. M.A. Quraishi, D.S. Chauhan, V.S. Saji, Journal of Molecular Liquids, 341 (2021) 117265.
8. A. Zarrouk, B. Hammouti, T. Lakhlifi, M. Traisnel, H. Vezin, F. Bentiss, Corrosion Science, 90 (2015) 572–584.
9. E.M. Kassim, I. Ibrahim, J. Jai, M. So'aib, N.A. Zamanhuri, H. Husin, M. Hashim, IOP Conference Series: Materials Science and Engineering, 358 (2018) 012045.
10. D. Gopi, E.S.M. Sherif, M. Surendiran, D. Angeline Sakila, L. Kavitha, Surface and Interface Analysis, 47 (2015) 618–625.
11. D. Gopi, K. Govindaraju, V. Collins Arun Prakash, V. Manivannan, L. Kavitha, Journal of Applied Electrochemistry, 39 (2009) 269–276.
12. C. Verma, E. Ebenso, I. Bahadur, I. Obot, M. Quraishi, Journal of Molecular Liquids, 212 (2015) 209–218.
13. A. Louroubi, A. Nayad, A. Hasnaoui, I. Hdoufane, R. Idouhli, M. Saadi, L. El Ammari, A. Abdessalam, M. Berraho, L. El Firdoussi, Journal of Molecular Structure, 1189 (2019) 240–248.
14. A. Louroubi, A. Hasnaoui, Y.A. Aicha, N. Abdallah, R. Idouhli, A. Benyaich, M.A. Ali, L. El Firdoussi, Chemical Data Collections, 32 (2021) 100662.
15. L. Guo, I.B. Obot, X. Zheng, X. Shen, Y. Qiang, S. Kaya, C. Kaya, Applied Surface Science, 406 (2017) 301–306.
16. A. Charles, K. Sivaraj, S. Thanikaikarasan, Materials Today: Proceedings, 33 (2020) 3135–3138.
17. A. Al-Mayout, A. Al-Suhybani, A. Al-Ameery, Desalination, 116 (1998) 25–33.
18. S. Abd El-Rehim, M.A. Ibrahim, K. Khaled, Journal of Applied Electrochemistry, 29 (1999) 593–599.
19. J. Bastidas, J. Polo, E. Cano, C. Torres, Journal of Materials Science, 35 (2000) 2637–2642.
20. G. Pandimuthu, S. Rameshkumar, K. Paramasivaganesh, A. Sankar, Oriental Journal of Chemistry, 37 (2021) 779–790.
21. A. Alamiery, E. Mahmoudi, T. Allami, International Journal of Corrosion and Scale Inhibition, 10 (2021) 749–765.
22. A. Gülden, A. Abdurrahman, Journal of the Turkish Chemical Society Section A: Chemistry, 7 151–154.
23. A. El Jaouhari, M. Laabd, E. Bazzaoui, A. Albourine, J. Martins, R. Wang, G. Nagy, M. Bazzaoui, Synthetic Metals, 209 (2015) 11–18.

24. F. Lallemand, D. Auguste, C. Amato, L. Hevesi, J. Delhalle, Z. Mekhalif, Electrochimica Acta, 52 (2007) 4334–4341.
25. R.A. Hameed, E. Aljuhani, A. Al-Bagawi, A. Shamroukh, M. Abdallah, The International Journal of Corrosion and Scale Inhibition, 9 (2020) 623–643.
26. H. Gadow, H. Dardeer, International Journal of Electrochemical Science, 12 (2017) 6137–6155.
27. L. Adamczyk, P.J. Kulesza, Electrochimica Acta, 56 (2011) 3649–3655.
28. M. Bagherzadeh, F. Jaberinia, Journal of Alloys and Compounds, 750 (2018) 677–686.
29. W. Qafsaoui, M. Kendig, H. Perrot, H. Takenouti, Electrochimica Acta, 87 (2013) 348–360.
30. X. Zhang, Q. Liao, K. Nie, L. Zhao, D. Yang, Z. Yue, H. Ge, Y. Li, Corrosion Science, 93 (2015) 201–210.
31. A.E. Al-Rawajfeh, E.M. Al-Shamaileh, Desalination, 206 (2007) 169–178.
32. J. Haque, M.A.J. Mazumder, M.A. Quraishi, S.A. Ali, N.A. Aljeaban, Journal of Molecular Liquids, 320 (2020) 114473.
33. M. Bouklah, A. Ouassini, B. Hammouti, A. El Idrissi, Applied Surface Science, 252 (2006) 2178–2185.
34. C. Verma, M. Quraishi, E. Ebenso, I. Obot, A. El Assyry, Journal of Molecular Liquids, 219 (2016) 647–660.
35. M. Jeeva, G.V. Prabhu, M.S. Boobalan, C.M. Rajesh, The Journal of Physical Chemistry C, 119 (2015) 22025–22043.
36. M. Jeeva, G.V. Prabhu, Research on Chemical Intermediates, 44 (2018) 425–454.
37. Y. Pan, F. Zhan, Z. Lu, Y. Lin, Z. Yang, Z. Wang, Anti-Corrosion Methods and Materials, 63 (2016) 153–159.
38. T. Karazehir, M.E. Mert, B.D. Mert, Journal of the Indian Chemical Society, 99 (2022) 100642.
39. J. Bouffard, R.F. Eaton, P. Müller, T.M. Swager, The Journal of Organic Chemistry, 72 (2007) 10166–10180.
40. H. Shirakawa, E.J. Louis, A.G. MacDiarmid, C.K. Chiang, A.J. Heeger, Journal of the Chemical Society, Chemical Communications, 16 (1977) 578–580.
41. S. Lin, Z. Liu, Y. Hu, Journal of Combinatorial Chemistry, 9 (2007) 742–744.
42. Y. Obata, Y. Senba, M. Koshika, Agricultural and Biological Chemistry, 27 (1963) 340–341.
43. S. Liu, X. Zhang, C. Ou, S. Wang, X. Yang, X. Zhou, B. Mi, D. Cao, Z. Gao, Acs Applied Materials & Interfaces, 9 (2017) 26242–26251.
44. M.E. Mashuga, L.O. Olasunkanmi, H. Lgaz, E.-S.M. Sherif, E.E. Ebenso, Journal of Molecular Liquids, 344 (2021) 117882.
45. L.O. Olasunkanmi, M.E. Mashuga, E.E. Ebenso, Surfaces and Interfaces, 12 (2018) 8–19.
46. L.O. Olasunkanmi, I.B. Obot, M.M. Kabanda, E.E. Ebenso, The Journal of Physical Chemistry C, 119 (2015) 16004–16019.
47. L.O. Olasunkanmi, M.M. Kabanda, E. Ebenso, Physica E: Low-Dimensional Systems and Nanostructures, 76 (2016) 109–126.
48. W. Luo, W. Li, J. Tan, J. Liu, B. Tan, X. Zuo, Z. Wang, X. Zhang, Journal of Molecular Liquids, 314 (2020) 113630.
49. W. Luo, Q. Lin, X. Ran, W. Li, B. Tan, A. Fu, S. Zhang, Journal of Molecular Liquids, 341 (2021) 117370.
50. M. El-Deeb, N. Abdel-Shafi, A. Shamroukh, International Journal of Electrochemical Science, 13 (2018) 5352–5369.
51. M.E. Mashuga, L.O. Olasunkanmi, E.E. Ebenso, Journal of Molecular Structure, 1136 (2017) 127–139.

52. P.B. Raja, M. Ismail, S. Ghoreishiamiri, J. Mirza, M.C. Ismail, S. Kakooei, A.A. Rahim, Chemical Engineering Communications, 203 (2016) 1145–1156.
53. F. Bentiss, M. Outirite, M. Traisnel, H. Vezin, M. Lagrenée, B. Hammouti, S. Al-Deyab, C. Jama, International Journal of Electrochemical Science, 7 (2012) 1699–1723.
54. L.O. Olasunkanmi, M.F. Sebona, E.E. Ebenso, Journal of Molecular Structure, 1149 (2017) 549–559.
55. A. Khadiri, R. Saddik, K. Bekkouche, A. Aouniti, B. Hammouti, N. Benchat, M. Bouachrine, R. Solmaz, Journal of the Taiwan Institute of Chemical Engineers, 58 (2016) 552–564.
56. A. Ghazoui, N. Bencaht, S. Al-Deyab, A. Zarrouk, B. Hammouti, M. Ramdani, M. Guenbour, International Journal of Electrochemical Science, 8 (2013) 2272.
57. A. Chetouani, B. Hammouti, A. Aouniti, N. Benchat, T. Benhadda, Progress in Organic Coatings, 45 (2002) 373–378.
58. A. Chetouani, A. Aouniti, B. Hammouti, N. Benchat, T. Benhadda, S. Kertit, Corrosion Science, 45 (2003) 1675–1684.
59. A. Ghazoui, N. Benchat, F. El-Hajjaji, M. Taleb, Z. Rais, R. Saddik, A. Elaatiaoui, B. Hammouti, Journal of Alloys and Compounds, 693 (2017) 510–517.
60. A. Nahle, F. El-Hajjaji, A. Ghazoui, N.-E. Benchat, M. Taleb, R. Saddik, A. Elaatiaoui, M. Koudad, B. Hammouti, Anti-Corrosion Methods and Materials, 65 (2018) 87–96.
61. M. Bouklah, N. Benchat, B. Hammouti, A. Aouniti, S. Kertit, Materials Letters, 60 (2006) 1901–1905.
62. A. Popova, M. Christov, S. Raicheva, E. Sokolova, Corrosion Science, 46 (2004) 1333–1350.
63. A. Zarrouk, T. Chelfi, A. Dafali, B. Hammouti, S. Al-Deyab, I. Warad, N. Benchat, M. Zertoubi, International Journal of Electrochemical Science, 5 (2010) 696–705.
64. F. Bentiss, F. Gassama, D. Barbry, L. Gengembre, H. Vezin, M. Lagrenée, M. Traisnel, Applied Surface Science, 252 (2006) 2684–2691.
65. X. Ren, J. Bai, X. Gu, H. Xu, B. Tan, S. Xu, J. Hao, F. Gao, X. Li, Journal of Industrial and Engineering Chemistry, 113 (2022) 348–359.
66. A. Fouda, M. El-Haddad, M. Ismail, A. Abd Elgyed, Journal of Bio-and Tribo-Corrosion, 5 (2019) 1–14.
67. F. El-Hajjaji, M. Messali, M.M. de Yuso, E. Rodríguez-Castellón, S. Almutairi, T.J. Bandosz, M. Algarra, Journal of Colloid and Interface Science, 541 (2019) 418–424.
68. L. Strekowski, M. Say, M. Henary, P. Ruiz, L. Manzel, D.E. Macfarlane, A.J. Bojarski, Journal of Medicinal Chemistry, 46 (2003) 1242–1249.
69. K. Barbade, K. Datar, Asian Journal of Pharmaceutical and Clinical Research, 8 (2015) 171–177.
70. Y.G. Avdeev, International Journal of Corrosion and Scale Inhibition, 7 (2018) 460–497.
71. Y.I. Kuznetsov, International Journal of Corrosion and Scale Inhibition, 4 (2015) 284–310.
72. S. Abd El-Maksoud, International Journal of Electrochemical Science, 3 (2008) 528–555.
73. K. Rasheeda, D. Vijaya, P. Krishnaprasad, S. Samshuddin, International Journal of Corrosion and Scale Inhibition, 7 (2018) 48–61.
74. H. Awad, S.A. Gawad, Anti-Corrosion Methods and Materials, 52 (2005) 328–336.
75. X. Li, S. Deng, T. Lin, X. Xie, G. Du, Corrosion Science, 118 (2017) 202–216.
76. M. El-Naggar, Corrosion Science, 49 (2007) 2226–2236.
77. X. Li, X. Xie, S. Deng, G. Du, Corrosion Science, 87 (2014) 27–39.
78. O.K. Abiola, Corrosion Science, 48 (2006) 3078–3090.
79. R. Solmaz, Corrosion Science, 81 (2014) 75–84.
80. D.M. Gurudatt, K.N.S. Mohana, European Journal of Chemistry, 5 (2014) 53–64.

10 Sulfur- and Sulfur–Nitrogen-containing Heterocycles-based Corrosion Inhibitors

Sheetal[1], Sheetal Kundu[1], Ashish Kumar Singh[1,2],
Manjeet Singh[3], Balaram Pani[4], and Sanjeeve Thakur[1]
[1]Department of Chemistry, Netaji Subhas
University of Technology, New Delhi, India
[2]Department of Chemistry, Hansraj College,
University of Delhi, New Delhi, India
[3]Department of Chemistry, School of Physical Sciences,
Mizoram University, Aizawl, Mizoram, India
[4]Department of Chemistry, Bhaskaracharya College of
Applied Sciences, University of Delhi, New Delhi, India

10.1 INTRODUCTION

Corrosion has been a trouble for metals ever since they were first discovered. Metallic materials have considerable mechanical and physical properties and are inexpensive. Various industrial processes, including oil production, refining, and other related industries, involve metals [1, 2]. However, aggressive acidic environments like HNO_3, H_2SO_4, and HCl during these industrial processes degrade metals leading to a sole phenomenon known as corrosion. Additionally, corrosion is a global problem resulting in economic and cultural losses in this period of rapid industrialisation. The current research aside from various other ways used to combat corrosion is more focused on the synthesis and usage of corrosion inhibitors [3]. There are a number of key benefits associated with corrosion inhibitors, including their affordability, effectiveness, and ease of supply. And, apart from these, they may attach to surfaces and thus are recognised for their capacity to suppress corrosion [4]. Three main aspects influence the inhibitor's adsorption onto metallic surfaces: type and charge over metal, characteristics of the electrolytic medium, and inhibitor's structure. The commercial use of corrosion inhibitors concerning inorganic moieties such as phosphates, silicates, and chromates are well documented in the literature. However, owing to the inconsistency of these inorganic inhibitors with the environment, more environment-friendly substitutes (organic moieties) are being used at their place. And, owing to their association with economy and effectivity, organic moieties are currently

DOI: 10.1201/9781003377016-10

regarded as the most well-executed strategy in the domain of fighting corrosion. Availability of heteroatoms (S, N, P, O) and polar functional assortments in organic corrosion inhibitors are accountable for their corrosion-combating potential. Rationally, every industrial area uses these organic molecules, though the usage is not without certain drawbacks. Low solubility, particularly in polar electrolytic solutions, is one of the major issues with organic molecules, particularly those with aromatic rings and non-polar hydrocarbon parts which offer restricted solubility owing to their hydrophobic nature, which has an impact on how effectively they inhibit corrosion [4]. With the aim of improvement among the connections of the organic corrosion inhibitors and the metal's topology, researchers are currently synthesising organic corrosion inhibitors with attached polar substituents and aromatic rings.

The adsorption of inhibitor moieties onto metal's surfaces with unoccupied d-orbitals, such as iron, is motivated by the presence of heteroatoms along with the existing electron cloud present in the inhibitor molecules [5, 6]. Extensive research has been published that demonstrate how the heteroatoms sulfur and nitrogen distinguish increased corrosion prevention efficacy in phenyl rings [7, 8]. This chapter provides a summary report of heterocyclic moieties having key atoms as sulfur (e.g. thiophene, thiane, etc.) and sulfur–nitrogen (e.g. thiazole, thiadiazole) which are being utilised to effectively suppress corrosion on various metals and their alloys. The main research used for experimental confirmation uses polarisation, electrochemical, and gravimetric techniques. SEM-EDS spectroscopy and other spectroscopic methods offer characteristics of the metal's morphology and the modification they have undergone during the corrosion-combating process, adding to the body of knowledge already known.

10.1.1 CORROSION AND ECONOMY

In celebration of Corrosion Awareness Day in 2019, the World Corrosion Organization (WCO) noted an estimated yearly expense of corrosion across the globe, i.e. \$2.5 trillion, or 3–4% of the Gross Domestic Income (GDI) of developed nations [9]. Additionally, for the United States solely, the cost of corrosion reported each year overreaches US\$500 annually [10]. There are many ways to reduce corrosion, which causes a huge economic loss. Due to the extreme corrosion in the locations where such metals have been employed, the significant financial estimates are not unexpected [11, 12]. The government as well as industrial setups should implement efficient analysis strategies, use tested sciences and technological procedures, and enact the necessary corrosion-combating solutions with an aim to reduce annual losses caused by corrosion [13]. The expense of corrosion is split into "avoidable costs" and "unavoidable costs" in several of these publications. "Avoidable costs" are expenses that can be reduced by using currently known corrosion control techniques, whereas "unavoidable costs" are expenses needed for new corrosion technologies and new materials.

10.1.2 Mechanism of Corrosion

Corrosion is an irreversible electrochemical degradation of metals when they are exposed to surrounding environment. Being an electrochemical phenomenon, corrosion is comprised of two principal reactions occurring over two prime sites: anodic (Equations 10.1–10.3) and cathodic (Equations 10.4–10.8):

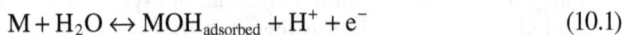

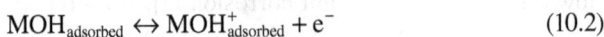

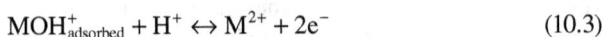

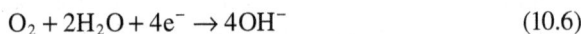

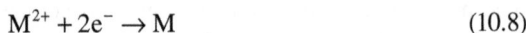

$$M + H_2O \leftrightarrow MOH_{adsorbed} + H^+ + e^-$$ (10.1)

$$MOH_{adsorbed} \leftrightarrow MOH^+_{adsorbed} + e^-$$ (10.2)

$$MOH^+_{adsorbed} + H^+ \leftrightarrow M^{2+} + 2e^-$$ (10.3)

$$O_2 + 4e^- + 2H_2O \rightarrow 4OH^-$$ (10.4)

$$H^+ + e^- \rightarrow \frac{1}{2}H_2$$ (10.5)

$$O_2 + 2H_2O + 4e^- \rightarrow 4OH^-$$ (10.6)

$$M^{3+} + e^- \rightarrow M^{2+}$$ (10.7)

$$M^{2+} + 2e^- \rightarrow M$$ (10.8)

It was remarked that corrosion phenomenon arises as a result of formation of various anodes as well as cathodes over metal surface out of which the one having more aeration belong to anode, whereas the comparatively less aerated were cathodes [14].

Reactions listed as Equations 10.1–10.3 were presented as anodic reactions, whereas (Equations 10.4–10.8) are referred to as cathodic reactions, depending on the availability of medium.

Consequently, the corresponding products of anodic and cathodic reactions react to yield corrosion (Equations 10.9–10.11) [15] :

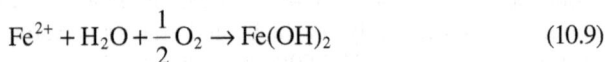

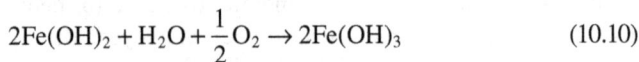

$$Fe^{2+} + H_2O + \frac{1}{2}O_2 \rightarrow Fe(OH)_2$$ (10.9)

$$2Fe(OH)_2 + H_2O + \frac{1}{2}O_2 \rightarrow 2Fe(OH)_3$$ (10.10)

$$2Fe(OH)_3 \rightarrow Fe_2O_3 (rust) + 3H_2O$$ (10.11)

10.2 HETEROCYCLIC COMPOUNDS AS CORROSION INHIBITORS

Corrosion inhibitors, known as chemical substances, when applied to corrosive conditions retard or slow down the rate of corrosion of metal exposed to surroundings. Research is continuously being focused on search and synthesis of such organic moieties which can efficiently retard corrosion kinetics. Heterocyclic compounds are formed when in carbocyclic ring compounds; carbon is being replaced with any other heteroatom like N, O, S, P, etc. [16] The cyclic rings have been seen more common to have five or six atoms in their structures compared to other having three or four atoms and are being seen to exhibit aromaticity as well as non-aromatic nature [17]. These compounds find their active usage almost in every industrial sector. Phytochemicals associated with various plant parts also constitute heterocyclic compounds. Additionally, certain amino acids like tryptophan, indole, etc. have heterocyclic structures as their building blocks [18–20]. Conventionally, carbocyclic ring when replaced with any other atoms apart from carbon is being referred to as heterocyclic compounds but the more common used are N, O, P, S, etc. The corrosion retarding property in these corrosion inhibitors is ascribed to the availability of heteroatoms, namely S, N, etc. apart from aromatic rings. These organic moieties protect metals from aggressive acid solutions by forming a protective layer onto them [21]. This adsorption is governed by two prime approaches; physical adsorption and chemical adsorption. Physisorption involves physical interactions in the form of van der Walls forces, whereas chemical adsorption involves emergence of chemical interactions among metal and the inhibitor surface [22]. This adsorption is regulated by a number of factors: electrolytic medium, inhibitors, and substituents attached along with immersion time and temperature [23]. And, this mode of interaction can be assessed utilising a number of theoretical approaches like DFT, molecular dynamics, etc.

10.2.1 SULFUR-CONTAINING HETEROCYCLES-BASED CORROSION INHIBITORS FOR FERROUS ALLOYS

These heterocyclic compounds have sulfur as heteroatom in their cyclic structure. Several classes of such compounds have been reported and evidenced in literature, e.g. thiophene, thiirane, thiolane, thiane, etc.

Literature supports ample evidence favouring these S-heterocyclic moieties as corrosion combaters for iron, mild steel, Q235 steel, and its certain other alloys. Daoud et al. explored the corrosion-mitigating behaviour of sulfur heterocycles–based Schiff's base, namely (E,E)-N,N'-dibenzo[b,d]thiene-2,8-diylbis[1-(thiophen-2-yl) methanimine] (SB) and its corresponding amine dibenzo[b,d]thiophene (DBTDA) over mild steel in the environment of 1 M HCl solution. Weight loss (WL) studies and EIS findings were incorporated for investigation of the same. The respective inhibition efficiency observed in both the cases were 92.95% for SB and 89.10% for DBTDA. The adsorption phenomenon for both of these over mild steel surface followed Langmuir adsorption isotherm, whereas PDP findings evaluated these mitigators decelerating both cathodic and anodic reactions corresponding to mixed behaviour [24].

Nyijime et al. reported the synthesis as well as anti-corrosive study of three sulfur-based heterocyclic compounds: THIO and its derivatives, including 2-TCA and

2-TCAH over iron metal. Quantum calculations were utilised in this regard and an inhibition efficiency order was seen to be THIO > 2-TCA > 2-TCAH. Fukui indices revealed that the interaction of S-heterocyclic compounds basically took place via the attached heteroatom in the moiety. THIO was seen to possess a comparative planer structure which, on the other hand, enhanced its inhibition efficiency to maximum of 3 [25].

Reeja et al. explored the anti-corrosive ability of S-based Schiff's base moiety, namely T2CDACH over carbon steel in the environment of 1 M HCl and the inhibitor outreached the corrosion-preventing efficacy above 95% owing to the presence of sulfur atoms. The adsorption of T2CDACH over carbon steel adheres to Langmuir adsorption isotherm and the same has been seen retarding the cathodic reactions taking place at carbon steel [26]. Ismail et al. investigated the corrosion-preventing ability of two derivatives of bithienyl fluorobenzamidine, MA-1615 and MA-1740, over carbon steel in the environment of 1 M HCl, respectively. EIS findings revealed the order of inhibition efficacy demarcated for both the inhibitors as MA-1740 (96.9%) > MA-1615 (95.6%), respectively. Mixed inhibition involving the slowing down of kinetics of anodic and cathodic reactions was seen in both the cases. Further, UV-visible spectroscopy in this regard affirmed the emergence of complexes between Fe^{2+} released from metal and inhibitor ions, thereby preventing corrosion [27].

Fouda et al. reported the synthesis and corrosion-combating ability testing of two naphthylbithiophene derivatives, namely MA-1341, and MA-1340, over carbon steel in the environment of 1 M HCl solution. Both the compounds were seen adsorbing chemically over carbon steel surface and adhered to Langmuir adsorption isothermal curves. The latter one possesses chlorine atom in the structure which basically hindered the electron donation to metallic vacant d-orbitals and decreased the inhibition efficiency for corrosion [28]. Issaadi et al. analysed the corrosion-protecting ability of two heterocyclic Schiff's bases, L_1 and L_2, over mild steel in the environment of 1 M HCl solution. Synthesis of above discussed compounds were performed using 3-carboxaldehydethiophene and amine relative to it. Polarisation studies indicated the mixed behaviour of both the inhibitors, whereas WL studies confirmed the enhancement in inhibition efficiency with concentration. Spontaneous adsorption of L_1 and L_2 over mild steel was observed following Langmuir adsorption isotherm [29]. Singh et al. remarked the effect of corrosion inhibition of various complexes of copper, zinc, cobalt, and nickle Hatbh over mild steel in 1 M HCl solution. Basically, here the structural aspects were confirmed by single-crystal XRD technique and electrochemical studies revealed that all complexes except the one having Cu^{2+} experienced a rise in charge transfer resistance along with a retardation in capacitance of double layer. This might be attributed to the fact that involvement of ligand, i.e. Hatbh, prevented the diminution of Cu^{2+} to Cu. Further, the redox behaviour of Cu complex was analysed with the involvement of cyclic voltammetry [30]. A summary report of sulfur-containing heterocyclic compounds having anti-corrosive abilities for ferrous alloys has been presented in Table 10.1.

TABLE 10.1
Sulfur-containing Heterocyclic Corrosion Inhibitors for Ferrous Alloys

Structure with Abbreviation	Metal/ Electrolytic Medium	Techniques Used	Inhibition Efficiency	References
 SB	Mild steel/ 1 M HCl	PDP, SEM, WL	92.95%	[24]
 DBTDA	Mild steel/ 1 M HCl	PDP, SEM, WL	89.10%	[24]
 THIO	Iron	Quantum chemical calculations	–	[25]
 2-TCA	Iron	Quantum chemical calculations	–	[25]
 2-TCAH	Iron	Quantum chemical calculations	–	[25]
 T2CDACH	Carbon steel/1 M HCl	PDP, EIS	95%	[26]

(Continued)

TABLE 10.1   (*Continued*)

Sulfur-containing Heterocyclic Corrosion Inhibitors for Ferrous Alloys

Structure with Abbreviation	Metal/ Electrolytic Medium	Techniques Used	Inhibition Efficiency	References
MA1615	Carbon steel/1 M HCl	WL, theoretical studies, UV-visible spectroscopy	95.6%	[27]
MA-1740	Carbon steel/1 M HCl	WL, theoretical studies, UV-visible spectroscopy	96.9%	[27]
MA-1341	Carbon steel/1 M HCl	WL, theoretical studies, PDP	93.55%	[28]
MA-1340	Carbon steel/1 M HCl	WL, theoretical studies, PDP	91.05%	[28]
L1	Mild steel/1 M HCl	WL, EIS, PDP	91.6%	[29]
L2	Mild steel/1 M HCl	EIS, WL	91.04%	[29]
[Co(atbh)2]	Mild steel/ 1 M HCl	EIS, cyclic voltammetry	99.4%	[30]

(*Continued*)

TABLE 10.1   (Continued)
Sulfur-containing Heterocyclic Corrosion Inhibitors for Ferrous Alloys

Structure with Abbreviation	Metal/ Electrolytic Medium	Techniques Used	Inhibition Efficiency	References
[Ni(atbh)2(H₂O)₂]	Mild steel/ 1 M HCl	EIS, cyclic voltammetry	96.4%	[30]
[Cu(atbh)₂]	Mild steel/ 1 M HCl	EIS, cyclic voltammetry	–	[30]
[Zn(atbh)₂]	Mild steel/1 M HCl	EIS, cyclic voltammetry	94.7%	[30]

10.2.2   SULFUR-CONTAINING HETEROCYCLES-BASED CORROSION INHIBITORS FOR NON-FERROUS ALLOYS

The present-day industries make inevitable usage of metals like copper, aluminium, and their alloys with altered compositions. However, copper is likely to get corroded in the presence of acidic and strong alkali media, especially when in proximity of oxygen and oxidants. In the environment of weak acidic or alkaline media, the copper surface is passivated by formation of an oxide layer. Therefore, inhibitors are used to combat the issue of corrosion as they are the most effective and economically viable solution. Organic heterocyclic compounds containing sulfur as their prime heteroatom are among one of the most promising inhibitors used for such metals [31].

Feng et al. primarily synthesised and reported the testing of two dithiane-based self-assembled monolayers: BD and HD over surface of copper in 0.5 M H₂SO₄. Basically, here it has been investigated that both the self-assembled layers were formed at copper surface at an optimum temperature and time period of 298 K and 18 hours, respectively. Self-assembled layer of BD was seen to exhibit more protection compared to the later derivative, i.e. 99% corrosion inhibition performance.

Potentiodynamic studies illustrated both the self-assembled moieties as mixed type of inhibitors [32].

Fouda et al. explored the corrosion-mitigating properties of thiophene derivatives: T, TC, and TE over copper surface in the environment of 2 M HNO_3 medium. WL studies affirmed a rise in inhibition efficacy with enhanced concentration of thiophene derivatives and negative valuate of enthalpy of adsorption confirmed the whole process of adsorption of these inhibitors over copper as exothermic one. Potentiodynamic studies confirmed all three inhibitors of mixed nature [33]. Allal et al. analysed the corrosion-preventing ability of varied thiophene derivatives: TPHCAL, ACTPH, TPHCAC, CLTPH, and TPH, respectively, for aluminium. Theoretical studies have been involved as an investigatory technique for the same. All the derivatives except thiophene adsorbed over aluminium surface using chemical interactions utilising the oxygen of carbonyl group present in their structures. On the other hand, comparatively weak forces of interactions were observed in case of TPH and CLTPH and this was further correlated with their weak inhibition efficiency of corrosion. The sequence for inhibition efficacy as remarked via varied quantum calculations followed: ACTPH = TPHCAL > TPHCAC > CLTPH >TPH [34].

Arrousse et al. investigated a thiophene compound OXM for its corrosion-mitigating ability for 2024-T3 aluminium alloy in 1 M HCl solution. Increase in concentration of OXM was seen to correlate with the rise in inhibition efficiency demarcated by the same. Polarisation study revealed the mixed-type nature of OXM. Barrier over aluminium surface was seen in the form of a lean protective layer. Adsorption of OXM over aluminium obeyed Temkin adsorption isotherm along with Flory–Huggins' isotherms [35]. A summary report of sulfur-containing heterocyclic compounds having anti-corrosive abilities for non-ferrous alloys has been presented in Table 10.2.

10.2.3   SULFUR–NITROGEN-CONTAINING HETEROCYCLES-BASED CORROSION INHIBITORS FOR FERROUS ALLOYS

Anti-corrosive studies of innumerable organic moieties having nitrogen as well as sulfur in their structures have been evidenced in literature for iron as well as its alloys like mild steel and carbon steel. Chugh et al. explored *N*-(benzo[*d*]thiazole-2-yl)-1-phenylethan-1-imines, namely BTPEI, BTCPEI, BTTEI, and BTPIA, for their corrosion-mitigating potential towards mild steel in the environment of 1 M HCl. BTPIA, BTPEI, BTCPEI, and BTTEI were reported to have a maximum inhibition efficacy of 90%, 85%, 79%, and 88%, respectively, for mild steel surface. Methods involving electrochemical and gravimetric analyses were incorporated for the exploration of the same. The inhibitors' mixed-type nature was unveiled using potentiodynamic polarisation [36]. Aoufir et al. in this regard have investigated the corrosion-preventing ability of T3 over mild steel surface in acidic medium of 1.0 M HCl. Adsorption-based evolution of a shielding layer via adsorption over mild steel surface was indicated via electrochemical findings and confirmed by the surface analysis technique, namely SEM–EDX. Other techniques employed in this study involved

TABLE 10.2

Sulfur-containing Heterocycles-based Corrosion Inhibitors for Non-Ferrous Alloys

Structure and Corrosion Inhibitor	Metal/Electrolyte	Techniques Used	IE%	Reference
BD	Copper/1 M H_2SO_4	WL, EIS, PDP	99.2%	[32]
HD	Copper/1 M H_2SO_4	WL, EIS, PDP	98.5%	[32]
T	Copper/ 2 M HNO_3	EIS, weight loss studies	82.6%	[33]
TC	Copper/ 2 M HNO_3	EIS, weight loss studies	74.2%	[33]
TE	Copper/ 2 M HNO_3	EIS, Weight loss studies	76.3%	[33]
TPHCAL	Aluminium	Quantum calculations	–	[34]
ACTPH	Aluminium	Quantum calculations	–	[34]

(Continued)

TABLE 10.2 *(Continued)*
Sulfur-containing Heterocycles-based Corrosion Inhibitors for Non-Ferrous Alloys

Structure and Corrosion Inhibitor	Metal/Electrolyte	Techniques Used	IE%	Reference
TPHCAC	Aluminium	Quantum calculations	–	[34]
CLTPH	Aluminium	Quantum calculations	–	[34]
TPH	Aluminium	Quantum calculations	–	[34]
OXM	Aluminium/1 M HCl	EIS	94%	[35]

molecular dynamics simulations and density fluctuation theory for gaining an insight into molecular interactions between the metal surface and the inhibitor [37].

Chugh et al. reported the findings with regard to evaluation of anti-corrosive potential of PMTTA, PATT, PMTA, and PTA over mild steel in 1 M HCl. These were found to have an inhibition efficiency of 92.7%, 91.1%, 87.8%, and 80.5%, respectively, towards mild steel in 1 M HCl with an optimum concentration of 175 ppm [38]. Later Salman et al. gave an account of the corrosion-inhibiting ability of DBTA over mild steel in 1 M HCl solution. The synthesised moiety was seen to exhibit the greatest inhibition efficacy of 91% at a concentration of 0.5 mM. The 1,3,4-thiadiazole ring performed a pivotal role in boosting the effectiveness of inhibition exhibited by DBTA. SEM affirmed the development of a shielding layer of DBTA over mild steel surface [39]. Sheetal et al. investigated the corrosion-combating properties of three aryl-substituted benzothiazoles, CBTA, TBTA, and MBTA, for mild steel in the environment of 1 M HCl. The maximum inhibition efficacies for CBTA, TBTA, and MBTA were 72.9%, 81.4%, and 96.4%, respectively. These inhibitors were remarked to follow Langmuir adsorption isotherm. PDP findings exhibited that these inhibitors have mixed kind of nature, slowing cathodic as well as anodic reactions [4]. Singh et al. further investigated BTDS for its anti-corrosive property towards mild steel in 0.5 M H_2SO_4 and 1 M HCl. It showcased a better corrosion combating efficacy in 0.5 M H_2SO_4 collated to that depicted in 1 M HCl. BTDS was also remarked to be a mixed kind of inhibitor [40].

Zhang et al. studied DSTA for its anti-corrosive potential and found it to have IE_{max} 99.4% towards mild steel in CO_2-saturated oilfield formation water. Thiadiazols are found to have quite a higher inhibition efficiency and are of ecofriendly nature; also, their efficiency makes them a promising range of corrosion inhibitors [41]. The substituted dithiazolidine family has some effective inhibitors for corrosion towards mild steel in varied acidic media. Quraishi et al. reported findings for evaluation of DPID, PCID, PTID, and PAID for their corrosion inhibition potential for mild steel in 1 M H_2SO_4 and 1 M HCl and at varied concentrations of 500 ppm, their corrosion inhibition efficiency was reported in respective order: PAID > PTID > PCID > DPID. The reason for the above order may be the existence of methoxy group ($-OCH_3$) in PAID molecule as it worked by increasing the electron density at nitrogen atoms which is enabled by resonance effect. The strong adsorption for these inhibitor molecules on metal's surface is ensured by the above-mentioned fact. A relatively negative and lower estimate of Gibbs free energy of adsorption was reported at varied temperature range of 30–50 °C, which explained inhibitor's spontaneous adsorption on surface of mild steel in the 1 M HCl and 1 M H_2SO_4 environment. These were remarked to obey Langmuir's adsorption isotherm. Also, potentiodynamic polarisation data supported mixed-type nature of above inhibitors in both the acidic media. Anodic nature was observed in 1 M H_2SO_4 medium for all the inhibitors [42]. Gong et al. explored the corrosion-mitigating ability of 2-amino-4-(4-methoxyphenyl)-thiazole (MPT) and 2-amino-(4-bromophenyl)-thiazole (BPT) towards corrosion of mild steel in acidic media of 0.5 M H_2SO_4 for a temperature variance of 30–60 °C. The effective corrosion inhibition for MS was successfully explained by various electrochemical measurements and surface analytical techniques in acidic media. Their adsorption on metal's surface meet Langmuir's adsorption isotherm [43]. Singh et al. analysed the corrosion-combating potential of bis-thiadiazole derivatives (BTDs), namely APT, APT-2, APT-4, and PAT, towards mild steel in the environment of 1 M HCl. These inhibitors adhere to Langmuir's adsorption isotherm and were remarked to exhibit a mixed kind of nature. The shield formed on the surface of ATP-4 was found to be less stable as compared to those of ATP, ATP-2, and PAT when studied for varied temperature ranges [44]. Abed et al. examined the inhibition efficiency of a recently discovered organic inhibitor named ANTD towards mild steel in the acidic media of 1 M HCl by using WL technique. ANTD's inhibition efficiency for mild steel increased by its concentration as well as increasing immersion time and decreased by increasing temperatures [45]. Attou et al. studied a novel 1,3,4-thiadiazole derivative 5-AMT for its corrosion-combating potential over mild steel. The various experimental studies implemented PDP, WL, EIS, and LPR techniques to examine the corrosion mitigation performance of 5-AMT. On the basis of experimental findings, an inhibition efficiency of 98% was attained utilising 5×10^{-4} M of studied compound at temperature 303 K after 1 hour of immersion. Also, potentiodynamic study focused that the examined 5-AMT is a mixed-type inhibitor. Furthermore, the adsorption of tested compound's molecules fits Langmuir isotherm model [46].

Ma et al. reported the findings for S3 and it was found to have IE_{max} of 99.2% towards carbon steel in 1 M HCl. PDP studies confirmed that inhibition efficacy increased, and I_{corr} decreased as the concentration of inhibitor was further increased. Also, no remarkable variation in the corrosion potential value was observed, which means that the above-discussed inhibitors have mixed kind of nature. The data for the PDP and WL study have been verified by the EIS investigation. It followed Langmuir's adsorption isotherm, and spontaneous process of inhibition was observed via the phenomena of chemisorption and physisorption with chemisorption as the paramount phenomenon [47]. Farag et al. analysed the corrosion-inhibiting ability of an ionic liquid named AMPMHMC towards carbon steel in the HCl solution of 1 M. The study disclosed that the efficacy of inhibition augmented with a rise in the concentration of the inhibitor, i.e. 10–40 ppm, and it outreached the upper limit of inhibition, i.e. 91.4% at 40 ppm concentration. Using the isotherms model, it was affirmed to obey Langmuir adsorption isotherm [48]. Khaled et al. investigated the corrosion-preventing ability of TB over iron in 1 M HNO_3 solution. Theoretical studies provided that corrosion-preventing ability enhanced with rise in values of energy of HOMO orbitals and high ΔN values [49]. The above-mentioned heterocyclic corrosion inhibitors containing both sulfur and nitrogen atoms hold good corrosion inhibition efficiency towards iron alloys, and this has been presented in Table 10.3.

10.2.4 SULFUR–NITROGEN-CONTAINING HETEROCYCLES-BASED CORROSION INHIBITORS FOR NON-FERROUS ALLOYS

Certain heterocyclic corrosion inhibitors incorporating the existence of both sulfur and nitrogen have enormously been utilised for corrosion prevention of varied metals and alloys like copper, aluminium, etc. Ting et al. reported the synthesis of an imidazole- and pyrazole-based corrosion inhibitor, namely LMS, which is among the efficient corrosion inhibitors for copper. It showed an IE_{max} of 93% towards copper in H_2SO_4 medium of 0.5 M. Langmuir adsorption isotherm was seen followed during adsorption of LMS over copper surface. Both the inhibitors were seen adsorbing over metal using physical as well as chemical modes of interactions [50]. Further, Farahati et al. investigated about PyTA (1 mM); it was found to have IE_{max} of 94% towards copper in 1 M HCl. Further, regarding investigation about the shielding layer and the explanation of the technical aspects of corrosion inhibitors, a variety of electrochemical methods and surface examination techniques were utilised [51]. Farahati et al. inspected about the corrosion inhibition ability of ATP on the copper surface. ATP adhered to Langmuir adsorption isotherm [51]. Qafsaoui et al. investigated the anti-corrosive potential of DMTD over bronze in 30 g L^{-1} of NaCl solution. It has been reported that high concentrations of DMTD basically blocked the oxide evolution which, on the other hand, formed a surface coverage over bronze surface. Formation of surface layer over bronze was seen to be more stable when the substrate was dissolved in 10 mM DMTD [52]. Raviprabha et al. basically investigated the

TABLE 10.3

Nitrogen–Sulfur-containing Heterocyclic Corrosion Inhibitors for Ferrous and Non-Ferrous Alloys

Structure and Corrosion Inhibitor	Metal/Electrolyte	Techniques Used	Inhibition Efficiency	Reference
BTPIA	Mild steel/1 M HCl	PDP, EDS, DFT	90%	[36]
BTPEI	Mild steel/1 M HCl	WL, FTIR	85%	[36]
BTCPEI	Mild steel/1 M HCl	PDP, EIS	79%	[36]
BTTEI	Mild steel/1 M HCl	EIS, SEM-EDX	88%	[36]
T3	Mild steel/1 M HCl	SEM-EDX, DFT	92.1%	[37]
PMTTA	Mild steel/1 M HCl	PDP, EIS	92.7%	[38]
PATT	Mild steel/1 M HCl	XRD, XPS	91.1%	[38]

(Continued)

TABLE 10.3 *(Continued)*
Nitrogen–Sulfur-containing Heterocyclic Corrosion Inhibitors for Ferrous and Non-Ferrous Alloys

Structure and Corrosion Inhibitor	Metal/Electrolyte	Techniques Used	Inhibition Efficiency	Reference
PMTA	Mild steel/1 M HCl	WL, XRD	87.8%	[38]
PTA	Mild steel/1 M HCl	WL, FTIR	80.5%	[38]
5-ATT	Mild steel/1 M HCl	EIS, XPS	92.0%	[54]
DBTA	Mild steel/1 M HCl	DFT, SEM	91.0%	[39]
CBTA	Mild steel/1 M HCl	PDP, XPS	72.9%	[4]
TBTA	Mild steel/1 M HCl	PDP, DFT	81.4%	[4]

(Continued)

TABLE 10.3 *(Continued)*
Nitrogen–Sulfur-containing Heterocyclic Corrosion Inhibitors for Ferrous and Non-Ferrous Alloys

Structure and Corrosion Inhibitor	Metal/Electrolyte	Techniques Used	Inhibition Efficiency	Reference
MBTA	Mild steel/1 M HCl	MD, AFM	96.4%	[4]
BTDS	Mild steel/0.5 M H₂SO₄	WL, SEM	99.29%	[40]
DSTA	Mild steel/CO₂-saturated brine solution	WL, MD	99.4%	[41]
DPID	Mild steel/1 N H₂SO₄	PDP	99.0%	[42]
PAID	Mild steel/1 N HCl	WL, EIS	99.5%	[42]
PCID	Mild steel/1 N H₂SO₄	WL, PDP	99.4%	[42]
PTID	Mild steel/ 1 N H₂SO₄	PDP	99.6%	[42]
MPT	Mild steel/0.5 M H₂SO₄	UV-vis, PDP, EIS	95.0%	[43]

(Continued)

TABLE 10.3 *(Continued)*

Nitrogen–Sulfur-containing Heterocyclic Corrosion Inhibitors for Ferrous and Non-Ferrous Alloys

Structure and Corrosion Inhibitor	Metal/Electrolyte	Techniques Used	Inhibition Efficiency	Reference
BPT	Mild steel/0.5 M H₂SO₄	WL, MD	95.4%	[43]
APT	Mild steel/1 M HCl	WL, AFM	95%	[44]
APT-2	Mild steel/1 M HCl	WL, PDP	96.8%	[44]
APT-4	Mild steel/1 M HCl	WL, EIS	92.7%	[44]
PAT	Mild steel/1 M HCl	EIS, AFM	98.6%	[44]
ANTD	Mild steel/1 M HCl	WL, PDP	82.0%	[45]
5-AMT	Mild steel/1 M HCl	LPR, EIS, XPS	98.0%	[46]

(Continued)

TABLE 10.3 *(Continued)*

Nitrogen–Sulfur-containing Heterocyclic Corrosion Inhibitors for Ferrous and Non-Ferrous Alloys

Structure and Corrosion Inhibitor	Metal/Electrolyte	Techniques Used	Inhibition Efficiency	Reference
S3	Carbon steel/1 M HCl	WL, PDP, EIS	99.2%	[47]
AMPMHMC	Carbon steel/1 M HCl	EDS, XRD, EIS	91.4%	[48]
MMPT	304L steel/3.0 M HCl	WL, EIS	90.5%	[55]
MAPT	304L steel/3.0 M HCl	WL, EIS	88.5%	[55]
MBT	API 5L X42 steel/ 3.5 wt.% NaCl	EIS, SEM	90.0%	[56]
TB	Iron/1.0 M HNO_3	WL, EIS, PDP, DFT	96.8%	[53]

(Continued)

TABLE 10.3 *(Continued)*

Nitrogen–Sulfur-containing Heterocyclic Corrosion Inhibitors for Ferrous and Non-Ferrous Alloys

Structure and Corrosion Inhibitor	Metal/Electrolyte	Techniques Used	Inhibition Efficiency	Reference
EMTC	AA6061 alloy/0.5 M HCl	WL, EIS, PDP	92.7%	[53]
CH₃-BOZ	Aluminium/0.1 M HCl	WL, PDP	73.7%	[57]
LMS	Copper/0.5 M H₂SO₄	EIS, XPS	93.0%	[50]
PyTA	Copper/1 M HCl	EIS, SEM, DFT	94.0%	[51]
ATP	Copper/1 M HCl	EIS, AFM, SEM	90.9%	[58]
DMTD	Bronze/NaCl	EIS, Raman, SEM	99.0%	[52]

corrosion-combating potential of EMTC for AA6061 alloy of aluminium in 0.05 M HCl solution. Polarisation techniques remarked that EMTC has a mixed kind of nature, retarding both cathodic and anodic reactions. Further, the adsorption of EMTC obeyed Langmuir adsorption isotherm over aluminium, and numeric figures for Gibbs free energy supported the whole process involving chemical interactions [53]. The above-mentioned heterocyclic corrosion inhibitors containing both sulfur and nitrogen that hold appreciable corrosion inhibition efficiency towards iron as well as other metals and associated alloys along with their maximum corrosion inhibition efficiency (IE_{max}) and techniques used are presented in Table 10.3.

10.3   CONCLUSION AND FUTURE PERSPECTIVES

Corrosion being a comprehensive approach prevails in every industrial sector from technology to engineering. It is almost next to impossible to completely vanish corrosion but it can be slowed down using different measures. This chapter has elaborated and discussed the most common approach being utilised to retard corrosion, i.e. use of heterocyclic corrosion inhibitors. In accordance with the recent trends in terms of economy and efficacy, organic moieties having heteroatoms are considered a valuable approach these days.

Certain literature involving investigation of anti-corrosive behaviour of sulfur- and nitrogen–sulfur-containing heterocyclic moieties has been incorporated here. Various methods of evaluating their anti-corrosive potential involving gravimetric and electrochemical techniques have been elaborated here. Also, the quantum chemical calculations along with surface studies have been utilised to get an insight into molecular interactions as well as comparative roughness of metallic counterparts has been discussed here. All heterocyclic corrosion inhibitors incorporated so far in literature for corrosion retardation do not add to environment; so, there is a need to develop such organic moieties which not only act as potent corrosion combaters but are also environment-friendly.

ABBREVIATIONS

ACTPH	2-acetylthiophene
AMPMHMC	3-((4-amino-2-methylpyrimidin-5-yl) methyl)-5-(2-hydroxyethyl)-4-methylthiazol-3-ium chloride
5-AMT	2-amino-5-(2-methoxyphenyl)-1,3,5-thiadiazole
ANTD	2-amino-5-(4-nitrophenyl)-1,3,4-thiadiazole
ATP	4-(2-aminothiazole-4-yl)phenol
BD	2-butyl-1,3-dithiane
BTCPEI	N-(benzo[d]thiazol-2-yl)-1-(3-chlorophenyl)ethan-1-imine
BTDS	2,20 benzothiazolyl disulfide
BTPEI	[N-(benzo[d]thiazol-2-yl)-1-phenylethan-1-imine
BTPIA	N-(benzo[d]thiazol-2-ylimino)ethyl)aniline
BTTE	N-(benzo[d]thiazol-2-yl)-1-(m-tolyl)ethan-1-imine
CBTA	6-(4-chlorophenyl)benzo[d]thiazol-2-amine
CLTPH	2-chlorothiophene

DBDTA	dibenzo[*b*,*d*]thiophene
DBTA	4-dimethylamino-benzylidene)-[1, 3, 4]thiadiazol-2-yl-amine
DMTD	2,5-dimercapto-1,3,4-thiadiazole
DPID	3,5-diphenyl-imino-1,2,4-dithiazolidine
DSTA	5,5′-disulfanediyl-bis-1,3,4-thiadiazol-2-amine
EMTC	ethyl-2-amino-4-methyl-1,3-thiazole-5-carboxylate
Hatbh	2-acetylthiophene benzoylhydrazone
HD	2-heptyl-1,3-dithiane
L_1	4,4′-bis(3-carboxaldehyde thiophene)diphenyl diimino ether
L_2	4,4′-bis(3-carboxaldehyde thiophene)diphenyl diimino ethane
LMS	levamisole
MA-1340	5′-(4-chlorophenyl)-2,2′-bithiophene-5-carboxamidine hydrochloride salt
MA-1341	5′-(naphthalen-2-yl)-[2,2′-bithiophene]-5-carboxamidine hydrochloride salt
MA-1615	4-([2,2′:5′,2″-terthiophen]-5-yl)-2-fluorobenzamidine hydrochloride salt
MA-1740	5′-(4-amidino-3-fluorophenyl)-[2,2′-bithiophene]-5-carboxamidine dihydrochloride salt
MBTA	6-(4-methoxyphenyl)benzo[*d*]thiazol-2-amine
OXM	(*E*)-thiophene-2-carbaldehyde oxime
PAID	3-phenyl-imiino-5-anisidylimino-1,2,4-dithiazolidine
PATT	5,5′-((1,4- phenylenebis(methanylylidene))bis(azanylylidene)) bis(1,3,4- thiadiazole-2-thiol)
PCID	3-phenyl-imino-5-chlorophenylimino-1,2,4-dithiazolidine
PMTA	*N*,*N*′-(1,4-phenylenebis-(methanylylidene)) bis(5-methyl-1,3,4-thiadizol-2-amine)
PMTTA	*N*,*N*′-(1,4-phenylenebis(methanylylidene)) bis-(5-methylthio)-1,3,4-thiadiazol-2-amine)
PTA	*N*,*N*′-(1,4-phenylenebis(methanylylidene)) bis-(1,3,4-thiadiazol-2-amine)
PTID	3-phenyl-imino-5-tolyl-imino-1,2,4-dithiazolidine
PyTA	4-(pyridin-3-yl)thiazol-2-amine
SB	(*E*,*E*)-*N*,*N*′-dibenzo[*b*,*d*]thiene-2,8-diylbis[1-(thiophen-2-yl) methanimine]
S3	6-((1,3,4-thiadiazol-2-yl)thio)-N_2-(3-(dimethylamino) propyl)-N_4-octyl-1,3,5-trizine-2,4-diamine
T	thiophene
TB	2-(4-thiazolyl)benzimidazole
TBTA	6-(*p*-tolyl)benzo[*d*]thiazol-2-amine
TC	2-thiophene carboxylic acid
2-TCA	2-thiophene carboxylic acid
2-TCAH	2-thiophene carboxylic acid hydrazide
TE	2-thienyl ethanol
THIO	thiophene
TPH	thiophene

TPHCAC	2-thiophenecarboxylic acid
TPHCAL	thiophenecarboxaldehyde
T2CDACH	$(13E)$-N_1,N_2-bis((thiophene-2-yl)methylene) cyclohexane-1,2-diamine
T3	methyl(E)-2-(((E)-4-chlorobenzylidene) hydrazono)-5-(4-chlorophenyl)-2,3-dihydrothiazole-4-carboxylate

REFERENCES

1. Guerraf, A. El; Titi, A.; Cherrak, K.; Mechbal, N.; Azzouzi, M. El; Touzani, R.; Hammouti, B.; Lgaz, H. The Synergistic Effect of Chloride Ion and 1,5-Diaminonaphthalene on the Corrosion Inhibition of Mild Steel in 0.5 M Sulfuric Acid: Experimental and Theoretical Insights. *Surf. Interfaces* 2018, *13*, 168–177. https://doi.org/10.1016/J. SURFIN.2018.09.004.

2. Al Hamzi, A. H.; Zarrok, H.; Zarrouk, A.; Salghi, R.; Hammouti, B.; Al-Deyab, S.; Bouachrine, S.; Amine, M.; Guenoun, A. The Role of Acridin-9(10H)-one in the Inhibition of Carbon Steel Corrosion: Thermodynamic, Electrochemical and DFT Studies. *Int. J. Electrochem. Sci.* 2013, *8*, 2586–2605.

3. Salghi, R.; Ben Hmamou, D.; Benali, O.; Jodeh, S.; Warad, I.; Hamed, O.; Ebenso, E. E.; Oukacha, A.; Tahrouch, S.; Hammouti, B. Study of the Corrosion Inhibition Effect of Pistachio Essential Oils in 0.5 M H_2SO_4. *Int. J. Electrochem. Sci.* 2015, *10*, 8403–8411.

4. Sheetal; Sengupta, S.; Singh, M.; Thakur, S.; Pani, B.; Banerjee, P.; Kaya, S.; Singh, A. K. An Insight about the Interaction of Aryl Benzothiazoles with Mild Steel Surface in Aqueous HCl Solution. *J. Mol. Liq.* 2022, *354*, 118890. https://doi.org/10.1016/J. MOLLIQ.2022.118890.

5. Daoud, D.; Douadi, T.; Issaadi, S.; Chafaa, S. Adsorption and Corrosion Inhibition of New Synthesized Thiophene Schiff Base on Mild Steel X52 in HCl and H_2SO_4 Solutions. *Corros. Sci.* 2014, *79*, 50–58. https://doi.org/10.1016/J.CORSCI.2013.10.025.

6. Bammou, L.; Belkhaouda, M.; Salghi, R.; Benali, O.; Zarrouk, A.; Zarrok, H.; Hammouti, B. Corrosion Inhibition of Steel in Sulfuric Acidic Solution by the *Chenopodium ambrosioides* Extracts. *J. Assoc. Arab Univ. Basic Appl. Sci.* 2014, *16*, 83–90. https://doi.org/10.1016/J.JAUBAS.2013.11.001.

7. El-Hajjaji Sidi Mohamed Ben, F.; Zerga Sidi Mohamed Ben, B.; Sfaira, M.; Taleb, M.; El-Hajjaji, F.; Belkhmima, R.; Zerga, B.; Ebn Touhami, M.; Hammouti, B.; Al-Deyab, S.; Ebenso, E. Temperature Performance of a Thione Quinoxaline Compound as Mild Steel Corrosion Inhibitor in Hydrochloric Acid Medium. *Artic. Int. J. Electrochem. Sci.* 2014, *9*, 4721–4731.

8. Lim, H. L. Assessing Level and Effectiveness of Corrosion Education in the UAE. *Int. J. Corros.* 2012, *2012*. https://doi.org/10.1155/2012/785701.

9. NAE Website – NACE International's IMPACT Study Breaks New Ground in Corrosion Management Research and Practice. https://www.nae.edu/19579/19582/21020/15526 6/155346/NACE-Internationals-IMPACT-Study-Breaks-New-Ground-in-Corrosion-Management-Research-and-Practice (accessed Dec 22, 2022).

10. Khan, M. A. A.; Irfan, O. M.; Djavanroodi, F.; Asad, M. Development of Sustainable Inhibitors for Corrosion Control. *Sustainability* 2022, *14* (15), 9502. https://doi. org/10.3390/SU14159502.

11. Koch, G. H.; Brongers, M. P. H.; Thompson, N. G.; Virmani, Y. P.; Payer, J. H.; CC Technologies, I.; International, N. Corrosion Cost and Preventive Strategies in the United States [Final Report]. 2002.

12. Bhaskaran, R.; Palaniswamy, N.; Rengaswamy, N. S.; Jayachandran, M. A Review of Differing Approaches Used to Estimate the Cost of Corrosion (and Their Relevance in the Development of Modern Corrosion Prevention and Control Strategies). *Anti-Corrosion Methods Mater.* 2005, *52* (1), 29–41.

13. Roberge, P. R. Corrosion Engineering Principles and Practice, McGraw-Hill Professional, 2008.

14. Aslam, J.; Aslam, R.; Verma, C. Imidazole and Its Derivatives as Corrosion Inhibitors. *Org. Corros. Inhib.* 2021, 95–122. https://doi.org/10.1002/9781119794516.CH6.

15. Hadfield, R.A. The Corrosion of Iron and Steel. *Proc. R. Soc. London. Ser. A* 1922, *101* (713), 472–486. https://doi.org/10.1098/RSPA.1922.0059.

16. Cook, M. J.; Katritzky, A. R.; Linda, P. Aromaticity of Heterocycles. *Adv. Heterocycl. Chem.* 1974, *17* (C), 255–356. https://doi.org/10.1016/S0065-2725(08)60910-1.

17. Heterocyclic Organic Corrosion Inhibitors. *Heterocycl. Org. Corros. Inhib.* (2020). ISBN-978-0-12-818558-2; https://doi.org/10.1016/C2018-0-04237-1.

18. Rogo, M. O. Modeling and Synthesis of Antiplasmodial Benzoxazines from Natural Products of Kenya. 2016.

19. Joule, J. A.; John, A.; Mills, K. Heterocyclic Chemistry at a Glance, Wiley–Blackwell, 2012.

20. Eicher, T.; Hauptmann, S.; Speicher, A. The Chemistry of Heterocycles : Structure, Reactions, Synthesis and Applications, Wiley-VCH Verlag GmbH, 2012, 632.

21. Bousskri, A.; Anejjar, A.; Messali, M.; Salghi, R.; Benali, O.; Karzazi, Y.; Jodeh, S.; Zougagh, M.; Ebenso, E. E.; Hammouti, B. Corrosion Inhibition of Carbon Steel in Aggressive Acidic Media with 1-(2-(4-Chlorophenyl)-2-oxoethyl)pyridazinium Bromide. *J. Mol. Liq.* 2015, *211*, 1000–1008. https://doi.org/10.1016/J.MOLLIQ.2015.08.038.

22. Arellanes-Lozada, P.; Olivares-Xometl, O.; Likhanova, N. V.; Lijanova, I. V.; Vargas-García, J. R.; Hernández-Ramírez, R. E. Adsorption and Performance of Ammonium-based Ionic Liquids as Corrosion Inhibitors of Steel. *J. Mol. Liq.* 2018, *265*, 151–163. https://doi.org/10.1016/J.MOLLIQ.2018.04.153.

23. Verma, C.; Olasunkanmi, L. O.; Ebenso, E. E.; Quraishi, M. A. Adsorption Characteristics of Green 5-Arylaminomethylene Pyrimidine-2,4,6-Triones on Mild Steel Surface in Acidic Medium: Experimental and Computational Approach. *Results Phys.* 2018, *8*, 657–670. https://doi.org/10.1016/J.RINP.2018.01.008.

24. Daoud, D.; Douadi, T.; Hamani, H.; Chafaa, S.; Al-Noaimi, M. Corrosion Inhibition of Mild Steel by Two New *S*-Heterocyclic Compounds in 1 M HCl: Experimental and Computational Study. *Corros. Sci.* 2015, *94*, 21–37. https://doi.org/10.1016/J.CORSCI.2015.01.025.

25. Nyijime, T. A.; Muhammad Ayuba, A.; Chahul, H. F. Computational Studies on Side Chain Effects of Five Membered Ring Sulphur Heterocycles on the Corrosion Inhibition of Iron Metal. *J. New Technol. Mater.* 2022, *12*, 80–88.

26. Reeja, J.; Thomas, K. J.; Ragi, K.; Binsi, M. P. Screening of Two Sulphur-containing Schiff's Bases Corrosion Inhibition Properties on CS: Gravimetric, Electrochemical and Quantum Chemical Studies. *Port. Electrochim. Acta* 2022, *40*, 223–241. https://doi.org/10.4152/pea.2022400401.

27. Ismail, M. A.; Shaban, M. M.; Abdel-Latif, E.; Abdelhamed, F. H.; Migahed, M. A.; El-Haddad, M. N.; Abousalem, A. S. Novel Cationic Aryl Bithiophene/Terthiophene Derivatives as Corrosion Inhibitors by Chemical, Electrochemical and Surface Investigations. *Sci. Reports 2022 121* 2022, *12* (1), 1–16. https://doi.org/10.1038/s41598-022-06863-8.

28. Fouda, A. E. A. S.; Etaiw, S. E. H.; Ismail, M. A.; Abd El-Aziz, D. M.; Eladl, M. M. Novel Naphthybithiophene Derivatives as Corrosion Inhibitors for Carbon Steel in 1 M HCl: Electrochemical, Surface Characterization and Computational Approaches. *J. Mol. Liq.* 2022, *367*, 120394. https://doi.org/10.1016/J.MOLLIQ.2022.120394.

29. Issaadi, S.; Douadi, T.; Zouaoui, A.; Chafaa, S.; Khan, M. A.; Bouet, G. Novel Thiophene Symmetrical Schiff Base Compounds as Corrosion Inhibitor for Mild Steel in Acidic Media. *Corros. Sci.* 2011, *53* (4), 1484–1488. https://doi.org/10.1016/J.CORSCI.2011.01.022.

30. Singh, V. P.; Singh, P.; Singh, A. K. Synthesis, Structural and Corrosion Inhibition Studies on Cobalt(II), Nickel(II), Copper(II) and Zinc(II) Complexes with 2-Acetylthiophene Benzoylhydrazone. *Inorganica Chim. Acta* 2011, *379* (1), 56–63. https://doi.org/10.1016/J.ICA.2011.09.037.

31. Zaid, A.K.; Merdas, S. M.; Hayal, M. Y. Heterocyclic Compounds Containing N Atoms as Corrosion Inhibitors: A Review. *J. Biosci. Appl. Res.* 2021, *7* (2), 93–103. https://doi.org/10.21608/JBAAR.2021.178505.

32. Feng, L.; Zheng, S.; Zhu, H.; Ma, X.; Hu, Z. Detection of Corrosion Inhibition by Dithiane Self-assembled Monolayers (SAMs) on Copper. *J. Taiwan Inst. Chem. Eng.* 2023, *142*, 104610. https://doi.org/10.1016/J.JTICE.2022.104610.

33. Fouda, A. S.; Wahed, H. A. A. Corrosion Inhibition of Copper in HNO_3 Solution Using Thiophene and Its Derivatives. *Arab. J. Chem.* 2016, *9*, S91–S99. https://doi.org/10.1016/J.ARABJC.2011.02.014.

34. Allal, H.; Belhocine, Y.; Zouaoui, E. Computational Study of Some Thiophene Derivatives as Aluminium Corrosion Inhibitors. *J. Mol. Liq.* 2018, *265*, 668–678. https://doi.org/10.1016/J.MOLLIQ.2018.05.099.

35. Arrousse, N.; Fernine, Y.; Al-Zaqri, N.; Boshaala, A.; Ech-Chihbi, E.; Salim, R.; El Hajjaji, F.; Alami, A.; Touhami, M. E.; Taleb, M. Thiophene Derivatives as Corrosion Inhibitors for 2024-T3 Aluminum Alloy in Hydrochloric Acid Medium. *RSC Adv.* 2022, *12* (17), 10321–10335. https://doi.org/10.1039/D2RA00185C.

36. Chugh, B.; Singh, A. K.; Thakur, S.; Pani, B.; Pandey, A. K.; Lgaz, H.; Chung, I. M.; Ebenso, E. E. An Exploration about the Interaction of Mild Steel with Hydrochloric Acid in the Presence of *N*-(Benzo[*d*]Thiazole-2-yl)-1-phenylethan-1-imines. *J. Phys. Chem. C.* 2019, *123* (37), 22897–22917.

37. El Aoufir, Y.; Zehra, S.; Lgaz, H.; Chaouiki, A.; Serrar, H.; Kaya, S.; Salghi, R.; AbdelRaheem, S. K.; Boukhris, S.; Guenbour, A.; Chung, I. M. Evaluation of Inhibitive and Adsorption Behavior of Thiazole-4-carboxylates on Mild Steel Corrosion in HCl. *Colloids Surf. A Physicochem. Eng. Asp.* 2020, *606*, 125351. https://doi.org/10.1016/J.COLSURFA.2020.125351.

38. Chugh, B.; Singh, A. K.; Thakur, S.; Pani, B.; Lgaz, H.; Chung, I. M.; Jha, R.; Ebenso, E. E. Comparative Investigation of Corrosion-mitigating Behavior of Thiadiazole-derived Bis-Schiff Bases for Mild Steel in Acid Medium: Experimental, Theoretical, and Surface Study. *ACS Omega.* 2020, *5* (23), 13503–13520.

39. Salman, T. A.; Zinad, D. S.; Jaber, S. H.; Al-Ghezi, M.; Mahal, A.; Takriff, M. S.; Al-Amiery, A. A. Effect of 1,3,4-Thiadiazole Scaffold on the Corrosion Inhibition of Mild Steel in Acidic Medium: An Experimental and Computational Study. *J. Bio- Tribo-Corrosion.* 2019, *5* (2), 1–11. https://doi.org/10.1007/S40735-019-0243-7/FIGURES/9.

40. Singh, A. K.; Quraishi, M. A. Effect of 2,2′ Benzothiazolyl Disulfide on the Corrosion of Mild Steel in Acid Media. *Corros. Sci.* 2009, *51* (11), 2752–2760. https://doi.org/10.1016/J.CORSCI.2009.07.011.

41. Zhang, Q. H.; Hou, B. S.; Xu, N.; Liu, H. F.; Zhang, G. A. Two Novel Thiadiazole Derivatives as Highly Efficient Inhibitors for the Corrosion of Mild Steel in the CO_2-saturated Oilfield Produced Water. *J. Taiwan Inst. Chem. Eng.* 2019, *96*, 588–598. https://doi.org/10.1016/J.JTICE.2018.11.022.

42. Quraishi, M. A.; Sardar, R. Dithiazolidines: A New Class of Heterocyclic Inhibitors for Prevention of Mild Steel Corrosion in Hydrochloric Acid Solution. *Corrosion.* 2002, *58* (2), 103–107. https://doi.org/10.5006/1.3277308.

43. Gong, W.; Yin, X.; Liu, Y.; Chen, Y.; Yang, W. 2-Amino-4-(4-methoxyphenyl)-thiazole as a Novel Corrosion Inhibitor for Mild Steel in Acidic Medium. *Prog. Org. Coatings.* 2019, *126*, 150–161. https://doi.org/10.1016/J.PORGCOAT.2018.10.001.

44. Singh, A. K.; Quraishi, M. A. The Effect of Some Bis-thiadiazole Derivatives on the Corrosion of Mild Steel in Hydrochloric Acid. *Corros. Sci.* 2010, *52* (4), 1373–1385. https://doi.org/10.1016/J.CORSCI.2010.01.007.

45. Abed, T. K.; Al-Azawi, K. F.; Jaber, S. H.; Al-Baghdadi, S. B.; Al-Amiery, A. A.; Abed, T. K.; Al-Azawi, K. F.; Jaber, H.; Al-Baghdadi, S. B.; Al-Amiery, A. A. Experimental Study of 2-Amino-5-(4-nitrophenyl)-1,3,4-thiadiazole for MS in HCl Solution. *Univ. Technol.* 2019, *37* (2), 214–218. https://doi.org/10.30684/ETJ.37.2C.3.

46. Attou, A.; Tourabi, M.; Benikdes, A.; Benali, O.; Ouici, H. B.; Benhiba, F.; Zarrouk, A.; Jama, C.; Bentiss, F. Experimental Studies and Computational Exploration on the 2-Amino-5-(2-methoxyphenyl)-1,3,4-thiadiazole as Novel Corrosion Inhibitor for Mild Steel in Acidic Environment. *Colloids Surf. A Physicochem. Eng. Asp.* 2020, *604*, 125320. https://doi.org/10.1016/J.COLSURFA.2020.125320.

47. Ma, X.; Wang, J.; Yu, S.; Chen, X.; Li, J.; Zhu, H.; Hu, Z. Synthesis, Experimental and Theoretical Studies of Triazine Derivatives with Surface Activity as Effective Corrosion Inhibitors for Medium Carbon Steel in Acid Medium. *J. Mol. Liq.* 2020, *315*, 113711. https://doi.org/10.1016/J.MOLLIQ.2020.113711.

48. Farag, A. A.; Migahed, M. A.; Badr, E. A. Thiazole Ionic Liquid as Corrosion Inhibitor of Steel in 1 M HCl Solution: Gravimetrical, Electrochemical, and Theoretical Studies. *J. Bio- Tribo-Corrosion.* 2019, *5* (3), 1–12. https://doi.org/10.1007/S40735-019-0246-4/FIGURES/11.

49. Khaled, K. F. Studies of Iron Corrosion Inhibition Using Chemical, Electrochemical and Computer Simulation Techniques. *Electrochim. Acta.* 2010, *55* (22), 6523–6532. https://doi.org/10.1016/J.ELECTACTA.2010.06.027.

50. Yan, T.; Zhang, S.; Feng, L.; Qiang, Y.; Lu, L.; Fu, D.; Wen, Y.; Chen, J.; Li, W.; Tan, B. Investigation of Imidazole Derivatives as Corrosion Inhibitors of Copper in Sulfuric Acid: Combination of Experimental and Theoretical Researches. *J. Taiwan Inst. Chem. Eng.* 2020, *106*, 118–129. https://doi.org/10.1016/J.JTICE.2019.10.014.

51. Farahati, R.; Behzadi, H.; Mousavi-Khoshdel, S. M.; Ghaffarinejad, A. Evaluation of Corrosion Inhibition of 4-(Pyridin-3-yl) Thiazol-2-amine for Copper in HCl by Experimental and Theoretical Studies. *J. Mol. Struct.* 2020, *1205*, 127658. https://doi.org/10.1016/J.MOLSTRUC.2019.127658.

52. Qafsaoui, W.; Et Taouil, A.; Kendig, M. W.; Heintz, O.; Cachet, H.; Joiret, S.; Takenouti, H. Corrosion Protection of Bronze Using 2,5-Dimercapto-1,3,4-thiadiazole as Organic Inhibitor: Spectroscopic and Electrochemical Investigations. *J. Appl. Electrochem.* 2019, *49* (8), 823–837. https://doi.org/10.1007/S10800-019-01329-8/FIGURES/13.

53. Raviprabha, K.; Bhat, R. S. Inhibition Effects of Ethyl-2-amino-4-methyl-1,3-thiazole-5-carboxylate on the Corrosion of AA6061 Alloy in Hydrochloric Acid Media. *J. Fail. Anal. Prev.* 2019, *19* (5), 1464–1474. https://doi.org/10.1007/S11668-019-00744-5/FIGURES/14.

54. Ouici, H.; Tourabi, M.; Benali, O.; Selles, C.; Jama, C.; Zarrouk, A.; Bentiss, F. Adsorption and Corrosion Inhibition Properties of 5-Amino 1,3,4-Thiadiazole-2-thiol on the Mild Steel in Hydrochloric Acid Medium: Thermodynamic, Surface and Electrochemical Studies. *J. Electroanal. Chem.* 2017, *803*, 125–134. https://doi.org/10.1016/J.JELECHEM.2017.09.018.

55. Verma, C.; Verma, D. K.; Ebenso, E. E.; Quraishi, M. A. Sulfur and Phosphorus Heteroatom-containing Compounds as Corrosion Inhibitors: An Overview. *Heteroat. Chem.* 2018, *29* (4), e21437. https://doi.org/10.1002/HC.21437.

56. Kartsonakis, I. A.; Stamatogianni, P.; Karaxi, E. K.; Charitidis, C. A. Comparative Study on the Corrosion Inhibitive Effect of 2-Mecraptobenzothiazole and Na$_2$HPO$_4$ on Industrial Conveying API 5L X42 Pipeline Steel. *Appl. Sci.* 2019, *10* (1), 290. https://doi.org/10.3390/APP10010290.

57. Goni, L. K. M. O.; Jafar Mazumder, M. A.; Quraishi, M. A.; Mizanur Rahman, M. Bioinspired Heterocyclic Compounds as Corrosion Inhibitors: A Comprehensive Review. *Chem. – An Asian J.* 2021, *16* (11), 1324–1364. https://doi.org/10.1002/ASIA.202100201.

58. Farahati, R.; Ghaffarinejad, A.; Mousavi-Khoshdel, S. M.; Rezania, J.; Behzadi, H.; Shockravi, A. Synthesis and Potential Applications of Some Thiazoles as Corrosion Inhibitor of Copper in 1 M HCl: Experimental and Theoretical Studies. *Prog. Org. Coatings.* 2019, *132*, 417–428. https://doi.org/10.1016/J.PORGCOAT.2019.04.005.

11 Nitrogen, Sulfur, and Nitrogen-based Heterocycles Corrosion Inhibitors

Humira Assad[1], Suresh Kumar[2], and Ashish Kumar[3]
[1]Department of Chemistry, School of Chemical Engineering and Physical Sciences, Lovely Professional University, Punjab, India
[2]Department of Chemistry, Chaudhary Devi Lal University, Haryana, India
[3]NCE, Department of Science and Technology, Government of Bihar, Bihar, India

11.1 INTRODUCTION

Metal corrosion is a problem in many industries, including mining, chemical industries, petrochemical engineering, material purification, and industrial cleaning. Metallic corrosion in a severe environment can lead to financial losses and safety and health concerns [1, 2]. External weathering is one of the main reasons for pipeline transportation malfunction for natural gases, chemicals, and petroleum, which can result in damage worth US$2.5 trillion, or 3.4% of the world GDP (2013) [3–5]. Therefore, in the realm of corrosion, researching ways to stop metal corrosion is a crucial duty that calls for scientists to concentrate on their research. There are many ways to slow down and halt metal corrosion, including creating materials, choosing appropriate materials for each atmosphere, covering metals to safeguard them, using electrochemical approaches, and more [6–8]. Regrettably, most are pricy, harmful to the environment, and inert. The creation of nontoxic, affordable, environment-friendly, and biodegradable compounds as inhibitors has become a focus of research due to the harmful chemicals and the current rise in environmental consciousness [9]. Therefore, corrosion inhibitors (CIs) help prevent metal corrosion. These compounds are unique because they are the best approach to preventing an aggressive medium from harming a metal surface. These goods are reasonably priced, so this strategy is simple and low-cost.

Over the past 50 years, numerous studies have generated specialized goods or product combinations suitable for various corrosion systems (metal–corrosive media). It is necessary to comprehend their fundamental functioning statistics, constraints of use, and toxicities to use them with sufficient assurance. Each case of corrosion continues to be a unique challenge [10]. Inhibitors additionally perform their functions by adhering to and forming a coating on the metal surface. They do this by

DOI: 10.1201/9781003377016-11

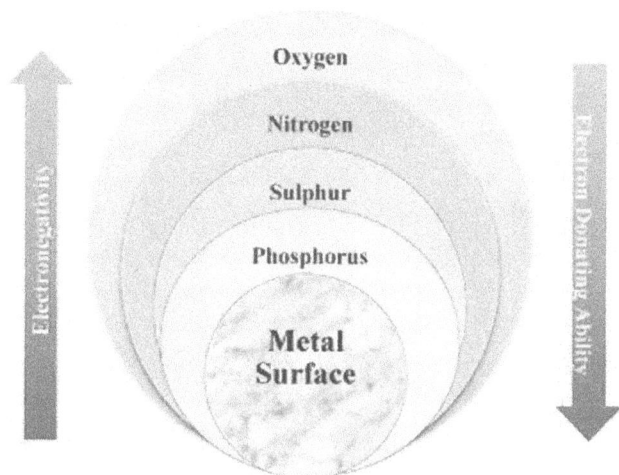

FIGURE 11.1 Schematic representation of heteroatoms as per electronegativity and electron-donating ability.

polarizing the anode or cathode more, restricting the flow of ions onto the surface of the metal, and boosting the impedance of the metallic substrate [11], which reduces corrosion. Steel is tested for its ability to resist corrosion in acid solutions by inorganic chemicals containing sulfur or nitrogen atoms.

Additionally, benzene rings in the inhibitor moiety play a significant role since they heighten the inhibitor's electrostatic contact with the metal surface. It improves metal inhibition's long-term ability as a result. Therefore, attractive candidates that should be considered while looking for potential CIs include organic molecules with HA and benzene rings. Organic molecules with unsaturated bonds and/or HAs like O, N, P, and S (Figure 11.1) are frequently effective CIs in acidic media [12–14]. As a result, various research has focused on heterocycles with inhibitory characteristics [15–17].

However, it is important to consider the possibility of an advantageous interaction between the "d" metal orbitals and the "π" ē's of the double bond. Additionally, when the ratio of double linkages in the chemical composition rises, the compounds' capacity to prevent growth likewise reaches its peak. Corrosion is reduced and the anodic/cathodic reactions are halted with improved corrosion inhibition effectiveness. Adsorption is heavily influenced by organic molecules' structural and chemical properties, including their size, electron density distribution, and interaction with electron orbital activity [18]. As a result, molecules can bind in four different ways (Figure 11.2) to the surface of a metal as follows:

- Via forming electrostatic connections between the metals charged surface and the inhibitor's charged surface;
- By interacting with the lone pair(s) in the inhibitor molecule's structure;
- By interacting with the "π"-electrons; or
- By combining the three previously mentioned methods [19, 20].

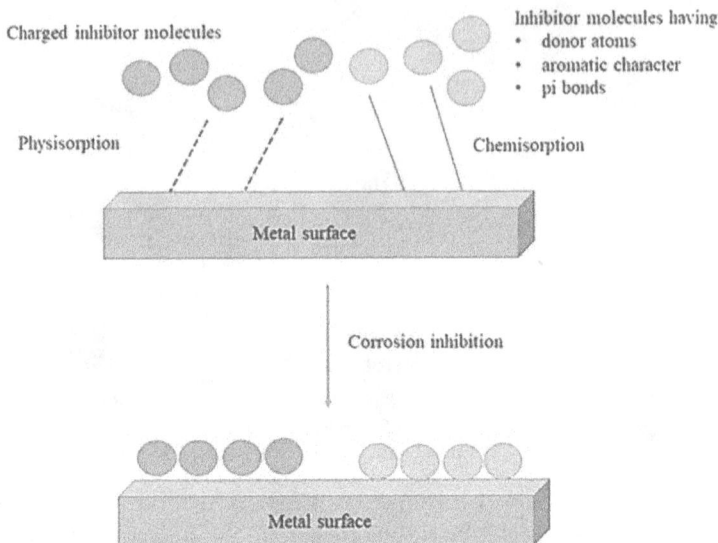

FIGURE 11.2   Corrosion protection mechanism of corrosion inhibitor [21].

Consequently, heterocyclic compounds are defined as organic compounds whose structures consist one or more rings of atoms with at minimum one HA (O, N, P, or S) being an element other than C [22–24]. As a result, heterocyclic compounds are a crucial and efficient class of CIs due to their cost-effectiveness, environment-friendliness, and superior inhibitory efficiency when combined with the properties of an inhibitor. But unlike the five- or six-membered heterocycles, the three- or four-membered heterocycles are significantly less stable due to ring strain. Hence they do not comprise the most significant families of heterocyclic CIs. Two or more N atoms are present in the most significant five-membered heterocycles (imidazoles, pyrazoles, azoles, etc.). Some of them, such as oxazoles, thiazoles, etc., also include O, S, and N atoms. Benzimidazoles, benzoxazoles, indoles, and other fused-ring-based heterocycles are also significant CIs. The most important class of HCIs is made up of six-membered heterocycles since they experience the minor ring strain. Interestingly, heterocycles with nitrogen act better as CIs in hydrochloric acid, but heterocycles with sulfur are favored for preventing corrosion in sulfuric acid [25]. Even better, though, are heterocycles with sulfur and nitrogen atoms [26]. This chapter will explore the potential for O- and S–N-based heterocyclic compounds to protect metal in various corrosive conditions.

11.2   HETEROCYCLES-BASED CORROSION INHIBITORS FOR DIFFERENT METALS AND THEIR ALLOYS

11.2.1   IRON

Iron is a high-tensile, reasonably priced metal. As a result, it is frequently used to produce machinery, cars, medical equipment, massive ship hulls, machine parts, and construction materials [27]. Iron combined with other elements offers sophisticated mechanical properties suitable for various uses and applications. The main problem, however, is that it has low corrosion impedance, particularly in hostile situations.

Acidic solutions are frequently used in industrial processes, including descaling, acid cleaning, stewing, and diggings in petroleum production, making these Fe or MS containers or frameworks more susceptible to corrosion under these circumstances [28]. As a result, steel corrosion is a severe problem that wastes resources, shortens the lifespan of equipment, and is bad for the environment [29]. Numerous studies have been conducted on corrosion prevention and corrosion rate reduction. Studies on Fe and MS corrosion in low-pH settings have taken much time [30]. The adsorption behavior and the relationship between the inhibitor compounds and their adsorption are listed in the literature on organic CIs [31, 32]. According to findings, the electronic and structural properties of the inhibitor molecule, such as aromaticity, steric factors, functional groups, electron density, and p-orbital behavior, play a significant role in adsorption [33]. In manufacturing operations, acid solutions are frequently utilized. Under these circumstances, iron corrosion and its prevention are dynamic phase issues.

Utilizing organic inhibitors to reduce MS deterioration in low-pH environments is typically very cost-effective. Numerous types of organic inhibitors have investigated steel inhibition in acidic solutions thoroughly over the past few years.

11.2.2 S–N-Heterocycles-based Corrosion Inhibitors

In contrast to those containing solely sulfur or nitrogen atoms, heterocycle-containing CIs with both S and N atoms in their configuration had excellent preventive actions [34]. In this regard, a number of N- and S-comprising azole molecules (Figure 11.3), such as thiazole and thiadiazole derivatives, etc., have been demonstrated to be effective CIs toward a variety of metal substrates, as shown in Figure 11.4, in a broad range of acidic conditions.

According to the current corrosion literature, thiadiazole-based compounds receive more focus than thiazole-based ones. This trend in attention is based on the idea that adding more HA (N atoms) to such heterocyclic compounds can increase their adsorption on the metallic outer layer and, as a result, increase their inhibition efficiency. In addition to the 1,3-thiazole ring's strong propensity to connect with the metallic outer layer when inhibited, it has also recently been reported that adding additional substituents to the ring increases the effectiveness of the inhibition. According to this theory, numerous 1,3-thiazole-centered derivatives are produced using various fabrication reaction techniques. The findings conclude that these novel compounds have a strong potential to slow down the dissolution of diverse metallic substrates.

FIGURE 11.3   Structure of thiazole.

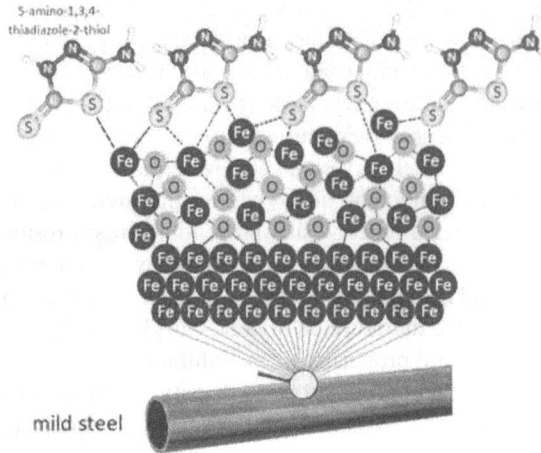

FIGURE 11.4   Schematic representation of the mild steel corrosion protection by thiazole derivative [35].

Using EIS (electrochemical impedance spectroscopy), weight loss (WL) techniques, and other techniques, Abdallah et al. demonstrate that the newly created bis-aminothiazole molecules had an inhibitory effect on the oxidation of C-steel in a 0.5 M sulfuric acid condition [36]. The outcomes show the sequence of the CI's inhibitory efficacy at numerous doses as assessed by polarization measurements: o-bis-MeAT > o-bis-AT > p-bis-MeAT > p-bis-AT, as shown in Figure 11.5.

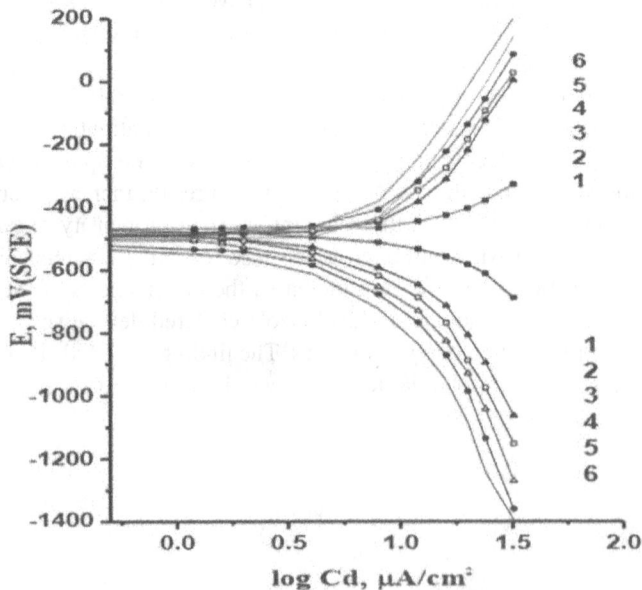

FIGURE 11.5   C-steel polarization curves at different concentrations in a 0.5 M sulfuric acid condition [36].

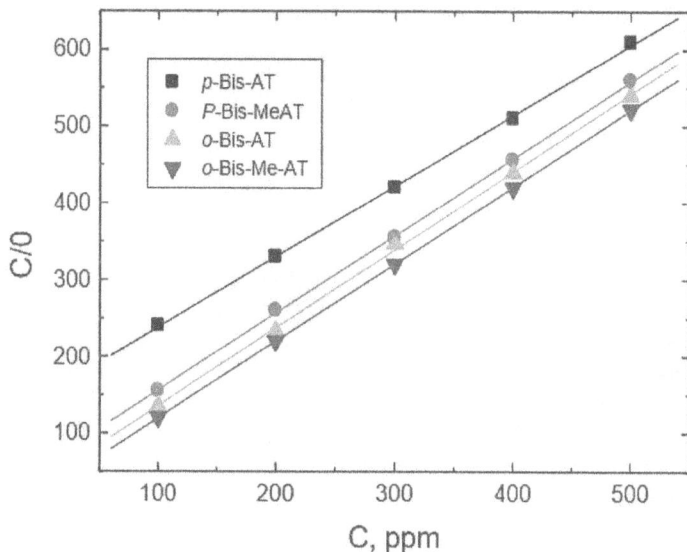

FIGURE 11.6   Langmuir adsorption isotherm [36].

Moreover, the development of an insoluble compound that adheres to the steel surface follows the Langmuir isotherm, as illustrated in Figure 11.6, which was attributed to the inhibitory effect.

Two monosubstituted 1,3-thiazole derivatives used to preserve the X65 steel alloy, primarily utilized in pipelines for natural gas transmission, recently exhibited similar behavior. X65 steel ten dissolutions were more effectively controlled by the ethenone-substituted 1,3-thiazole derivative than by the isobutyl one, with observed prevention efficiencies at 0.005 M being approximately 90% and 70% for C-3 and C-4, respectively (Figure 11.7) [37].

The inhibition execution of 1,3-thiazole-centered molecules is enhanced not only by lateral functional groups, which encompass additional ē- contributing centers (such as functional groups, aromatic, and azole rings), but also by enhancing their ē-contributing competency through connection with a C_6H_6 fragment. In this context, Chugh and coworkers [38] have created four new derivatives centered on the

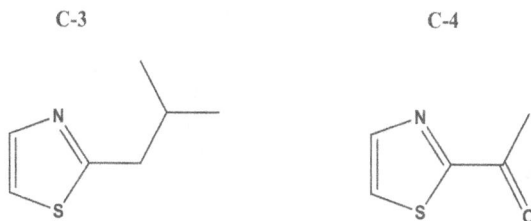

FIGURE 11.7   Structures of C-3 and C-4.

benzo[d]thiazole basic framework, which displayed improved anticorrosion properties by substituting Cl atom (IE = 85%), CH₃ moiety (IE = 88%), and finally dNH₂
fragment (IE = 90%) for H atom (IE = 79%) of R-substituent (on the lateral C_6H_6
fragment).

Furthermore, the use of 2-amino-4-methyl-thiazole (2A4MT) as a CI for MS
prevention in 0.5 M acidic solutions was investigated by Ongun et al. [39]. The
findings show that 2A4MT's inhibition efficiency rises with inhibitor concentration but falls with rising temperature. Its high inhibition efficiency against corrosion was also linked to its full adherence as a protective coating on the outer
metal layer. Additionally, several intricate 1,3-thiazole derivative compounds
were assessed and shown to be effective CIs at minimum doses. For instance,
the ceftriaxone 1,3-thiazole derivative showed a 95% inhibition percentage for
MS at a low-pH solution of 400 ppm [39]. Moreover, Chaouiki et al. used a variety of approaches to study the effects of two thiazolidinediones, MeOTZD and
MeTZD, as carbon steel (CS) CIs in 1.0 M HCl solution. Inhibitors can significantly improve the corrosion resistance of CS, as demonstrated by the study,
which exhibits the effectiveness of 95% and 87% at five 103 mol/L of MeOTZ and
MeTZD, respectively. Additionally, first-principles DFT simulations' projected
densities of states and interaction energies demonstrate that the investigated compounds connect with iron atoms via charge transfer to create covalent bonds, as
shown in Figure 11.8.

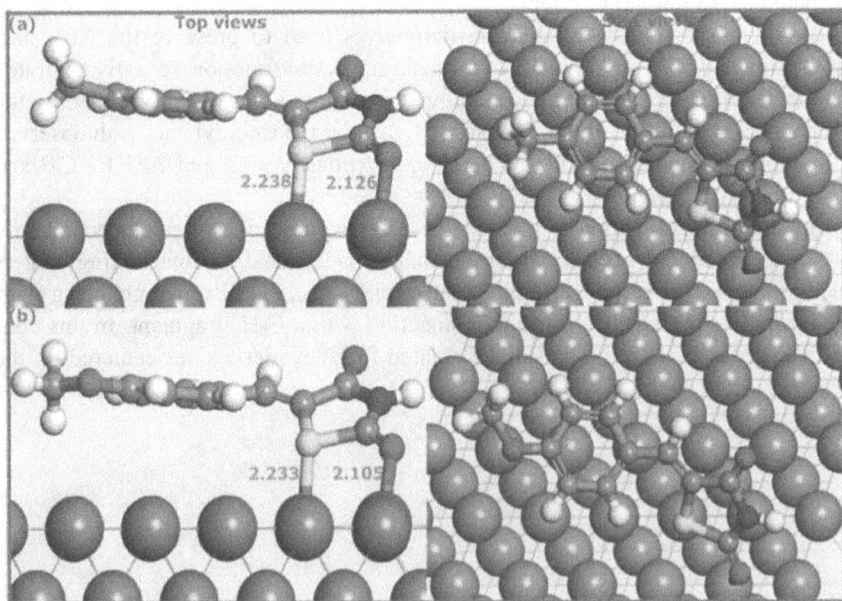

FIGURE 11.8 First-principles DFT simulations exhibiting adsorption geometries of (a)
MeTZD and (b) MeOTZD on Fe(110) surface [40].

FIGURE 11.9 Basic structure of furan.

11.2.3 OXYGEN-BASED CORROSION INHIBITORS

In addition to substances containing N and S, O-heterocycles-based CIs such as furans (Figure 11.9) have attracted much interest in reducing metallic corrosion.

One of the best O-heterocycles-based CIs that has recently provided remarkable preventative activities against a number of metallic corrosions is furan derivatives, which are heterocyclic five-membered aromatic complexes. Furan derivatives' aromatic properties and the existence of O atoms in their structures with two pairs of unshared electrons enable them to function as effective CIs [41]. In a research by Ahmed and his associates, the effectiveness of 2-(5-amino-1,3,4-oxadiazol-2-yl)-5-nitrofuran (AONF) as a CI for MS in a solution of HCl was explored [42]. Surface morphology was assessed by using scanning electron microscopy (SEM) techniques in addition to WL. The outcomes show that the tested inhibitor's highest inhibition effectiveness was 0.005 M, with a maximum IE of 79.49%. The inclusion of N and O atoms in the framework of CI was found to be the primary root of adsorption on an MS outer layer.

Additionally, Fakih et al. examined the effectiveness of 2-furanmethanethiol (FMT) and 2-furonitrile (FN) as inhibitors of MS corrosion in 1 M acidic solution utilizing WL, adsorption isotherms, and EIS [43]. The findings demonstrate that both CIs prevent MS deterioration, and that the potency of their inhibition (IE) rises with rise in inhibitor dosages. The inhibitory impact of FMT is stronger, with an IE of 94.54% at 0.005 M. Additionally, the findings of quantum chemical calculations using the DFT indicate that the reaction cores on FN are the furan ring, the O HA and its adjacent C atom in the furan fragment, the C and S atoms of the substitute methane thiol $-C-SH$ group, and the N HA of the cyano $-C\equiv N$ unit.

11.2.4 ALUMINUM

Aluminum and its alloys are widely utilized in the aircraft, electrical engineering, packaging, home appliances, transportation, and building industries as well as in a variety of items. One rationale is that aluminum has excellent thermal and electrical conductivity, has a light mass (2.6 g/cm^3), is relatively inexpensive, and is nearly twice as efficient as iron. Aluminum also exhibits great corrosion impedance owing to the development of a resistant oxide film, even when exposed to the environment and other unfavorable circumstances that can cause corrosion [44]. According to a literature review, certain organic CIs are frequently applied to avert an aluminum breakdown in basic and acidic conditions. Thermodynamic parameters were used to determine the thiazole as well as furan derivatives' ability to reduce corrosion.

For instance, in a 0.05 M HCl solution, Raviprabha and Bhat [45] assessed the anticorrosion performance of the ethyl-2-amino-4-methyl-1,3-thiazole-5-carboxylate derivative for the AA6061 aluminum alloy. Based on the projected thermodynamic features matching the adsorption progression of inhibitor chemicals, the chemisorption progression of derivative fragments is recommended as a possible route of inhibition. Additionally, it was revealed that a rise in temperature indicates an increase in the inhibitory action of the examined 1,3-thiazole derivative, with a 93% success rate at 333 K and 100 ppm of inhibitor. Using density functional theory (DFT), molecular dynamics (MD), and Monte Carlo (MC) simulation, Zhang et al. conducted a series of methodical theoretical investigations on the usage of N-thiazolyl-2-cyanoacetamide derivatives as CIs of aluminum [46] in a basic condition. Due to its low surface energy and numerous adsorption spots, the results demonstrate that Al(111) face is a good substrate for adsorption. However, solvation effects can alter the CI's molecular structure, which in turn changes how effectively it inhibits corrosion. The findings also indicate that the free volume and the film's self-diffusion are the two key determinants of how effectively the CI layer prevents the passage of corrosive particulates.

11.2.5 COPPER

Although copper scores very high among construction materials that experience significant corrosion rates, Cu and its alloys have a broad array of implementations in industry and technology due to their accessibility and economic viability. Conventional sectors seek new uses for copper and its alloy [47]. Copper is susceptible to acid corrosion and strong basic solutions, especially when O or other oxidants are present. In the pH range between 2.5 and 5, Cu deteriorates exceedingly quickly, and stable surface oxide coatings that can passively protect metal surfaces cannot form. Copper corrosion can be prevented for low-acid or alkaline solutions by creating a barrier on the metal's surface [48]. Some thiazole and furan derivatives were successful in preventing copper corrosion. For example, 1,3-thiazole derivative containing a pyridinium ring has shown a strong ability to control Cu disbanding in an acidic environment, with the highest prevention efficacy of 94% at 10^{-3} M. A 1,3-thiazole ring-centered CI's ability to inhibit can be significantly influenced by the kind and position of added substituents [49].

By employing EIS, AFM (atomic force microscopy), and other techniques, Farahati et al. created and examined the corrosion inhibitions of three thiazole derivatives on the copper metal in 1 M HCl. The findings show that PTA, ATP, and TATD had corrosion inhibition efficiencies at optimal concentrations of about 90%, and these compounds were recommended for use as efficient CIs [50]. Nevertheless, using 1,3,4-thiadiazole (TD)-centered complexes as CI also helped to lessen the amount of metal deterioration brought about by the hostile environment around it. Numerous TD derivatives are effective anticorrosion agents under various operating circumstances. A supplementary N atom is incorporated into the 1,3-thiazole fragment in site 4, giving this five-atom ring type its

distinctive chemical structure. Incongruously, the protective properties of conjugated TD-based compounds depend on the presence of additional HAs. The
most recent feature results from the maximum propensity of HAs with concurrent
multi-bonds to promote these moieties adsorption onto the outer layer of metal,
and the consequently produced defending layer separates the metal from solution
constituents. Numerous TD derivatives with various connected HC chains were
created and tested for their ability to prevent corrosion. It was discovered that the
created TD derivatives' ability to delay metal disbanding could be affected by
the dimension and morphology of the introduced functional moieties and their
chemical characteristics.

For instance, copper corrosion's inhibition efficacy in the PAO base oil environment increased from 82.4% to 88.1% at 0.4 mM of inhibitors when mercapto
groups at positions 2 and 5 of the thiadiazole nucleus were replaced with ethyl-
disulfanyl. These compounds' protective properties were linked to their physical
adsorption on CuO surfaces, which was predicted theoretically and experimentally [51]. In deaerated, aerated, and oxygenated 3% NaCl solutions, another
derivative of 1,3,4-thiadiazol-2-amine has also been reported to function as an
effective copper inhibitor, with the highest IE of 94% attained at 5 mM of the CI
[52]. Additionally, Nakomčić et al. studied the effects of two thiazole complexes,
MFDT and TEBOT, on the suppression of Cu corrosion in $0.1 \ mol/dm^3 \ Na_2SO_4$
solution (pH 3). The findings show that the IE of the inspected thiazole derivatives, MFDT and TEBOT, increases as inhibitor concentration does, preventing
copper from corroding in the following sequence: TEBOT > MFDT. As demonstrated in Figure 11.10, the existence of O, N, and S, carbonyl, >C=S, −CH=C
faction, and furan and benzene rings was found to be accountable for the adsorption of the scrutinized complexes [53].

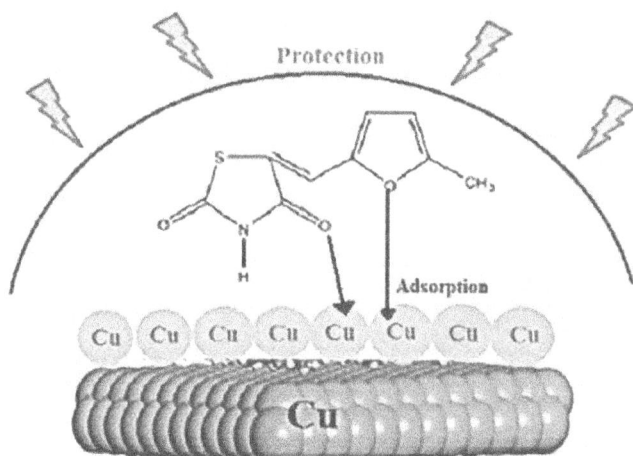

FIGURE 11.10 Schematic representation of thiazole adsorption on copper surface [54].

11.3 CONCLUSION

The use of CI chemicals is a well- acknowledged strategy to combat metallic corro-sion and its associated unwanted effects. In this sense, heterocyclic compounds can be utilized to prevent metal corrosion in a diversity of harsh situations efficiently. Based on their chemical composition, electrostatic interaction, steric influences, aromaticity, charge density, p-orbital feature, electronic structure, and accessibil-ity, heterocyclic compounds have been proven efficient metal CIs. Such chemicals block active sites and slow the rate of corrosion by adhering to metal surfaces. Heterocyclic compound inhibition may be anodic, cathodic, or mixed type. The preponderance of the CI's inhibitory effectiveness tended to rise with increasing inhibitor concentration. For various metal/medium systems, numerous research-ers have implemented a broad array of organic heterocyclic compounds worldwide as anticorrosion substances, and more are continuously being studied. Especially, heterocyclic structures with N, O, and/or S atoms, like azole, oxazole, and thia-zole substances or their analogues, have displayed exceptional protective efficacy against metallic corrosion in various challenging environments. In this regard, using these compounds to prevent metallic corrosion is a well-researched academic and industrial subject.

The use of O- and S–N-heterocycle-based CIs as retarders for the safeguard of various metals, namely iron, aluminum, and copper, in diverse harsh environments is covered in this chapter. Compared to O-heterocycle-based CIs, thiazole derivatives (N-, S-heteroatoms) are intensively researched and reported as CIs. Therefore, it is crucial to pay special attention and focus on synthesizing innovative O-heterocycles-based CIs and their substituted analogues with improved preventative competencies and steadiness for various metal–solution groupings.

ABBREVIATIONS

AFM	atomic force microscopy
CIs	corrosion inhibitors
DFT	density functional theory
EIS	electrochemical impedance spectroscopy
HA	heteroatoms
IE	inhibitory efficiency
MC	Monte Carlo
MD	molecular dynamics
SEM	scanning electron microscopy
WL	weight loss

REFERENCES

1. Assad, H., I. Fatma, and A. Kumar, An overview of the application of graphene-based materials in anticorrosive coatings. Materials Letters, 2022. **330**: p. 133287.
2. Ganjoo, R., et al., Experimental and theoretical study of sodium cocoyl glycinate as corrosion inhibitor for mild steel in hydrochloric acid medium. Journal of Molecular Liquids, 2022. **364**: p. 119988.

3. Carranza, M.S.S., et al., Electrochemical and quantum mechanical investigation of various small molecule organic compounds as corrosion inhibitors in mild steel. Heliyon, 2021. **7**(9): p. e07952.

4. Sharma, S., et al., Investigation of inhibitive performance of betahistine dihydrochloride on mild steel in 1 M HCl solution. Journal of Molecular Liquids, 2022. **347**: p. 118383.

5. Thakur, A., et al., Plant extracts as environmentally sustainable corrosion inhibitors II. In: *Eco-friendly Corrosion Inhibitors*, 2022, Elsevier, pp. 283–310.

6. Huong, D.Q., et al., Pivotal role of heteroatoms in improving the corrosion inhibition ability of thiourea derivatives. ACS Omega, 2020. **5**(42): pp. 27655–27666.

7. Thakur, A., et al., Coordination polymers as corrosion inhibitors. In: *Functionalized Nanomaterials for Corrosion Mitigation: Synthesis, Characterization, and Applications*, 2022, ACS Publications, pp. 231–254.

8. Thakur, A., et al., Computational and experimental studies on the efficiency of *Sonchus arvensis* as green corrosion inhibitor for mild steel in 0.5 M HCl solution. Materials Today: Proceedings, 2022. **66**: p. 609–621.

9. Mobin, M., et al., Biopolymer from tragacanth gum as a green corrosion inhibitor for carbon steel in 1 M HCl solution. ACS Omega, 2017. **2**(7): pp. 3997–4008.

10. Saady, A., et al., Chemical, electrochemical, quantum, and surface analysis evaluation on the inhibition performance of novel imidazo[4,5-*b*] pyridine derivatives against mild steel corrosion. Corrosion Science, 2021. **189**: p. 109621.

11. Malinowski, S., J. Jaroszyńska-Wolińska, and T. Herbert, Theoretical predictions of anti-corrosive properties of THAM and its derivatives. Journal of Molecular Modeling, 2018. **24**(1): pp. 1–12.

12. Daoud, D., et al., Adsorption and corrosion inhibition of new synthesized thiophene Schiff base on mild steel X52 in HCl and H_2SO_4 solutions. Corrosion Science, 2014. **79**: pp. 50–58.

13. Issaadi, S., T. Douadi, and S. Chafaa, Adsorption and inhibitive properties of a new heterocyclic furan Schiff base on corrosion of copper in HCl 1 M: Experimental and theoretical investigation. Applied Surface Science, 2014. **316**: pp. 582–589.

14. Assad, H., and A. Kumar, Understanding functional group effect on corrosion inhibition efficiency of selected organic compounds. Journal of Molecular Liquids, 2021. **344**: p. 117755.

15. Zhang, W., et al., Tetrahydroacridines as corrosion inhibitor for X80 steel corrosion in simulated acidic oilfield water. Journal of Molecular Liquids, 2019. **293**: p. 111478.

16. El-Hajjaji, F., et al., Effect of 1-(3-phenoxypropyl) pyridazin-1-ium bromide on steel corrosion inhibition in acidic medium. Journal of Colloid and Interface Science, 2019. **541**: pp. 418–424.

17. Zhang, G., et al., Benzimidazole derivatives as novel inhibitors for the corrosion of mild steel in acidic solution: Experimental and theoretical studies. Journal of Molecular Liquids, 2019. **278**: pp. 413–427.

18. Chakravarthy, M., and K. Mohana, Adsorption and corrosion inhibition characteristics of some nicotinamide derivatives on mild steel in hydrochloric acid solution. International Scholarly Research Notices, 2014. **9**: pp. 1–13.

19. Naderi, E., et al., Effect of carbon steel microstructures and molecular structure of two new Schiff base compounds on inhibition performance in 1 M HCl solution by EIS. Materials Chemistry and Physics, 2009. **115**(2–3): pp. 852–858.

20. Bentiss, F., et al., The corrosion inhibition of mild steel in acidic media by a new triazole derivative. Corrosion Science, 1999. **41**(4): pp. 789–803.

21. Eddy, N.O., et al., A brief review on fruit and vegetable extracts as corrosion inhibitors in acidic environments. Molecules, 2022. **27**(9): p. 2991.

22. Quraishi, M.A., D.S. Chauhan, and V.S. Saji, *Heterocyclic Organic Corrosion Inhibitors: Principles and Applications*, 2020, Elsevier.
23. Sharma, S., et al., Multidimensional analysis for corrosion inhibition by isoxsuprine on mild steel in acidic environment: Experimental and computational approach. Journal of Molecular Liquids, 2022. **357**: p. 119129.
24. Thakur, A., et al., Computational and experimental studies on the corrosion inhibition performance of an aerial extract of *Cnicus benedictus* weed on the acidic corrosion of mild steel. Process Safety and Environmental Protection, 2022. **161**: pp. 801–818.
25. Quraishi, M.A., 2-Amino-3,5-dicarbonitrile-6-thio-pyridines: New and effective corrosion inhibitors for mild steel in 1 M HCl. Industrial & Engineering Chemistry Research, 2014. **53**(8): pp. 2851–2859.
26. Döner, A., et al., Experimental and theoretical studies of thiazoles as corrosion inhibitors for mild steel in sulphuric acid solution. Corrosion Science, 2011. **53**(9): pp. 2902–2913.
27. Bashir, S., et al., Potential of venlafaxine in the inhibition of mild steel corrosion in HCl: Insights from experimental and computational studies. Chemical Papers, 2019. **73**(9): pp. 2255–2264.
28. Bashir, S., et al., Electrochemical behavior and computational analysis of phenylephrine for corrosion inhibition of aluminum in acidic medium. Metallurgical and Materials Transactions A, 2019. **50**(1): pp. 468–479.
29. Solmaz, R., Investigation of adsorption and corrosion inhibition of mild steel in hydrochloric acid solution by 5-(4-dimethylaminobenzylidene) rhodanine. Corrosion Science, 2014. **79**: pp. 169–176.
30. Liu, Y., et al., β-Cyclodextrin modified natural chitosan as a green inhibitor for carbon steel in acid solutions. Industrial & Engineering Chemistry Research, 2015. **54**(21): pp. 5664–5672.
31. Thakur, R., et al., Thermodynamic and transport studies of some aluminium salts in water and binary aqueous mixtures of tetrahydrofuran. Journal of Materials and Environmental Science, 2015. **6**(5): pp. 1330–1336.
32. Sharma, V., et al., Effect of cosolvents DMSO and glycerol on the self-assembly behavior of SDBS and CPC: An experimental and theoretical approach. Journal of Chemical & Engineering Data, 2018. **63**(8): pp. 3083–3096.
33. Singh, A., et al., A combined electrochemical and theoretical analysis of environmentally benign polymer for corrosion protection of N80 steel in sweet corrosive environment. Results in Physics, 2019. **13**: p. 102116.
34. Assad, H., R. Ganjoo, and S. Sharma. A theoretical insight to understand the structures and dynamics of thiazole derivatives. Journal of Physics: Conference Series. 2022, 2267: 012063.
35. Kartsonakis, I.A., and C.A. Charitidis, Corrosion protection evaluation of mild steel: The role of hybrid materials loaded with inhibitors. Applied Sciences, 2020. **10**(18): p. 6594.
36. Abdallah, M., et al., Inhibition effects and theoretical studies of synthesized novel bisaminothiazole derivatives as corrosion inhibitors for carbon steel in sulphuric acid solutions. International Journal of Electrochemical Science, 2014. **9**(5): pp. 2186–2207.
37. Tan, B., et al., Corrosion inhibition of X65 steel in sulfuric acid by two food flavorants 2-isobutylthiazole and 1-(1, 3-thiazol-2-yl)ethanone as the green environmental corrosion inhibitors: Combination of experimental and theoretical researches. Journal of Colloid and Interface Science, 2019. **538**: p. 519–529.
38. Chugh, B., et al., An exploration about the interaction of mild steel with hydrochloric acid in the presence of *N*-(benzo[*d*]thiazol-2-yl)-1-phenylethan-1-imines. The Journal of Physical Chemistry C, 2019. **123**(37): pp. 22897–22917.

39. Ongun Yüce, A., B. Doğru Mert, and G. Kardaş, Electrochemical and quantum chemical studies of 2-amino-4-methyl-thiazole as corrosion inhibitor for mild steel in HCl solution. Corrosion Science, 2014. **83**: pp. 310–316.

40. Chaouiki, A., et al., Adsorption mechanism of eco-friendly corrosion inhibitors for exceptional corrosion protection of carbon steel: Electrochemical and first-principles DFT evaluations. Metals, 2022. **12**(10): p. 1598.

41. Khaled, K., and N. Al-Mobarak, A predictive model for corrosion inhibition of mild steel by thiophene and its derivatives using artificial neural network. International Journal of Electrochemical Science, 2012. **7**(2): pp. 1045–1059.

42. Al-Amiery, A.A., et al., Exploration of furan derivative for application as corrosion inhibitor for mild steel in hydrochloric acid solution: Effect of immersion time and temperature on efficiency. Materials Today: Proceedings, 2021. **42**: pp. 2968–2973.

43. Al-Fakih, A.M., H.H. Abdallah, and M. Aziz, Experimental and theoretical studies of the inhibition performance of two furan derivatives on mild steel corrosion in acidic medium. Materials and Corrosion, 2019. **70**(1): pp. 135–148.

44. Bashir, S., G. Singh, and A. Kumar, An investigation on mitigation of corrosion of aluminium by *Origanum vulgare* in acidic medium. Protection of Metals and Physical Chemistry of Surfaces, 2018. **54**(1): pp. 148–152.

45. Raviprabha, K., and R.S. Bhat, Inhibition effects of ethyl-2-amino-4-methyl-1,3-thiazole-5-carboxylate on the corrosion of AA6061 alloy in hydrochloric acid media. Journal of Failure Analysis and Prevention, 2019. **19**(5): pp. 1464–1474.

46. Zhang, X., Q. Kang, and Y. Wang, Theoretical study of *N*-thiazolyl-2-cyanoacetamide derivatives as corrosion inhibitor for aluminum in alkaline environments. Computational and Theoretical Chemistry, 2018. **1131**: pp. 25–32.

47. Khaled, K., Studies of the corrosion inhibition of copper in sodium chloride solutions using chemical and electrochemical measurements. Materials Chemistry and Physics, 2011. **125**(3): pp. 427–433.

48. Singh, A., A. Kumar, and T. Pramanik, A theoretical approach to the study of some plant extracts as green corrosion inhibitor for mild steel in HCl solution. Oriental Journal of Chemistry, 2013. **29**(1): pp. 277–283.

49. Mistry, B.M., and S. Jauhari, Corrosion inhibition of mild steel in 1 N HCl solution by mercapto-quinoline Schiff base. Chemical Engineering Communications, 2014. **201**(7): pp. 961–981.

50. Farahati, R., et al., Synthesis and potential applications of some thiazoles as corrosion inhibitor of copper in 1 M HCl: Experimental and theoretical studies. Progress in Organic Coatings, 2019. **132**: pp. 417–428.

51. Xiong, S., et al., Adsorption behavior of thiadiazole derivatives as anticorrosion additives on copper oxide surface: Computational and experimental studies. Applied Surface Science, 2019. **492**: pp. 399–406.

52. Zucchi, F., G. Trabanelli, and C. Monticelli, The inhibition of copper corrosion in 0.1 M NaCl under heat exchange conditions. Corrosion Science, 1996. **38**(1): pp. 147–154.

53. Nakomčić, J., et al., Effect of thiazole derivatives on copper corrosion in acidic sulphate solution. International Journal of Electrochemical Science, 2015. **10**: pp. 5365–5381.

54. Thakur, A., and A. Kumar, A review on thiazole derivatives as corrosion inhibitors for metals and their alloys. European Journal of Molecular & Clinical Medicine, 2020. **7**(7): p. 2020.

12 Naturally Derived Polymers Modified with N/S-Containing Heterocyclic Compounds for Corrosion Inhibition of Steel in Acidic Media

Demian I. Njoku[1,2,3,9], Ini-Ibehe N. Etim[3,4,9], Okpo O. Ekerenam[5,9], Chigoziri N. Njoku[3,6,9], Sharafadeen K. Kolawole[7,9], and Wilfred Emori[8,9]

[1]Department of Applied Science, School of Science and Technology, Hong Kong Metropolitan University, Ho Man Tin, Hong Kong SAR
[2]Center for Corrosion and Protection of Materials, Institute of Metal Research, Chinese Academy of Sciences, Shenyang, China
[3]Africa Centre of Excellence in Future Energies and Electrochemical Systems (ACE-FUELS), Federal University of Technology, Owerri, Nigeria
[4]Marine Chemistry and Corrosion Research Group, Department of Marine Biology, Akwa Ibom State University, Mkpat-Enin, Nigeria
[5]Department of Pure & Applied Chemistry, Veritas University, Abuja, Nigeria
[6]Department of Chemical Engineering, Federal University of Technology, Owerri, Nigeria
[7]Mechanical Engineering Department, School of Engineering and Technology, Federal Polytechnic, Offa, Nigeria
[8]School of Materials Science and Engineering, Sichuan University of Science and Engineering, Zigong, Sichuan, PR China
[9]Nigerian Alumni Association of the Institute of Metal Research, Chinese Academy of Sciences (NAAIMCAS), Nigeria

DOI: 10.1201/9781003377016-12

12.1 INTRODUCTION

The degradation of steel in an acidic environment is inevitable. But owing to their outstanding mechanical properties and affordability, they are the materials of choice in various facilities. Several industrial processes expose the steel to degradation, like acid-pickling, oil-well acidization, cleaning, and descaling. The cheapest and most effective method to mitigate steel corrosion in acidic systems is the addition of corrosion-inhibiting additives. But due to environmental, ecological, and health concerns, nontoxic additives are preferred as a replacement for toxic inorganic-based inhibiting additives. Consequently, using naturally derived polymeric additives as green corrosion inhibitors has attracted severe interest.

Natural polymers are often water based and are easily extracted. Examples are wool, silk, DNA, cellulose (chitosan, alginate, gelatin, polyaspartic acid, hyaluronic acid, etc.), and proteins. Furthermore, natural polymers (carbohydrates and biopolymers) such as chitosan [1, 2], sodium alginate [3, 4], glucose [5], gellan gum [5], hydroxypropyl cellulose [6], aspartic acid [7, 8], hyaluronic acid [9, 10], carboxymethyl cellulose [11–13], etc. have been investigated as corrosion inhibitors. Other examples of natural polymers that have been studied are dextran (Dex) [14, 15], pectin (PEC) [16–18], gum Arabic (GA) [19, 20], agarose [21], guar gum/guaran [22, 23], agar, carrageenan [24, 25], gelatin [26], and xanthan [27, 28].

Some of these natural polymers did not possess sufficient heteroatoms and conjugated multiple bond systems, which require strong interactions with metal surfaces and lead to compelling inhibition performances. When applied as corrosion inhibitors, the intra/inter-modification of natural polymers with heterocyclic compounds can include the required heteroatom and enhanced adsorption/inhibition efficacy. As such, several natural polymers have been modified with different organic molecules or compounds, including the heterocyclic, which is the focus of this chapter. A heterocyclic compound is a cyclic compound with atoms of at least two different elements as part of its ring structure. It can be aromatic or aliphatic heterocyclic compounds. Hence, in this chapter, we will discuss using some natural products modified with the heterocyclic class of organic molecules for corrosion inhibition of steel in acidic systems.

12.2 CARBOHYDRATE-BASED BIOPOLYMERS

12.2.1 CHITOSAN (CTS) MODIFIED WITH AROMATIC HETEROCYCLIC COMPOUNDS

Among the natural polymers that have enjoyed investigations as corrosion inhibitors in acidic mediums, chitosan has attracted the most attention [29–32]. This is likely because it is nontoxic, cheap, and readily available [29–32]. It is a polysaccharide with β-(1-4)-linked N-acetyl-D-glucosamine units in its chain. Chitosan is a hydrophilic biomacromolecule obtained through the N-deacetylation of chitin, which exists in the exoskeletons of arthropods, especially crustaceans, after sodium hydroxide treatment [29–32]. Chitosan is sparingly soluble in an aqueous system and performs only slightly as a corrosion inhibitor in various aggressive systems.

The weak performance can be linked to the absence of such heterocyclic ring/s with heteroatoms like N and S, which have been established as the primary binding sites for organic molecules in corrosion-inhibiting systems. Nevertheless, scientists have been blending and modifying CTS with many compounds, including heterocyclic compounds.

Chauhan et al. modified chitosan triazole using a two-step treatment with formaldehyde and 4-amino-5-methyl-1,2,4-triazole-3-thiol to produce modified chitosan as a new environmentally benign corrosion inhibitor for carbon steel in HCl solution [33]. Heightened inhibition performance approaching 97% was recorded with the revised CTS, better than previous records for chitosan alone in acidic systems. Modification can improve other properties, such as water solubility and inhibition performance. Soluble inhibitors have wider applications than insoluble ones. Chauhan et al. employed aminotriazolethiol to modify CTS to obtain a macromolecule with improved solubility and inhibition performance following the scheme shown in Figure 12.1 [34]. Zhang et al. [35] synthesized two novel chitosan derivatives using two heterocyclic compounds to obtain highly effective environment-friendly corrosion inhibitors for steel protection in acidic systems [35]. The synthetic route is illustrated in Figure 12.1c. The authors believed that obtaining an inhibition efficacy as high as 98% for the compounds is quite epochal compared with previously modified chitosan molecules. Electrochemical data revealed the disparity in the inhibition behaviors of the two fabricated derivatives. Polarization results showed that both derivatives impacted the anodic and cathodic branches of the polarization curve of Q235 steel in an acid environment. Consequently, the adsorption of the derivatives on the steel surface followed the Langmuir adsorption isotherm, comprising both physical- and chemical-type adsorptions. Moreover, computational tools (quantum chemical parameters and molecular dynamic simulation) affirmed the experimental results.

Furthermore, Eduak et al. [36] compared the anticorrosion effect of chitosan, carboxymethyl chitosan, and carboxymethyl chitosan grafted to poly(N-vinyl imidazole) via a series of chemical reactions [36]. The synthesis was probed with combined TGA, FTIR, and XPS. At the same time, the corrosion inhibition assessments were performed with electrochemical impedance spectroscopy (EIS) and potentiodynamic polarization measurements, and the experimental outcome showed that the composite massively controlled the corrosion action of acids on the studied steel (X70) more than either chitosan or carboxymethyl chitosan. These results affirm that the modification of polymers influences some critical parameters such as improved solubility and adsorption -sites as well as surface coverage by the molecules leading to better inhibition performances.

12.2.2 Hyaluronic Acid and Its Derivatives or Related Compound

Hyaluronic acid (HA) and its derivatives have been thoroughly investigated for the possibility of using them as corrosion inhibitors in acidic conditions. This is due to their capacity to form protective layers on metallic surfaces, effectively blocking corrosion processes. Several techniques have been used to assess HA's corrosion-inhibiting qualities and its derivatives, including electrochemical techniques like EIS, gravimetry, and surface analytical methods. A recent

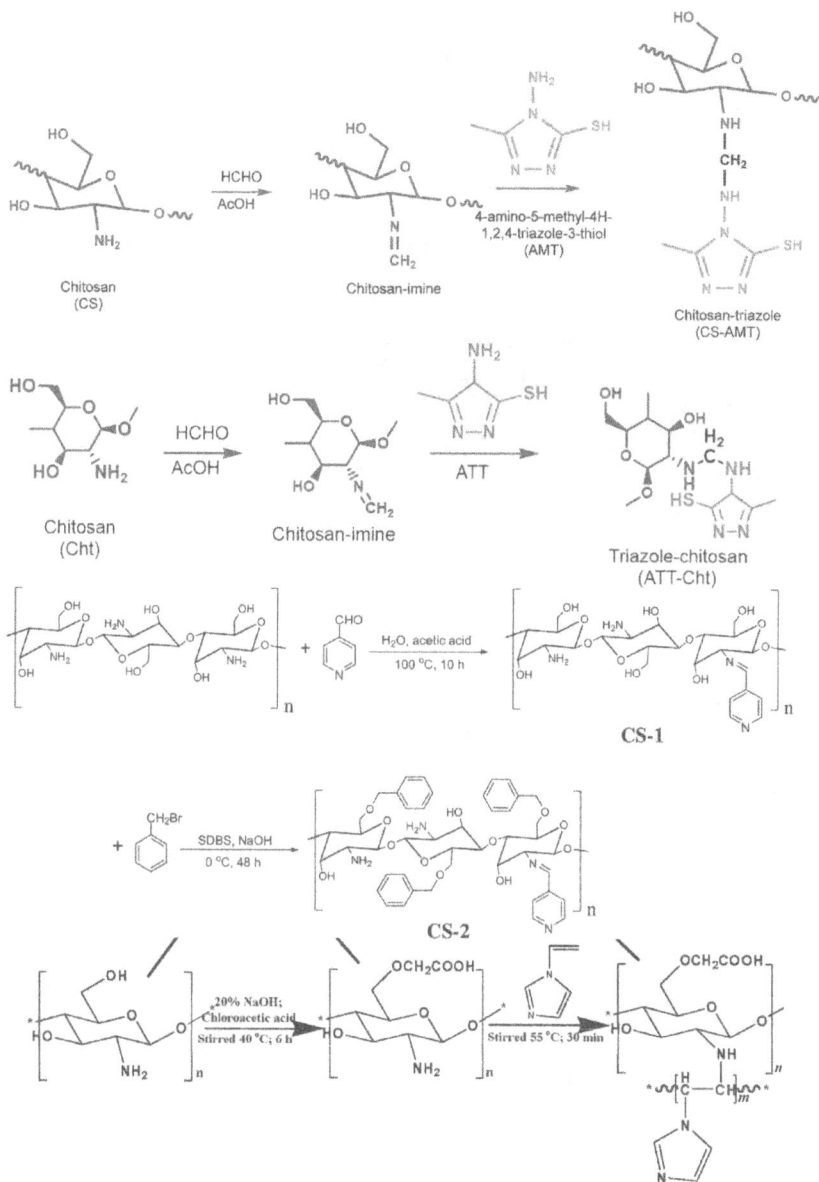

FIGURE 12.1 The modification of chitosan with (a) 4-amino-5-methyl-1,2,4-triazole-3-thiol [33], (b) aminitriazolethiol [34], (c) isonicotinaldehyde and benzylbromide [35], and (d) poly(N-vinyl imidazole) [36].

investigation by Zhou et al. [37] through potentiodynamic polarization and immersion shows that upon HA mobilization on AZ31 alloy through a polydopamine (PDA) anchor layer, HA concealed the cracks created by the PDA layer, thereby decreasing the corrosion density of the polydopamine/HA coating. In a different investigation, HA derivatives were used to prevent corrosion in sulfuric acid solutions using electrochemical techniques. They found that the derivatives were effective in inhibiting corrosion, with inhibition efficiencies ranging from 68% to 95% depending on the concentration of the inhibitor used. In their study, Kim et al. [38] discovered that HA coating through hydrothermal treatment on a Ce-containing polymer with a stable, thick layer of MgO was the most effective self-healing and corrosion resistance.

Recently, He et al. [39] explored the modification of HA with hydroxyl treatment in alkali. The researchers adopted the hydroxyl of polydopamine or the bonds between the carboxyl group of HA and amine to form magnesium/OH/polydopamine/HA. They found that the hydroxyl significantly improved the corrosion resistance, whereas the polydopamine and HA layers further acted as barriers to inhibit corrosion. In addition to its effectiveness, HA has also been shown to have good stability and low toxicity, making it an attractive alternative to traditional chemical inhibitors. Despite the promising results obtained with HA as a corrosion inhibitor, its use still has some limitations. One of the main challenges is its high cost, which discourages large-scale applications. Additionally, its ability to form a protective layer on metallic surfaces can make it difficult to remove, which can be an issue in specific applications.

The modification of HA with heterocyclic organic compounds is an area of active research, as it could enhance HA's ability to prevent corrosion since they have a variety of chemical and physical properties. Several methods can be used to modify HA with heterocyclic organic compounds. One standard method is a covalent modification, in which a covalent bond is formed between the HA molecule and the heterocyclic organic compounds. This can be achieved through chemical reactions such as esterification or amide formation, which involve adding functional groups (e.g., carboxyl) to the HA molecule. Another method for modifying HA with heterocyclic organic compounds is non-covalent modification. Without forming a covalent bond, the heterocyclic organic compound is adsorbing onto the surface of the HA molecule. Non-covalent change can be achieved through a variety of mechanisms.

For example, a study found that modifying HA with pyridine increased its corrosion inhibition efficacy on aluminum in acidic systems. The researchers used electrochemical methods to assess the corrosion suppression properties of the modified HA compound and found that it could inhibit corrosion at low concentrations. Li et al. [40] obtained a polydopamine/hyaluronic acid composite coating on the ZE21B alloy using the dip-coating method. They discovered that the coating samples strengthened the capacity of Mg alloy to resist corrosion. According to Carmela et al. [41], improved bioavailability and stability were recorded in producing an acetylated HA derivative compared to the HA-free form. They discovered that HA-Acet demonstrated low cytotoxicity in all three cell lines, at least at the drug dosages used in the experiments, and a slight but significantly improved anti-inflammation property, dependent on dosage. This was determined by evaluating the NO release inhibition from murine monocyte/macrophage cell lines. Modifying HA with heterocyclic organic compounds is a promising approach for improving its corrosion inhibition properties and enhancements for other uses, especially in biotechnology and bioengineering.

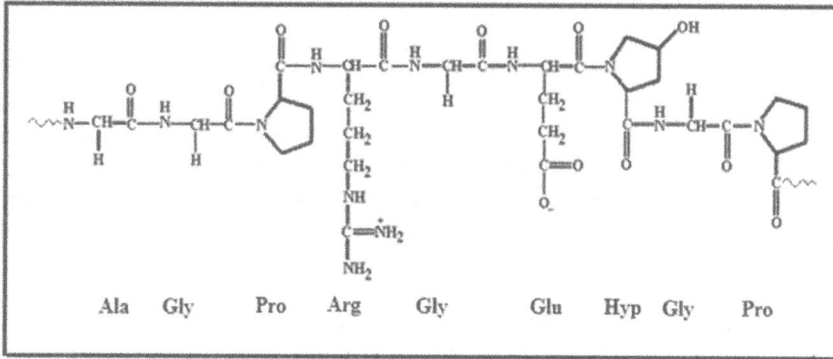

FIGURE 12.2 Chemical structure of gelatin [42].

12.2.3 GLUCOSE, ASPARTIC ACID, GELATIN, THEIR DERIVATIVES, AND COMBINATION WITH HETEROCYCLIC COMPOUNDS

Gelatin was investigated by Haruna et al. as an eco-friendly inhibitor against carbon steel corrosion in acidifying oil-well environments [26]. In 15% HCl at 25 °C, gelatin inhibited corrosion of the carbon steel well, and the inhibition efficacy increased as gelatin concentration increased. The structure of gelatin is given in Figure 12.2. Inhibiting both the cathodic and anodic reactions, gelatin was a mixed-type inhibitor. Gelatin molecules impeded steel dissolution by developing a metal–gelatin complex, which efficaciously stopped the corrosive solution from directly contacting the steel surface, following its adsorption on the metal.

The gelatin molecules were adsorbed on the steel surface through interactions with its O and N atoms, accounting for the metal–gelatin complex, as depicted in Figure 12.3.

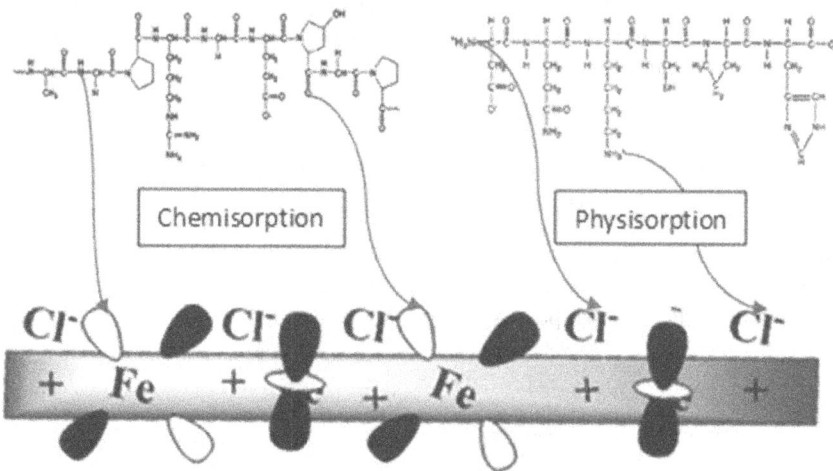

FIGURE 12.3 Representative mechanism for gelatin adsorption on steel surface in the HCl environment [26].

In a related study by Farag et al. [43], a squid by-product (SBP) gelatin polymer employed as an environmentally benign corrosion suppressor for steel in the sulfuric acid environment at 298 K exhibited strong inhibition properties, and the recorded efficacy increased with SBP concentration increments and decreased with temperature increments. Mild steel corrosion was effectively impeded by aspartic acid at 301 K, 305 K, 309 K, and 313 K following gravimetry, acidimetry, and electrochemistry, according to a study by Adejo et al. [44]. The resistance of the mild steel to corrosion was adjudged to be dependent on the aspartic acid concentration and temperature, and chemisorption dominated the adsorption process. The chemical adsorption mechanism proved that aspartic acid's inhibition efficiency and activation energy increased with temperature and concentration. In a similar way, Masroor et al. [45] investigated a biodegradable amino acid–derived surfactant (aspartic di-dodecyl ester hydrochloride acid) and its ZnONPs derivative and aspartic acid derivatives as green corrosion inhibitors for carbon steel. The aspartic di-dodecyl ester hydrochloride acid (Figure 12.4) was discovered to be a powerful corrosion suppressor for steel in 1 M HCl, with its potency increasing as the inhibitor concentration was raised and decreasing when the temperature was raised from 303 to 333 K, indicative of a physisorption-type adsorption mechanism. According to thermodynamic parameters, the adsorption process was discovered to be spontaneous. At 303.15 K, the aspartic di-dodecyl ester hydrochloride acid and ZnO nanocomposite demonstrated the best performance and most potent corrosion inhibition.

Being fully soluble in aqueous media, glucose is a high-purity, cheap, nontoxic organic compound employed as a corrosion inhibitor. For a long time, glucose has been used to stop different metals from corroding in acidic solutions. Glucose has an oxide-phase inhibitory effect in acidic solutions, particularly for copper and copper-containing alloys. Its ability to prevent corrosion in X70 steel exposed to 1 M HCl was assessed in a comparative study with other carbohydrate-based inhibitors [46]. With 87% efficiency at 50 ppm, methyl-4,6-O-benzylidene-D-glucopyranosae exhibited optimum corrosion inhibition. Notably, D(+)-glucose performed better as a corrosion inhibitor than D(+)-galactose.

12.2.4 HETEROCYCLIC-MODIFIED CELLULOSE AND ITS CARBOXYMETHYLCELLULOSE DERIVATIVE AS ANTICORROSION MATERIALS

Using corrosion inhibitor–based polymers became popular in deploying organic and inorganic corrosion inhibitors of small-sized particles for different situations. Polymers are materials that possess impressive adhesion abilities on the surfaces of

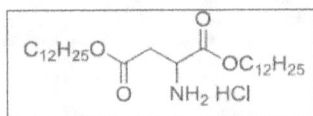

FIGURE 12.4   Chemical structure of aspartic di-dodecyl ester hydrochloride acid.

metals. Various of them have been singled out for their corrosion-resisting properties, including serving as pre-covers on metals or acting as an inhibitor in a mixture
of corrosive solutions [47].

Moreover, the application of carbohydrate-based polymers in corrosion protection
has birthed the utilization of eco-friendly, artificially stable, biodegradable materials with distinctive inhibitive properties, renewability, and low cost [48]. Molecular
weights usually dictate the effectiveness of carbohydrate polymers in inhibition
studies and, invariably, the closeness of cyclic rings in the molecular structures of
such carbohydrates. The anticorrosion properties of these polymeric materials are
directly related to their ability to frame compound structures and complexes with
metal entities on the exterior of such metals via their assembly. Such complexes are
constructed to protect the outer parts of the metals due to their considerable surface
areas, thereby protecting the metals from the aggressive ions found in their environments. This inhibition ability depends on the polymer's heteroatoms (oxygen and
nitrogen) and cyclic rings naturally found [49].

Cellulose is a carbohydrate polymer with inherent nontoxicity, biocompatibility, biodegradability, and the limitless presence of reactive functional groups,
which aids its ability to be modified when there is a need to incorporate another
polymer(s) or in adjusting their properties toward achieving novel applications [50].
An essential type of cellulose is carboxymethyl cellulose (CMC), and sodium salt
is its most common occurrence. Its structural configurations are the same as conventional cellulose; however, it possesses reactive carboxymethyl groups linked to
the −OH groups of its cellulosic glucopyranyl component. CMC is usually produced through an alkali-catalyzed cellulose/$C_2H_3ClO_2$ reaction. The corrosion
suppression behavior of CMC on steel in acidic systems is believed to result from
the possible physical adsorption of protonated CMC through electrostatic interactions with the negatively charged weakly adsorbed hydrated ions on the metal surface. The protonation of CMC is generally known to occur at the carbonyl oxygen,
thus yielding polycations. Meanwhile, its corrosion-resisting abilities have been
extensively investigated for steel materials in varying acidic media, and it was
discovered that the compound forms a relatively thin adsorbed film layer on steel
surface, consequently decreasing its rate of corrosion by impeding one or both the
Tafel reactions [51].

The mechanism in which this occurs is that polymers can transfer electrons to
the unfilled d-orbitals of metals to form coordinate covalent bonds. Moreover, they
can also receive free electrons from the metals via their antibonding orbitals, forming feedback bonds. They can quickly form CMC–metal complexes with extensive
surface coverages, thus constituting excellent corrosion inhibitors. It is noteworthy
that the formation of halide ions usually promotes these complexes. Moreover, the
physisorption of cellulose at the metal surface typically complies with the Langmuir
adsorption isotherm. Its corrosion-inhibiting efficacy in varying environments stems
from both its bulky molecular size and the special functional groups (for example,
−COOH and −OH) on its cellulose framework. Together, these factors present CMC
with inhibitive abilities [47, 49, 52].

It is also worthy of note that quite a large number of the existing effective anticorrosion materials for acid pickling are organic, particularly heterocyclic compounds. For

their organic nature, therefore, adding these corrosion inhibitors not only diminishes the amount of acid used but also reduces the metallic corrosion to a great extent, subsequently extending the service life of the metallic equipment. Organic compounds such as alkaloids, organic amines, rosin amine, urea and thiourea derivatives, quaternary ammonium salt, acetylenic compounds, Mooney alkali, etc. are commonly employed as corrosion inhibitors for sulfuric acid–based pickling. Conversely, the acid pickling of several alloys, especially steel materials in an HCl medium, primarily uses heterocyclic organic compounds consisting of nitrogen, oxygen, phosphorus, and sulfur atoms [50]. Also, ammonia and thiourea-based organic compounds are common anticorrosion materials in this medium. Since metallic oxides and boiler scales are highly soluble in nitric acid (HNO_3) media, and because of the highly oxidizing nature of nitric acid solutions, there are very few existing corrosion-inhibiting materials for pickling in nitric acid environments. However, combinations of hydrazine (C_8H_7N) and sodium sulfide (Na_2S) or ammonium thiocyanate (NH_4SCN), and thiourea (CH_4N_2S) with Na_2S are extensively employed nitric acid–based anticorrosion materials [50]. In the case of pickling processes in phosphoric acid media, the use of various heterocyclic compounds such as triazole ($C_2H_3N_3$), benzotriazole ($C_6H_5N_3$), and urea derivatives, polyvinylpyrrolidone (PVP), sulfonated imidazoline, and polyethyleneimine (PEI), and so on, in combination with the inorganic and their mixed compound formulations is common. Nevertheless, organic compound corrosion inhibitors are also applied in so many other pickling acidic media, and their adsorption on metallic surfaces inspires the effectiveness of these compounds via their electron-rich centers. The important point is that descaling processes involve acidic solutions at lower concentrations, while acid-pickling methods use highly concentrated ones [47, 52, 53].

12.2.5 Polymeric Protein-based Corrosion Inhibitors Modified with Heterocyclic Compounds

Polymers are reported to occupy vast surface areas which supply ample surface protection and media from corrosive attack [54]. In recent years, the adoption of natural polymers, for example, gum Arabic, starch, cellulose, chitin, proteins, and chitosan, among others, as anticorrosive compounds have gained significant attention due to their respective environment-friendly, nonbioaccumulative, biodegradable, bioresistant nature, safer, and no accumulation of toxic materials or heavy metals [54]. The polar-related functional entities aid the polymer material within its electrolytes. Based on the Langmuir adsorption isotherm theory, research has shown that naturally derived polymers mitigate corrosion via adhesion primarily within the metal surface [54]. Among the known natural polymeric compounds exist protein-based polymers.

The protein-based polymers constitute mixed-type and interface corrosion inhibitors. These polymer components have an exceptional anticorrosive function because of the enormous electron-abundance adsorption areas occurring as broad conjugation and polar functional compounds. The polymer occurs in different types, such as silk sericin, gelatin, wool keratin, soy protein collagen, and silk fibroin [55].

Interestingly, these polymers are commercially and naturally accessible in large quantities. It is imperative to note that the polymers, as stated in the preceding discussion, are often merged with heterocyclic compounds.

Adopting heterocyclic compounds was introduced as an essential aspect of practical and economical corrosion mitigation methods. The heterocyclic constituents are often synthesized for use in research and industry by adsorbing within the substrate–media surface adopting their electron centers which are regarded as adsorption media. Notwithstanding, due to their known reduced size, they cause insufficient substrate surface safety and protection. In addition, the development of the compound is linked to the use of toxic and expensive solvents, catalysts, and components. Thus, the focus should be tilted towards developing an eco-friendly substitute of essentially more significant molecular parts. Adopting synthetic and natural polymers as anticorrosive compounds is essential in this aspect. Based on its polymeric characteristics and the occurrence of polar functional groups within its outer margin, these polymeric materials adequately adhere to the substrate surface, enhancing the metal's protectiveness. Furthermore, the natural polymer substitute, especially proteins, and its by-products enable a clear pathway that ousts locally non-eco-friendly corrosion inhibitors [56], as shown in Figure 12.5.

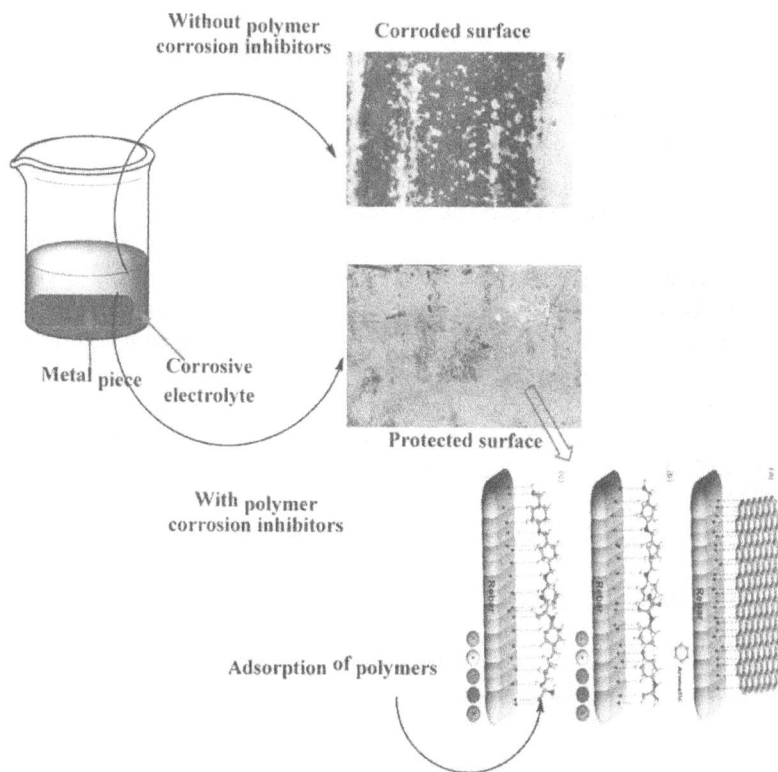

FIGURE 12.5   Mechanism of the protein-based polymer [56].

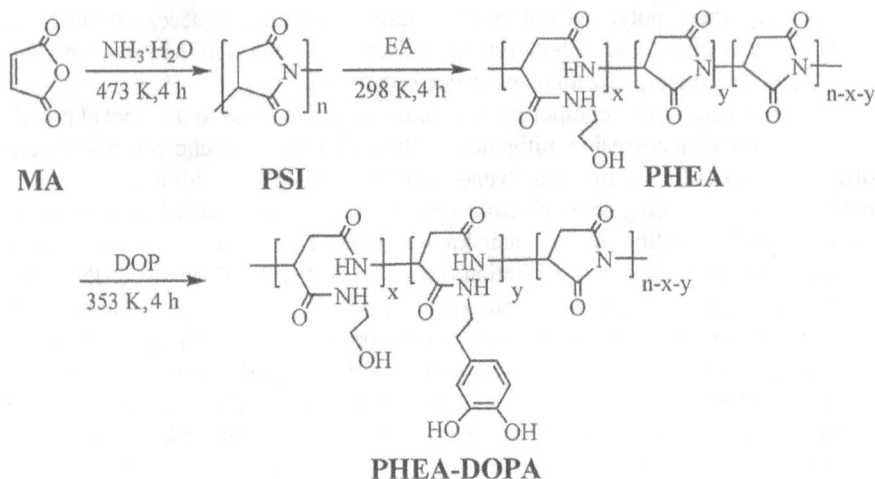

FIGURE 12.6   Illustration of PHEA-DOPA critical synthesis [58].

Another study also reported that protein derived from natural polymers was able to mitigate corrosion inhibition. Umoren et al. [57] presented some of the natural polymers (chitosan, carboxymethyl cellulose, dextran, pectin, sodium alginate, gum Arabic, and hydroxylethyl cellulose) that were screened for anticorrosion characteristics using AZ31 Mg metal in solutions. Chitosan and other natural polymers enhanced the metal corrosion process compared to sodium alginate and hydroxylethyl cellulose. The results were confirmed using surface characterization and electrochemical methods [57]. In a related study by Mingliang Yang et al. [58], it was observed that a mussel-adopted adhesive polymer with the 3,4-dihydroxyphenylalanine functional component was efficient in corrosion mitigation as a polymer with heterocyclic derivatives. The synthesis of the polymer is shown in Figure 12.6. The study was primarily conducted using electrochemical impedance spectroscopy (EIS).

The findings showed that the mussel-adopted adhesive polymer improved the corrosion inhibition of the coating. The products from the corrosion process indicated the adherence of the 3,4-dihydroxyphenylalanine-Fe compound within the substrate exterior. These compounds function as a passive film, thus inhibiting the corrosion processes. Another analysis using the differential scanning calorimeter showed that mussel-adopted adhesive polymer increases the linking density area of coating materials. The influence of oxygen within material protection activities of the mussel-adopted adhesive polymer coating in a medium was carried out using EIS. The findings showed that more compact layers were achieved, thus enhancing the corrosion protection governing the substrate. The mechanism of the process is presented in Figure 12.7.

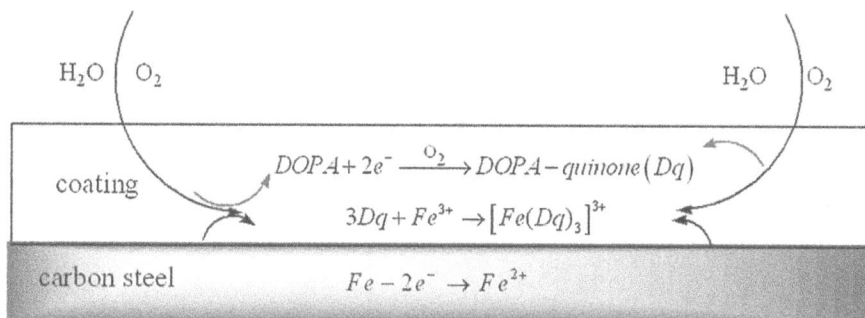

FIGURE 12.7 Illustration of the protective mechanism involving polymer and O_2 within the coating matrix. Note that the polymer is referred to as DOPA in the figure [58].

A patent developed by Panduranga et al. [59] showed a novel protein-based inhibitor with heterocyclic amide constituent of pharmaceutically based salt, ester, solvate, stereoisomer, and prodrug. The invention also presented the compositions of the established compounds, with their unique methods on why they can be used for treatment or preventive purposes. An example of a heterocyclic amide compound is the flavopiridolas shown in Figure 12.8.

A protein kinase inhibitor was reported to serve as an inhibitor [60]. The authors showed the inhibition tendency of olomoucine within the medium containing *Xenopus* egg components. The *in vitro* DNA synthesized within *Xenopus* egg components also inhibited the licensing factor, a vital replication factor that

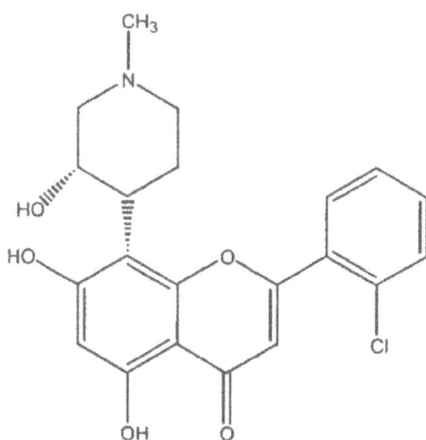

FIGURE 12.8 Flavopiridol as a protein-based corrosion inhibitor with heterocyclic derivative [59].

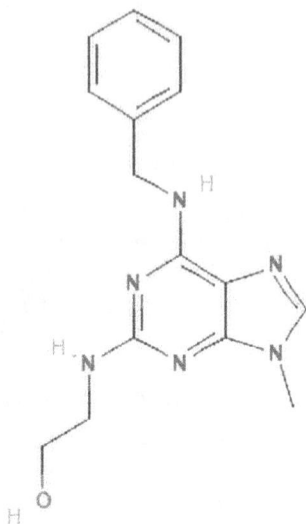

FIGURE 12.9   An olomoucine structure showing its heterocyclic components [60].

ensured DNA was duplicated directly via each cell membrane. Olomoucine inhibited the starfish female gametocyte media within the *in vivo* DNA. Olomoucine's distinctive characteristics enable a corresponding antimitotic compound that, conversely, inhibits most cell process stages. Figure 12.9 shows an olomoucine structure.

An important polymer-based compound with heterocyclic derivatives that can act as an inhibitor is the vincristine compound, as shown in Figure 12.10. It is a heterocyclic anticancer derivative that finds application in nanomedicines. The vincristine compound is often adopted as a combinatorial treatment primarily for intense lymphoblastic leukemia, non-Hodgkin's lymphoma, and Hodgkin's [60].

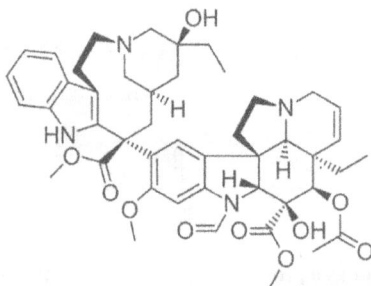

FIGURE 12.10   Vincristine compound [60].

12.3 CONCLUSIONS

The continued search for nontoxic anticorrosion materials for steel protection in acidic systems has led to the discovery of biopolymers as excellent nontoxic options compared with organic anticorrosion materials with toxic tendencies. Given this, many naturally derived polymers like carbohydrates (alginate, chitosan, hyaluronic acid, etc.) and proteins have been tested as anticorrosion materials for steel in acid environments. The corrosion protection is linked to the extensive surface coverage of the polymers. However, in most scenarios, especially in aggressive systems, moderate inhibition efficiencies have been reported probably due to the insolubility and insufficient heteroatoms (i.e., S, O, N, etc.) and double bonds in conjugated systems in the monomers. But improved protection efficacies have been achieved by introducing functional groups into the molecular chains of eco-friendly polymers. Interestingly, heterocyclic compounds are cyclic compounds with atoms of at least two different elements as part of their ring structure. The aromatic type comprises heteroatoms and double bonds in conjugated systems, which are strong adsorption sites for inhibitor adsorption. Thus, the present report identified reports where heterocyclic compounds had been employed to modify natural polymers. Adjusting natural polymers with heterocyclic compounds improves inhibitive properties, stability, and solubility. The essence of this report is not to provide an exhaustive survey of the words but to expose the advantages and methods of using heterocyclic compounds to modify naturally developed polymers. The reaction methods and experimental methods are highlighted. The reported naturally derived polymeric compounds were classified into carbohydrate and protein based. They are cheap, readily available, biodegradable, nontoxic, and renewable. In spite of these, there are still very few existing reports on the modification of natural polymers using heterocyclic compounds, which is the motivation for this topic and the relevance of the information being the first of its kind. The improved inhibition performances revealed the need to intensify research efforts in this research area.

CONFLICTS OF INTEREST

The authors declare no conflict of interest.

ACKNOWLEDGMENT

The authors acknowledge the Institute of Metal Research, Chinese Academy of Sciences.

REFERENCES

1. H. Liu, Z. Zhu, J. Hu, X. Lai, J. Qu, Inhibition of Q235 corrosion in sodium chloride solution by chitosan derivative and its synergistic effect with ZnO, Carbohydr. Polym. 296 (2022) 119936. https://doi.org/10.1016/j.carbpol.2022.119936.
2. H. Ashassi-Sorkhabi, A. Kazempour, Chitosan, its derivatives and composites with superior potentials for the corrosion protection of steel alloys: A comprehensive review, Carbohydr. Polym. 237 (2020) 116110. https://doi.org/10.1016/j.carbpol.2020.116110.

3. I.B. Obot, I.B. Onyeachu, A.M. Kumar, Sodium alginate: A promising biopolymer for corrosion protection of API X60 high strength carbon steel in saline medium, Carbohydr. Polym. 178 (2017) 200–208. https://doi.org/10.1016/j.carbpol. 2017.09.049.

4. W. Zhang, B. Nie, H.-J. Li, Q. Li, C. Li, Y.-C. Wu, Inhibition of mild steel corrosion in 1 M HCl by chondroitin sulfate and its synergistic effect with sodium alginate, Carbohydr. Polym. 260 (2021) 117842. https://doi.org/10.1016/j.carbpol. 2021.117842.

5. V. Rajeswari, D. Kesavan, M. Gopiraman, P. Viswanathamurthi, Physicochemical studies of glucose, gellan gum, and hydroxypropyl cellulose: Inhibition of cast iron corrosion, Carbohydr. Polym. 95 (2013) 288–294. https://doi.org/10.1016/j. carbpol.2013.02.069.

6. K. Liu, H. Du, T. Zheng, H. Liu, M. Zhang, R. Zhang, H. Li, H. Xie, X. Zhang, M. Ma, C. Si, Recent advances in cellulose and its derivatives for oilfield applications, Carbohydr. Polym. 259 (2021) 117740. https://doi.org/10.1016/j.carbpol.2021.117740.

7. D.J. Kalota, D.C. Silverman, Behavior of aspartic acid as a corrosion inhibitor for steel, Corrosion. 50 (1994) 138–145. https://doi.org/10.5006/1.3293502

8. B. El Ibrahimi, A. Jmiai, L. Bazzi, S. El Issami, Amino acids and their derivatives as corrosion inhibitors for metals and alloys, Arab. J. Chem. 13 (2020) 740–771. https:// doi.org/10.1016/j.arabjc.2017.07.013.

9. S. Radice, J. Yao, J. Babauta, M.P. Laurent, M.A. Wimmer, The effect of hyaluronic acid on the corrosion of an orthopedic CoCrMo-alloy in simulated inflammatory conditions, Materialia. 6 (2019) 100348. https://doi.org/10.1016/j.mtla.2019.100348.

10. Y.-K. Kim, S.-Y. Kim, Y.-S. Jang, I.-S. Park, M.-H. Lee, Bio-corrosion behaviors of hyaluronic acid and cerium multi-layer films on degradable implant, Appl. Surf. Sci. 515 (2020) 146070. https://doi.org/10.1016/j.apsusc.2020.146070.

11. R.G.M. de Araújo Macedo, N. do Nascimento Marques, J. Tonholo, R. de Carvalho Balaban, Water-soluble carboxymethylchitosan used as corrosion inhibitor for carbon steel in saline medium, Carbohydr. Polym. 205 (2019) 371–376. https://doi.org/10.1016/j. carbpol.2018.10.081.

12. R. Aslam, M. Mobin, J. Aslam, H. Lgaz, I.-M. Chung, Inhibitory effect of sodium carboxymethylcellulose and synergistic biodegradable Gemini surfactants as effective inhibitors for MS corrosion in 1 M HCl, J. Mater. Res. Technol. 8 (2019) 4521–4533. https://doi.org/10.1016/j.jmrt.2019.07.065.

13. S.A. Umoren, A.A. AlAhmary, Z.M. Gasem, M.M. Solomon, Evaluation of chitosan and carboxymethyl cellulose as ecofriendly corrosion inhibitors for steel, Int. J. Biol. Macromol. 117 (2018) 1017–1028. https://doi.org/10.1016/j.ijbiomac.2018.06.014.

14. Q.H. Zhang, B.S. Hou, Y.Y. Li, G.Y. Zhu, Y. Lei, X. Wang, H.F. Liu, G.A. Zhang, Dextran derivatives as highly efficient green corrosion inhibitors for carbon steel in CO_2-saturated oilfield produced water: Experimental and theoretical approaches, Chem. Eng. J. 424 (2021) 130519. https://doi.org/10.1016/j.cej.2021.130519.

15. M.M. Solomon, A. Umoren, I.B. Obot, A.A. Sorour, H. Gerengi, Exploration of dextran for application as corrosion inhibitor for steel in strong acid environment: Effect of molecular weight, modification, and temperature on efficiency, ACS Appl. Mater. Interfaces. 10 (2018) 28112–28129. https://doi.org/10.1021/acsami.8b09487.

16. J. Halambek, I. Cindrić, A. Ninčević Grassino, Evaluation of pectin isolated from tomato peel waste as natural tin corrosion inhibitor in sodium chloride/acetic acid solution, Carbohydr. Polym. 234 (2020) 115940. https://doi.org/10.1016/j.carbpol.2020.115940.

17. P.R. Prabhu, P. Hiremath, D. Prabhu, M.C. Gowrishankar, B.M. Gurumurthy, Chemical, electrochemical, thermodynamic and adsorption study of EN8 dual-phase steel with ferrite–martensite structure in 0.5 M H_2SO_4 using pectin as inhibitor, Chem. Pap. 75 (2021) 6083–6099. https://doi.org/10.1007/s11696-021-01773-x.

18. D. Prabhu, S. Sharma, P.R. Prabhu, J. Jomy, R.V. Sadanand, Analysis of the inhibiting action of pectin on corrosion of AISI1040 dual-phase steel with ferrite–martensite and ferrite–bainite structure: A comparison in 0.5 M sulphuric acid, J. Iran. Chem. Soc. 19 (2022) 1109–1128. https://doi.org/10.1007/s13738-021-02368-9.

19. M.A. Asaad, G.F. Huseien, M.H. Baghban, P.B. Raja, R. Fediuk, I. Faridmehr, F. Alrshoudi, Gum Arabic nanoparticles as green corrosion inhibitor for reinforced concrete exposed to carbon dioxide environment, Mater. (Basel, Switzerland) 14 (2021). https://doi.org/10.3390/ma14247867.

20. M.-O.M. Danyliak, Y.Y. Rizun, Gum Arabic as an environmentally friendly inhibitor for corrosion protection of 09G2S steel in neutral media, Mater. Sci. 58 (2022) 47–53. https://doi.org/10.1007/s11003-022-00629-3.

21. R.S. Nathiya, S. Perumal, V. Murugesan, P.M. Anbarasan, V. Raj, Agarose as an efficient inhibitor for aluminium corrosion in acidic medium: An experimental and theoretical study, J. Bio-Tribo-Corrosion 3 (2017) 44. https://doi.org/10.1007/s40735-017-0103-2.

22. G. Palumbo, K. Berent, E. Proniewicz, J. Banaś, Guar gum as an eco-friendly corrosion inhibitor for pure aluminium in 1-M HCl solution, Mater. (Basel, Switzerland) 12 (2019). https://doi.org/10.3390/ma12162620.

23. A. Singh, H. Samih Mohamed, S. Singh, H. Yu, Y. Lin, Corrosion inhibition using guar gum grafted 2-acrylamido-2-methylpropanesulfonic acid (GG-AMPS) in tubular steel joints, Constr. Build. Mater. 258 (2020) 119728. https://doi.org/10.1016/j.conbuildmat.2020.119728.

24. A. Golshirazi, N. Golafshan, M. Kharaziha, Multilayer self-assembled kappa carrageenan/chitosan: Heparin coating on Mg alloys for improving blood compatibility, Mater. Today Commun. 32 (2022) 104085. https://doi.org/10.1016/j.mtcomm.2022.104085.

25. M.H. Maleki, M. Rezaie, M. Dinari, Facile synthesis of green and efficient magnetic nanocomposites of carrageenan/copper for the reduction of nitrophenol derivatives, Int. J. Biol. Macromol. 220 (2022) 954–963. https://doi.org/10.1016/j.ijbiomac.2022.08.138.

26. K. Haruna, I.B. Obot, N.K. Ankah, A.A. Sorour, T.A. Saleh, Gelatin: A green corrosion inhibitor for carbon steel in oil well acidizing environment, J. Mol. Liq. 264 (2018) 515–525. https://doi.org/10.1016/j.molliq.2018.05.058.

27. K. Vimal Kumar, B.V. Appa Rao, Phosphorylated xanthan gum, an environment-friendly, efficient inhibitor for mild steel corrosion in aqueous 200 ppm NaCl, Mater. Today Proc. 15 (2019) 155–165. https://doi.org/10.1016/j.matpr.2019.05.038.

28. Y. Cao, C. Zou, C. Wang, W. Chen, H. Liang, S. Lin, Green corrosion inhibitor of β-cyclodextrin modified xanthan gum for X80 steel in 1 M H_2SO_4 at different temperature, J. Mol. Liq. 341 (2021) 117391. https://doi.org/10.1016/j.molliq.2021.117391.

29. M.M. Solomon, H. Gerengi, T. Kaya, S.A. Umoren, Enhanced corrosion inhibition effect of chitosan for St37 in 15% H_2SO_4 environment by silver nanoparticles, Int. J. Biol. Macromol. 104 (2017) 638–649. https://doi.org/10.1016/j.ijbiomac.2017.06.072.

30. A.M. Fekry, R.R. Mohamed, Acetyl thiourea chitosan as an eco-friendly inhibitor for mild steel in sulphuric acid medium, Electrochim. Acta 55 (2010) 1933–1939. https://doi.org/10.1016/j.electacta.2009.11.011.

31. J. Haque, V. Srivastava, D.S. Chauhan, H. Lgaz, M.A. Quraishi, Microwave-induced synthesis of chitosan Schiff bases and their application as novel and green corrosion inhibitors: Experimental and theoretical approach, ACS Omega 3 (2018) 5654–5668. https://doi.org/10.1021/acsomega.8b00455.

32. M. Li, J. Xu, R. Li, D. Wang, T. Li, M. Yuan, J. Wang, Simple preparation of aminothiourea-modified chitosan as corrosion inhibitor and heavy metal ion adsorbent, J. Colloid Interface Sci. 417 (2014) 131–136. https://doi.org/10.1016/j.jcis.2013.11.053.

33. D.S. Chauhan, M.A. Quraishi, A.A. Sorour, S.K. Saha, P. Banerjee, Triazole-modified chitosan: A biomacromolecule as a new environmentally benign corrosion inhibitor for carbon steel in a hydrochloric acid solution, RSC Adv. 9 (2019) 14990–15003. https://doi.org/10.1039/c9ra00986h.

34. D.S. Chauhan, K.E.L. Mouaden, M.A. Quraishi, L. Bazzi, Aminotriazolethiol-functionalized chitosan as a macromolecule-based bioinspired corrosion inhibitor for surface protection of stainless steel in 3.5% NaCl, Int. J. Biol. Macromol. 152 (2020) 234–241. https://doi.org/10.1016/j.ijbiomac.2020.02.283.

35. Q.H. Zhang, B.S. Hou, Y.Y. Li, G.Y. Zhu, H.F. Liu, G.A. Zhang, Two novel chitosan derivatives as high efficient eco-friendly inhibitors for the corrosion of mild steel in acidic solution, Corros. Sci. 164 (2020) 108346. https://doi.org/10.1016/j.corsci.2019.108346.

36. U. Eduok, E. Ohaeri, J. Szpunar, Electrochemical and surface analyses of X70 steel corrosion in simulated acid pickling medium: Effect of poly(N-vinyl imidazole) grafted carboxymethyl chitosan additive, Electrochim. Acta 278 (2018) 302–312. https://doi.org/10.1016/j.electacta.2018.05.060.

37. Z. Zhou, B. Zheng, H. Lang, A. Qin, J. Ou, Corrosion resistance and biocompatibility of polydopamine/hyaluronic acid composite coating on AZ31 magnesium alloy, Surf. Interfaces 20 (2020) 100560. https://doi.org/10.1016/j.surfin.2020.100560.

38. S. Kim, Y. Jang, M. Jang, A. Lim, J.G. Hardy, H.S. Park, J.Y. Lee, Versatile biomimetic conductive polypyrrole films doped with hyaluronic acid of different molecular weights, Acta Biomater. 80 (2018) 258–268. https://doi.org/10.1016/j.actbio.2018.09.035.

39. X. He, G. Zhang, Y. Pei, H. Zhang, Layered hydroxide/polydopamine/hyaluronic acid functionalized magnesium alloys for enhanced anticorrosion, biocompatibility and antithrombogenicity in vascular stents, J. Biomater. Appl. 34 (2020) 1131–1141. https://doi.org/10.1177/0885328219899233.

40. J. Li, L. Chen, X. Zhang, S. Guan, Enhancing biocompatibility and corrosion resistance of biodegradable Mg-Zn-Y-Nd alloy by preparing PDA/HA coating for potential application of cardiovascular biomaterials, Mater. Sci. Eng. C 109 (2020) 110607. https://doi.org/10.1016/j.msec.2019.110607.

41. C. Saturnino, M.S. Sinicropi, O.I. Parisi, D. Iacopetta, A. Popolo, S. Marzocco, G. Autore, A. Caruso, A.R. Cappello, P. Longo, F. Puoci, Acetylated hyaluronic acid: Enhanced bioavailability and biological studies, BioMed Res. Int. 2014 (2014) 921549. https://doi.org/10.1155/2014/921549.

42. N. Devi, M. Sarmah, B. Khatun, T.K. Maji, Encapsulation of active ingredients in polysaccharide–protein complex coacervates, Adv. Colloid Interface Sci. 239 (2017) 136–145. https://doi.org/10.1016/j.cis.2016.05.009.

43. A.A. Farag, A.S. Ismail, M.A. Migahed, Squid by-product gelatin polymer as an eco-friendly corrosion inhibitor for carbon steel in 0.5 M H_2SO_4 solution: Experimental, theoretical, and Monte Carlo simulation studies, J. Bio-Tribo-Corrosion. 6 (2019) 16. https://doi.org/10.1007/s40735-019-0310-0.

44. S.O. Adejo, S.G. Yiase, A. Gbertyo, E.O. Ojah, Aspartic acid as corrosion inhibitor of mild steel corrosion using weight loss, acidimetry and EIS measurement, J. Adv. Chem. 15 (2018) 6262–6274.

45. S. Masroor, M. Mobin, A.K. Singh, R.A.K. Rao, M. Shoeb, M.J. Alam, Aspartic didodecyl ester hydrochloride acid and its ZnO-NPs derivative, as ingenious green corrosion defiance for carbon steel through theoretical and experimental access, SN Appl. Sci. 2 (2020) 144. https://doi.org/10.1007/s42452-019-1515-z.

46. A.E. Vázquez, Carbohydrates as corrosion inhibitors of API 5L X70 steel immersed in acid medium, Int. J. Electrochem. Sci. 14 (2019) 1–36.

47. R.K. Farag, A.A. Farag, Cellulose-based hydrogels as smart corrosion inhibitors. In: Cellulose-based Superabsorbent Hydrogels, Springer, 2019, pp. 979–1014. https://doi.org/10.1007/978-3-319-77830-3_32.

48. M. Gumienna, B. Górna, Antimicrobial food packaging with biodegradable polymers and bacteriocins, Molecules 26 (2021) 3735. https://doi.org/10.3390/molecules26123735.

49. M. Gouda, H.M. Abd El-Lateef, Novel cellulose derivatives containing metal (Cu, Fe, Ni) oxide nanoparticles as eco-friendly corrosion inhibitors for C-steel in acidic chloride solutions, Molecules 26 (2021) 7006. https://doi.org/10.3390/molecules26227006.

50. C. Verma, E.E. Ebenso, M.A. Quraishi, C.M. Hussain, Recent developments in sustainable corrosion inhibitors: Design, performance and industrial scale applications, Mater. Adv. 2 (2021) 3806–3850. https://doi.org/10.1039/d0ma00681e.

51. N. Manimaran, S. Rajendran, M. Manivannan, J.A. Thangakani, A.S. Prabha, Corrosion inhibition by carboxymethyl cellulose, Eur. Chem. Bull. 2 (2013) 494–498.

52. S.A. Umoren, U.M. Eduok, Application of carbohydrate polymers as corrosion inhibitors for metal substrates in different media: A review, Carbohydr. Polym. 140 (2016) 314–341. https://doi.org/10.1016/j.carbpol.2015.12.038.

53. I.O. Arukalam, I.C. Madufor, O. Ogbobe, E. Oguzie, Experimental and theoretical studies of hydroxyethyl cellulose as inhibitor for acid corrosion inhibition of mild steel and aluminium, Open Corros. J. 6 (2014) 1–10. https://doi.org/10.2174/1876503301406010001.

54. D.E. Arthur, A. Jonathan, P.O. Ameh, C. Anya, A review on the assessment of polymeric materials used as corrosion inhibitor of metals and alloys, Int. J. Ind. Chem. 4 (2013) 2. https://doi.org/10.1186/2228-5547-4-2.

55. C. Verma, Natural polymers as green corrosion inhibitors. In: Handbook of Science & Engineering of Green Corrosion Inhibitors, Elsevier, 2022, pp. 207–224.

56. C. Verma, M.A. Quraishi, K.Y. Rhee, Aqueous phase polymeric corrosion inhibitors: Recent advancements and future opportunities, J. Mol. Liq. 348 (2022) 118387. https://doi.org/10.1016/j.molliq.2021.118387.

57. S.A. Umoren, M.M. Solomon, A. Madhankumar, I.B. Obot, Exploration of natural polymers for use as green corrosion inhibitors for AZ31 magnesium alloy in saline environment, Carbohydr. Polym. 230 (2020) 115466. https://doi.org/10.1016/j.carbpol.2019.115466.

58. M. Yang, J. Wu, D. Fang, B. Li, Y. Yang, Corrosion protection of waterborne epoxy coatings containing mussel-inspired adhesive polymers based on polyaspartamide derivatives on carbon steel, J. Mater. Sci. Technol. 34 (2018) 2464–2471. https://doi.org/10.1016/j.jmst.2018.05.009.

59. M.A.U.S.A.M. (Newton M.A.U.W.T.T. Belmont M.A.U. Reddy Panduranga Adulla P.) Walpole, Heterocyclic Amide Compounds as Protein Kinase Inhibitors (2012). https://www.freepatentsonline.com/y2012/0208826.html.

60. J. Veselý, L. Havlicek, M. Strnad, J.J. Blow, A. Donella-Deana, L. Pinna, D.S. Letham, J. Kato, L. Detivaud, S. Leclerc, Inhibition of cyclin-dependent kinases by purine analogues, Eur. J. Biochem. 224 (1994) 771–786. https://doi.org/10.1111/j.1432-1033.1994.00771.x.

13 Pyridine-, Imidazopyridine-, and Pyrimidine-based Corrosion Inhibitors

Walid Daoudi[1], Omar Dagdag[2], Salma Lamghafri[3], Rajesh Haldhar[4], and Abdelmalik El Aatiaoui[1]
[1]Laboratory of Molecular Chemistry, Materials and Environment (LCM2E), Department of Chemistry, Multidisciplinary Faculty of Nador, University Mohamed I, Nador, Morocco
[2]Centre for Materials Science, College of Science, Engineering and Technology, University of South Africa, Johannesburg, South Africa
[3]Laboratory of Applied Sciences, National School of Applied Sciences Al-Hoceima, Abdelmalek Essaâdi University, Tetouan, Morocco
[4]School of Chemical Engineering, Yeungnam University, Gyeongsan, Republic of Korea

13.1 INTRODUCTION

Metallic resources are employed in several businesses for household and building purposes [1]. However, the interactions of the elements in their environment easily cause them to corrode [2]. Due to corrosion failure, there are a great deal of incidents recorded across the world, which results in significant health, financial, and fatality losses. The NACE estimates that corrosion causes a 3.5% loss in global GDP [3]. As a result, several efforts have been undertaken to decrease corrosion damage. According to evidence, the previously employed corrosion-monitoring techniques can reduce corrosion losses by 15–35% [3]. Utilizing chemical molecules is among the most efficient, well-liked, and cost-effective choices for lowering the risks and losses associated with corrosion [4]. At the boundary between metal and a harsh environment, organic compounds, particularly heterocyclic compounds with electron-rich centers, adsorb and create protective coatings that separate the metal surfaces from the electrolyte [5].

These substances create an adsorbed protective coating on the metallic structures in this item, shielding the metals from the harmful acid mixture [6]. There are two methods for determining this adsorption: chemisorption and electrostatic adsorption. Strong electronic bonds are formed by a chemical interaction between

DOI: 10.1201/9781003377016-13

the metallic structure and the inhibitor chemicals during chemisorption. In contrast, the van der Waals force generates the electrostatic contact [7]. Numerous variables, including the type and amount of charge on the metal surface, the nature of the aggressive environment, the electronic structure of the inhibitor compounds, the type of substituents, the aggressive environment, temperature, immersion time, etc., have an impact on the adsorption of these compounds [8]. Pyridine, imidazopyridine, and pyrimidine, as well as their derivatives, are crucial in many industries, particularly biological and pharmaceutical ones [9, 10]. However, recently, several researchers have investigated this derivative as a metal corrosion inhibitor and found that it exhibits high inhibition efficacy in a variety of acid solutions.

The collection of pyridines, imidazopyridine, and pyrimidine and its by-products as anticorrosion compounds are described in this chapter. For a variety of metallic element and their alloys, diverse classes of pyridine, imidazopyridine, and pyrimidine-based compounds are often utilized. This is the first complete chapter on pyridine-, imidazopyridine-, and pyrimidine-based molecules as corrosion inhibitors.

13.2 PYRIDINE-BASED CORROSION INHIBITORS

13.2.1 PYRIDINE

Pyridine is very important class of compounds that is widely used in pharmaceuticals conception. Due to their presence in natural products and exceptional biological properties, the heterocyclic compounds have received the greatest research attention. These heterocyclic compounds rule the fields of biochemistry and medicinal chemistry, and they are becoming more significant in a variety of other fields.

13.2.2 THEORETICAL UNDERSTANDING OF THE ABILITY OF PYRIDINE-BASED COMPOUNDS TO SUPPRESS CORROSION

Recently, computer simulations and models have attracted a lot of attention for researching how organic chemicals limit corrosion [11–13]. DFT-based studies offer a number of theoretical factors that may be used to characterize the adsorption behavior of pyridine, imidazopyridine, and pyrimidine, as well as their derivatives, on metal surfaces. These parameters' significance, computation, and detailed description are covered elsewhere [14–16]. The chemical locations involved for electron donation as well as acceptance may be simply seen using the DFT approach in the configuration of FMOs. It will facilitate the development of efficient anticorrosion materials. Many reviews of the literature suggest that DFT-based computer analyses are frequently employed to determine how well organic chemicals control corrosion [17–20].

The ability of organic compounds to reduce corrosion is frequently tested using MD and MC simulations [21–23]. These simulations enable measurement of the location of the corrosion inhibitors on the steel substrate. The efficiency of corrosion inhibitors may be determined by measuring how they are oriented on metallic surfaces. The most important variable is adsorption energy (E_{ads}). Spontaneous adsorption is compatible with a negative value of E_{ads}, and vice versa [24–27]. Six-membered heterocyclic pyridine (C_5H_5H) has a nitrogen atom in lieu of one of the benzene ring's carbons. Although it is colorless in its purest form, contaminants cause it to turn yellow. Several organic compounds with pyridine rings in their chemical structures are helpful for a variety of purposes [28]. A lot of agrochemicals, vitamins, and chemical medications include pyridine [29]. Pyridine, in contrast to benzene, has a 2.2 Debye dipole moment, including a polar heterocyclic structure [30]. Pyridine possesses an extra nonbonding pair of electrons of nitrogen compared to benzene. As a result, pyridine and its derivatives ought to be more effective corrosion inhibitors than benzene and its analogues [31–35]. Figure 13.1 displays the frontier molecular orbital images of pyridine, 2-aminopyridine, and 2,6-diaminopyridine. It is evident that the presence of electron-donating —NH_2 substituent(s) significantly improves LUMO and HOMO electron distribution, which is attributable to the +R effect of the —NH_2. Additionally, the energy bandgaps are seen to decrease sequentially when one transitions from benzene to 2,6-diaminopyridine. This finding shows that these chemicals' chemical and corrosion resistance capabilities grow in the same order.

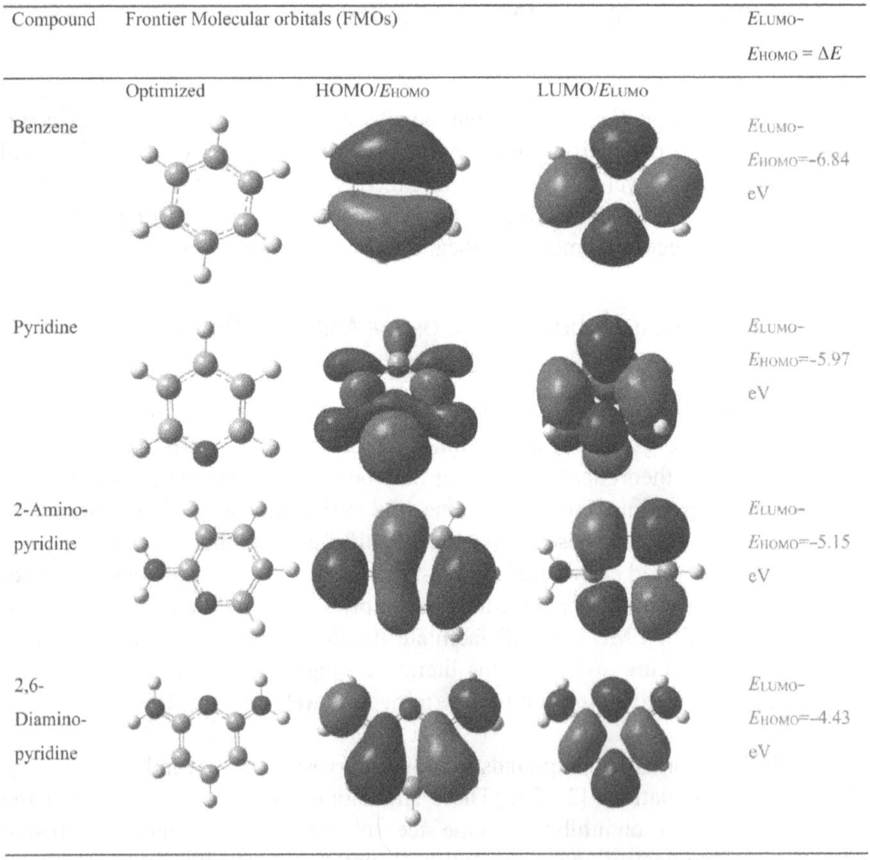

Compound	Frontier Molecular orbitals (FMOs)			$E_{LUMO}-$ $E_{HOMO} = \Delta E$
	Optimized	HOMO/E_{HOMO}	LUMO/E_{LUMO}	
Benzene				$E_{LUMO}-$ $E_{HOMO}=-6.84$ eV
Pyridine				$E_{LUMO}-$ $E_{HOMO}=-5.97$ eV
2-Amino-pyridine				$E_{LUMO}-$ $E_{HOMO}=-5.15$ eV
2,6-Diamino-pyridine				$E_{LUMO}-$ $E_{HOMO}=-4.43$ eV

FIGURE 13.1 FMOs pictures of benzene, pyridine, 2-aminopyridine, and 2,6-diaminopyridine. (Reprinted with permission from Ref. [36] © 2020 Elsevier Publications.)

The presence of substituents, particularly at the *ortho*-position in the aromatic ring, is crucial for determining how well pyridine derivatives suppress corrosion. 2-Aminopyridines (amino-substituted), due to their significant anticorrosive efficacy are frequently used as corrosion inhibitors. The inclusion of extra electron-releasing $-NH_2$ substituents in 2-aminopyridine as opposed to pyridine is thought to be the cause of its superior efficiency. "2-Aminopyridine as well as its derivatives function as bidentate ligands which create chelating compounds with the metal atoms by having two electron donating nitrogen atoms. Corrosion inhibitors based on substituted pyridine that have 2-hydroxy, 2-mercapto, and 2-methoxy groups have a similar mode of action. The pyridine derivatives can function as tri-dentate ligands for a slightly more stable and efficient chelating compound than the 2-substituted pyridine derivatives. These substituents include $-NHMe$, $-NMe_2$, $-NH_2$, and $-OMe$"etc. [36]. Figure 13.2 illustrates a potential mechanism

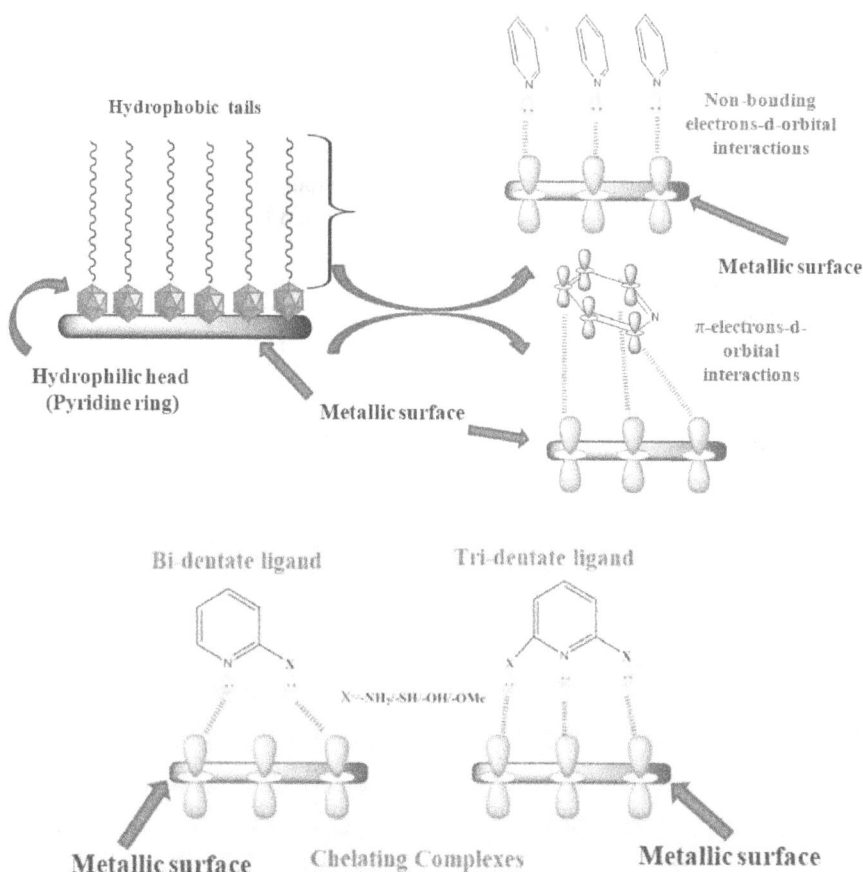

FIGURE 13.2 Pyridine, 2-aminopyridien, and 2,6-diaminopyrdine's behavior upon adsorption on a metallic surface. (Reprinted with permission from Ref. [36] © 2020 Elsevier Publications.)

of correlations involving pyridine, 2-aminopyridine, and 2,6-diaminopyrine and metallic surfaces.

13.2.3 Pyridine Compounds as Corrosion Inhibitors

Pyridine is an organic molecule with the chemical formula C_5H_5N. It resembles benzene structure with one methine group (=CH) substituted by a nitrogen atom. As corrosion inhibitors, pyridine and its derivatives are widely used; according to the literature, their derivatives prove to be more effective against corrosion than unsubstituted pyridine. The structure of the pyridine derivatives influences their inhibitory efficacy. The substitute effect is extensively researched when evaluating the ability of pyridine derivatives. These substances can generate a protective hydrophobic layer while adsorbing at the metal surface; in fact, pyridine has six π-electrons delocalized over the ring allowing the formation of bonds with the metallic surface. Also, electrophilic attack is promoted by the lone pair of electrons in nitrogen atom. Pyridine derivatives can be classified as interface-type inhibitors; they are generally used in acidic two-dimensional that block the reaction sites, and prevent anodic and cathodic reactions.

13.2.4 Corrosion Materials and Methods

Many developed methodologies are employed to study inhibitory efficiency, but recently, experimenters' research has been focused on WL studies, which is a simple method that does not demand a sophisticated equipment; also, electrochemical experiments, namely stationary methods such as PDP and EIS, allow studying the adsorption mechanism. These techniques used potentiostat which impose a potential for which the current density is measured: the galvanostat is connected, on one hand, to a computer associated with an appropriate software and, on the other hand, to a three-electrode cell containing working electrode whose corrosion is evaluated, reference electrode, against which the metal's or alloy's corrosion potential is determined, and a counter-auxiliary electrode which maintains the reference electrode stable and protects it against current flow.

13.2.5 Adsorption Isotherm

To better understand the mechanism of reaction of pyridine toward metallic surfaces, it is necessary to study the adsorption isotherm. In fact, pyridine is an organic compound that is more likely used in acidic media. Their adsorption depends on several factors, namely the nature of the metal and the condition of the surface on which the molecule adsorb. Also, knowing the optimal concentration of molecule increases the ability of adsorption of the nature of medium and the aggressivity of ions influence negatively the processes. Also, the presence of dissolved oxygen can change the adsorption mechanism; this latter is highly affected by temperature which provoke molecule desorption from the surface, sometimes it can lead to its total destruction.

Literature showed that these heterocyclic compounds adsorb strongly into the steel surface forming potent covalent chemical bonds. This is due to the presence of heteroatoms N providing unshared pair of electrons, the adsorption ability increase with the tendency of donating electron to the metallic surface. Also the π-electron of the aromatic ring interacts physically to improve their affinity of adsorption. Generally, pyridine adsorbed by displacement of water molecule was first adsorbed; in fact, a quasi-substitution process occurs between water molecules at the electrode surface and organic compounds in the aqueous phase. The following reaction illustrates the phenomenon:

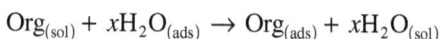

$$Org_{(sol)} + xH_2O_{(ads)} \rightarrow Org_{(ads)} + xH_2O_{(sol)}$$

Here x is the number of water molecules that replace one organic inhibitor. Because of the energy of interaction between organic compounds and metal surfaces is greater than that between water molecule and interfaces, substitution took place. A further explanation of the relationship between an organic molecule and a metal surface: to find the most appropriate kind of adsorption isotherm, a number of isotherms, including those explored by Langmuir, Frumkin, Temkin, Freundlich, and Flory-Huggins, have been examined. The literature claims that pyridine adsorbs via the Langmuir isotherm. Table 13.1 includes a variety of active compounds, the kind of metal/medium utilized in anticorrosion experiments, the adsorption technique applied, and the greatest level of shielding efficacy for pyridine derivatives recently reported [36–38].

Yıldız et al. [39] utilized microstructural studies and electrochemical techniques to investigate the anticorrosive impact of PCN for mild steel in a 1 M HCl medium. At 10 mM, PCN showed a maximum effectiveness of 94.7%. Adsorption worked in accordance with the Langmuir adsorption isotherm. PCN acted as an inhibitor of a mixed kind. Investigations on the anticorrosion capabilities of nitro, amino, methyl, mercapto, and hydroxyl substituted pyridine for steel metals [40, 41] and Al [42]. These pyridine derivatives primarily function as mixed-type inhibitors, and the Langmuir isotherm model was followed during their adsorption. The most often used compounds based on 2-aminopyridine among the studied pyridine derivatives are corrosion inhibitors. In this system, the impact of substituents is extensively researched. In general, electron-donating substituents improve protection efficiency, and vice versa.

Ansari et al. [43] used theoretical and practical techniques to show that three pyridine derivatives (PCs) with various substituents inhibited the oxidation of mild steel. They discovered that the pyridine derivative's inhibitory efficacy followed the pattern: PC-1, PC-2, and PC-3. All of the examined PCs exhibited mostly inhibitory cathodic behavior, and their adsorption followed the Langmuir isotherm model. On the basis of the findings of DFT research, it was determined that the engagement of PCs and metallic surfaces was of the donor–acceptor kind. These authors found that electron-donating —Ome- and —Me-modified pyridine derivatives demonstrated higher protection effectiveness than nitro-substituted pyridine derivative when examining the corrosion protection impact of naphthyridine derivatives (ANCs) for N80 steel in a 15% of HCl solution [44]. Their inhibitory power was distributed as follows: ANC-1, ANC-2, and ANC-3. They act as inhibitors of the

TABLE 13.1
Corrosion Properties of Some Pyridines Molecules

Structures	Metal/ Medium	Adsorption Behavior	EI(%)	Reference
P1	MS/0.1 M HCl	Langmuir/ mixed-type	94.7% at 10 mM	[39]
P2	MS/0.5 M HCl	Langmuir/ mixed-type	94.4% at 50 mM	[45]
P3	MS/1 M HCl	Langmuir	92.8% at 1 mM	[46]
P4	CS/6 M H$_2$SO$_4$	Langmuir/ mixed-type	86.6% at 2.66 × 10^{-3} M	[47]
P5	MS/1 M HCl	Langmuir	96.2% at 0.005 M	[41]
P6 P7	Copper/0.1 M HNO$_3$	Langmuir/ mixed-type	90.14% at 1 mM (P6) 26.9% at 1 mM (P7)	[48]

(Continued)

TABLE 13.1   (*Continued*)
Corrosion Properties of Some Pyridines Molecules

Structures	Metal/ Medium	Adsorption Behavior	EI(%)	Reference
P8	CS/2 M HCl	Langmuir/ mixed-type	91% at 10^{-4} M	[49]
P9 P10	Al/1 M KOH	-	53.35% at 5.10^{-4} M (P9) 46.31 at 5.10^{-4} M (P10)	[50]
P11	Al/H_2O	Langmuir	86% at 0.0187 M	[42]
P12 P13 P14	MS/1 M HCl	Langmuir/ mixed-type	97.4% (P12) at 400 mg/L 95.7% (P13) at 400 mg/L 88.0% (P14) at 400 mg/L	[43]
P15	MS/1 M HCl	Langmuir/ anodic-type	90.0% at 0.5 g/L	[51]
P16	N80 steel/ 15% HCl	Langmuir/ cathodic-type	92.31% at 200 mg/L	[52]

(Continued)

TABLE 13.1   (*Continued*)
Corrosion Properties of Some Pyridines Molecules

Structures	Metal/ Medium	Adsorption Behavior	EI(%)	Reference
P17 P18 P19	MS/1 M HCl	Langmuir/ mixed-type	95.8% (P17) at 20.2×10^{-5} M 96.2% (P18) at 20.2×10^{-5} M 96.6% (P19) at 20.2×10^{-5} M	[53]
P19 P20 P21	MS/1 M HCl	Langmuir/ mixed-type	97.6% (P19) at 1.52 mM 96.1% (P20) at 1.52 mM 94.7% (P21) at 1.52 mM	[54]
P22	Q235 steel/1 M HCl	Langmuir/ mixed-type	95.41% at 3.24 mM	[55]
P23	Q235 MS/1 M HCl	Langmuir/ mixed-type	91.9% at 1 mM	[56]
P24 P25	MS/1 M HCl	Langmuir/ mixed-type	93.0% (P24) at 5 mM 88.9% (P25) at 5 mM	[57]

(*Continued*)

TABLE 13.1 *(Continued)*
Corrosion Properties of Some Pyridines Molecules

Structures	Metal/ Medium	Adsorption Behavior	EI(%)	Reference
(structure P26)	MS/1 M HCl	Langmuir/ mixed-type	81% at 2 mM	[58]
(structures P27, P28)	N80 steel/ 15% HCl	Langmuir/ mixed-type	95.0% (P27) at 100 ppm 94.6% (P28) at 100 ppm	[59]

cathodic type and adsorb utilizing the Langmuir isotherm model. According to DFT calculations, ANCs interact and adsorb on metallic surfaces through interactions that share electrons.

Utilizing both experimental and theoretical techniques, our research team recently documented the inhibitory impact of three pyridine compounds with different types of substituents on mild steel in a 1 M HCl medium The results demonstrated that the pyridine derivatives' inhibitory efficiency followed the following pattern: DHPN, DMPN, and DPPN. By adhering to a mild steel sample, all the chemicals are rendered useful. According to DFT calculations, the torsion angle between the pyridine ring and the −CH substituent determines how well the pyridine derivatives protect. The high level of protection efficacy was associated with a larger torsion angle value. Figure 13.3 displays the FMOs of the analyzed pyridine derivatives.

The MC simulation demonstrated that the pyridine derivatives, in both their neutral and protonated forms, assume essentially flat orientations on metallic surfaces, as shown in Figure 13.4. Pyridine derivatives' negative E_{ads} values indicated that they spontaneously interact with metallic surfaces.

13.3 IMIDAZOPYRIDINE-BASED CORROSION INHIBITORS

13.3.1 IMIDAZOPYRIDINES

One chemical, imidazopyridine, contains a number of physiologically active compounds that are found in plants, medications, and human enzymes. It is regarded as one of the substances that are comparable to purine and benzimidazole. Utilizing

HOMO isosurfaces

Nucleophilic Fukui indices (f^-) isosurfaces

DPPN-H⁺   DHPN-H⁺   DMPN-H⁺

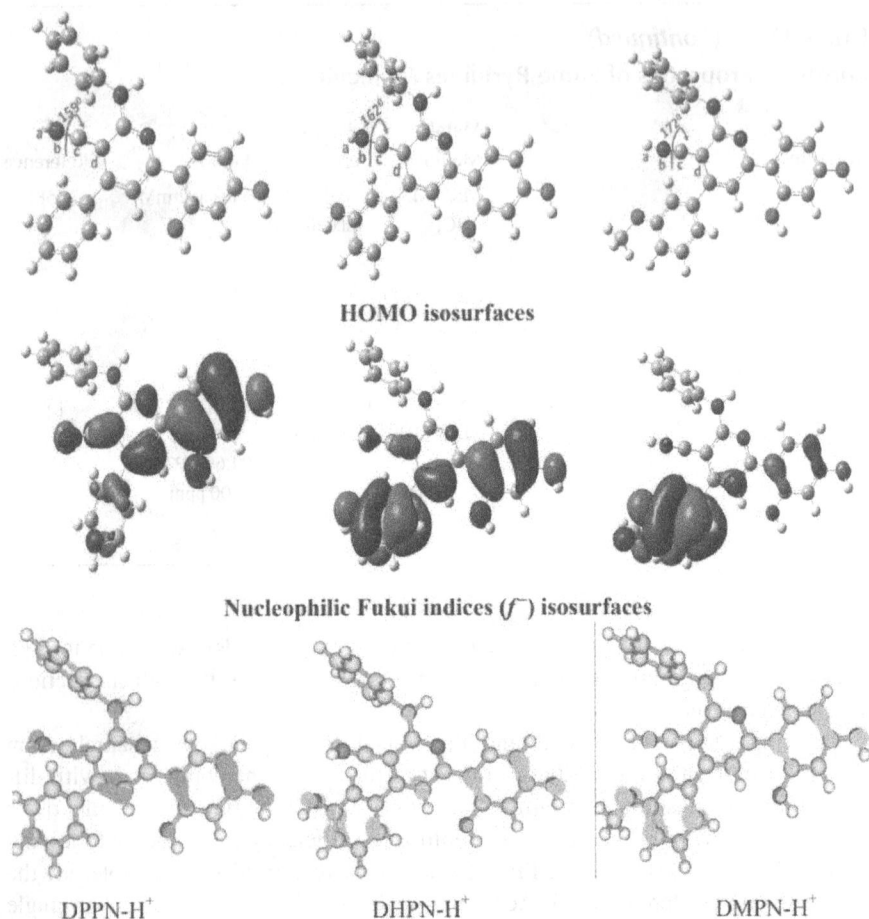

FIGURE 13.3   FMO images of the protonated forms of DHPN-H⁺, DMPN-H⁺, and DPPN-H⁺. (Reprinted with permission from Ref. [36] © 2020 Elsevier Publications.)

this organic heterocyclic structure, many chemical–biological instruments and medicinal substances are synthesized [60, 61]. There are a variety of derivatives of imidazopyridine; however, in this chapter, we are only concerned in imidazo[1,2-a] pyridine. Anti-inflammatory, antibacterial, and anti-ulcer effects have been demonstrated for this structure [62]. Additionally, it has cyclin-dependent kinase inhibitor characteristics [62], cardiotonic substances, GABA, and benzodiazepine receptor agonists [62]. On the other hand, there are currently marketed drug formulations incorporating imidazo[1,2-a]pyridine, including alpidem (hypnotic), zolpidem (hypnotic), zolimidine (anti-ulcer), and olprinone (PDE-3 inhibitor) (anxiolytic) [62]. A summary of imidazopiridines as corrosion inhibitors is presented in Table 13.2.

Neutral molecules

DPPN/Fe(110)	DHPN/Fe(110)	DMPN/Fe(110)
$E_{binding}$ = 881.73 kJ/mol	$E_{binding}$ = 926.33 kJ/mol	$E_{binding}$ = 964.38 kJ/mol

Protonated molecules

DPPN-H+/Fe(110)	DHPN-H+/Fe(110)	DMPN-H+/Fe(110)
$E_{binding}$ = 1231.03 kJ/mol	$E_{binding}$ = 1287.57 kJ/mol	$E_{binding}$ = 1324.55 kJ/mol

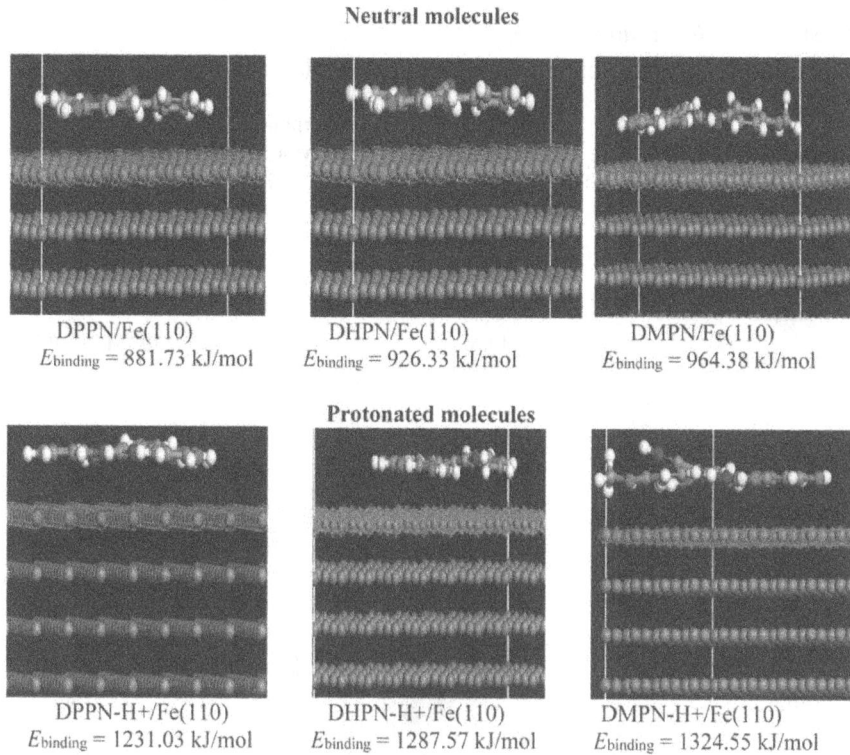

FIGURE 13.4 DHPN, DMPN, and DPPN's optimal orientation on the surface of Fe(110) in their protonated forms. (Reprinted with permission from Ref. [36] © 2020 Elsevier Publications.)

TABLE 13.2
Corrosion Properties of Some Imidazopyridine Molecules

Structures	Metal/Medium	Adsorption Behavior	EI(%)	Reference
  2-(3-Nitrophenyl)-6-methylimidazo[1,2-a]pyridine	C38/1 M HCl	Langmuir/mixed-type	90% at 10^{-3} M	[63]
  7-Methyl-2-phenylimidazo[1,2-α]pyridine	CS/2 M H_3PO_4	Langmuir/mixed-type	92.6% at 10^{-3} M	[64]

(Continued)

TABLE 13.2   (*Continued*)
Corrosion Properties of Some Imidazopyridine Molecules

Structures	Metal/Medium	Adsorption Behavior	EI(%)	Reference
 M3 6-Chloro-2-(4-fluorophenyl) imidazo[1,2-*a*]pyridine	C38/1 M HCl	Langmuir/ mixed-type	91.2% at 10^{-3} M	[63]
 NH₂ M4 3-Amino-2-phenylimidazo[1,2-*a*] pyridine	C38/1 M HCl	Langmuir/ mixed-type	86.1% at 10^{-3} M	[65]
 3-Bromo-2-phenylimidazol[1,2-*α*] pyridine	C38/0.5 M H₂SO₄	Langmuir/ mixed-type	87.5% at 10^{-3} M	[66]
 HO M6 2-Phenylimidazo[1,2-*a*]pyridin-3-yl) methanol	MS/1 M HCl	Langmuir	92% at 0.5 mM	[67]
 O₂N M7 6-Nitroso-2-phenylimidazo[1,2-*a*] pyridine-3 carbaldehyde	MS/1 M HCl	Langmuir	96% at 0.5 mM	[67]
 M8 (2*E*)-3-[2-(4-chlorophenyl) imidazo[1,2-*a*]pyridin-3-yl]-1-(4-methoxyphenyl)prop-2-en-1-one	C38/1 M HCl	Langmuir/ mixed-type	93% at 5.10^{-5} M	[68]

13.4   PYRIMIDINE-BASED CORROSION INHIBITORS

13.4.1   PYRIMIDINE

Pyrimidine is a six-membered heterocyclic aromatic chemical molecule with two nitrogen atoms located at positions 1 and 3. In the fields of medicine, agrochemicals, and numerous biological processes, the chemical composition of pyrimidine derivatives is crucial. A variety of pyrimidine derivatives have biological and pharmaceutical effects, including antifungal, antibacterial, anticonvulsant, antiviral, and anticancer capabilities [69]. Several pyrimidine derivatives have already been created, and their potential for preventing corrosion in a range of steel samples in an acidic media has been researched. Numerous pyrimidine derivatives' effects have been investigated utilizing a variety of methodologies (Table 13.3). According to studies, the key factors influencing the pyrimidine derivative's adsorption on steel surfaces are its physical and chemical characteristics, the donor atom's electronic density, and any potential interactions between its p-orbitals and the d-orbitals of the surface atom. The findings demonstrated that the concentration and type of the inhibitors had an impact on the studied compounds' ability to inhibit.

TABLE 13.3
List of Pyrimidine Derivatives Used to Prevent Steel Samples from Corroding in Various Acid Media

Structures	−R	Medium/Metal and Adsorption Behavior	EI (%)	Reference
	−H −NH$_2$	CS/2 M HCl Langmuir/ mixed-type	81% at 10^{-2} M 88%% at 10^{-2} M	[70]
	−H −OH −SH −NH$_2$		91% at 10^{-2} M 96.50% at 10^{-2} M 98.50% at 10^{-2} M 94% at 10^{-2} M	[70]
	–	Austenitic stainless steel/1 M HCl Langmuir/ mixed-type	90% at 5.10^{-3}M	[71]
	–	Austenitic stainless steel/0.5 M H$_2$SO$_4$ Langmuir/ mixed-type	90% at 2.10^{-3} M	[72]

(Continued)

TABLE 13.3 (*Continued*)
List of Pyrimidine Derivatives Used to Prevent Steel Samples from Corroding in Various Acid Media

Structures	−R	Medium/Metal and Adsorption Behavior	EI (%)	Reference
		CS//1 M HCl Langmuir/ mixed-type	98.7% at 10^{-2} M	[73]
			98.9% at 10^{-2} M	
			99.3% at 10^{-2} M	
		CS/0.1 M HCl Langmuir/ anodic-type	839.3% at 10^{-4} M	[74]
			71.8% at 10^{-4} M	
		CS/2 M HCl	93% at 5×10^{-3} M	[75]
		CS/1 M HCl Langmuir/ mixed-type	44% at 17×10^{-6} M	[69]
			52.8% at 17×10^{-6} M	
			32.3% at 17×10^{-6} M	

13.5 CONCLUSION AND FUTURE PERSPECTIVES

Recently published studies on the effectiveness of pyridine, imidazopyridine, and pyrimidine derivatives in inhibiting metal corrosion are included in in this chapter. The structure exhibits good corrosion resistance, which demonstrates its relevance in the fields of medicine, biology, biochemistry, and medicinal chemistry, due to the presence of nitrogen heteroatoms in the pyridine and pyrimidine rings. Research published in the literature shows that pyrimidine derivatives are more efficient than pyridine derivatives due to the presence of two nitrogen atoms in the pyrimidine molecular entity compared to the presence of a single nitrogen atom in the pyridine ring. According to the Langmuir isotherm, it prevents metal corrosion by creating an adsorbed monolayer on the steel substrate, forming an energetic barrier that prevents the attack of aggressive ions. In addition, it has a lesser impact on the corrosion potential of the metal, acting as mixed-type inhibitors with prominent cathodic and anodic properties at different times. Finally, theoretical calculations by different techniques show the importance of these molecular entities in the biological and industrial field, especially in corrosion. In perspectives, we intend to delimit some physicochemical properties for a better valorization of pyridine and pyrimidine compounds. This will be made possible by trying to establish a more advanced quantum study in order to target the different parameters of our structures for further use in the prediction of the activity of different products, and to evaluate the pharmacological and physicochemical potentialities of these compounds, in order to widen this research field for the synthesis of new structures and complexes.

ABBREVIATIONS

EIS   electrochemical impedance spectroscopy
FMO   frontier molecular orbital
GDP   gross domestic product
MC   Monte Carlo
MD   molecular dynamics
PDP   potentiodynamic polarization
WL   weight loss
PCN   2-pyridinecarbonitrile

REFERENCES

1. N. Baddoo, Journal of Constructional Steel Research, 64 (2008) 1199–1206.
2. K. Suzumura, S.-I. Nakamura, Journal of Materials in Civil Engineering, 16 (2004) 1–7.
3. C. Verma, E.E. Ebenso, M. Quraishi, Journal of Molecular Liquids, 233 (2017) 403–414.
4. L.T. Popoola, Corrosion Reviews, 37 (2019) 71–102.
5. S.K. Ahmed, W.B. Ali, A.A. Khadom, International Journal of Industrial Chemistry, 10 (2019) 159–173.
6. C. Verma, E.E. Ebenso, I. Bahadur, M. Quraishi, Journal of Molecular Liquids, 266 (2018) 577–590.
7. A. Bousskri, A. Anejjar, M. Messali, R. Salghi, O. Benali, Y. Karzazi, S. Jodeh, M. Zougagh, E.E. Ebenso, B. Hammouti, Journal of Molecular Liquids, 211 (2015) 1000–1008.
8. C. Verma, L.O. Olasunkanmi, E.E. Ebenso, M. Quraishi, Results in Physics, 8 (2018) 657–670.
9. R.A. El-Awady, M.H. Semreen, M.M. Saber-Ayad, F. Cyprian, V. Menon, T.H. Al-Tel, DNA Repair, 37 (2016) 1–11.
10. C. Li, L. Chen, D. Steinhuebel, A. Goodman, Tetrahedron Letters, 57 (2016) 2708–2712.
11. O. Dagdag, A. Berisha, Z. Safi, O. Hamed, S. Jodeh, C. Verma, E. Ebenso, A. El Harfi, Journal of Applied Polymer Science, 137 (2020) 48402.
12. O. Dagdag, Z. Safi, N. Wazzan, H. Erramli, L. Guo, A.M. Mkadmh, C. Verma, E. Ebenso, L. El Gana, A. El Harfi, Journal of Molecular Liquids, 302 (2020) 112535.
13. O. Dagdag, R. Hsissou, A. El Harfi, Z. Safi, A. Berisha, C. Verma, E.E. Ebenso, M. Quraishi, N. Wazzan, S. Jodeh, Journal of Molecular Liquids, 315 (2020) 113757.

14. O. Dagdag, Z. Safi, H. Erramli, O. Cherkaoui, N. Wazzan, L. Guo, C. Verma, E. Ebenso, A. El Harfi, RSC Advances, 9 (2019) 14782–14796.
15. R. Haldhar, D. Prasad, N. Mandal, F. Benhiba, I. Bahadur, O. Dagdag, Colloids and Surfaces A: Physicochemical and Engineering Aspects, 614 (2021) 126211.
16. O. Dagdag, Z. Safi, H. Erramli, N. Wazzan, L. Guo, C. Verma, E. Ebenso, S. Kaya, A. El Harfi, Materials Today Communications, 22 (2020) 100800.
17. O. Dagdag, Z. Safi, H. Erramli, N. Wazzan, I. Obot, E. Akpan, C. Verma, E. Ebenso, O. Hamed, A. El Harfi, Journal of Molecular Liquids, 287 (2019) 110977.
18. R. Haldhar, S.-C. Kim, D. Prasad, M. Bedair, I. Bahadur, S. Kaya, O. Dagdag, L. Guo, Journal of Molecular Structure, 1242 (2021) 130822.
19. R. Haldhar, D. Prasad, L.T. Nguyen, S. Kaya, I. Bahadur, O. Dagdag, S.-C. Kim, Materials Chemistry and Physics, 267 (2021) 124613.
20. R. Haldhar, D. Prasad, I. Bahadur, O. Dagdag, S. Kaya, D.K. Verma, S.-C. Kim, Journal of Molecular Liquids, 335 (2021) 116184.
21. O. Dagdag, A. El Harfi, L. El Gana, Z.S. Safi, L. Guo, A. Berisha, C. Verma, E.E. Ebenso, N. Wazzan, M. El Gouri, Journal of Applied Polymer Science, 138 (2021) 49673.
22. O. Dagdag, A. Berisha, V. Mehmeti, R. Haldhar, E. Berdimurodov, O. Hamed, S. Jodeh, H. Lgaz, E.-S.M. Sherif, E.E. Ebenso, Journal of Molecular Liquids, 346 (2022) 117886.
23. E. Berdimurodov, A. Kholikov, K. Akbarov, L. Guo, S. Kaya, K.P. Katin, D.K. Verma, M. Rbaa, O. Dagdag, R. Haldhar, Colloids and Surfaces A: Physicochemical and Engineering Aspects, 637 (2022) 128207.
24. E. Berdimurodov, A. Kholikov, K. Akbarov, L. Guo, S. Kaya, K.P. Katin, D.K. Verma, M. Rbaa, O. Dagdag, R. Haldhar, Journal of Electroanalytical Chemistry, 901 (2021) 115794.
25. R. Haldhar, D. Prasad, D. Kamboj, S. Kaya, O. Dagdag, L. Guo, SN Applied Sciences, 3 (2021) 1–13.
26. N. Bhardwaj, P. Sharma, L. Guo, O. Dagdag, V. Kumar, Colloids and Surfaces A: Physicochemical and Engineering Aspects, 632 (2022) 127707.
27. M.H. Abdellattif, S.H. Alrefaee, O. Dagdag, C. Verma, M. Quraishi, Journal of Molecular Liquids, 337 (2021) 116954.
28. S. Pal, Pyridine, 57 (2018) 57–74.
29. S. Shimizu, N. Watanabe, T. Kataoka, T. Shoji, N. Abe, S. Morishita, H. Ichimura, Ullmann's Encyclopedia of Industrial Chemistry, 2000, Wiley-VCH Verlag GmbH.
30. G. Brownson, J. Yarwood, Journal of Molecular Structure, 10 (1971) 147–153.
31. A. Hassan, R. Hussein, M. Abou-krisha, M.I. Attia, International Journal of Electrochemical Science, 15 (2020) 4274–4286.
32. M. Benabdellah, A. Ousslim, B. Hammouti, A. Elidrissi, A. Aouniti, A. Dafali, K. Bekkouch, M. Benkaddour, Journal of Applied Electrochemistry, 37 (2007) 819–826.
33. A. Donya, M. Pakter, M. Shalimova, V. Lambin, Protection of Metals, 38 (2002) 216–219.
34. M. Lashgari, M. Arshadi, G.A. Parsafar, Corrosion, 61 (2005) 778–783.
35. T. Chaitra, K. Mohana, H. Tandon, International Journal of Corrosion, 2016 (2016) 9532809.
36. C. Verma, K.Y. Rhee, M. Quraishi, E.E. Ebenso, Journal of the Taiwan Institute of Chemical Engineers, 117 (2020) 265–277.
37. P. Dohare, M. Quraishi, I. Obot, Journal of Chemical Sciences, 130 (2018) 1–19.
38. Y. Qiang, L. Guo, S. Zhang, W. Li, S. Yu, J. Tan, Scientific Reports, 6 (2016) 1–14.
39. R. Yıldız, A. Döner, T. Doğan, İ. Dehri, Corrosion Science, 82 (2014) 125–132.

40. R. Karthik, G. Vimaladevi, S.-M. Chen, A. Elangovan, B. Jeyaprabha, P. Prakash, International Journal of Electrochemical Science, 10 (2015) 4666–4681.
41. A. Al-Amiery, L. Shaker, Koroze a ochrana materiálu, 64 (2020) 59–64.
42. R. Padash, E. Jamalizadeh, A.H. Jafari, Anti-Corrosion Methods and Materials, 64 (2017) 550–554.
43. K. Ansari, M. Quraishi, A. Singh, Journal of Industrial and Engineering Chemistry, 25 (2015) 89–98.
44. K. Ansari, M. Quraishi, Physica E: Low-Dimensional Systems and Nanostructures, 69 (2015) 322–331.
45. B.D. Mert, A.O. Yüce, G. Kardaş, B. Yazıcı, Corrosion Science, 85 (2014) 287–295.
46. M. Dawood, Z. Alasady, M. Abdulazeez, D. Ahmed, G. Sulaiman, A. Kadhum, L. Shaker, A. Alamiery, The International Journal of Corrosion and Scale Inhibition, 10 (2021) 1766–1782.
47. A.A. Farag, E.A. Mohamed, G.H. Sayed, K.E. Anwer, Journal of Molecular Liquids, 330 (2021) 115705.
48. C. Varghese, K. Thomas, V. Raphael, K. Shaju, Current Chemistry Letters, 8 (2019) 1–12.
49. S.A.E.-M.A. Fouda, Chemical Physics, 93 (2005) 84–90.
50. D. Patil, A. Sharma, International Scholarly Research Notices, 2014 (2014) 154285.
51. A. Al-Amiery, T.A. Salman, K.F. Alazawi, L.M. Shaker, A.A.H. Kadhum, M.S. Takriff, International Journal of Low-Carbon Technologies, 15 (2020) 202–209.
52. K. Ansari, M. Quraishi, A. Singh, Measurement, 76 (2015) 136–147.
53. C. Verma, L.O. Olasunkanmi, T.W. Quadri, E.-S.M. Sherif, E.E. Ebenso, The Journal of Physical Chemistry C, 122 (2018) 11870–11882.
54. M.A. Quraishi, Industrial & Engineering Chemistry Research, 53 (2014) 2851–2859.
55. R. Chaudhary, T. Namboodhiri, I. Singh, A. Kumar, British Corrosion Journal, 24 (1989) 273–278.
56. Y. Ji, B. Xu, W. Gong, X. Zhang, X. Jin, W. Ning, Y. Meng, W. Yang, Y. Chen, Journal of the Taiwan Institute of Chemical Engineers, 66 (2016) 301–312.
57. M. Murmu, S.K. Saha, N.C. Murmu, P. Banerjee, Corrosion Science, 146 (2019) 134–151.
58. B. Xu, Y. Ji, X. Zhang, X. Jin, W. Yang, Y. Chen, RSC Advances, 5 (2015) 56049–56059.
59. M. Yadav, S. Kumar, Surface and Interface Analysis, 46 (2014) 254–268.
60. L. Dymińska, Bioorganic & Medicinal Chemistry, 23 (2015) 6087–6099.
61. W.M. El-Sayed, W.A. Hussin, Y.S. Al-Faiyz, M.A. Ismail, European Journal of Pharmacology, 715 (2013) 212–218.
62. R. Salim, E. Ech-Chihbi, H. Oudda, F. El Hajjaji, M. Taleb, S. Jodeh, Journal of Bio- and Tribo-Corrosion, 5 (2019) 1–10.
63. R. Salim, E. Ech-Chihbi, H. Oudda, Y. Aoufir, F. El-Hajjaji, A. Elaatiaoui, A. Oussaid, B. Hammouti, H. Elmsellem, M. Taleb, Der Pharma Chemica, 8 (2016) 200–213.
64. D.B. Hmamou, R. Salghi, A. Zarrouk, H. Zarrok, B. Hammouti, S. Al-Deyab, A. El Assyry, N. Benchat, M. Bouachrine, International Journal of Electrochemical Science, 8 (2013) 11526–11545.
65. A. Ghazoui, R. Saddik, N. Benchat, B. Hammouti, M. Guenbour, A. Zarrouk, M. Ramdani, Der Pharma Chemica, 4 (2012) 352–364.
66. R. Salghi, A. Anejjar, O. Benali, S. Al-Deyab, A. Zarrouk, M. Errami, B. Hammouti, N. Benchat, International Journal of Electrochemical Science, 9 (2014) 3087–3098.
67. T.A. Salman, D.S. Zinad, S.H. Jaber, M. Al-Ghezi, A. Mahal, M.S. Takriff, A.A. Al-Amiery, Journal of Bio-and Tribo-Corrosion, 5 (2019) 1–11.

68. R. Salim, A. Elaatiaoui, N. Benchat, E. Ech-Chihbi, Z. Rais, H. Oudda, F. El Hajjaji, Y. ElAoufir, M. Taleb, Journal of Mechanical Engineering Sciences, 8 (2017) 3747–3758.
69. K. Rasheeda, D. Vijaya, P. Krishnaprasad, S. Samshuddin, International Journal of Corrosion and Scale Inhibition, 7 (2018) 48–61.
70. H. Awad, S.A. Gawad, Anti-Corrosion Methods and Materials, 52 (2005) 328–336.
71. N. Caliskan, E. Akbas, Materials and Corrosion, 63 (2012) 231–237.
72. N. Caliskan, E. Akbas, Materials Chemistry and Physics, 126 (2011) 983–988.
73. H. Ashassi-Sorkhabi, B. Shaabani, D. Seifzadeh, Electrochimica Acta, 50 (2005) 3446–3452.
74. A. Yurt, A. Balaban, S.U. Kandemir, G. Bereket, B. Erk, Materials Chemistry and Physics, 85 (2004) 420–426.
75. G. Elewady, International Journal of Electrochemical Science, 3 (2008) 1149.

14 N-Heterocyclics as Corrosion Inhibitors: Miscellaneous

Amarpreet K. Bhatia[1] and Shippi Dewangan[2]
[1]Department of Chemistry, Bhilai Mahila
Mahavidyalaya, Chhattisgarh, India
[2]Department of Chemistry, SW Pukeshwar Singh
Bhardiya Govt. College, Chhattisgarh, India

14.1 INTRODUCTION

The expression "corrosion" generally denotes the progressive degradation of metallic substances from exposure to the surrounding environment. The corrosion process can be conceptualized as the manifestation of an electrochemical or a biochemical interaction concerning a metallic and its immediate surroundings, with the outcome being the gradual deterioration of the metal in question. The grave ramifications of corrosion have emerged as a vexing issue of worldwide significance. The detrimental effects of corrosion on global infrastructure and its substantial economic and ecological impact have been widely acknowledged. The financial cost of this phenomenon is significant, accounting for 3–4% of the nation's output domestic product of manufacturing homelands [1, 2].

Choosing and utilizing appropriate consumption counteraction techniques are thus exceptionally fundamental for the assurance and effective utilization of metallic designs. Many industries, such as those involved in oil and gas, water desalination, and chemical production, encounter diverse corrosion-related challenges resulting in substantial economic losses. The righteousness is that the reception of reasonable erosion counteraction methodologies can avoid a considerable degree of misfortune. Among the various strategies for regulating consumption, the practice of anticorrosion agents is a relatively uncomplicated, financial, and practical approach frequently employed within industrial settings. A consumption inhibitor can be characterized as a substance that, when included reasonably in a destructive climate, fundamentally reduces the erosion rate. An example of inhibitor classification is separated into two sorts, explicitly inorganic and organic inhibitors [1, 2].

In contrast to traditional external passivation techniques utilizing inorganic inhibitors, organic inhibitors, specifically those of the adsorption type, present an attractive alternative given their high efficacy and environmentally conscious properties. Organic inhibitors have been widely utilized across multiple industries to counteract the effects of aggressive environments. The performance of inhibitory substances depends on their chemical composition and physicochemical characteristics, such as functional group properties, electron density at donor atoms, p-orbital traits, and the molecular electronic arrangement. The impediment primarily results from the adsorption mechanism and the subsequent creation of a shielding layer on the exterior [1, 2].

Nevertheless, most extant inhibitors take time to prepare, are costly, and are toxic. The release and buildup of said corrosion inhibitors into terrestrial and aquatic environments pose a significant environmental threat. The inhibitory action of organic compounds occurs through their adsorption onto the metallic constituent. The adsorption process may be expedited through electrostatic fascination, which happens through charge allocation or chemoadsorption, whereby the back-donation of lone-pair electrons is brought about. The comprehension of the adsorption mechanism can be attained by disclosing diverse kinetic and thermodynamic limitations [3, 4].

The effectiveness of an organic anticorrosion agent may be quantified by its exterior coverage, which is impacted by various factors such as molecular geometry, the presence of electron-donating or electron-withdrawing functional groups, steric hindrance, aromaticity, and planarity of the molecular structure. In contemporary existence, several inhibitor chemistries have been suggested to tackle this problem. The heterocyclic compounds belonging to the class of organic anticorrosion agents have been demonstrated to exhibit remarkable inhibitory properties. The molecules above consist of heteroatoms, namely nitrogen (N), sulfur (S), and oxygen (O), as well as phenyl rings, p-bonds, and diverse functional groups. These components contribute substantially to the coverage of metal surfaces and consequent corrosion protection [5–7].

Furthermore, nitrogen-containing heterocyclic compounds have extensive applications as anticorrosion agents because they can make strong coordination bonds with metals and metallic ions. This study investigates the anticorrosive efficacy of compounds derived from pyrrole, pyrrolidine, pyridazine, and pyrimidine. Furthermore, this chapter provides an intricate portrayal of the various types of nitrogen-based heterocyclic organic corrosion inhibitors [3, 6].

14.1.1 CORROSION INHIBITION AND SHIELDING APPROACHES

Corrosion inhibitors serve as a means to prevent the corrosion of metals. Inhibitors refer to compounds incorporated into corrosive environments with the primary objective of impeding or arresting the corrosion of metals [8]. Erosion forms cause numerous casualties, for the most part, within the manufacturing segment. As it were, a way to bargain with it is to debilitate it. Of the distinctive procedures for dodging or anticipating the devastation or oxidation of metal surfaces, erosion inhibitors are one of the foremost well-known and compelling within the industry. Inhibitors have long been utilized in the industry because of their excellent anticorrosive properties. However, numerous side effects happened, causing natural harm. As a result, researchers have begun seeking out naturally neighborly operators, such as organic inhibitors [8].

Tentatively and/or hypothetically utilizing calculation chemistry, the inhibitor's effectiveness can be inspected. Different strategies are regularly used to perform exploratory corrosion restraint productivity estimations and screen the hindrance instrument. These techniques include weight misfortune, linear polarization, potentiodynamic polarization, electrochemical impedance spectroscopy (EIS), UV-visible spectroscopy, filtering electron-magnifying lens (SEM), X-ray spectroscopy (EDX), and cyclic voltammetry. Utilizing fair exploratory strategies, on the other hand, is expensive, time devouring, and unsafe for the environment [9–12]. The progression of preventing and controlling corrosion extends beyond the mere choice of constituents. However, its duration spans the entire life cycle of a product, encompassing the phases from attainment to maintenance. Specific factors that contribute to corrosion all over the lifespan of a system are visually represented in Figure 14.1.

The initiation of corrosion anticipation ought to commence through the formulation of the design conception and persist throughout the material assortment procedure. During the phase dedicated to detail designing, it is imperative to formulate preventive maintenance, monitoring, and inspection plans. Confirming that these plans are consistently conserved and updated when indispensable throughout the sustainment phase is significant. [13]. Corrosion administration requires the execution of designing standards and methods to moderate erosion to a suitable degree by the foremost temperate strategy. Various techniques and methodologies can be employed to effectively manage or govern the erosion process (Figure 14.2) [12].

14.1.2 IMPACT OF MOLECULAR ASSEMBLY ON INHIBITION

The mechanism underlying the effectiveness of organic corrosion inhibitors typically involves surface adsorption phenomena. At suboptimal inhibitor concentrations, adsorption of inhibitor molecules occurs primarily in a planar orientation, while at concentrations surpassing the optimal threshold, vertical adsorption of inhibitors becomes predominant. Henceforth, the metal–inhibitor interplay is subject to the

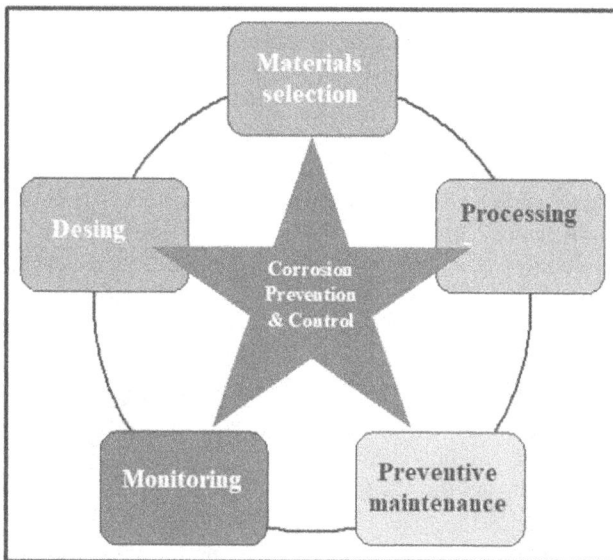

FIGURE 14.1   Some observations for corrosion anticipation and control.

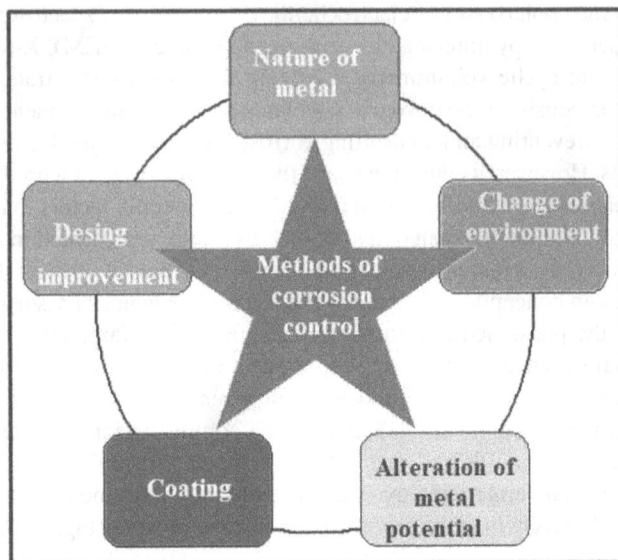

FIGURE 14.2 A diagram of different strategies utilized in corrosion control modification.

supremacy of attractive forces below the optimal inhibitor concentration. By contrast, surpassing the threshold of optimal adsorption elicits the dominance of intramolecular repulsive forces within the inhibitor. The comportment of metal in the incidence of organic inhibitors is significantly affected by various factors, including the properties and category of the organic anticorrosion agents, the properties and configuration of the metal, and the operational settings, such as acid concentration and temperature. Organic anticorrosion agents could possess chains of hydrocarbon of different dimensions. The disparity may significantly influence the hydrophobicity of a corrosion inhibitor in the size of the involved alkyl chains [3, 14, 15].

The adsorption strength is significantly influenced by various factors, including the number of alkyl chains, chain dimension, and the existence of π-bonds featuring involved heteroatoms. Although the elongation of alkyl chains can enhance hydrophobicity, it can concurrently lead to a crowding effect, thereby impeding the adsorption of inhibitors [16]. The exterior adsorption process could be significantly inclined by numerous aspects comprising heteroatoms' identity, quantity, location, and the nature of the phenyl rings and attached functional groups [17–20]. This phenomenon can potentially provoke perturbations in the steadiness of the inhibitor layer, inducing bending, twisting, and variations in the alignment of the inhibitor molecules. The organic compounds employed as anticorrosion agents typically comprise heterocycles with five or six carbon atoms and phenyl rings [21]. Furthermore, beyond the ring above systems, the inhibitors could potentially contain dimensions of rings that differ, for instance, macrocyclic compounds.

The enhanced availability of dynamic positions for adsorption and improved exterior coverage in organic molecules can be attributed to the incidence of augmented numbers of π-bonds, heteroatoms, and larger ring sizes. A range of substituent groups may be appended to the inhibitor structure to constitute an organic corrosion inhibitor. The presence of substituent groups on aromatic rings may exert either an electron-withdrawing or electron-donating effect, thereby modifying the electron density of the ring. The electron density of a given ring can be modified by functional groups possessing electron-withdrawing properties, which decrease said density, or by functional groups with electron-donating properties which, conversely, increase it. Hence, the existence of electron-withdrawing moieties may potentially diminish the effectiveness of inhibition. On the contrary, introducing electron-donating groups can potentially increase efficacy [22–24].

The impact of substituent groups on the inhibition behavior is subject to variation based on molecular dimension, solubility, and electronic effects. The impact of substituent assemblies on the inhibitor's molecular size may result in surface coverage variations and consequent fluctuations in protective efficacy. The incorporation of polar substituent moieties has the potential to augment the solubility of a compound in an aqueous environment. In contrast, including non-polar substituent groups can reduce its aqueous solubility. The electron-donating properties of certain substituent groups, including but not limited to $-CH_3$, $-N(CH_3)_2$, $-NHCH_3$, $-NH_2$, $-OCH_3$, and halogens, have been acknowledged in academic literature. These clusters are postulated to facilitate the movement of electrons toward the interior of the phenyl ring, thereby preventing erosion. In contrast, functional groups such as $-CN$, $-NO_2$, $>C=O$, and $-COOH$ exhibit an electron-withdrawing nature, reducing steric hindrance efficacy. The relationship between inhibition efficiency and functional groups can be quantified using the Hammett substitution constant (r) [3, 21].

14.1.3 CONCEPTUALIZATION OF HETEROCYCLIC COMPOUNDS

According to its definition, a heterocyclic compound is an organic cyclic compound that features the replacement of one or more carbon atoms with other atoms, including, but not limited to, nitrogen (N), sulfur (S), oxygen (O), phosphorus (P), and so forth [3, 25]. The illustration depicted in Figure 14.3 delineates the configurations and nomenclatures of select essential heterocyclic compounds.

A diverse array of heterocyclic compounds has been documented in the literature as exhibiting corrosion-inhibiting capabilities. These compounds include, but are not limited to, pyridines [26], pyrimidines [27], pyrazole [28], imidazoles [29], triazoles [30], tetrazoles [31], indoles [32], thiazoles [33, 34], oxadiazoles [35], benzotriazoles [36, 37], benzimidazoles [38, 39], and macrocyclic compounds [40, 41]. Additionally, specific biomolecules have also demonstrated inhibitory effects on corrosion. Heterocyclic compounds serve as the foundational backbone of several medication molecules. The plant extracts are known to contain phytochemicals that are constituted by intricate heterocyclic structures. Numerous amino acids [42], proteins, and carbohydrates [43], among other biomolecules, are present. The building blocks of these compounds consist of heterocycles, as corroborated by various academic sources [3, 25, 44, 45].

Consequently, all categories above of molecules can act as prospective anticorrosion agents. The heteroatoms nitrogen (N), sulfur (S), and oxygen (O) possess non-bonding

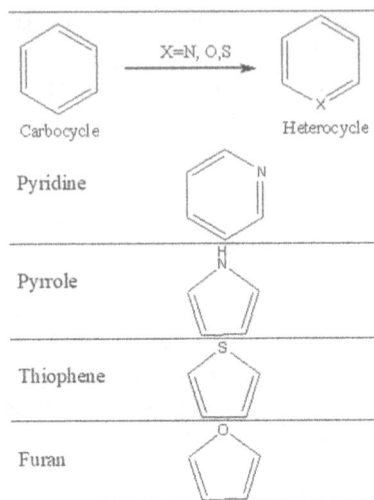

FIGURE 14.3 Examples of heterocyclic compounds.

pairs of electrons, rendering these molecules capable of coordinating with metal exteriors through the formation of chemical bonds. The pH of the harsh electrolyte influences the protonation/deprotonation of heteroatoms and enables inhibitor adsorption through electrostatic interactions. These heteroatoms facilitate such electrostatic interactions. Incorporating the p-electron cloud within the heterocyclic ring can lead to adsorption in a planar orientation, which facilitates interactions with the highest possible surface zone. Furthermore, the incidence of polar functional assemblies with electron-withdrawing or electron-donating properties may promote surface adsorption [3].

The heteroatoms within a heterocyclic corrosion inhibitor possess a readily available lone pair of electrons that may be contributed to the targeted metal. This, in turn, facilitates the effective chemical adsorption of the inhibitor molecule. Moreover, in instances where heteroatoms exist within an environment that is either acidic or alkaline, these atoms can undergo a process of protonation or deprotonation, leading to the emergence of a positive or negative charge on said atoms. This phenomenon can facilitate electron acceptance through a back-donation from the metal atoms or the donation of electrons to the metal surface [1, 3].

14.2 N-BASED HETEROCYCLES AS CORROSION INHIBITORS

Hydrochloric acid (HCl) is a frequently employed chemical agent in eradicating rust and undesired scale from metallic surfaces. Additionally, it is commonly used to clean heat exchangers and boilers. The occurrence of scorched covering is attributed to the excessive pickling of metals. Establishing a defensive layer on the steel exterior and the consequent representation of the metal exterior have emerged as crucial focus areas. Carbon steel is a prominent engineering material that finds extensive usage in numerous applications, including but not limited to marine operations, fossil fuel and nuclear power plants, metal-processing machinery, chemical-processing industries, transportation activities, pipeline systems, mining, and construction ventures. Using iron alloys as structural components in the industrial field presents a significant challenge for corrosion scientists and engineers in modern-day contexts. Corrosion inhibitors are regarded as critical for safeguarding many metals and alloys. Consequently, an investigation into utilizing organic compounds as anti-corrosion agents is warranted. The adsorption phenomenon of organic molecules on the metal exterior is contingent upon their molecular architecture and surface charge density. A significant figure of literature attests to the usefulness of organic compounds that carry nitrogen (N) as part of an aromatic or a heterocyclic ring in serving as anticorrosion agents for steel in corrosive settings when adsorbed onto the surface of a metal. The optimal selection of inhibitors is contingent upon several factors: the variety and concentration of acid, ambient temperature, dissolved organic and/or inorganic constituents, even in trace quantities, and the composition of the metallic material being safeguarded [46].

14.2.1 PYRROLE-BASED CORROSION INHIBITORS

The utilization of pyrrole and pyrrole-based derivatives may encompass various applications in producing medicinally active compounds comprising cholesterol-reducing agents, antifungals, and antibiotics [46]. Pyrrole derivatives have been employed in diverse applications, including usage as solvents for resin, luminescence chemistry, catalysts for polymerization processes, synthesis of alkaloids, and corrosion inhibitors [46]. The contemporary research systematically investigates the inhibition effect of novel pyrrole derivatives (Table 14.1) [46] on the corrosion of carbon steel in a 1.0 M HCl solution, employing electrochemical techniques such as potentiodynamic polarization measurements, impedance spectroscopy, and electrochemical frequency modulation. The presented findings illustrate that compound (A) exhibited the most effective inhibition activity of the analyzed compounds, measuring 82.8% at a concentration of 21×10^{-6} M [46].

14.2.1.1 Adsorption Isotherm

The inhibition results in highly acidic media are believed to result from adsorption occurring at the edge of the metal and solution. How adsorption occurs is contingent upon variables, including the constitution of the compounds, modifications to the compounds at a molecular level, and the inherent surface charge of the metallic materials in question. The preferential adsorption of anions is governed by the presence of a positive exterior charge, and conversely, a negative external charge promotes cation adsorption. This phenomenon indicates the electrostatic interaction between the surface and ionic species in solution [46, 47].

To derive the adsorption isotherm, it is necessary to determine the degree of exterior coverage (θ) for the involved compounds. Several scientific formulations have been developed to consider nonideal consequences for determining the adsorption mode. The Frumkin, De Boer, Parsons, Temkin, Flory–Huggins, and Bockris–Swinkels isotherms are the most commonly utilized models [47–50]. The inhibition phenomenon adheres to the Langmuir adsorption isotherm model, described by the corresponding formula [51]:

$$C/\theta = 1/K_{ads} + C \tag{14.1}$$

TABLE 14.1

Structure of Pyrrole-Based Corrosion Inhibitors [46]

Name of Pyrrole-based Corrosion Inhibitors	Structure
(A) 13-(4-Aminophenyl)sulfonyl)-10,11-dihydro-9H-9,10-[3,4]epipyrrolo anthracene-12,14(13H,15H)-dione	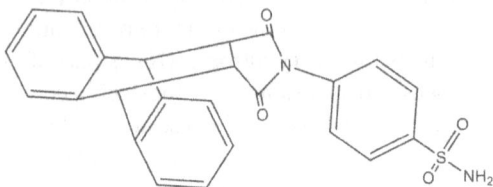
(B) Anthracen-9(10H)- ylidenehydrazine	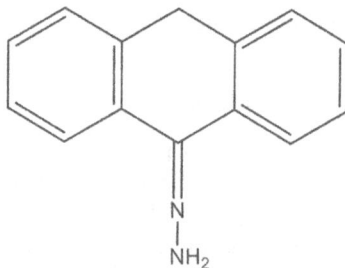
(C) (11S,15R)-13-(Anthracen-9(10)-ylideneamino)10,11-dihydro-9H-9,10[3,4]epipyrroloanthracene-12,14-(13H,15H)-dione	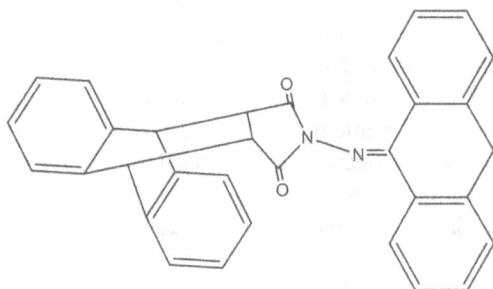
(D) 1,6-Bis-(N-hexachloro-5-norbornene-2,3-dicarboximidyl) hexane	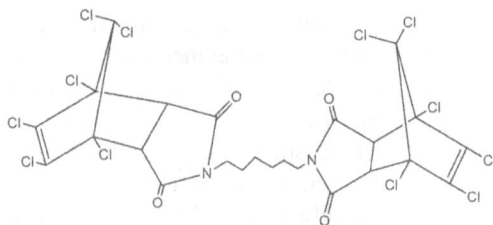
(E) N-Phenylsulfonyloxy-hexachloro-5-norbornene-2,3-dicarboximide	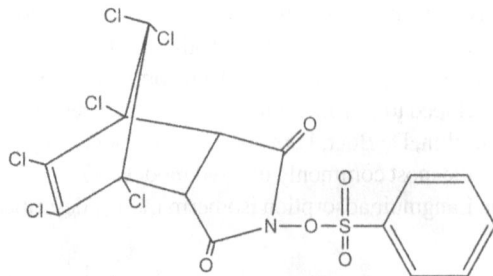

In this study, the molar concentration of inhibitors is represented by C, while K_{ads} denotes the equilibrium constant of the adsorption procedure. Additionally, the extent to which inhibitors' molecules cover the exterior of metal is expressed as θ. The straight line is attained by plotting the ratio of C to θ against C, and the degree of linear correlation observed in the adjusted data approaches unity. The K_{ads} value may be exploited to assess the adsorption forces' potency between the metallic exterior and the inhibitor molecules. The elevated K_{ads} values indicate the inhibitor's enhanced capability to adsorb onto the exterior of carbon steel. Through the utilization of Equation 14.2, it is possible to derive the numerical values of the standard Gibbs free energy (ΔG_{ads}) about adsorption [52]:

$$K_{ads} = \frac{1}{55.5} \exp^{\left(-\Delta G_{ads}/RT\right)} \tag{14.2}$$

Equation 14.2 may be more academic: R represents the universal gas constant, T denotes the thermodynamic temperature, and the molar concentration of water in the solution is expressed as 55.5 mol L^{-1}. The negative values of Gibbs free energy of adsorption (ΔG_{ads}) denote the process of adsorption's inherent spontaneity and further signal the steadfastness of the adsorbed film on the exterior of carbon steel. Theoretically, a range of values for the adsorption-free energy, denoted as ΔG_{ads}, that fall within the vicinity of -20 kJ mol^{-1} or exhibit positivity could indicate physisorption. Conversely, values approaching or surpassing -40 kJ mol^{-1} represent chemisorption [53]. The range of ΔG_{ads} values, from -40.1 kJ mol^{-1} to -42.6 kJ mol^{-1}, reveals a chemisorption mechanism for the adsorption of complexes on carbon steel in a 1.0 M HCl solution [46].

14.2.1.2 Inhibition Mechanism

The phenomenon of pyrrole derivatives compounds being adsorbed is explicable by polar units, characterized by atoms of N, S, and O, as well as aromatic or heterocyclic rings. Henceforth, the feasible response sites entail the unshared electron pair of heteroatoms and the aromatic ring's π-electrons. In 1.0 M HCl solution, pyrrole derivatives compounds exhibit an adsorption and inhibition impact that can be elucidated in the following manner: through a chemisorption mechanism, neutral molecules can adhere to the alloy surface. This process involves water molecules displaced from the alloy surface and electrons shared between the heteroatoms and metal. The amalgam exterior can attract inhibitor molecules over a donor–acceptor interaction, where the aromatic ring's π-electrons bind with the unoccupied d-orbitals of Fe atoms on the surface. The hierarchy of inhibition is reduced in the subsequent order: A > C > E > D > B. The observed phenomenon is attributed to variations in the quantity of nitrogen, oxygen, and sulfur atoms present within the compound, resulting in an augmented electron charge density within the molecular structure [46].

14.3 FUNCTIONAL DENSITY THEORY (DFT) MODELLING FOR 1*H* PYRROLE INHIBITOR

The corrosion-inhibiting abilities of organic corrosion inhibitors are explicated using deploying quantum chemical computations to elucidate their structural features. The density functional theory (DFT) is a significant quantum chemical scheme employed to determine the electronic components of molecular species. Moreover, implementing DFT is frequently used to analyze the correlations between inhibitors and surfaces in studies related to corrosion prevention. In theoretical investigations concerning corrosion inhibition and support of empirical measurements, electronic parameters garnered through quantum chemical methodologies are commonly employed [54].

This investigation employs a computational quantum approach to investigate the erosion reticence attributes of $1H$-pyrrole molecules over the application of DFT analysis, employing a 6-31G (d, p) basis set. The parameters that directly correlate with the inhibition efficacy level are accurately characterized. In the preliminary stage of computational research, an essential aspect is establishing the optimized molecular configuration. The quantum chemical properties of the molecules under investigation were evaluated by determining various parameters. These parameters include the energies of the highest occupied molecular orbital (E_{HOMO}) and the lowest unoccupied molecular orbital (E_{LUMO}), which were found to be -5.503 eV and 1.346 eV, respectively. The energy gap (ΔE) between these two orbitals was also calculated to be 6.849 eV. The dipole moment of the molecules (μ) was determined to be 1.902, while the absolute electronegativity (χ) and chemical hardness (η) were calculated as 2.079 and 3.424, respectively. Furthermore, the global electrophilicity index (ω) and chemical softness (S) were evaluated and found to be 0.631 and 0.289, respectively. Finally, the fraction of electrons transferred (ΔN) between the molecules was calculated to be 0.825. According to the obtained data, the inhibitor of molecule 1 ($1H$-pyrrole) exhibits significant inhibition activity, suggesting its potential use as an effective anticorrosion agent. This finding was determined by applying DFT [8, 54].

14.3.1 PYRROLIDINE-BASED CORROSION INHIBITORS

The utilization of inhibitors represents the most viable erosion fortification method, predominantly in corrosive situations. Including these entities is imperative in protecting metals from potentially corrosive media, such as chemical cleaning and pickling procedures commonly employed to eliminate mill or oxide scales from metallic surfaces. Inhibitors ought to exhibit efficacy under challenging conditions such as concentrated acid concentrations of 20% and temperatures within the range of 333 K to 368 K [55]. The inhibitive properties of organic compounds containing heteroatoms (nitrogen, sulfur, or oxygen) that possess a lone pair of electrons and π-electrons present in conjugated double or triple bonds are attributed to the functional motifs displayed by these compounds. This could be elucidated by the interaction of electrons with the exterior of metal, leading to effective inhibition [3]. The facile adsorption of N-heterocyclic molecules onto metal surfaces via N heteroatom has been established, and they have been found to manifest superior inhibitive performance in reducing corrosion levels under acidic conditions [4, 5]. According to a review of relevant literature, pyrrolidone compounds have demonstrated efficacy as corrosion inhibitors at temperatures reaching 353 K. The amalgamation of novel organic molecules presents many molecular structures encompassing multiple heteroatoms and substituents. The phenomenon of adsorption is often explicated through the formation of a layer that is adsorptive in nature and which may possess both chemical and physical attributes on the substrate made of metal [55].

A set of corrosion inhibitors based on pyrrolidone were synthesized, including N,N-dipropargylpyrrolidium bromide (DPPB), N-dodecyl, N-propargylpyrrolidium bromide (DDPPB), and *N*-hexadecyl, *N*-propargylpyrrolidium bromide (HDPPB). These inhibitors demonstrated significant efficacy in inhibiting corrosion, even at low concentrations of 60.3 μmol L^{-1}, with efficiency values of 92.6%, 93.7%, and 96.2%, respectively, in aqueous acidic media [55, 56]. Table 14.2 shows the structures of pyrrolidine-based corrosion inhibitors [55, 56].

TABLE 14.2

Structure of Pyrrolidine-Based Corrosion Inhibitors [55, 56]

Name of Pyrrolidine-based Corrosion Inhibitors	Structure
{[2-(2-Oxopyrrolidin-1-yl) ethyl]thio}acetic acid	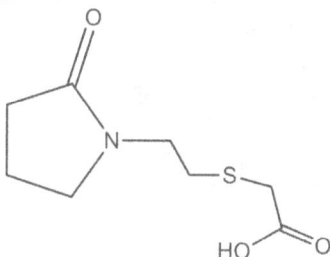
1-{2-[(2-Hydroxyethyl)thio] ethyl}pyrrolidone-2-one	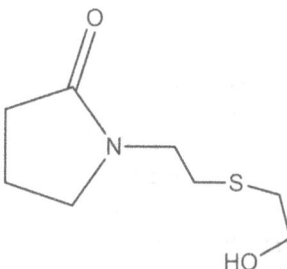
N,N-Dipropargylpyrrolidium bromide (DPPB)	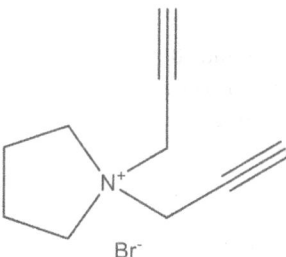
N-Dodecyl, *N*-propargylpyrrolidium bromide (DDPPB)	
N-Hexadecyl, *N*-propargylpyrrolidium bromide	

14.3.1.1   Inhibition Mechanism

The contemporary research investigated the effect of varying concentrations of anticorrosion agents, namely DPPB, DDPPB, and HDPPB, on the erosion mitigation of mild steel in a 1 M HCl solution. The experiment outcomes demonstrated that HDPPB is the optimal inhibitive effect surrounded by the three anticorrosion agents under scrutiny. The capacity to use adsorption on the metallic exterior over physical and chemisorptive processes is present. Physisorption and chemisorption are distinct phenomena linking an inhibitor and a metallic surface at the interface. Electrostatic interactions between the ionic charges in the inhibitive agent and the metal surface mainly control the physisorption mechanism. This statement denotes a scientific perspective on the nature of physisorption and the factors that influence this interfacial phenomenon. The chemisorption process is commonly explained as the result of charge transfer or electron allocation between the inhibitive agent's lone electron pairs and the Fe surface [56, 57]. This research report presents the ΔG_{ads} concerning the anticorrosion agents DPPB, DDPPB, and HDPPB. The obtained data exhibited values within the −39.4 to −43.6 kJ mol^{-1} ranges. These results suggest that the adsorption route potentially follows a diverse process, with chemisorption as the primary controlling factor. The investigation revealed that the anticorrosion agent, HDPPB, demonstrated the uppermost value of ΔG_{ads} (−43.6 kJ mol^{-1}), suggesting that the inclusion of an extended dimension having alkyl group led to an enhancement of the chemisorption interface concerning the inhibitive agent and mild steel. Such findings suggest a potential strategy for improving corrosion inhibition efficacy. The inhibitor HDPPB comprises a cationic moiety containing a pyrrolidine ring, a prop-1-yne, and an alkyl chain (C-16), alongside a counteranion in the form of a bromide ion. Determining the E_{corr} value for mild steel submerged in a 1 M HCl solution yields a result of −459 mVSCE. According to the findings, the value of the potential at zero charges is deemed below zero (PZC > 0). This consequently creates a negatively charged metal surface [58]. The steric hindrance exerted by cyclic quaternary amine nitrogen is comparatively lower than that of acyclic quaternary amine. Consequently, it can be positioned near the metal exterior and, subsequently, interact more efficiently with the adsorbed negative chloride ions [59].

Furthermore, it was observed that the bromide anion exhibited a greater degree of synergistic activity compared to the chloride ion, thereby facilitating a more substantial interaction between cationic-charged ions [60]. The inhibitor molecules containing a terminal alkyne group possess four π-electrons, which can contribute an electron to the vacant d-orbital of Fe, facilitating chemisorption. Concurrently, the anti-bonding π-orbital of the alkyne moiety can accept an electron from the occupied orbital of Fe, thereby augmenting the degree of chemisorption of HDPPB on the metallic substrate through retro-donation. About alkynes, a possibility has been recommended that it undergoes polymerization, leading to the creation of a defensive film or coating on the exterior of the metal [61]. The resulting layer functions to impede mass transport and consequently delays the process of corrosion [62]. Furthermore, it is worth noting that the inhibition enactment is heightened by the reduction of alkyl chain length to less than 18 carbon atoms, which may be attributed to the densification of chain packing and subsequent decrease in the geometric volume of the chain. This observation holds significant implications for designing and optimizing inhibition strategies [63]. Additionally, introducing longer alkyl chains enhances the hydrophobic characteristics exhibited by the exterior of metal and mitigates the rate of diffusion of harsh ions toward it. The current investigation has determined that the HDPPB inhibitor containing the C-16 atom displays outstanding inhibitory performance at elevated temperatures and extended duration of immersion when compared to the hydrophobic alkyl chain DDPPB containing the C-12 atom and the corrosion inhibitor DPPB containing the alkyne group. Experimental findings provide evidence for this conclusion [56].

14.3.2 PYRIDAZINE-BASED CORROSION INHIBITORS

The pyridazine moiety is a heterocyclic system with a six-membered ring with two neighboring nitrogen atoms. Substitution of a single CH unit in a pyridine moiety forms three distinct types of aromatic diazines, specifically pyridazine, pyrimidine, and pyrazine. The presence of an extra nitrogen atom compared to pyridine augments pyridine chemistry's fundamental characteristics. The pyridazine functional group is present in several pharmacological agents, including cefozopran, cadralazine, minaprine, pipofezine, and hydralazine. Chetouani et al. [64] conducted an intriguing investigation to compare the impact of $=S$ and $=O$ groups as a moiety in two pyridazines concerning the behavior of pure iron in a 1 M HCl solution [64]. The derivative containing S demonstrated a significantly greater efficacy of 98% compared to the product containing O, which exhibited an efficacy of 80%. The study revealed a decrease in efficiency as the temperature was raised from 298 K to 353 K.

A consequent analysis assessed a carboxylic acid derivative with an S atom and ester derivatives with two N- and $-O$ of the aforementioned inhibitive agents [65]. The result of carboxylic acid exhibited a performance rate of 93% when present at a concentration of 10^{-4} M [66]. Including the S atom significantly heightened the usefulness of the inhibitive phenomena. The replacement of oxygen with sulfur in a distinct subset of pyridazines has been found to significantly improve the efficacy of steel corrosion inhibition, with an inhibition efficiency of 99% observed in a 0.5 M H_2SO_4 solution. The intensification in temperature from 298 K to 323 K did not result in significant changes to the level of inhibition. However, when the temperature rose further to 343 K, there was a noticeable decrease in the level of inhibition. This study investigated four pyridazine derivatives incorporating functional groups such as $=O$, $=S$, and substituted $-CH_3$, $-OH$ [67]. The high proficiency of 92% was endorsed to the presence of both S and $-CH_3$. The corrosion inhibition performance of 3,6-bis(3-pyridyl)pyridazine on carbon steel in a 1 M HCl solution was investigated [68]. EIS investigations publicized that charge relocation processes determine the observed behavior and involve a combination of inhibitory mechanisms. The X-ray analysis revealed the occurrence of N—C bond formation, thus providing evidence of the chemical interface concerning the anticorrosion agents and the metallic substrate. A subsequent investigation indicated that replacing oxygen with sulfur yielded a favorable outcome [69]. The existence of the S atom was also validated in other studies [70, 71]. The impact of the $-Cl$ substituent in pyridazine thione on the inhibitive effect of steel in a 0.5 M H_2SO_4 solution was investigated. At a concentration of 5×10^{-4} M, the inhibitor featuring a 2-chlorophenyl substituent exhibited nearly complete efficiency [72].

This study examined four pyridazine derivatives in the presence of copper ions within a solution containing 2 M HNO_3 [73]. It was found that including chlorine groups at positions 2 and 6 significantly enhanced the efficiency of the reaction, resulting in a remarkable 96.2% yield at a concentration of 10^{-3} M. The computational investigations conducted using the DFT/B3LYP level facilitated the association of reactivity parameters with the molecular architecture [74]. Four pyridazines were created and subsequently subjected to analysis for their effect on mild steel [75]. The computational analysis findings demonstrate the pyridazine ring's interface mechanism with a metallic substrate, in both the frontward and reverse directions, during adsorption in a planar conformation. The compound 3,6-di(pyridin-2-yl)pyridazine derivative was subjected to assessment in a solution containing 1 M HCl [76]. The existence of the isatin group within the pyridazine ring resulted in a significantly elevated inhibitive capacity at a concentration of 10^{-4} M. The study encompassed an investigation of five distinct sulfanyl pyridazines [77]. The recent findings suggest that the inhibitive consequence of the substrate under investigation was positively associated with the concentration of said substrate. Conversely, an inverse correlation was observed between the inhibitory effect and temperature [3].

14.3.3 PYRIMIDINE-BASED CORROSION INHIBITORS

Pyrimidine represents an additional constituent of the diazine family, distinguished by the occurrence of two nitrogen (N) elements at the 1 and 3 sites. They comprise a substantial member of nucleic acids. The pyrimidine pharmacophore is essential in various biological processes and possesses significant pharmacological potential as antibiotics, cardiovascular, antibacterial, and veterinary products. Several pyrimidine derivatives, including but not limited to piritrexim, uramustine, isethionate, tegafur, and fluorouracil, have garnered significant commercial popularity as pharmaceutical drugs. The occurrence of two pairs of electrons in isolation on each N atom endows the structure with exceptional stability, rendering it an exemplary heterocyclic entity [78–84]. Pyrimidine derivatives have been observed to exhibit remarkable corrosion inhibition properties. Elewady et al. [85] have presented a scholarly contribution in their work. The compound 2,6-dimethylpyrimidine-2-amine and its two derivatives were examined in a study and found that the efficiency of protection demonstrated an uninterrupted correlation with the amount of the inhibitive agent [85, 86].

Caliskan and Akbas reported using 5-benzoyl-4-(substituted phenyl)-6-phenyl-3,4-dihydropyrimidine-2 as an effective compound for austenitic stainless steel [87]. The nature of the inhibitive agent was identified as diverse, as confirmed by a prominently negative magnitude of ΔG_{ads} [88, 89]. Another investigation involved analyzing a limited number of amino-pyrimidines under exposure to 1018 steel in an environment containing 0.05 M HNO_3 [86]. The incorporation of potassium iodide (KI) yielded a beneficial synergistic effect on the prevention of corrosion. A range of 2-mercaptopyrimidine derivatives was explored for their impending as inhibitive agents for ARMCO Fe and low amalgam steels in a solution of 1 M H_2SO_4 [90]. Table 14.3 displays illustrative pyrimidine-based anticorrosion agents [3, 91–100].

TABLE 14.3

Structure of Some Pyrimidine-based Corrosion Inhibitors [3, 91–100]

Material	Medium	Name Pyrimidine-based Corrosion Inhibitors	Structure	Inhibition Competence (%)
Mild steel	1 M HCl	Dihydropyrimidinone		98.8
Mild steel	1 M HCl	Condensed uracil		96.1
Mild steel	1 M HCl	Fused pyrimidines		96.52
Mild steel	1 M HCl	Dihydrodipyrimidine		97.82
N80 steel	15% HCl	Pyrimidine derivatives		89.1

(Continued)

TABLE 14.3   (Continued)
Structure of Some Pyrimidine-based Corrosion Inhibitors [3, 91–100]

Material	Medium	Name Pyrimidine-based Corrosion Inhibitors	Structure	Inhibition Competence (%)
Mild steel	1 M HCl	Thiopyrimidines		98.58
Mild steel	1 M H$_2$SO$_4$	Thiazolopyrimidine		99
Mild steel	1 M HCl	Pyrimidopyrimidine		97.1
Al steel	1 M HCl	Pyridine–pyrimidine derivative		95.6
Mild steel	1 M HCl	Pyrazole–pyrimidine derivative		92

14.4 CONCLUSION

The erosion of metals and amalgams is a significant deleterious occurrence prevalent in industrial applications. A general prevention method involves incorporating organic inhibitive agents as supplementary agents. One of the greatest efficacious categories of organic compounds is the heterocyclic compounds. The *N*- heterocyclic compounds, such as pyrrole derivatives, pyrrolidine, pyrazines, and pyrimidines, are comprehensively employed as inhibitive agents. However, most extant corrosion inhibitors present considerable difficulties in production, are inherently toxic, and entail significant costs. Based on the current discourse, it is apparent that compounds found in pyrrole, pyrrolidine, pyradiazines, and pyrimidines exhibit potent anticorrosion properties, rendering them desirable materials for inhibiting corrosion.

REFERENCES

1. M.A. Quraishi, D.S. Chauhan, V.S. Saji, Heterocyclic Organic Corrosion Inhibitors: Principles and Applications, Elsevier, 2020.
2. V.S. Sastri, Corrosion Inhibitors: Principles and Applications, John Wiley & Sons, Inc., New York, 1998.
3. M.A. Quraishi, D.S. Chauhan, V.S. Saji, Heterocyclic biomolecules as green corrosion inhibitors, J. Mol. Liq. 341 (2021) 117265.
4. M. Murmu, S.K. Saha, P. Bhaumick, N.C. Murmu, H. Hirani, P. Banerjee, Corrosion inhibition property of azomethine functionalized triazole derivatives in 1 mol L^{-1} HCl medium for mild steel: Experimental and theoretical exploration, J. Mol. Liq. 313 (2020) 113508.
5. A. Suhasaria, M. Murmu, S. Satpati, P. Banerjee, D. Sukul, Bis-benzothiazoles as efficient corrosion inhibitors for mild steel in aqueous HCl: Molecular structure–reactivity correlation study, J. Mol. Liq. 313 (2020) 113537.
6. C. Verma, L. Olasunkanmi, E.E. Ebenso, M.A. Quraishi, Substituents effect on corrosion inhibition performance of organic compounds in aggressive ionic solutions: A review, J. Mol. Liq. 251 (2018) 100–118.
7. L. Guo, I.B. Obot, X. Zheng, X. Shen, Y. Qiang, S. Kaya, C. Kaya, Theoretical insight into an empirical rule about organic corrosion inhibitors containing nitrogen, oxygen, and sulfur atoms, Appl. Surf. Sci. 406 (2017) 301–306.
8. L. Ahmed, R. Omer, 1*H*-Pyrrole, furan, and thiophene molecule corrosion inhibitor behaviors, J. Phy.Chem. Funct. Mater. 4:2 (2021) 1–4.
9. D.K. Yadav, M.A. Quraishi, Application of some condensed uracils as corrosion inhibitors for mild steel: Gravimetric, electrochemical, surface morphological, UV–visible, and theoretical investigations, Ind. Eng. Chem. Res. 51:46 (2012) 14966–14979.
10. M. Bobina, et al., Corrosion resistance of carbon steel in weak acid solutions in the presence of L-histidine as corrosion inhibitor, Corr. Sci. 69 (2013) 389–395.
11. H. El Sayed, S.A. Senior, QSAR of lauric hydrazide and its salts as corrosion inhibitors by using the quantum chemical and topological descriptors, Corr. Sci. 53:3 (2011) 1025–1034.
12. S. Zehra, M. Mobin, R. Aslam, Corrosion Prevention and Protection Methods, in: L. Guo, C. Verma, D. Zhang (Eds.) Eco-Friendly Corrosion Inhibitors, Principles, Designing and Applications, Elsevier, 2022.
13. X. Han, Y. Zhu, X. Yang, C. Li, Electrocatalytic activity of Pt doped TiO$_2$ nanotubes catalysts for glucose determination, J. Alloys Compd. 500:2 (2010) 247–251. https://doi.org/10.1016/j.jallcom.2010.04.019.

14. J. Haque, V. Srivastava, D.S. Chauhan, M.A. Quraishi, A.M. Kumar, H. Lgaz, Electrochemical and surface studies on chemically modified glucose derivatives as environmentally benign corrosion inhibitors, Sustainable Chem. Pharm. 16 (2020) 100260.

15. S. Cao, D. Liu, H. Ding, J. Wang, H. Lu, J. Gui, Task-specific ionic liquids as corrosion inhibitors on carbon steel in 0.5 M HCl solution: An experimental and theoretical study, Corros. Sci. 153 (2019) 301–313.

16. M.A. Mazumder, H.A. Al-Muallem, S.A. Ali, The effects of N-pendants and electron-rich amidine motifs in 2-(p-alkoxyphenyl)-2-imidazolines on mild steel corrosion in CO_2-saturated 0.5 M NaCl, Corros. Sci. 90 (2015) 54–68.

17. B. Tan, S. Zhang, Y. Qiang, W. Li, H. Li, L. Feng, L. Guo, C. Xu, S. Chen, G. Zhang, Experimental and theoretical studies on the inhibition properties of three diphenyl disulfide derivatives on copper corrosion in acid medium, J. Mol. Liq. 298 (2020) 111975.

18. Y. Qiang, H. Li, X. Lan, Self-assembling anchored film basing on two tetrazole derivatives for application to protect copper in sulfuric acid environment, J. Mater. Sci. Technol. 52 (2020) 63–71.

19. Y. Qiang, S. Zhang, H. Zhao, B. Tan, L. Wang, Enhanced anticorrosion performance of copper by novel N-doped carbon dots, Corros. Sci. 161 (2019) 108193.

20. B. Tan, S. Zhang, H. Liu, Y. Guo, Y. Qiang, W. Li, L. Guo, C. Xu, S. Chen, Corrosion inhibition of X65 steel in sulfuric acid by two food flavorants 2- isobutylthiazole and 1-(1,3-thiazol-2-yl)ethanone as the green environmental corrosion inhibitors: Combination of experimental and theoretical researches, J. Colloid Interface Sci. 538 (2019) 519–529.

21. M.A. Quraishi, D.S. Chauhan, V.S. Saji, Heterocyclic Organic Corrosion Inhibitors: Principles and Applications, Elsevier Inc., Amsterdam, 2020.

22. Y.G. Skrypnik, T. Doroshenko, S.Y. Skrypnik, On the influence of the nature of substituents on the inhibiting activity of meta- and para-substituted pyridines, Mater. Sci. 31 (1996) 324–330.

23. E. Schwoegler, L. Berman, The evaluation of certain organic nitrogen compounds as corrosion inhibitors, Corrosion 15 (1959) 44–46.

24. E. Ebenso, U. Ekpe, B. Ita, O. Offiong, U. Ibok, Effect of molecular structure on the efficiency of amides and thiosemicarbazones used for corrosion inhibition of mild steel in hydrochloric acid, Mater. Chem. Phys. 60 (1999) 79–90.

25. J.A. Joule, K. Mills, Heterocyclic Chemistry at a Glance, John Wiley & Sons, Inc., New York, 2012.

26. C.T. Ser, P. Z˘uvela, M.W. Wong, Prediction of corrosion inhibition efficiency of pyridines and quinolines on an iron surface using machine learning–powered quantitative structure–property relationships, Appl. Surf. Sci. 512 (2020) 145612.

27. K. Rasheeda, D. Vijaya, P. Krishnaprasad, S. Samshuddin, Pyrimidine derivatives as potential corrosion inhibitors for steel in acid medium: An overview, Int. J. Corros. Scale Inhib. 7 (2018) 48–61.

28. M. Yadav, R.R. Sinha, T.K. Sarkar, N. Tiwari, Corrosion inhibition effect of pyrazole derivatives on mild steel in hydrochloric acid solution, J. Adhes. Sci. Technol. 29 (2015) 1690–1713.

29. M.B.P. Mihajlovic´, M.B. Radovanovic´, Z˘.Z. Tasic´, M.M. Antonijevic´, Imidazole based compounds as copper corrosion inhibitors in seawater, J. Mol. Liq. 225 (2017) 127–136.

30. M. ElBelghiti, Y. Karzazi, A. Dafali, B. Hammouti, F. Bentiss, I. Obot, I. Bahadur, E.-E. Ebenso, Experimental, quantum chemical and Monte Carlo simulation studies of 3,5-disubstituted-4-amino-1,2,4-triazoles as corrosion inhibitors on mild steel in acidic medium, J. Mol. Liq. 218 (2016) 281–293.

31. J. Seetharaman, E.A. Reny, D.A. Johnson, K.B. Sawant, V. Sivaswamy, Tetrazole based corrosion inhibitors, US 9771336 B2, 2017.
32. C. Verma, M.A. Quraishi, E. Ebenso, I. Obot, A. El Assyry, 3-Amino alkylated indoles as corrosion inhibitors for mild steel in 1 M HCl: Experimental and theoretical studies, J. Mol. Liq. 219 (2016) 647–660.
33. M.A. Quraishi, W. Khan, M. Ajmal, The influence of some condensation products of aminobenzothiazoles and salicylaldehyde on corrosion inhibition and hydrogen permeation in sulphuric acid solution, J. Electrochem. Soc. India 46 (1997) 133–138.
34. M.A. Quraishi, W. Khan, M. Ajmal, S. Muralidharan, S.V. Iyer, Influence of substituted benzothiazoles on corrosion in acid solution, J. Appl. Electrochem. 26 (1996) 1253–1258.
35. M.A. Quraishi, F.A. Ansari, Fatty acid oxadiazoles as corrosion inhibitors for mild steel in formic acid, J. Appl. Electrochem. 36 (2006) 309–314.
36. M. Albini, P. Letardi, L. Mathys, L. Brambilla, J. Schröter, P. Junier, E. Joseph, Comparison of a bio-based corrosion inhibitor versus benzotriazole on corroded copper surfaces, Corros. Sci. 143 (2018) 84–92.
37. C. Gattinoni, A. Michaelides, Understanding corrosion inhibition with van der Waals DFT methods: The case of benzotriazole, Faraday Discuss. 180 (2015) 439–458.
38. I. Obot, U.M. Edouk, Benzimidazole: Small planar molecule with diverse anticorrosion potentials, J. Mol. Liq. 246 (2017) 66–90.
39. I. Obot, A. Madhankumar, S. Umoren, Z. Gasem, Surface protection of mild steel using benzimidazole derivatives: Experimental and theoretical approach, J. Adhes. Sci. Technol. 29 (2015) 2130–2152.
40. K.R. Ansari, S. Ramkumar, D.S. Chauhan, M. Salman, D. Nalini, V. Srivastava, M.A. Quraishi, Macrocyclic compounds as green corrosion inhibitors for aluminium: Electrochemical, surface and quantum chemical studies, Int. J. Corros. Scale Inhib. 7 (2018) 443–459.
41. S. Hadisaputra, S. Hamdiani, M.A. Kurniawan, N. Nuryono, Influence of macrocyclic ring size on the corrosion inhibition efficiency of dibenzo crown ether: A density functional study, Indones. J. Chem. 17 (2017) 431–438.
42. B. El Ibrahimi, A. Jmiai, L. Bazzi, S. El Issami, Amino acids and their derivatives as corrosion inhibitors for metals and alloys, Arabian J. Chem. 13 (2020) 740–771.
43. S.A. Umoren, U.M. Eduok, Application of carbohydrate polymers as corrosion inhibitors for metal substrates in different media: A review, Carbohydr. Polym. 140 (2016) 314–341.
44. D.S. Chauhan, K.R. Ansari, A.A. Sorour, M.A. Quraishi, H. Lgaz, R. Salghi, Thiosemicarbazide and thiocarbohydrazide functionalized chitosan as ecofriendly corrosion inhibitors for carbon steel in hydrochloric acid solution, Int. J. Biol. Macromol. 107 (2018) 1747–1757.
45. S. Kirchhecker, M. Antonietti, D. Esposito, Hydrothermal decarboxylation of amino acid derived imidazolium zwitterions: A sustainable approach towards ionic liquids, Green Chem. 16 (2014) 3705–3709.
46. H.S. Gadow, H.M. Dardeer, The corrosion inhibition effect of pyrrole derivatives on carbon steel in 1.0 M HCl, Int. J. Electrochem. Sci. 12 (2017) 6137–6155.
47. Z.M. Hadi, J. Al-Sawaad, Thermodynamic and quantum chemistry study for dimethylol-5-methyl hydantoin and its derivatives as corrosion inhibitors for carbon steel N-80 in raw water (cooling water system). Mater. Environ. Sci. 2:2 (2011) 128.
48. P.N.G. Shankar, K.I. Vasu, J. Electrochem. Soc. India. 32 (1983) 47-51.
49. A. Amin, K.F. Khaled, Q. Mohsen, A. Arida, A study of the inhibition of iron corrosion in HCl solutions by some amino acids, Corros. Sci. 52 (2010) 1684.

50. S.A. Umoren, O. Ogbobe, I.O. Igwe, E.E. Ebenso, Inhibition of mild steel corrosion in acidic medium using synthetic and naturally occurring polymers and synergistic halide additives, Corros. Sci. 50 (2008) 1998.

51. R. Solmaza, G. Kardas, M. Culha, B. Yazıcı, M. Erbil, Investigation of adsorption and inhibitive effect of 2-mercaptothiazoline on corrosion of mild steel in hydrochloric acid media. Electrochim. Acta 53 (2008) 5941.

52. M.N. El-Haddad, Inhibitive action and adsorption behavior of cefotaxime drug at copper/hydrochloric acid interface: Electrochemical, surface and quantum chemical studies. RSC Adv. 6 (2016) 57844.

53. A.K. Singh, M.A. Quraishi, Investigation of the effect of disulfiram on corrosion of mild steel in hydrochloric acid solution. Corros. Sci. 53 (2011) 1288.

54. H. Zhao, et al., Quantitative structure–activity relationship model for amino acids as corrosion inhibitors based on the support vector machine and molecular design, Corros. Sci., 83 (2014) 261–271.

55. M. Bouklah, A. Ouassini, B. Hammouti, A. El Idrissi, Corrosion inhibition of steel in sulphuric acid by pyrrolidine derivatives, Appl. Surf. Sci. 252 (2006) 2178–2185.

56. J. Haque, M.A.J. Mazumder, M.A. Quraishi, S.A. Ali, N.A. Aljeaban, Pyrrolidine-based quaternary ammonium salts containing propargyl and hydrophobic C-12 and C-16 alkyl chains as corrosion inhibitors in aqueous acidic media, J. Mol. Liq. 320:15 (2020) 114473.

57. M.A.J. Mazumder, New, amino acid based zwitterionic polymers as promising corrosion inhibitors of mild steel in 1 M HCl, Coatings. 9 (2019) 675.

58. R.S. Erami, M. Amirnasr, S. Meghdadi, M. Talebian, H. Farrokhpour, K. Raeissi, Carboxamide derivatives as new corrosion inhibitors for mild steel protection in hydrochloric acid solution, Corros. Sci. 151 (2019) 190–197.

59. I. Fernández, G. Frenking, E. Uggerud, The interplay between steric and electronic effects in S_N2 reactions, Chem. Eur. J. 15 (2009) 2166–2175.

60. O. Olivares-Xometl, E. Álvarez-Álvarez, N.V. Likhanova, I.V. Lijanova, R.E. Hernández-Ramírez, P. Arellanes-Lozada, J.L. Varela-Caselis, Synthesis and corrosion inhibition mechanism of ammonium-based ionic liquids on API 5L X60 steel in sulfuric acid solution, J. Adhes. Sci. Technol. 32 (2018) 1092–1113.

61. F.B. Growcock, W.W. Frenier, V.R. Lopp, in: 6th European symposium on corrosion inhibitors, Ann. Univ. Ferrara, N.S., Sez, V. Suppl. No. 7, 1980, p. 1185.

62. D. Jayaperumal, S. Muralidharan, P. Subramanian, G. Venkatachari, S. Senthilvel, Propargyl alcohol as hydrochloric acid inhibitor for mild steel: Temperature dependence of critical concentration, Anti-Corros. Method. Mater. 44 (1997) 265–268.

63. Z. Yang, C. Qian, W. Chen, M. Ding, Y. Wang, F. Zhan, M.U. Tahir, Synergistic effect of the bromide and chloride ion on the inhibition of quaternary ammonium salts in haloid acid, corrosion inhibition of carbon steel measured by weight loss, Colloid Interface Sci. Commun. 34 (2020) 100228.

64. A. Chetouani, B. Hammouti, A. Aouniti, N. Benchat, T. Benhadda, New synthesised pyridazine derivatives as effective inhibitors for the corrosion of pure iron in HCl medium, Prog. Org. Coat. 45 (2002) 373–378.

65. A. Chetouani, A. Aouniti, B. Hammouti, N. Benchat, T. Benhadda, S. Kertit, Corrosion inhibitors for iron in hydrochloride acid solution by newly synthesised pyridazine derivatives, Corros. Sci. 45 (2003) 1675–1684.

66. M. Bouklah, N. Benchat, A. Aouniti, B. Hammouti, M. Benkaddour, M. Lagrenée, H. Vezin, F. Bentiss, Effect of the substitution of an oxygen atom by sulphur in a pyridazinic molecule towards inhibition of corrosion of steel in 0.5 M H_2SO_4 medium, Prog. Org. Coat. 51 (2004) 118–124.

67. B. Zerga, B. Hammouti, M. Ebn Touhami, R. Touir, M. Taleb, M. Sfaira, M. Bennajeh, I. Forssal, Comparative inhibition study of new synthesised pyridazine derivatives towards mild steel corrosion in hydrochloric acid. Part-II: Thermodynamic proprieties, Int. J. Electrochem. Sci. 7 (2012) 471–483.

68. F. Bentiss, M. Outirite, M. Traisnel, H. Vezin, M. Lagrenée, B. Hammouti, S. Al Deyab, C. Jama, Improvement of corrosion resistance of carbon steel in hydrochloric acid medium by 3,6-bis(3-pyridyl)pyridazine, Int. J. Electrochem. Sci. 7 (2012) 1699–1723.
69. Z. El Adnani, M. Mcharfi, M. Sfaira, A. Benjelloun, M. Benzakour, M. Ebn Touhami, B. Hammouti, M. Taleb, Investigation of newly pyridazine derivatives as corrosion inhibitors in molar hydrochloric acid. Part III: Computational calculations, Int. J. Electrochem. Sci. 7 (2012) 3982–3996.
70. A. Ghazoui, N. Bencaht, S. Al-Deyab, A. Zarrouk, B. Hammouti, M. Ramdani, M. Guenbour, An investigation of two novel pyridazine derivatives as corrosion inhibitor for C38 steel in 1.0 M HCl, Int. J. Electrochem. Sci. 8 (2013) 2272–2292.
71. A. Khadiri, R. Saddik, K. Bekkouche, A. Aouniti, B. Hammouti, N. Benchat, M. Bouachrine, R. Solmaz, Gravimetric, electrochemical and quantum chemical studies of some pyridazine derivatives as corrosion inhibitors for mild steel in 1 M HCl solution, J. Taiwan Inst. Chem. Eng. 58 (2016) 552–564.
72. M. Bouklah, N. Benchat, B. Hammouti, A. Aouniti, S. Kertit, Thermodynamic characterisation of steel corrosion and inhibitor adsorption of pyridazine compounds in 0.5 M H_2SO_4, Mater. Lett. 60 (2006) 1901–1905.
73. A. Zarrouk, T. Chelfi, A. Dafali, B. Hammouti, S. Al-Deyab, I. Warad, N. Benchat, M. Zertoubi, Comparative study of new pyridazine derivatives towards corrosion of copper in nitric acid: Part-1, Int. J. Electrochem. Sci. 5 (2010) 696–705.
74. A. Zarrouk, B. Hammouti, H. Zarrok, R. Salghi, M. Bouachrine, F. Bentiss, S. AlDeyab, Theoretical study using DFT calculations on inhibitory action of four pyridazines on corrosion of copper in nitric acid, Res. Chem. Intermed. 38 (2012) 2327–2334.
75. M.E. Mashuga, L.O. Olasunkanmi, E.E. Ebenso, Experimental and theoretical investigation of the inhibitory effect of new pyridazine derivatives for the corrosion of mild steel in 1 M HCl, J. Mol. Struct. 1136 (2017) 127–139.
76. M. Filali, E. El Hadrami, A. Ben-Tama, B. Hafez, I. Abdel-Rahman, A. Harrach, H. Elmsellem, B. Hammouti, M. Mokhtari, S. Stiriba, 3,6-Di(pyridin-2-yl) pyridazine derivatives as original and new corrosion inhibitors in support of mild steel: Experimental studies and DFT investigational, Int. J. Corros. Scale Inhib. 8 (2019) 93–109.
77. R.A. Hameed, E. Aljuhani, A. Al-Bagawi, A. Shamroukh, M. Abdallah, Study of sulfanyl pyridazine derivatives as efficient corrosion inhibitors for carbon steel in 1.0 M HCl using analytical techniques, Int. J. Corros. Scale Inhib. 9 (2020) 623–643.
78. M. Abdallah, Rhodanine azosulpha drugs as corrosion inhibitors for corrosion of 304 stainless steel in hydrochloric acid solution, Corros. Sci. 44 (2002) 717–728.
79. S. Bilgic, N. Caliskan, An investigation of some Schiff bases as corrosion inhibitors for austenitic chromium–nickel steel in H_2SO_4, J. Appl. Electrochem. 31 (2001) 79–83.
80. A. Fouda, M. Abdallah, S. Al-Ashrey, A. Abdel-Fattah, Some crown ethers as inhibitors for corrosion of stainless steel type 430 in aqueous solutions, Desalination. 250 (2010) 538–543.
81. V. Reznik, V. Akamsin, Y.P. Khodyrev, R. Galiakberov, Y.Y. Efremov, L. Tiwari, Mercaptopyrimidines as inhibitors of carbon dioxide corrosion of iron, Corros. Sci. 50 (2008) 392–403.
82. S. Vega, J. Alonso, J.A. Diaz, F. Junquera, Synthesis of 3-substituted-4-phenyl2-thioxo-1,2,3,4,5,6,7,8-octahydrobenzo[4,5]thieno [2,3-*á*]pyrimidines, J. Heterocycl. Chem. 27 (1990) 269–273.
83. C. Shishoo, K. Jain, Synthesis of some novel azido/tetrazolothienopyrimidines and their reduction to 2,4-diamino thieno[2,3-*d*] pyrimidines, J. Heterocycl. Chem. 29 (1992) 883–893.
84. N.A. Hassan, Syntheses of furo[3,2-*e*][1,2,4]triazolo[1, 5-*c*]pyrimidines and furo[2,3:5,6]-pyrimido[3,4-*b*][2,3-*e*] indolo[1,2,4]triazine as a new ring system, Molecules 5 (2000) 826–834.

85. G. Elewady, Pyrimidine derivatives as corrosion inhibitors for carbon-steel in 2 M hydrochloric acid solution, Int. J. Electrochem. Sci. 3 (2008) 1149–1161.

86. M. Abdallah, E. Helal, A. Fouda, Aminopyrimidine derivatives as inhibitors for corrosion of 1018 carbon steel in nitric acid solution, Corros. Sci. 48 (2006) 1639–1654.

87. N. Caliskan, E. Akbas, Corrosion inhibition of austenitic stainless steel by some pyrimidine compounds in hydrochloric acid, Mater. Corros. 63 (2012) 231–237.

88. A.Y. Musa, A.A.H. Kadhum, A.B. Mohamad, M.S. Takriff, A.R. Daud, S.K. Kamarudin, On the inhibition of mild steel corrosion by 4-amino-5-phenyl-4H-1,2,4-trizole-3-thiol, Corros. Sci. 52 (2010) 526–533.

89. W.-H. Li, Q. He, S.-T. Zhang, C.-L. Pei, B.-R. Hou, Some new triazole derivatives as inhibitors for mild steel corrosion in acidic medium, J. Appl. Electrochem. 38 (2008) 289–295.

90. F. Zucchi, G. Trabanelli, G. Brunoro, C. Monticelli, G. Rocchini, Corrosion inhibition of carbon and low alloy steels in sulphuric acid solutions by 2-mercaptopyrimidine derivatives, Mater. Corros. 44 (1993) 264–268.

91. D.K. Yadav, M.A. Quraishi, Application of some condensed uracils as corrosion inhibitors for mild steel: Gravimetric, electrochemical, surface morphological, UV–visible, and theoretical investigations, Ind. Eng. Chem. Res. 51 (2012) 14966–14979.

92. C. Verma, L.O. Olasunkanmi, E.E. Ebenso, M.A. Quraishi, I. Obot, Adsorption behavior of glucosamine-based, pyrimidine-fused heterocycles as green corrosion inhibitors for mild steel: Experimental and theoretical studies, J. Phys. Chem. C 120 (2016) 11598–11611.

93. P. Singh, D.S. Chauhan, S.S. Chauhan, G. Singh, M.A. Quraishi, Bioinspired synergistic formulation from dihydropyrimidinones and iodide ions for corrosion inhibition of carbon steel in sulphuric acid, J. Mol. Liq. 298 (2019) 112051.

94. J. Haque, K. Ansari, V. Srivastava, M.A. Quraishi, I. Obot, Pyrimidine derivatives as novel acidizing corrosion inhibitors for N80 steel useful for petroleum industry: A combined experimental and theoretical approach, J. Ind. Eng. Chem. 49 (2017) 176–188.

95. P. Singh, A. Singh, M.A. Quraishi, Thiopyrimidine derivatives as new and effective corrosion inhibitors for mild steel in hydrochloric acid: Electrochemical and quantum chemical studies, J. Taiwan Inst. Chem. Eng. 60 (2016) 588–601.

96. S. Hejazi, S. Mohajernia, M.H. Moayed, A. Davoodi, M. Rahimizadeh, M. Momeni, A. Eslami, A. Shiri, A. Kosari, Electrochemical and quantum chemical study of thiazolo-pyrimidine derivatives as corrosion inhibitors on mild steel in 1 M H_2SO_4, J. Ind. Eng. Chem. 25 (2015) 112–121.

97. K.R. Ansari, A. Sudheer, M.A. Singh, Quraishi, some pyrimidine derivatives as corrosion inhibitor for mild steel in hydrochloric acid, J. Dispersion Sci. Technol. 36 (2015) 908–917.

98. N. Abdelshafi, Electrochemical and molecular dynamic investigation of some new pyrimidine derivatives as corrosion inhibitors for aluminium in acid medium, Prot. Met. Phys. Chem. 56 (2020) 1066–1080.

99. X. Li, S. Deng, H. Fu, Three pyrazine derivatives as corrosion inhibitors for steel in 1.0 M H_2SO_4 solution, Corros. Sci. 53 (2011) 3241–3247.

100. N. Arrousse, R. Salim, Y. Kaddouri, D. Zahri, F. El Hajjaji, R. Touzani, M. Taleb, S. Jodeh, The inhibition behavior of two pyrimidine-pyrazole derivatives against corrosion in hydrochloric solution: Experimental, surface analysis and *in silico* approach studies, Arabian J. Chem. 13 (2020) 5949–5965.

15 Heterocyclic as Corrosion Inhibitors for Petrochemical Industries

Jasdeep Kaur and Akhil Saxena
Department of Chemistry, Chandigarh
University Mohali, Punjab, India

15.1 INTRODUCTION

With a total income of US$2 trillion, the oil and gas extraction business is a crucial contributor to the world's economic growth, contributing 2–3% of it at present [1]. By 2040, crude oil and natural gas are expected to hold the highest shares of all fuels, with approximately 27% and 25%, respectively [2]. Serious corrosion issues during manufacture, transit, storage, and processing are challenging for the industry [3]. According to a recent research [4], a significant gas and oil business spent $1.372 billion annually on corrosion-related issues. Corrosion was described by Fontana and Greene [5] as an unexpected degradation of a material due to interaction with the surroundings; this process considerably affects the properties of the materials. It has been determined that there are several types of corrosion in petrochemical plants, including uniform corrosion, galvanic corrosion, pitting, crevice corrosion, corrosion fatigue, stress corrosion cracking, intergranular corrosion, erosion–corrosion, and fretting corrosion [6]. Uniform, erosion, and pitting corrosion are the types of corrosion that occur most frequently in the petroleum industry [7, 8]. Various factors are [9] responsible for the corrosion in the pipeline, as shown in Figure 15.1. Corrosion is typically an electrochemical reaction [10]. In the gas and oil industries, three primary forms of corrosion can occur: corrosion brought on by CO_2 (sweet corrosion), corrosion brought on by H_2S (sour corrosion), and corrosion brought on by oxygen dissolved in injection water [11, 12]. In this chapter, we will discuss sweet and sour corrosion inhibition in petrochemical industries.

15.1.1 SWEET CORROSION

In addition to being naturally present in gas and oil wells, carbon dioxide is also purposefully pumped into the wells to improve oil extraction. Carbon dioxide corrosion, commonly called "sweet corrosion," is the biggest challenge in the oil and gas sector and has a yearly economic impact of billions of dollars. Coronation prevention requires significant effort for environmental, commercial, and safety concerns. Sweet corrosion is brought on by carbonic acid, which is created when CO_2 is dissolved in water (H_2CO_3). As the system's temperature, pressure, and CO_2 concentrations rise, corrosion accelerates. Pitting corrosion is the expected result of this slow,

DOI: 10.1201/9781003377016-15

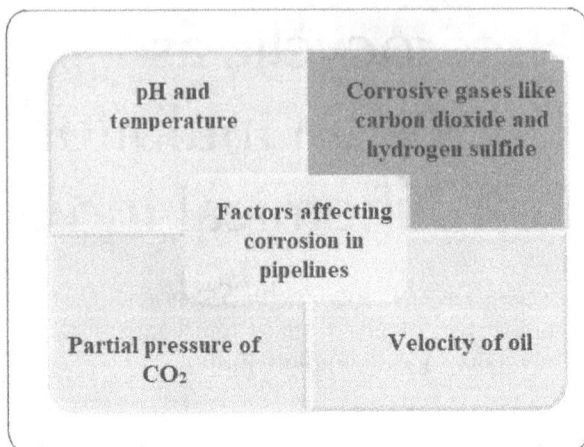

FIGURE 15.1   Factors affecting corrosion in pipelines.

localized corrosion. Pits are exceedingly challenging to find because of their small size and product of corrosion, which hides them [13].

More than 25% of problems in the oil and gas industries are attributed to corrosion, and more than 28% of these failures are attributed to CO_2 corrosion, according to literature studies [14]. Such shortcomings are incredibly concerning not only because of their financial consequences but also because they could result in environmental issues due to leaks from corroded transportation pipelines. Because of this, this industry must reduce the effects of corrosion. According to Zhao et al. [15], 30–40% of the economic loss caused by corrosion might be prevented if effective anticorrosion measures were applied. In the gas and oil business, corrosion inhibitors have effectively prevented corrosion. This is because adding small amounts of corrosion inhibitors, ranging from organic to inorganic substances and also natural products, lowers the pace of corrosion by getting adsorbed on the surface of the metal, creating a coating that shields the metal from the corrosive environments.

15.1.2   MECHANISM OF CORROSION: SWEET CORROSION

As mentioned, sweet corrosion is a significant issue for the oil and gas sector. It can occur at any point in the production process and lead to problems, including good failure, tank and tubing leaks, and well failure. It has been stated that most equipment subjected to a CO_2 environment is damaged in about six months. Oil and gas companies suffer significant financial losses due to pipe corrosion brought by sweet metal corrosion.

Based on Figure 15.2, sweet corrosion is commonly associated with anodic reactions involving iron oxidation. Several cathodic responses are critical in CO_2 corrosion, but hydrogen evolution is the most prominent. As a result of the hydrating action of the dissolved CO_2 molecules in aqueous media, carbonic acid is formed. The reactions on the cathode side cause H_2CO_3 to dissociate, whereas the responses on the anode side produce solid $FeCO_3$. Carbon dioxide is a mild acidic gas that becomes acidic when dissolved in water. Carbon dioxide has been

$$Fe + OH^- \longrightarrow FeOH + e^-$$
$$FeOH \longrightarrow FeOH^+ + e^-$$
$$FeOH^+ \longrightarrow Fe^{2+} + 2e^-$$

$$2H_2CO_3 + 2e^- \longrightarrow H_2 + 2HCO_3^-$$
$$2H^+ + 2e^- \longrightarrow H_2$$
$$2HCO_3^- + 2e^- \longrightarrow H_2 + 2CO_3^{2-}$$

$Fe(CO_3)_2$

$$Fe + CO_2 + H_2O \longrightarrow FeCO_3 + H_2$$
overall reaction

Anode side

$$Fe \longrightarrow Fe^{2+} + 2e^-$$

Cathode side

e^-

Pipeline

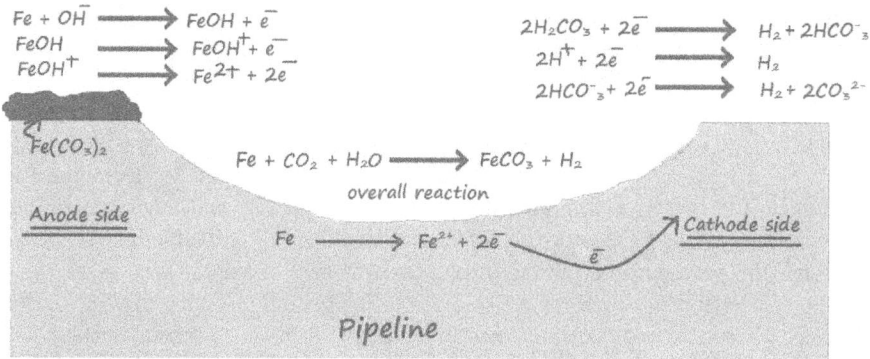

FIGURE 15.2 Mechanism of sweet corrosion.

dissolved in water to create carbonic acid, which mainly promotes sweet corrosion. As the system pressure, temperature, and CO_2 concentration rise, corrosion will also increase. Often, this localized, gradual corrosion leads to pitting corrosion. Pits are tough to find due to their small size and the corrosion products that hide them. The by-products of CO_2 corrosion include iron carbonate, iron oxide, and magnetite. Corrosion products are formed in various shades, from green to tan or brown to black. Steel is unaffected by dry CO_2, whether in gaseous, liquid, or solid form. But in water, CO_2 dissolves to form H_2CO_3, a mild acidic solution corrosive to steel. The strength of an acid is determined by its capacity to release a hydrogen ion.

Strong acids like hydrochloric acid completely break down in the water, but weak acids like carbonic acid only partially dissociate. Carbonic acid is a steady source of H^+ ions, resulting in a greater corrosion rate than strong acid solutions at the same pH.

Temperature, pH, pressure, contaminants, and salinity are just a few variables that might determine the pace of sweet corrosion. The solubility of O_2 and CO_2 reduces as the temperature rises, thus lowering their concentrations in solution. Despite this, corrosion rates typically rise as temperature increases because reaction rates do so more quickly than solubility drops. Corrosion products also change morphology and phase identity with temperature. The corrosion product layer may become protective and reduce corrosion rates if flow, pH, and ferrous iron concentration are sufficient. Generally, corrosion rates increase with increasing temperature at a low pH, around pH 4, and corrosion products provide less protection than at a high pH. This is because of the nature of corrosion products, which are porous and weak and do not offer steel surface protection.

15.2 HETEROCYCLIC AS CORROSION INHIBITION FOR SWEET CORROSION

In the production and processing of gas and oil, corrosion inhibitors have generally been regarded as the primary protective barrier against internal corrosion. These compounds can provide corrosion protection by adsorbing on the metal surface due to their chemical structure's many heteroatoms and multiple bonds. Initially, inorganic inhibitors like sodium arsenite and sodium ferrocyanide were utilized to prevent CO_2 corrosion in oil wells, but the performance of those treatments could have been better. Consequently, several organic chemicals were invented, most of which included film-forming amines and the salts of those amines [13].

Fouda et al. [13] used EIS, SEM, EDX spectroscopy, and XRD techniques to investigate the corrosion inhibition capacity of itraconazole and fluconazole compounds on steel in CO_2-saturated 3.5% NaCl solutions. Temperature's influence on corrosion pace and the inhibition process was also explored. Itraconazole and fluconazole effectively inhibit steel in a 3.5% NaCl solution saturated with carbon dioxide. These two molecules formed a protective film on the metal surface and blocked reaction sites, reducing exposure to the corrosive medium. The maximum inhibition efficiency of these two inhibitors was reported as 92% and 90%. Ambrish et al. [16] explained the inhibition of two porphyrin compounds in a 3.5% NaCl and CO_2 environment. Researchers used techniques like weight loss, EIS, SEM, and AFM to find inhibition efficiency. The heterocyclic compound gave 93% efficiency at 400 ppm with these studies. Metal protection was observed due to the adsorption of heterocyclic compounds on the metal surface as heteroatoms present in porphyrin compounds donate electrons to the vacant d-orbitals of the Fe atoms (Figure 15.3).

Zhang et al. [17] used EIS and PDP to examine the inhibitory effect of an imidazoline derivative for X65 steel in CO_2 and NaCl solution. This imidazoline derivative functions as a mixed-type inhibitor, inhibiting both cathodic and anodic processes by adsorbing onto the surface of the electrode. Imidazoline [18] compounds are the best

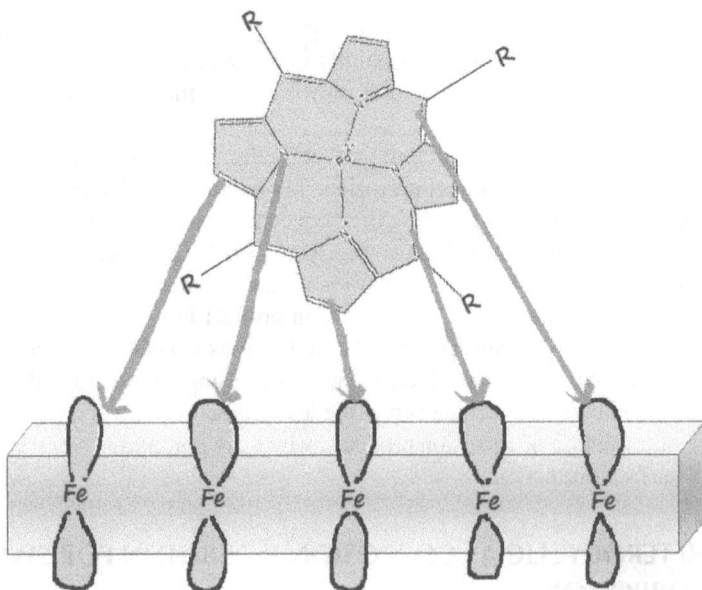

FIGURE 15.3 Mechanism of adsorption of heterocyclic compounds on the surface of the metal.

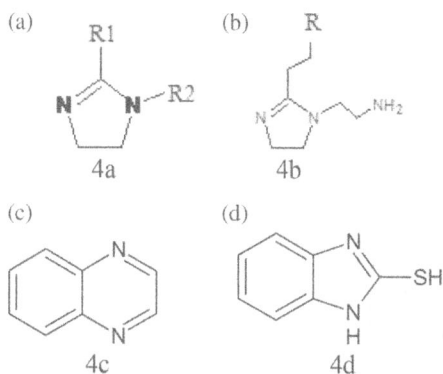

FIGURE 15.4 Chemical structures of heterocyclic compounds that act as sweet corrosion inhibitors; (a) 11,2-dialkyl-2,5-dihydro-1H-imidazole; (b) 2-(2-alkyl-2,5-dihydro-1H-imid-azol-1-yl)ethan-1-amine; (c) Quinoxaline and (d) 1H-benzo[d]imidazole-2-thiol

surface protectors as these compounds contain two nitrogen atoms (Figure 15.4a). These heterocyclic compounds act as mixed corrosion inhibitors that may affect the cathodic and anodic reactions. As shown in Figure 15.4b, the hydrocarbon chain length also affects imidazoline derivative inhibition behavior. When the temperature was 150 °C, the partial pressure of CO_2 was 15 psi, and the rotation speed was 6000 rpm, 50 ppm of imidazoline could reduce corrosion.

The quinoxaline [19] (Figure 15.4c) molecule is a heterocyclic chemical found in various medicines and natural compounds. Quinoxalines are extensively used in industry as alloy and metal corrosion inhibitors. Because of the existence of two aromatic rings and two nitrogen atoms, the quinoxaline molecule can effectively cover the metal. By using electrochemical studies, weight loss studies, surface studies, and other calculations, the inhibitory properties of 2-mercapto benzimidazole [20] (Figure 15.4d) and its synergism using oleic imidazoline for the steel in CO_2-saturated brine solution were examined. The carbon steel is efficiently guarded against CO_2 corrosion by 2-mercapto benzimidazole, which adsorbs iron through a bidentate binding N–S bridge link.

In sodium chloride at 3% (m/v), saturated with carbon dioxide, the 1-butyl-3-ethyl imidazolate [21] is investigated as a corrosion inhibitor of steel. Electrochemical techniques, polarization curves, and other techniques were used to measure the inhibitory effect. The highest efficiency observed was 94.9% when the concentration was 50 ppm. The chemical interaction between the anionic portion of 1-butyl-3-ethyl imidazolate and the metal surface explains the mixed physisorption–chemisorption.

15.3 SOUR CORROSION

When a metal surface is exposed to a highly acidic condition that contains hydrogen sulfide, it can deteriorate or rust; this process is known as sour corrosion. Sour corrosion is frequent in the petroleum industry because a large amount of H_2S is a primary chemical component of crude oil and refinery processes.

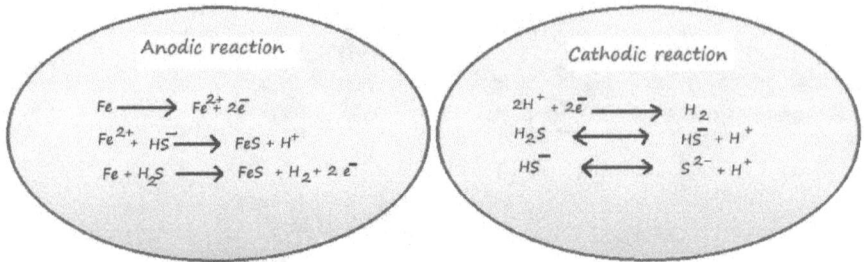

FIGURE 15.5   Mechanism of sour corrosion.

15.3.1   MECHANISM OF CORROSION: SOUR CORROSION

Elemental sulfur is routinely co-produced by sour gas wells, which can seriously corrode mild steel. Sulfur corrosion is predicted to become more common as producing
wells become more acidic. Despite the seriousness of the issue, little mechanistic
data is available in the literature regarding this process. The primary mechanism of
corrosion is shown in Figure 15.5.

   The mechanism can be explained by the fact that H_2S performs dual functions:
both a corrosive species that speeds up corrosion and a protected species that interacts with steel to form a scale containing the "kinetically favored" mackinawite
phase [22]. As a result, during short-term tests at high H_2S and higher temperatures,
the initial corrosion rate may be increased because the scale's protectiveness has yet
to develop fully and is therefore dominated by the rapid corrosion rate. The development of a more protective mackinawite structure is encouraged by extending the
reaction time to several days.

15.3.2   HETEROCYCLIC AS CORROSION INHIBITION FOR SOUR CORROSION

Figure 15.6 represents the structures of several heteromolecules described as sour
corrosion inhibitors [23]. Investigations have been conducted on the sour corrosion of low-carbon steel in a carbon dioxide–saturated 3.5% sodium chloride +
100 ppm hydrogen sulfide solution and its inhibition by the low-toxic compound

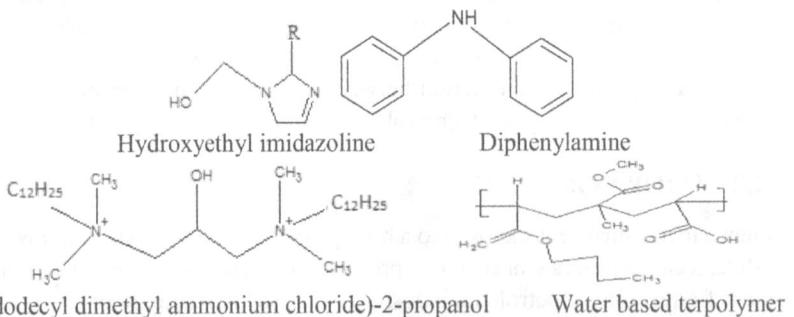

FIGURE 15.6   Examples of sour corrosion inhibitors.

1-benzylimidazole. Results from electrochemical impedance spectroscopy demonstrate that 1-benzylimidazole can create a protective inhibitor layer that is significantly adsorbed to the surface of the steel. Results from the PDP show that 1-benz is a mixed-type inhibitor with a slightly stronger anodic tendency. Up to 83% inhibitory efficiency was achieved at an ideal concentration of 150 ppm. According to FTIR analysis, 1-benzylimidazole utilizes its nitrogen atom and $C=C$ π-electrons to interact with the steel surface [24].

Corrosion inhibitors are essential for preventing corrosion caused by sour oil and gas. Polysuccinimide and lactobionic acid derivatives are utilized for mild steel inhibition in sour conditions. The effectiveness of the results is compared to that of a well-known conventional sour gas inhibitor [25]. Electrochemical tests examined corrosion inhibition of a water-base acrylic terpolymer (ATP), methyl methacrylate/acrylic acid, for steel in a sour petroleum corrosive solution. Although increased rotational speed speeds up corrosion, it also makes corrosion inhibitors more effective. A decrease in corrosion attacks is seen when ATP is present. Additionally, thermodynamic calculations revealed that ATP adheres to the Langmuir adsorption isotherm and chemically adsorbs to the surface [26].

The influence of quaternary ammonium salts with various concentrations on the mild steel in the sour brine solution at various temperatures was investigated using weight loss, electrochemical tests, surface examination, and theoretical calculations. The corrosion pace decreased as the inhibitor concentration decreased, and the inhibition efficiency was close to 98%. The Langmuir adsorption isotherm was followed when inhibitors were adsorbed onto the metal surface. Additionally, inhibitors performed well at high temperatures. The computational studies well supported the experimental data [27].

The development of corrosion products on the metal is typically facilitated by an increase in temperature, and pH, which slows the speed of corrosion. However, organic acids and the fast flow of liquids have the reverse effect [28].

15.4 CONCLUSION

For the oil and gas sector, the problem of sweet and sour corrosion is inevitable, having comparable effects to natural disasters. Therefore, it is impossible to resolve this problem, but taking precautions to protect the metal surface from corroding is more cost-effective. The use of corrosion inhibitors is the most feasible and cost-effective method for addressing the issues caused by CO_2 corrosion (sweet corrosion) and H_2S corrosion (sour corrosion). Its key protective mechanism relies on its capacity to adsorb on metal surfaces, creating a barrier between the metal surface and the aggressive medium.

REFERENCES

1. IBISWorld, Global Oil & Gas Exploration & Production US Industry Market Research Report (March 2018) https://www.ibisworld.com/industry-trends/global-industry-reports/mining/oil-gas-exploration-production.html (accessed 12 January 2019).
2. Organization of Petroleum Exporting Countries, World Oil Outlook 2040. https://www.opec.org/opec_web/flflipbook/WOO2017/WOO2017/assets/common/downloads/WOO%202017 (accessed 12 January 2019).

3. Menendez CM, Jardine J, Mok WY, Ramachandran S, Jovancicevic V, Bhattacharya A. New Sour Gas Corrosion Inhibitor Compatible with Kinetic Hydrate Inhibitor. Paper presented at the International Petroleum Technology Conference, Doha, Qatar. OnePetro. 2014 Jan 19.

4. Perez TE. Corrosion in the oil and gas industry: An increasing challenge for materials. JOM. 2013 Aug;65(8): 1033–42.

5. Fontana MG, Greene ND, Klerer J. Corrosion engineering. Journal of the Electrochemical Society. 1968 May 1;115(5): 142C.

6. Bahadori A. Cathodic Corrosion Protection Systems: A Guide for Oil and Gas Industries. Gulf Professional Publishing. 2014 Jul 5.

7. Aribo S, Olusegun SJ, Ibhadiyi LJ, Oyetunji A, Folorunso DO. Green inhibitors for corrosion protection in acidizing oilfield environment. Journal of the Association of Arab Universities for Basic and Applied Sciences. 2017 Oct 1;24: 34–8.

8. Balan KP. Metallurgical Failure Analysis: Techniques and Case Studies. Elsevier. 2018 Jan 3.

9. Al-Moubaraki AH, Obot IB. Top of the line corrosion: Causes, mechanisms, and mitigation using corrosion inhibitors. Arabian Journal of Chemistry. 2021 May 1;14(5): 103116.

10. Shreir LL. 1.05 Basic Concepts of Corrosion. Oxford: Elsevier. 2010:89–100.

11. Holloway MD, Nwaoha C, Onyewuenyi OA, editors. Process Plant Equipment: Operation, Control, and Reliability. New York: John Wiley & Sons, Inc. 2012 Aug 20.

12. Finšgar M, Jackson J. Application of corrosion inhibitors for steels in acidic media for the oil and gas industry: A review. Corrosion Science. 2014 Sep 1;86: 17–41.

13. Ibraheem MA, El Sayed Fouda AE, Rashad MT, Nagy Sabbahy F. Sweet corrosion inhibition on API 5L-B pipeline steel. International Scholarly Research Notices. 2012 Dec 25; 2012: 892385.

14. Kermani MB, Harrop D. The impact of corrosion on the oil and gas industry. SPE Production & Facilities. 1996 Aug 1;11(03): 186–90.

15. Xhanari K, Wang Y, Yang Z, Finšgar M. A review of recent advances in the inhibition of sweet corrosion. The Chemical Record. 2021 Jul; 21(7): 1845–75.

16. Singh A, Talha M, Xu X, Sun Z, Lin Y. Heterocyclic corrosion inhibitors for J55 steel in a sweet corrosive medium. ACS Omega. 2017 Nov 20;2(11): 8177–86.

17. Zhang G, Chen C, Lu M, Chai C, Wu Y. Evaluation of inhibition efficiency of an imidazoline derivative in CO_2-containing aqueous solution. Materials Chemistry and Physics. 2007 Oct 15;105(2–3): 331–40.

18. Obot IB, Onyeachu IB, Umoren SA, Quraishi MA, Sorour AA, Chen T, Aljeaban N, Wang Q. High temperature sweet corrosion and inhibition in the oil and gas industry: Progress, challenges and future perspectives. Journal of Petroleum Science and Engineering. 2020 Feb 1;185: 106469.

19. Chauhan DS, Singh P, Quraishi MA. Quinoxaline derivatives as efficient corrosion inhibitors: Current status, challenges and future perspectives. Journal of Molecular Liquids. 2020 Dec 15;320: 114387.

20. Wang X, Yang J, Chen X, Ding W. Synergism of 2-mercaptobenzimidazole and oleic imidazoline on corrosion inhibition of carbon steel in CO_2-saturated brine solutions. Journal of Molecular Liquids. 2022 Dec 15;368: 120645.

21. Ontiveros-Rosales M, Espinoza-Vázquez A, Gómez FR, Valdez-Rodríguez S, Miralrio A, Acosta-Garcia BA, Castro M. Imidazolate of 1-butyl-3-ethyl imidazole as corrosion inhibitor on API 5L X52 steel in NaCl saturated with CO_2. Journal of Molecular Liquids. 2022 Oct 1;363: 119826.

22. Sun W, Pugh DV, Smith SN, Ling S, Pacheco JL, Franco RJ. A Parametric Study of Sour Corrosion of Carbon Steel. CORROSION. 2010 Mar 14. OnePetro.

23. Obot IB, Solomon MM, Umoren SA, Suleiman R, Elanany M, Alanazi NM, Sorour AA. Progress in the development of sour corrosion inhibitors: Past, present, and future perspectives. Journal of Industrial and Engineering Chemistry. 2019 Nov 25;79: 1–8.

24. Onyeachu IB, Njoku DI, Kaya S, El Ibrahimi B, Nnadozie CF. Sour corrosion of C1018 carbon steel and its inhibition by 1-benzylimidazole: Electrochemical, SEM, FTIR and computational assessment. Journal of Adhesion Science and Technology. 2022;36(7): 774–94.

25. Schmitt G, Saleh AO. Evaluation of Environmentally Friendly Corrosion Inhibitors for Sour Service. CORROSION 2000 Mar 26. OnePetro.

26. Azghandi MV, Davoodi A, Farzi GA, Kosari A. Water-base acrylic terpolymer as a corrosion inhibitor for SAE1018 in simulated sour petroleum solution in stagnant and hydrodynamic conditions. Corrosion Science. 2012 Nov 1;64: 44–54.

27. Iravani D, Esmaeili N, Berisha A, Akbarinezhad E, Aliabadi MH. The quaternary ammonium salts as corrosion inhibitors for X65 carbon steel under sour environment in NACE 1D182 solution: Experimental and computational studies. Colloids and Surfaces A: Physicochemical and Engineering Aspects. 2023 Jan 5;656: 130544.

28. Obot IB, Sorour AA, Verma C, Al-Khaldi TA, Rushaid AS. Key parameters affecting sweet and sour corrosion: Impact on corrosion risk assessment and inhibition. Engineering Failure Analysis. 2022 Dec 19: 107008.

16 Heterocyclic Organic Derivatives as Acidizing Corrosion Inhibitors

Sourav K. Saha[1] and Manilal Murmu[2]
[1]Department of Materials Science and Engineering, Pusan National University, Busan Republic of Korea
[2]Department of Nuclear and Quantum Engineering, Korea Advanced Institute of Science and Technology (KAIST), Daejeon, South Korea

16.1 INTRODUCTION

Acidizing is one of the most effective approaches to stimulating oil well to enhance oil recovery from the reversers [1–5]. During this process, the acid or its solutions are forced to flow through the borehole pipeline under high pressure into the porous regions of the rock made of calcite, dolomite, and limestones. The injected acids chemically react with the rocks compelling them to dissolve. Such dissolution helps to expand the existing bore well channels and open new channels simultaneously [6, 7]. In 1895, based on chemical reactions, Harman Frasch conceived the procedure for acidizing treatments of oil wells to increase the flow of oil wells in a limestone formation. In this procedure, 30–40 wt% of HCl was recommended for acidizing oil wells, followed by neutralizing reacted acid after the stimulation was completed [8]. Generally, highly concentrated acid (5–28%) are used during oil-well acidizing, and these acids creates highly corrosive environments for the metallic materials used in the pipelines [9, 10]. In the HCl medium, the corrosion mechanism has been proposed for iron- and steel-made materials [11, 12]. The anodic dissolution reaction has been shown as follows:

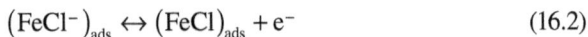

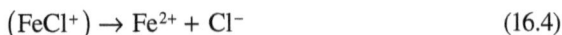

$$Fe + Cl^- \rightarrow \left(FeCl^-\right)_{ads} \tag{16.1}$$

$$\left(FeCl^-\right)_{ads} \leftrightarrow \left(FeCl\right)_{ads} + e^- \tag{16.2}$$

$$\left(FeCl\right)_{ads} \rightarrow \left(FeCl^+\right) + e^- \tag{16.3}$$

$$\left(FeCl^+\right) \rightarrow Fe^{2+} + Cl^- \tag{16.4}$$

The cathodic reactions which cause the hydrogen evolution are shown as follows:

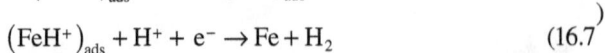

$$Fe + H^+ \rightarrow \left(FeH^+\right)_{ads} \tag{16.5}$$

$$\left(FeH^+\right)_{ads} + e^- \rightarrow \left(FeH\right)_{ads} \tag{16.6}$$

$$\left(FeH^+\right)_{ads} + H^+ + e^- \rightarrow Fe + H_2 \tag{16.7}$$

DOI: 10.1201/9781003377016-16

The electrochemical techniques, density functional theory (DFT), and molecular dynamics simulations are also essential strategies for exploring the insight of corrosion and its inhibition mechanisms [13–17].

16.2 HETEROCYCLIC ORGANIC MOLECULES AS ACIDIZING CORROSION INHIBITOR

An organic compound comprising a cyclic ring of carbon atoms is known as a carbocyclic compound. When the carbocyclic compound contains heteroatoms like N, O, S, P, etc. [4–6], the constituents ring is called a heterocyclic ring. Several heterocyclic compounds are explored for their corrosion inhibition capability in the acidizing medium. For instance, azoles, including pyrazoles, imidazoles, benzimidazoles, benzotriazoles, etc., are five-membered heterocyclic rings used as corrosion inhibitors in the acidic medium. Indoles are benzene rings infused with five-membered pyrrole rings. Pyrimidines are six-membered heterocyclic rings in which an N atom replaces one CH. Diazines are heterocycles in which two N atoms replace the two CH groups of the benzene ring. Similarly, quinolines are the heterocyclic compound in which the benzene ring is infused with the pyridine ring. The presence of heteroatoms in the heterocyclic ring facilitates its adsorption onto the metal surface. It leads to the form of a thin protective layer on the metal surface, reducing the protection of the vulnerable metal surface from severe corrosion in the acidic solution [3, 14, 18–20].

16.2.1 FIVE-MEMBERED HETEROCYCLIC ORGANIC MOLECULES AS ACIDIZING CORROSION INHIBITOR

Recently, Yadav et al. synthesized amino acid–based organic compounds, namely, OYAA and OYPA (Figure 16.1), as acidizing corrosion inhibitors to protect mild steel 15% HCl solution. The electrochemical studies revealed that 50 ppm of OYAA

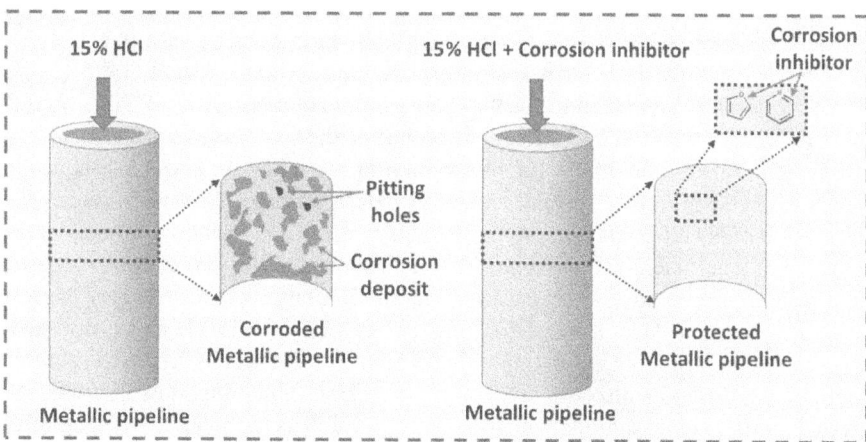

FIGURE 16.1 Schematic representation for corrosion and corrosion inhibition by applied corrosion inhibitors.

and OYPA adsorbed onto mild steel following Langmuir adsorption isotherm; thereby, these inhibitors exhibited mixed-type corrosion inhibition with 93.2% and 97.9% corrosion inhibition efficiency (CIE), respectively, at 303 K. These corrosion inhibitors' adsorption is primarily attributed to the electrostatic interaction between the protonated inhibitors with previously adsorbed chloride ions. The adsorption of these corrosion inhibitors is facilitated through physisorption. Second, the donor–acceptor interaction between the π-electrons of the aromatic ring and the vacant orbital of the metal surface atoms. Third, the interaction between the unshared electrons of the heteroatoms (N, O) of inhibitor molecules with vacant d-orbitals of the iron surface atoms. Thus, the adsorption of these corrosion inhibitors is also facilitated through the chemisorption pathway. The high CIE of OYPA is attributed to the higher energy of the highest occupied molecular orbital, the lower value of the lowest unoccupied molecular orbital, and the smaller value of the energy gap for OYPA than that of OYAA [21]. Etuen et al. reported 78.1% CIE of 10×10^{-5} M concentration of 5-HTP at 30 °C for mild steel in acidizing conditions. The CIE decreased with the increase in temperature and showed 39.8% at 90 °C. Furthermore, when 5-HTP is used by blending it with potassium iodide and glutathione, the CIE increases to 80.7% and 82.5%, respectively. This implies the stability of the corrosion inhibitor at 90 °C, and the blend could be used in the oil-well acidizing procedure at elevated temperatures. It has been reported that the negatively charged group of 5-HTP, i.e., $-COO^-$ interacts electrostatically with the positively charged iron surface facilitating physical adsorption. The acidic condition is also sufficient to protonate 5-HTP at amino functional ($-NH_2$ site). The protonated amino group (i.e., $-NH^{3+}$) may interact with the previously adsorbed chloride ions on the iron surface, facilitating physical adsorption. The physisorption property of 5-HTP was also confirmed from the free energy of adsorption in the ranges of -12.67 kJ mol^{-1} to -13.53 kJ mol^{-1} [22].

Recently, Solomon et al. synthesized imidazole-based corrosion inhibitors with variations of aliphatic chain ($-C_{13}H_{27}$, $-C_{15}H_{31}$, and $-C_{17}H_{35}$) length, namely, NTETD [23], NIMP [24], and QSI [25] to be employed as corrosion inhibitors in oil-well acidizing. NTETD possesses strong bonding capability with the steel surface, as revealed by the adsorption–desorption equilibrium constant (1.015×10^3). The adsorption is attributed to the N, O, and imidazoline ring of NTETD through electron donation to the low energy 3d-orbital of the iron atom. Furthermore, the strong acidic media facilitates the protonation of the inhibitor, and it subsequently gets adsorbed onto chloride ions previously adsorbed onto the steel surface. The adsorption of NTETD has been reported to be of chemisorption as confirmed by the determined coefficient of adsorption. At the same time, the hydrophobic myristic acid pendant group covers the steel surface and inhibits corrosive species' ingress toward the steel surface. It was found that 300 mg L^{-1} of NTETD acted as mixed-type corrosion inhibition exhibiting 93% for St37-2 steel sheet in 15% HCl solution. Similarly, NIMP was also explored for its corrosion inhibition capability for N80 steel in similar acidic solution with low (25 °C) as well as high (60 °C) temperatures which showed 97.92% and 95.59% CIE, correspondingly, with the use of 300 ppm of NIMP. The N and O heteroatoms of the palmitic acid pendant group of NIMP can be protonated easily in the acidic media. This cationic NIMP facilitates electrostatic interaction with the previously adsorbed chloride ions on the steel surface $((FeCl^-)_{ads})$ as shown by the equations and gets adsorbed thereby following the physisorption:

$$(FeCl^-)_{ads} + NIMP^+ \rightarrow (FeCl^-NIMP^+)_{ads} \qquad (16.8)$$

The adsorption of NIMP is possible through N, O, and heterocyclic imidazoline ring with the iron surface atoms (Figure 16.2). The chemisorption, considered the major pathway for the adsorption of NIMP on the metal substrate, was confirmed by the determined standard enthalpy of adsorption (100.34 kJ mol^{-1}). The hydrophobic aliphatic chain covers the surface and shifts its barrier effect by retarding those penetrating corrosive species toward metal surface atoms.

Similarly, QSI also exhibited mixed-type corrosion inhibitory action showing less than 50% corrosion inhibition at 400 ppm addition into 15% HCl solution for carbon steel. The QSI exhibited high CIE with the co-addition of potassium iodide, which upgraded CIE to 90%. The chemisorption of QSI has been deduced from the calculated corrosion kinetic parameters. Furthermore, it was found that QSI performed better in hydrodynamic conditions than in static conditions up to 1000 rpm. The presence of the hydrocarbon chain as the pendant group has been found to affect the CIE of imidazoline derivatives more than the tail group. The CIE decreases with an increased hydrocarbon chain length of the pendant active functional group.

Furthermore, Ituen et al. also reported a synergistic increase of CIE of 5-HTP upon blending it with GLU. This 5-HTP + GLU blend [10×10^{-5} M] showed 96.3% CIE at 30 °C and 82.5% CIE at 90 °C for mild steel in 15% HCl (Table 16.1). This acted as a mixed-type corrosion inhibitor and underwent physisorption obeying Langmuir adsorption isotherm [22]. Gerengi et al. reported CIE of DNPP to be 87% at 4 mM in SAE 1012 carbon steel (Table 16.1). The DNPP followed a mixed pathway of adsorption, thereby revealing mixed-type corrosion inhibition properties [26]. Yadav et al. reported the CIE of amino acid–based compounds, namely, MPIT was 92% and MPII exhibited 97.6% at 200 ppm at 303 K for N80 steel in 15% HCl

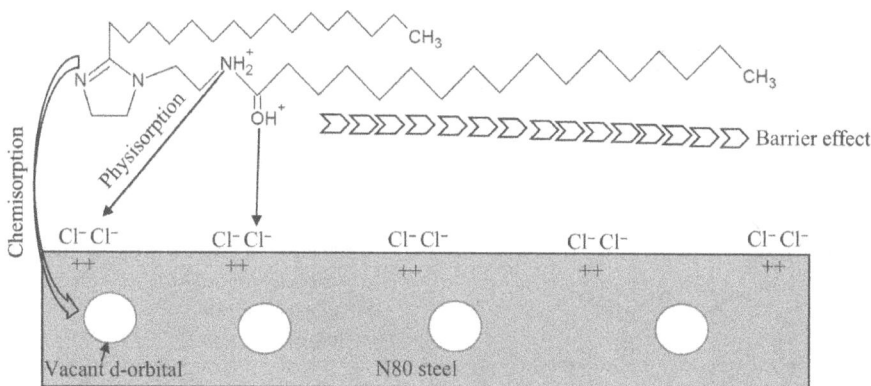

FIGURE 16.2 Schematic representation for adsorption of NIMP on N80 steel. (Reproduced with permission from Ref. [24], Copyright 2019, Elsevier.)

TABLE 16.1
Heterocyclic Organic Compounds for Acidizing Corrosion Inhibitors in 15 wt% HCl

Corrosion Inhibitors	Corrosion Inhibition Properties	Reference
OYPA OYAA	OYPA 97.9% at 50 ppm OYAA 93.2% at 50 ppm at 303 K, mild steel	[21]
5-HTP	10×10^{-5} M 5-HTP showed 78.1% at 30 °C and 39.8% at 90 °C for mild steel 5-HTP + GLU [10×10^{-5} M] 96.3% at 30 °C and 82.5% at 90 °C	[22]
NIMP	97.48% and 94.42% at 25 °C and 60 °C N80 steel	[24]
NTETD	96.51% or >93% at 300 ppm (corrosion rate 0.33 mm/year) at 25 °C, St37-2 steel sheet	[23]
QSI	400 mg L^{-1} afforded η of <50%. QSI +1 mM KI afforded 90% for low-carbon steel	[25]

(Continued)

TABLE 16.1   (*Continued*)
Heterocyclic Organic Compounds for Acidizing Corrosion Inhibitors in 15 wt% HCl

Corrosion Inhibitors	Corrosion Inhibition Properties	Reference
DNPP	87% by 4 mM DNPP at 25 °C In SAE 1012, carbon steel	[26]
MPIT MPII	MPIT 92%, MPII 97.6% at 200 ppm at 303 K	[27]
MBI PBI	98.90% and 97.74% for MBI and PBI at 200 ppm at 303 K	[28]
a b c d e	95.3% when 9% recycled cigarette butt extract is used at 90 °C to inhibit corrosion of N80 steel	[29]

(*Continued*)

TABLE 16.1   (Continued)
Heterocyclic Organic Compounds for Acidizing Corrosion Inhibitors in 15 wt% HCl

Corrosion Inhibitors	Corrosion Inhibition Properties	Reference
PZ1 PZ2	92.0% and 85.1% for PZ-1 and PZ-2 at 300 ppm at 308 K	[30]
SAHMT	53.17% and 90% upon the addition of 1000 ppm at 28 ± 2 °C N80 steel and mild steel	[31]
BDMA BDMT	BDMT 95.4% and BDMA 93.7% at 50 ppm at 303 K	[32]
PZ-1 PZ-2	98.4% (PZ-1) and 94.3% (PZ-2) at 308 K, N80 steel	[33]

(Continued)

TABLE 16.1   (*Continued*)
Heterocyclic Organic Compounds for Acidizing Corrosion Inhibitors in 15 wt% HCl

Corrosion Inhibitors	Corrosion Inhibition Properties	Reference
AMPC · ACPC	AMPC and ACPC at 300 ppm showed 98.26% and 96.21% inhibition efficiency, respectively, at 303 K for mild steel	[34]
n + m = 4EO,4 n + m = 8EO,8 n + m = 12EO,12	1×10^{-3} M (EO, 12) shows maximum inhibition of 95.5% at 25 °C X65 steel	[35]
PZ	88.7% at 200 ppm at 60 °C for mild steel	[36]
FACI	95.4% at 333 K [136×10^{-5} M] 90.8% at 353 K [136×10^{-5} M]	[37]
DAMA-ran-DAMTDB	DAMA-ran-DAMTDB showed 87% IE at 500 ppm; and DAMA-ran-DAMTDB + 1 mM KI 93% at 25 °C for st37-2 steel	[38]

(*Continued*)

TABLE 16.1   (Continued)

Heterocyclic Organic Compounds for Acidizing Corrosion Inhibitors in 15 wt% HCl

Corrosion Inhibitors	Corrosion Inhibition Properties	Reference
DMIC	68.61% at 400 mg L^{-1}. 90.15% upon adding KI (0.5 mM) + DMIC (400 mg L^{-1}).	[39]
NDTHDC	86.1% with 100 mg L^{-1} poly-NDTHDC at 25 °C, and 90.2% at 60 °C, 87.5% at 90 °C	[40]

(Table 16.1). The CIE increased with the increase in concentration, reaching the optimum concentration of 200 pp. The experimentally determined CIE corroborated with quantum chemical parameters specified by the density functional study. Both these corrosion inhibitors are of a mixed-type nature which gets adsorbed following Langmuir isotherm [27]. Kumari et al. explored the CIE of indolines derivatives: MBI and PBI. They found it a mixed-type corrosion inhibitor with 98.90% and 97.74% at 200 ppm at 303 K for mild steels in 15% HCl medium, respectively (Table 16.1). It has been reported that the CIE get enhanced with the increase in the concentration of these corrosion inhibitors studied up to 200 ppm. At the same time, it decreased with the elevation of temperature studied up to 333 K. The adsorption of these corrosion inhibitors obeyed Langmuir adsorption isotherm. The adsorption of these corrosion inhibitors is attributed to the interaction of the lone pairs of electrons in nitrogen and oxygen atoms and the delocalized π-electrons in the aromatic rings in their neutral forms. In such an acidic medium, the protonated forms may be adsorbed through electrostatic interaction with the pre-adsorbed chloride anions on the metal surface atoms. The determined quantum chemical parameters supported the adsorption capability of these corrosion inhibitors [28].

It has also been reported that the heterocyclic compounds present in recycled cigarette butts can act as corrosion inhibitor by Zhang et al. It showed 95.3% CIE when 9% recycled cigarette butts extract at 90 °C. The primary chemical constituents which acted as corrosion inhibitors might be (a) nicotine, (b) 3-(pyridin-3-yl)pyrrolidine-1-carbaldehyde, (c) 5-(pyridin-3-yl)pyridine-1(2H)-carboxamide, (d)

1-(3-pyridyl)-4-(methylnitrosoamino)-1-butanone, and (e) quercitrin. These major constituents possess heterocyclic moieties [29]. Ansari et al. reported corrosion inhibitory action of PZ1 and PZ2. They found the CIE as 92% and 85.1% for PZ1 and PZ2 at 300 ppm at 308 K exhibiting mixed-type corrosion inhibition with predominant cathodic inhibition and following Langmuir type of adsorption isotherm. The enhancement in the surface coverage by PZ1 compared to PZ2 is attributed to the higher CIE. The experimental observations were also validated by DFT and Monte Carlo (MC) simulation [30]. Quraishi and Jamal synthesized SAHMT and explored its corrosion inhibitory action in the boiling condition in 15% HCl. The CIE of SAHMT was 53.17% and 90% at 1000 ppm at 28 ± 2 °C for N80 steel and mild steel, respectively. It showed mixed-type inhibitory action by blocking active sites of metal. Adsorption of SAHMT was governed by Temkin adsorption isotherm [31].

Yadav et al. explored and reported thiazole derivatives, namely, BDMT and BDMA, exhibiting Langmuir adsorption isotherm and mixed-type corrosion inhibition behavior. The BDMT showed 95.4% and BDMA showed 93.7% CIE at 50 ppm at 303 K for oil-well tubular steel exposed in 15% HCl. The experimental observations were also validated by DFT [32]. Singh et al. synthesized ultrasonically pyrazole derivatives, namely, PZ-1 and PZ-2, to inhibit corrosion. These corrosion inhibitors showed mixed-type corrosion inhibitors with predominant cathodic nature. The CIE increased with the increase in concentration. These inhibitors established high CIE as 98.4% (PZ-1) and 94.3% (PZ-2) at 308 K for N80 steel in 15% HCl. The DFT study revealed the strong adsorption capability of PZ-1 compared to PZ-2. The MD simulation results also showed more muscular binding energy of PZ-1 compared to PZ-2 [33]. Paul et al. investigated and revealed the corrosion inhibition efficacy of carbohydrazide-pyrazole compounds, namely, AMPC and ACPC. Only 300 ppm of AMPC and ACPC exhibited 98.26% and 96.21% CIE, correspondingly at 303 K for mild steel. Both of these compounds followed acted as mixed corrosion inhibitors retarding both the cathodic and anodic corrosion of the metal substrate. Its fitting with Langmuir adsorption isotherm suggested the monolayer adsorption. The substitution on the phenyl moiety by the phenyl-releasing methyl group in AMPC contributed to more electron donation leading to enhanced adsorption and corrosion inhibition than ACPC. The DFT and MC simulation results corroborated the experimental findings [34].

Basiony et al. reported the newly synthesized ethoxylated aminothiazole compounds, [4, EO, 4], [8, EO, 8], and [12, EO, 12] varying the carbon chain lengths. The adsorption of [12, EO, 12] is more stable on the X65 steel surface and exhibited the highest CIE. However, these corrosion inhibitors exhibited physiochemisorption following Langmuir adsorption isotherm. A 1×10^{-3} M of [12, EO, 12] showed maximum inhibition of 95.5% at 25 °C in 15% HCl. The DFT and MC simulations were also compatible with experimental results [35]. Yaagoob and Ali reported the high CIE of homo-copolymers containing symmetrical motifs of PZ ions compared to that of the co-copolymers motif of PZ ions. It exhibited 88.7% CIE at 200 ppm at 60 °C for mild steel. The specific polymers (PZ 13, i.e., 13 carbon units and 100 ppm) upon addition with potassium iodide (400 ppm) can also show 93% CIE for mild steel in 15% HCl at 60 °C for 6 hours [36]. Rahimi et al. reported a high corrosion inhibition capability of furfuryl alcohol-based corrosion inhibitor, FACI, with 95.4% at 333 K [136×10^{-5} M] and 90.8% at 353 K [136×10^{-5} M] in 15% HCl acting as mixed-type corrosion inhibitor following Langmuir adsorption isotherm [37]. A random polymer containing diallylmethylamine and DAMA-ran-DAMTDB was synthesized and revealed its corrosion inhibition capability by Umoren et al. This DAMA-ran-DAMTDB showed 87% IE at 500 ppm; and DAMA-ran-DAMTDB + 1 mM KI 93% at 25 °C for st37-2 steel in a simulated acidizing environment. It gets easily chemisorbed onto the metal substrate, and the synergistic increase in adsorption get facilitated by the co-addition of iodides. The CIE enhanced to 92.99% with the co-addition of 1 mM of potassium iodide into 500 ppm of the corrosion inhibitor at 25 °C [38]. Singh et al. reported the corrosion inhibitory action of DMIC on P110 steel in static and dynamic conditions of acidic solutions. It exhibited increasing CIE with simultaneous concentration increments and showed mixed-type inhibitory activity. The maximum CIE of 68.61% was found at 400 mg L^{-1}. The co-addition of KI as a corrosion inhibitor improved its CIE. DFT suggested that the adsorption of DMIC occurred through both the anionic and cationic parts of DMIC. The MD simulation revealed the horizontal disposition of neutral DMIC, while the protonated DMIC get adsorbed onto the metal surface twistedly [39].

Some polymers of heterocyclic compounds were synthesized, and their corrosion-inhibitory actions have been validated. Odewunmi et al. synthesized a dicationic monomer: NDTHDC, and its polymer poly-NDTHDC. It was found that the addition of 100 mg L^{-1} poly-NDTHDC into 15% HCl showed 86.1% at 25 °C to 90.2% at 60 °C but decreased to 87.5% at 90 °C. The poly-NDTHDC is thermally and chemically stable. It exhibited mixed-type inhibition behavior, with its adsorption mainly governed by the chemisorption pathway [40]. Similarly, Haruna et al. revealed gelatin's efficient corrosion inhibition property for oil-well acidizing. Typically, the gelatin comprises glycine residues with heterocyclic amino acid residues such as proline and hydroxyl proline, along with other amino acids like arginine, alanine, and glutamic acid. The gelatin showed CIE of 88.35% at 2.5 (w/v%) at 25 °C for X60 steel in 15% HCl medium. Its adsorption occurs through physisorption and chemisorption on the metal surface, forming a metal–gelatin complex that blocks the steel surface and acts as a mixed-type corrosion inhibitor inhibiting cathodic and anodic dissolution [41].

16.2.2 FIVE- AND SIX-MEMBERED HETEROCYCLIC RING–INFUSED ORGANIC MOLECULES AS ACIDIZING CORROSION INHIBITOR

Similarly, there are also five-membered heterocyclic rings infused with other aromatic or nonaromatic rings for designing the efficient corrosion inhibitor. Five- and six-membered heterocyclic ring–infused organic molecules have found splendid applications as an acidizing corrosion inhibitors. For oil-well acidizing purposes, 5–15% HCl solution is mainly used. Singh et al. synthesized and reported pyridine derivative, PD, as a corrosion inhibitor. The CIE of 93% at 400 mg L^{-1} for Q235 steel in 15% HCl. The CIE increased to 97% with KI addition at 308 K. It exhibited Freundlich adsorption isotherm and its performances as dual-natured corrosion inhibition. DFT showed that the neutral form PD was capable of better electron donation facilitating adsorption. The MD simulation revealed the parallel adsorption of protonated PD on metal surface atoms [42].

Wang et al. synthesized and explored the corrosion inhibition behavior of phena-cyl quinolinium bromide, DaQBr, and its indoline derivative DiDaQBr. The CIE was reported to be 97.4% and 97% for PaQBr and DiPaQBr, respectively, at 25 °C for N80 steel in 15% HCl. PaQBr and DiPaQBr were mixed-type corrosion inhibitors and obeyed Langmuir-type adsorptions while getting adsorbed onto the N80 steel sur-face. The high corrosion inhibition property of DiDaQBr is attributed to converting the active methylene group of DaQBr to corresponding inhibitive imidazoline derivatives, suggesting the dimer imidazoline derivatives to be novel corrosion inhibitors [43].

Yang et al. reported the novel benzyl quinolinium chloride derivative, BQD, exhibiting mixed-type corrosion inhibiting capability with CIE of 97.5% and 94.6% for its precursor benzyl quinolinium chloride, BQC, for N80 steel in 15% HCl. The corrosion rate is 1.98 g m^{-2} h^{-1} when the concentration of BQD is 0.744 mmol L^{-1} at 25 °C (Table 16.2). The CIE increases with concentration,

TABLE 16.2
Heterocyclic Organic Compounds as Acidizing Corrosion Inhibitors in 15 wt% HCl

Corrosion Inhibitors	Corrosion Inhibition Properties	Reference
PD	93% at 400 mg L^{-1}; 97% with KI (0.5 mM) + PD (400 mg L^{-1}) at 308 K, Q235 steel	[42]
PaQBr DiPaQBr	97.4% and 97% for PaQBr and DiPaQBr, respectively, 25 °C, N80 Steel	[43]
BQC BQD	97.5% and 94.6% for BQD and BQC, respectively, 1.98 g m^{-2} h^{-1} when the concentration of BQD is 0.744 mmol L^{-1} at 25 °C, N80 steel	[44]

followed by Langmuir adsorption isotherm. The standard free energy of adsorption suggested the chemisorption of BQD on the steel surface [44].

16.2.3 SIX-MEMBERED HETEROCYCLIC ORGANIC MOLECULES AS ACIDIZING CORROSION INHIBITOR

Six-membered heterocyclic organic molecules are acidizing corrosion inhibitors in a 15% hydrochloric acid solution. Furthermore, a cleaner corrosion inhibition strategy of carbon steel using *Peumus boldus* Molina extract has also been explored by Furtado et al. The major constituent, namely, boldine (B), played a major role in the adsorption and corrosion protection of the steel substrate. The sole B and binary combination with propargyl alcohol (B + PA) showed 41.11% and 97.42%, respectively, at 333 K for API P110 carbon steel acidic solution (Table 16.3) [45]. Moreover, the corrosion inhibition property of pyrimidine derivatives, namely, PP-1 and PP-2, have also been reported by Haque et al. to protect N80 steel in 15% HCl. These derivatives showed CIE of 89.1% (250 mg L^{-1}, PP-1) and 73.1% (250 mg L^{-1}, PP-2), respectively, at 308 K. PP-1 and PP-2 also acted as mixed-type corrosion inhibitors with predominant cathodic inhibition capability by adsorbing on the active sites of the N80 steel. These inhibitors also followed Langmuir adsorption isotherm [46].

Ansari et al. revealed environment-friendly chromenopyrimidine derivatives, namely, PPC-1 and PPC-2 exhibiting Langmuir-type adsorption following physisorption and possessing cathodic corrosion inhibitory action with 92.4% and 82.1% CIE, respectively, at 200 mg L^{-1} at 308 K for N80 steel in 15% HCl. The order of corrosion inhibitory action was well supported by electron density distribution in the frontiers molecular orbitals exhibiting more reactivity of the protonated forms of chromenopyrimidine derivatives in the acidic medium [47].

Moreover, benzoxanthone derivatives are also acting as good corrosion inhibitors. Recently, Singh et al. reported CIE of TBX to be 92.3% at 200 mg L^{-1} to protect Q235 steel in 15% HCl. It revealed that temperature increment leads to enhance corrosion. It followed Langmuir's adsorption isotherm. Both the DFT and MD results corroborated the experimental outcomes [48]. Singh et al. also synthesized BX-Cl and BX-NO$_2$ through microwave irradiation techniques. The CIE reaches 92.21% for BX-Cl and around 84.21% CIE for BX-NO$_2$ of 200 mg L^{-1} concentration at 308 K, P110 steel. The synergistic corrosion inhibition behavior increases with the addition of potassium iodide. These corrosion inhibitors followed Freundlich adsorption isotherm and exhibited anodic-type corrosion inhibitor. DFT and MD revealed BX-Cl as a better corrosion inhibitor than BX-NO$_2$ [49]. Furthermore, Singh et al. synthesized pyran derivatives, namely, AP-1, AP-2, and AP-3. The AP-1, AP-2, and AP-3 showed 97.7%, 88.5%, and 75.1% at 300 mg L^{-1} at 308 K for N80 steel acting as mixed-type corrosion inhibitors with cathodic dominance. These pyran derivatives followed Langmuir adsorption isotherm. Furthermore, DFT revealed more electron donation from the neutral form of derivatives to the metal surface atoms leading to adsorption. The MD showed the strong adsorption of these pyran derivatives in their neutral forms [50].

Similarly, 8-HQ was revealed as a corrosion inhibitor for an oil-well acidizing environment by Obot et al. It exhibited CIE of 65.13%, 0.4 wt% of 8-HQ at 25 °C X60 steel in 15% HCl in hydrodynamic condition. It showed increased corrosion

TABLE 16.3

Heterocyclic Organic Derivatives as Acidizing Corrosion Inhibitors in 15 wt% HCl

Corrosion Inhibitors	Corrosion Inhibition Properties	Reference
Boldine	B and B + PA showed 41.11% and 97.42% at 333 K for API P110 carbon steel	[45]
PP-1 PP-2	89.1% (PP-1) and 73.1% (PP-2) at 250 mg L^{-1}, 308 K for N80 steel	[46]
PPC-1 PPC-2	PPC-1 and PPC-2 at 200 mg L^{-1} are 92.4% and 82.1%, respectively, at 308 K	[47]
Benzoxanthone	92.3% at 200 mg L^{-1}, Q235 steel (15% HCl hydrodynamic condition)	[48]

(Continued)

TABLE 16.3   *(Continued)*
Heterocyclic Organic Derivatives as Acidizing Corrosion Inhibitors in 15 wt% HCl

Corrosion Inhibitors	Corrosion Inhibition Properties	Reference
BX-Cl BX-NO₂	92.21% (BX-Cl); 84.21% (BX-NO$_2$) at 200 mg L^{-1}, 308 K, P110 steel	[49]
AP-1 AP-2 AP-3	AP-1, AP-2, and AP-3 showed 97.7%, 88.5%, and 75.1%, respectively, at 300 mg L^{-1} at 308 K for N80 steel	[50]
8-HQ	65.13%, 0.4 wt% of 8-HQ at 25 °C X60 steel	[51]
HA-1 HA-2 HA-3	89.2% at 1.00 mM HA-1 at 45 °C 92.4% at 1.00 mM HA-1 at 60 °C 91.6% at 1.00 mM HA-2 at 45 °C 94.1% at 1.00 mM HA-2 at 60 °C 96.2% at 1.00 mM HA-3 at 45 °C 97.5% at 1.00 mM HA-3 at 60 °C, carbon steel C1018 alloy analyzed using potentiodynamic polarization	[52]

(Continued)

TABLE 16.3 *(Continued)*

Heterocyclic Organic Derivatives as Acidizing Corrosion Inhibitors in 15 wt% HCl

Corrosion Inhibitors	Corrosion Inhibition Properties	Reference
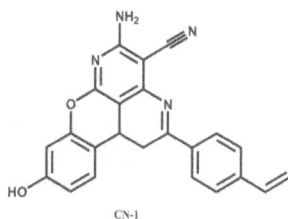 CN-1	CN-1, CN-2, and CN-3 showed 98.3%, 97.9%, and 95.5%, respectively, at only 300 mg L^{-1} at 35 °C for N80 steel	[53]
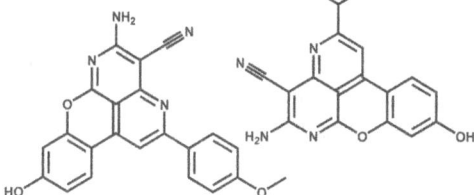 CN-2 CN-3		
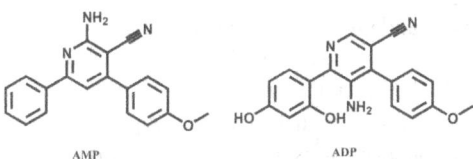 AMP ADP	86.94% for AMP, 90.24% for ADP at 200 mg L^{-1} at 308 K for N80 steel	[54]
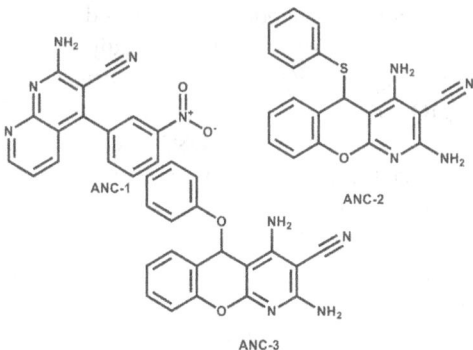 ANC-1 ANC-2 ANC-3	AANC-1, ANC-2, and ANC-3 showed 93.9%, 91.48%, and 85.07%, respectively, at 200 mg L^{-1} at 308 K for N80 steel	[55]
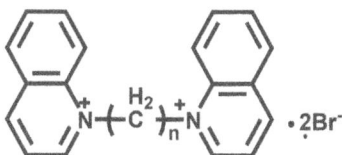 n = 4, Q4Q n = 6, Q6Q n = 8, Q8Q	88.7%, 92.9%, and 95.5% for Q4Q, Q6Q, and Q8Q, respectively, at 25 °C [10×10^{-2} M] in 15% HCl for N80 steel	[56]

inhibition upon increasing concentrations which further improved synergistically upon adding potassium iodide. It acted as a mixed-type corrosion inhibitor [51]. El-Lateef et al. revealed the corrosion inhibition activity of three Schiff base derivatives, HA-1, HA-2, and HA-3, for carbon steel C1018 alloy used in acidizing oil wells. The CIE of HA-3 was superior to other derivatives, which was 96.2% and 97.5% at 1.00 mM at 45 °C and 60 °C, respectively, for carbon steel. These corrosion inhibitors followed Langmuir adsorption isotherm and got chemisorbed on the metal surface. These Schiff bases–based heterocycles acted as mixed-type corrosion inhibitors; their CIE improved with concentration or temperature increment until the optimum concentration of 1.00 mM and temperature of 60 °C [52].

Salman et al. synthesized chromeno naphthyridines–based heterocyclic compounds having nontoxic and environment-friendly nature, namely, CN-1, CN-2, and CN-3. The CN-1, CN-2, and CN-2 showed CIE of 98.3%, 97.9% and 95.5%, respectively, at only 300 mg L^{-1} at 35 °C for N80 steel. The corrosion inhibitory action was controlled by charge transfer, exhibited mixed-type corrosion inhibition, and followed Langmuir-type adsorption isotherm. DFT and MD simulation results corroborated experimentally determined results [53]. Similarly, Ansari et al. reported the adsorption and corrosion inhibitory action of ADP and AMP for N80 steel in 15% HCl. The CIE was 86.94% for AMP and 90.24% for ADP at 200 mg L^{-1} at 308 K for N80 steel. The corrosion inhibitory action was mixed, predominantly suppressing the cathodic reaction. ADP and AMP obeyed the Langmuir adsorption isotherm. The DFT study also corroborated the experimentally determined corrosion inhibitory action [54]. Furthermore, Ansari and Quraishi synthesized ANC-1, ANC-2, and ANC-3. The ANC-1, ANC-2, and ANC-3 showed 93.9%, 91.48%, and 85.07% at 200 mg L^{-1} at 308 K for N80 steel in 15% HCl with CIE order: ANC-1 > ANC-2 > ANC-3. These corrosion inhibitors were found to act as mixed-type corrosion-mitigating nature and followed Langmuir adsorption isotherm during their adsorption. Additionally, the quantum chemical parameters supported the CIE order [55].

Zhang et al. revealed the effect of corrosion inhibitory action by alkanediyl spacers in the diquarternary ammonium salts, namely, Q4Q, Q6Q, and Q8Q. The determined CIE was 88.7%, 92.9%, and 95.5% for Q4Q, Q6Q, and Q8Q, respectively, at 25 °C $[10 \times 10^{-2}$ M] in 15% HCl for N80 steel. Additionally, the CIE of Q8Q reached 91% at 90 °C for 0.01 mol L^{-1}. It has been reported that the CIE increases with concentration, and hydrophobic alkanediyl spacer length increase in the molecular skeleton of these corrosion inhibitors. These corrosion inhibitors followed Langmuir-type adsorption; the strong adsorption is attributed to bromine and nitrogen in its chemical structure. The adsorption of these diquarternary ammonium salts was attributed to (a) electrostatic attraction, (b) interaction of cationic N^+ ion with the metal surface, and (c) the interaction of π-electrons present in the quinolone rings with the metal surface atoms [56].

16.2.4   SEVEN-MEMBERED HETEROCYCLIC ORGANIC MOLECULES AS ACIDIZING CORROSION INHIBITOR

Some seven-membered heterocyclic organic molecules are also used as acidizing corrosion inhibitors. Recently, Ituen et al. reported the corrosion inhibition efficiency of clomipramine (CLO) on J55 steel in 15% HCl which showed 73.2% at

TABLE 16.4

Heterocyclic Organic Derivatives as Acidizing Corrosion Inhibitors in 15 wt% HCl

Corrosion Inhibitors	Corrosion Inhibition Properties	Reference
3CDA	74.8% for 3CDA at 10×10^{-5} M; 3CDA-NaG, 3CDA-KI, 3CDA-PEG, and 3CDA-GLU IE is 78.1%, 88.6%, 90.3%, and 91.4%, respectively, at 30 °C for X80 steel	[58]
BPBD BMBD	95.6% (BMBD), 93.4% (BPBD), 200 ppm at 303 K	[59]

10×10^{-5} M at 30 °C [57]. In addition, Ituen et al. also reported the synergistic corrosion inhibitory action of 3CDA on X80 steel in 15% HCl to be 74.8% at 10×10^{-5} M; this corrosion inhibitor with sodium gluconate (3CDA-NaG), with potassium iodide (3CDA-KI), with polyethylene glycol (3CDA-PEG), and with glutathione (3CDA-GLU) exhibiting CIE of 78.1%, 88.6%, 90.3%, and 91.4%, respectively, at 30 °C (Table 16.4) [58]. Furthermore, other corrosion inhibitors containing seven-membered heterocycles along with the aromatic benzene rings, namely, BPBD and BMBD, were also found to exhibit CIE of 95.6% and 93.4%, respectively, with 200 ppm addition at 303 K [59].

16.3 SUMMARY

This chapter outlines the heterocyclic corrosion inhibitor explored for effective corrosion inhibition of metal substrates exposed to oil-well acidizing using a 15% hydrochloric acid solution to maximize the production from oil reservoirs. Adding an anticorrosive additive is one of the practical approaches to minimizing the direct and indirect damages owing to corrosion and cleansing the tubing materials during acidizing. Consequently, the emergence of the heterocyclic organic corrosion inhibitor as additives has significantly impacted oil-well acidizing industries owing to its cost-effective synthesis technique, eco-friendly nature, adsorption capability, high corrosion inhibition capability, etc. The active adsorption capabilities and the corrosion inhibition mechanism revealed that these corrosion inhibitors undergo cathodic, anodic, cathodic and anodic combined, i.e., mixed-type, corrosion inhibitors. Usually, the heterocyclic corrosion inhibitors used in the acidizing solution obey Langmuir, Temkin, Freundlich, etc., adsorption isotherms. It has also been found that blending the heterocyclic corrosion inhibitors with iodide salts, like potassium iodides, etc., or organic co-additive, like glutathione, etc., synergistically enhances the corrosion inhibition efficiency of the corrosion inhibitors. In most cases, the DFT-based quantum chemical parameters and MD and MC simulation results effectively corroborate with the experimental corrosion inhibition order of the studied heterocyclic corrosion inhibitors.

ABBREVIATIONS

ACPC	(*E*)-5-amino-*N'*-(4-chlorobenzylidene)-3-(4-chlorophenyl)-1*H*-pyrazole-4-carbohydrazide
ADP	2-amino-6-(2,4-dihydroxyphenyl)-4-(4-methoxyphenyl) nicotinonitrile
AMP	2-amino-4-(4-methoxyphenyl)-6-phenylnicotinonitrile
AMPC	(*E*)-5-amino-3-(4-methoxyphenyl)-*N'*-(1-(4-methoxyphenyl) ethylidene)-1*H*-pyrazole-4-carbohydrazide
ANC-1	2-amino-4-(4-methoxyphenyl)-1,8-naphthyridine-3-carbonitrile
ANC-2	2-amino-4-(4-methylphenyl)-1,8-naphthyridine-3-carbonitrile
ANC-3	2-amino-4-(3-nitrophenyl)-1,8-naphthyridine-3-carbonitrile
AP-1	2-amino-4-(4-methoxyphenyl)-7,7-dimethyl-5-oxo-5,6,7,8-tetrahydro-4*H*-chromene-3-carbonitrile
AP-2	2-amino-7,7-dimethyl-5-oxo-4-phenyl-5,6,7,8-tetrahydro-4*H*-chromene-3-carbonitrile
AP-3	2-amino-7,7-dimethyl-4-(4-nitrophenyl)-5-oxo-5,6,7,8-tetrahydro-4*H*-chromene-3-carbonitrile
BDMA	benzo[*d*]thiazol-2-yl)-2-(3,5-dichlorophenyl)-2,5-dimethylthiazolidin-4-one
BDMT	1-(benzo[*d*]thiazol-2-yl)-3-chloro-4-(3,5-dichlorophenyl)-4-methylazetidin-2-one
BMBD	2,4-bis(methoxyphenyl)-1*H*-benzodiazepin
BPBD	2,4-bis(phenyl)-1*H*-benzodiazepine
BX-Cl	12-(4-chlorophenyl)-9,9-dimethyl-8,9,10,12-tetrahydro-11*H*-benzo[*a*]xanthen-11-one

BX-NO$_2$	9,9-dimethyl-12-(4-nitrophenyl)-8,9,10,12-tetrahydro-11*H*-benzo[*a*]xanthen-11-one
3CDA	3-(2-chloro-5,6-dihydrobenzo[*b*][1]benzazepin-11-yl)-*N*,*N*-dimethylpropan-1-amine
CIE	corrosion inhibition efficiency
CN-1	5-amino-9-hydroxy-2-(4-vinylphenyl)-1,11*b*-dihydrochromeno[4,3,2-de][1, 6]naphthyridine-4-carbonitrile
CN-2	5-amino-9-hydroxy-2-(4-methoxyphenyl)chromeno[4,3,2-de][1, 6]naphthyridine-4-carbonitrile
CN-3	5-amino-9-hydroxy-2-phenylchromeno[4,3,2-de][1, 6]naphthyridine-4-carbonitrile
DAMA-ran-DAMTDB	N_1,N_1-diallyl-N_1-methyl-N_6,N_6,N_6-tripropylhexane-1,6-diammonium dibromide
DMIC	1-decyl-3-methylimidazolium chloride
DNPP	(*E*)-1,5-dimethyl-4-((1-(3-nitrophenyl)ethylidene)amino)-2-phenyl-1,2-dihydro-3*H*-pyrazol-3-one
GLU	glutathione
HA-1	4-bromo-2-{(*Z*)-[(3,5-dimethylphenyl)imino]methyl}phenol
HA-2	1-{(*Z*)-[(4,6-dimethylpyridin-2-yl)imino]methyl}naphthalen-2-ol
HA-3	1-{(*Z*)-[(2-methoxy-4-nitrophenyl)imino]methyl}naphthalen-2-ol
8-HQ	8-hydroxyquinoline
5-HTTP	5-hydroxytryptophan
MBI	2-((1-((3-methoxyphenylamino)methyl)-1-benzoimidazol-2-yl)methyl)isoindoline-13-dione
MPII	1-(4-(1-acetyl-5-(4-methoxyphenyl)-4,5-dihydro-1*H*-pyrazol-3-yl)phenyl)spiro[imidazolidine-2,3′-indoline]-2′,5-dione
MPIT	3′-(4-(1-acetyl-5-(4-methoxyphenyl)-4,5-dihydro-1*H*-pyrazol-3-yl)phenyl)spiro[indoline-3,2′-thiazolidine]-2,4′-dione
NDTHDC	N_1,N_1-diallyl-N_6,N_6,N_6-tripropylhexane-1,6-diaminium chloride
NIMP	*N*-(2-(2-pentadecyl-4,5-dihydro-1*H*-imidazol-1-yl)ethyl)palmitamide
NTETD	*N*-(2-(2-tridecyl-4,5-dihydro-1*H*-imidazol-1-yl)ethyl)tetradecanamide
OYAA	(*Z*)-2-(2-oxoindolin-3-ylideneamino)acetic acid
OYPA	2-(2-oxoindolin-3-ylideneamino)-3-phenylpropanoic acid
PBI	2-((1-(3-methylphenylamino)methyl)-1-benzoimidazol-2-yl)methyl)isoindoline-13-dione

PD	4-(3-methoxy-4-hydroxy-phenyl)-2-phenyl-1,4-dihydro-benzo[4, 5]imidazo [1,2-*a*]pyrimidine-3-carboxylicacid ethyl ester
poly-NDTHDC	poly(N_1,N_1-diallyl-N_6,N_6,N_6-tripropylhexane-1,6-diaminium chloride)
PP-1	5-styryl-2,7-dithioxo-2,3,5,6,7,8-hexahydropyrimido[4,5-*d*] pyrimidin-4(1*H*) one
PP-2	5-(2-hydroxyphenyl)-2,7-dithioxo-2,3,5,6,7,8-hexahydropyrimido[4,5-*d*]-pyrimidin-4(1*H*) one
PPC-1	2,4-diamino-3,5-dihydro-5-(phenylthio)-2*H*-chromeno[2,3-*b*]pyridine-3-carbonitrile
PPC-2	2,4-diamino-3,5-dihydro-5-phenoxy-2*H*-chromeno[2,3-*b*]pyridine-3-carbonitrile
PZ	(diallylamino)diacetate polyzwitter
PZ1	2-(3-amino-5-oxo-4,5-dihydro-1*H*-pyrazol-1-yl)(*p*-tolyl) methyl)malononitrile
PZ2	2-((3-amino-5-oxo-4,5-dihydro-1*H*-pyrazol-1-yl) (phenyl)methyl)malononitrile
PZ-1	4,4′-((4-methoxyphenyl)methylene) bis(3-methyl-1-phenyl-1*H*-pyrazol-5-ol)
PZ-2	4,4′-((4-nitrophenyl)methylene) bis(3-methyl-1-phenyl-1*H*-pyrazol-5-ol)
Q4Q	*N*,*N*′-butane-1,4-diyl-bisquinolinium dibromide
Q6Q	*N*,*N*′-hexane-1,6-diyl-bisquinolinium dibromide
Q8Q	*N*,*N*′-octane-1,8-diyl-bisquinolinium dibromide
QSI	2-heptadecyl-1-[2-(octadecanoylamino) ethyl]-2- imidazoline
SAHMT	4-salicylideneamino-3-hydrazino-5-mercapto-1,2,4-triazole
TBX	9,9-dimethyl-12-*p*-tolyl-8,9,10,12-tetrahydrobenzo[*a*] xanthen-1-one

REFERENCES

1. S. K. Saha, P. Ghosh, A. Hens, N. C. Murmu, and P. Banerjee, "Density functional theory and molecular dynamics simulation study on corrosion inhibition performance of mild steel by mercapto-quinoline Schiff base corrosion inhibitor," *Physica E: Low-dimensional Systems and Nanostructures*, vol. 66, pp. 332–341, 2015. doi: 10.1016/j.physe.2014.10.035.
2. M. Murmu, S. K. Saha, P. Bhaumick, N. C. Murmu, H. Hirani, and P. Banerjee, "Corrosion inhibition property of azomethine functionalized triazole derivatives in 1 mol L^{-1} HCl medium for mild steel: Experimental and theoretical exploration," *Journal of Molecular Liquids*, vol. 313, p. 113508, 2020. doi: 10.1016/j.molliq.2020.113508.
3. S. K. Saha, M. Murmu, N. C. Murmu, and P. Banerjee, "Benzothiazolylhydrazine azomethine derivatives for efficient corrosion inhibition of mild steel in acidic environment: Integrated experimental and density functional theory cum molecular dynamics simulation approach," *Journal of Molecular Liquids*, vol. 364, p. 120033, 2022. doi: 10.1016/j.molliq.2022.120033.

4. S. K. Saha, A. Dutta, P. Ghosh, D. Sukul, and P. Banerjee, "Adsorption and corrosion inhibition effect of Schiff base molecules on the mild steel surface in 1 M HCl medium: A combined experimental and theoretical approach," *Physical Chemistry Chemical Physics*, vol. 17, no. 8, pp. 5679–5690, 2015. doi: 10.1039/C4CP05614K.

5. S. K. Saha, A. Dutta, P. Ghosh, D. Sukul, and P. Banerjee, "Novel Schiff-base molecules as efficient corrosion inhibitors for mild steel surface in 1 M HCl medium: Experimental and theoretical approach," *Physical Chemistry Chemical Physics*, vol. 18, no. 27, pp. 17898–17911, 2016. doi: 10.1039/C6CP01993E.

6. M. A. Quraishi, D. S. Chauhan, and V. S. Saji, *Heterocyclic Organic Corrosion Inhibitors*, Amsterdam, Elsevier, 2020. doi: 10.1016/C2018-0-04237-1.

7. M. Murmu, S. Kr. Saha, C. Murmu, and P. Banerjee, "Effect of stereochemical conformation into the corrosion inhibitive behaviour of double azomethine based Schiff bases on mild steel surface in 1 mol L^{-1} HCl medium: An experimental, density functional theory and molecular dynamics simulation study," *Corrosion Science*, vol. 146, pp. 134–151, 2019. doi: 10.1016/j.corsci.2018.10.002.

8. H. Frash, "Increasing the flow of oil well," 1985 (accessed Feb 16, 2023). Available at: https://patents.google.com/patent/US556651A/en

9. A. Singh and M. A. Quraishi, "Acidizing corrosion inhibitors: A review," *Journal of Materials and Environmental Science*, vol. 6, no. 1, pp. 224–235, 2015.

10. H. Li, Y. Qiang, and C. Verma, "Acidizing corrosion inhibitors," in *Eco-Friendly Corrosion Inhibitors*, Elsevier, 2022, pp. 45–54. doi: 10.1016/B978-0-323-91176-4.00022-2.

11. X. Ma et al., "New corrosion inhibitor acrylamide methyl ether for mild steel in 1 M HCl," *Applied Surface Science*, vol. 371, pp. 248–257, 2016. doi: 10.1016/j.apsusc.2016.02.212.

12. A. Yurt, A. Balaban, S. U. Kandemir, G. Bereket, and B. Erk, "Investigation on some Schiff bases as HCl corrosion inhibitors for carbon steel," *Materials Chemistry Frontiers*, vol. 85, no. 2–3, pp. 420–426, 2004. doi: 10.1016/j.matchemphys.2004.01.033.

13. S. K. Saha, A. Hens, N. C. Murmu, and P. Banerjee, "A comparative density functional theory and molecular dynamics simulation studies of the corrosion inhibitory action of two novel *N*-heterocyclic organic compounds along with a few others over steel surface," *Journal of Molecular Liquids*, vol. 215, pp. 486–495, 2016. doi: 10.1016/j.molliq.2016.01.024.

14. S. K. Saha and P. Banerjee, "Introduction of newly synthesized Schiff base molecules as efficient corrosion inhibitors for mild steel in 1 M HCl medium: An experimental, density functional theory and molecular dynamics simulation study," *Materials Chemistry Frontiers*, vol. 2, no. 9, pp. 1674–1691, 2018. doi: 10.1039/C8QM00162F.

15. S. K. Saha and P. Banerjee, "A theoretical approach to understand the inhibition mechanism of steel corrosion with two aminobenzonitrile inhibitors," *RSC Advances*, vol. 5, no. 87, pp. 71120–71130, 2015. doi: 10.1039/c5ra15173b.

16. A. Dutta, S. K. Saha, P. Banerjee, and D. Sukul, "Correlating electronic structure with corrosion inhibition potentiality of some bis-benzimidazole derivatives for mild steel in hydrochloric acid: Combined experimental and theoretical studies," *Corrosion Science*, vol. 98, pp. 541–550, 2015. doi: 10.1016/j.corsci.2015.05.065.

17. S. K. Saha, M. Murmu, N. C. Murmu, and P. Banerjee, "Synthesis, characterization and theoretical exploration of pyrene based Schiff base molecules as corrosion inhibitor," *Journal of Molecular Structure*, vol. 1245, p. 131098, 2021. doi: 10.1016/j.molstruc.2021.131098.

18. M. Murmu, S. Kr. Saha, N. C. Murmu, and P. Banerjee, "Amine cured double Schiff base epoxy as efficient anticorrosive coating materials for protection of mild steel in 3.5% NaCl medium," *Journal of Molecular Liquids*, vol. 278, pp. 521–535, 2019. doi: 10.1016/j.molliq.2019.01.066.

19. S. K. Saha, M. Murmu, N. C. Murmu, and P. Banerjee, "Evaluating electronic structure of quinazolinone and pyrimidinone molecules for its corrosion inhibition effectiveness on target specific mild steel in the acidic medium: A combined DFT and MD simulation study," *Journal of Molecular Liquids*, vol. 224, pp. 629–638, 2016. doi: 10.1016/j. molliq.2016.09.110.

20. S. K. Saha, M. Murmu, N. C. Murmu, I. B. Obot, and P. Banerjee, "Molecular level insights for the corrosion inhibition effectiveness of three amine derivatives on the carbon steel surface in the adverse medium: A combined density functional theory and molecular dynamics simulation study," *Surfaces and Interfaces*, vol. 10, pp. 65–73, 2018. doi: 10.1016/j.surfin.2017.11.007.

21. M. Yadav, L. Gope, and T. K. Sarkar, "Synthesized amino acid compounds as eco-friendly corrosion inhibitors for mild steel in hydrochloric acid solution: Electrochemical and quantum studies," *Research on Chemical Intermediates*, vol. 42, no. 3, pp. 2641–2660, 2016. doi: 10.1007/s11164-015-2172-5.

22. E. Ituen, O. Akaranta, and A. James, "Electrochemical and anticorrosion properties of 5-hydroxytryptophan on mild steel in a simulated well-acidizing fluid," *Journal of Taibah University for Science*, vol. 11, no. 5, pp. 788–800, 2017. doi: 10.1016/j. jtusci.2017.01.005.

23. M. M. Solomon, S. A. Umoren, M. A. Quraishi, and M. Salman, "Myristic acid based imidazoline derivative as effective corrosion inhibitor for steel in 15% HCl medium," *Journal of Colloid and Interface Science*, vol. 551, pp. 47–60, 2019. doi: 10.1016/j. jcis.2019.05.004.

24. M. M. Solomon, S. A. Umoren, M. A. Quraishi, and M. A. Jafar Mazumder, "Corrosion inhibition of N80 steel in simulated acidizing environment by *N*-(2-(2-pentadecyl-4,5-dihydro-1*H*-imidazol-1-yl)ethyl)palmitamide,"*Journal of Molecular Liquids*, vol. 273, pp. 476–487, 2019. doi: 10.1016/j.molliq.2018.10.032.

25. M. M. Solomon, S. A. Umoren, M. A. Quraishi, D. B. Tripathy, and E. J. Abai, "Effect of alkyl chain length, flow, and temperature on the corrosion inhibition of carbon steel in a simulated acidizing environment by an imidazoline-based inhibitor," *Journal of Petroleum Science and Engineering*, vol. 187, p. 106801, 2020. doi: 10.1016/j. petrol.2019.106801.

26. H. Gerengi, R. Cakmak, B. Dag, M. M. Solomon, A. Tuysuz, and E. Kaya, "Synthesis and anticorrosion studies of 4-[(2-nitroacetophenonylidene)-amino]-antipyrine on SAE 1012 carbon steel in 15 wt.% HCl solution," *Journal of Adhesion Science and Technology*, vol. 34, no. 22, pp. 2448–2466, 2020. doi: 10.1080/01694243.2020.1766400.

27. M. Yadav, T. K. Sarkar, and T. Purkait, "Studies on adsorption and corrosion inhibitive properties of indoline compounds on N80 steel in hydrochloric acid," *Journal of Materials Engineering and Performance*, vol. 24, no. 12, pp. 4975–4984, 2015. doi: 10.1007/s11665-015-1765-x.

28. N. Kumari, P. K. Paul, L. Gope, and M. Yadav, "Studies on anticorrosive action of synthesized indolines on mild steel in 15% HCl solution," *Journal of Adhesion Science and Technology*, vol. 31, no. 14, pp. 1524–1544, 2017. doi: 10.1080/01694243.2016.1263473.

29. J. Zhang, J. Zhao, N. Zhang, C. Qu, and X. Zhang, "Synergized action of CuCl on recycled cigarette butts as corrosion inhibitor for N80 steel at 90 °C in 15% HCl," *Industrial & Engineering Chemistry Research*, vol. 50, no. 12, pp. 7264–7272, 2011. doi: 10.1021/ ie102288b.

30. K. R. Ansari, M. A. Quraishi, A. Singh, S. Ramkumar, and I. B. Obote, "Corrosion inhibition of N80 steel in 15% HCl by pyrazolone derivatives: Electrochemical, surface and quantum chemical studies," *RSC Advances*, vol. 6, no. 29, pp. 24130–24141, 2016. doi: 10.1039/C5RA25441H.

31. M. A. Quraishi and D. Jamal, "Corrosion inhibition of N-80 steel and mild steel in 15% boiling hydrochloric acid by a triazole compound: SAHMT," *Materials Chemistry and Physics*, vol. 68, no. 1–3, pp. 283–287, 2001. doi: 10.1016/S0254-0584(00)00369-2.

32. M. Yadav, D. Sharma, and S. Kumar, "Thiazole derivatives as efficient corrosion inhibitor for oil-well tubular steel in hydrochloric acid solution," *Korean Journal of Chemical Engineering*, vol. 32, no. 5, pp. 993–1000, 2015. doi: 10.1007/s11814-014-0275-0.

33. A. Singh, K. R. Ansari, M. A. Quraishi, and S. Kaya, "Theoretically and experimentally exploring the corrosion inhibition of N80 steel by pyrazol derivatives in simulated acidizing environment," *Journal of Molecular Structure*, vol. 1206, p. 127685, 2020. doi: 10.1016/j.molstruc.2020.127685.

34. P. K. Paul, M. Yadav, and I. B. Obot, "Investigation on corrosion protection behavior and adsorption of carbohydrazide-pyrazole compounds on mild steel in 15% HCl solution: Electrochemical and computational approach," *Journal of Molecular Liquids*, vol. 314, p. 113513, 2020. doi: 10.1016/j.molliq.2020.113513.

35. N. M. el Basiony et al., "Synthesis, characterization, experimental and theoretical calculations (DFT and MC) of ethoxylated aminothiazole as inhibitor for X65 steel corrosion in highly aggressive acidic media," *Journal of Molecular Liquids*, vol. 297, p. 111940, 2020. doi: 10.1016/j.molliq.2019.111940.

36. I. Y. Yaagoob and S. A. Ali, "Homo- and co-cyclopolymers containing symmetrical motifs of (diallylammonio)diacetate,"*Reactive & Functional Polymers*, vol. 179, p. 105379, 2022. doi: 10.1016/j.reactfunctpolym.2022.105379.

37. A. Rahimi, M. Abdouss, A. Farhadian, L. Guo, and J. Neshati, "Development of a novel thermally stable inhibitor based on furfuryl alcohol for mild steel corrosion in a 15% HCl medium for acidizing application," *Industrial & Engineering Chemistry Research*, vol. 60, no. 30, pp. 11030–11044, 2021. doi: 10.1021/acs.iecr.1c01946.

38. S. A. Umoren, M. M. Solomon, S. A. Ali, and H. D. M. Dafalla, "Synthesis, characterization, and utilization of a diallylmethylamine-based cyclopolymer for corrosion mitigation in simulated acidizing environment," *Materials Science and Engineering: C*, vol. 100, pp. 897–914, 2019. doi: 10.1016/j.msec.2019.03.057.

39. A. Singh, K. R. Ansari, M. A. Quraishi, and P. Banerjee, "Corrosion inhibition and adsorption of imidazolium based ionic liquid over P110 steel surface in 15% HCl under static and dynamic conditions: Experimental, surface and theoretical analysis," *Journal of Molecular Liquids*, vol. 323, p. 114608, 2021. doi: 10.1016/j.molliq.2020.114608.

40. A. Odewunmi, M. M. Solomon, A. Umoren, and A. Ali, "Comparative studies of the corrosion inhibition efficacy of a dicationic monomer and its polymer against API X60 steel corrosion in simulated acidizing fluid under static and hydrodynamic conditions," *ACS Omega*, vol. 5, no. 42, pp. 27057–27071, 2020. doi: 10.1021/acsomega.0c02345.

41. K. Haruna, I. B. Obot, N. K. Ankah, A. A. Sorour, and T. A. Saleh, "Gelatin: A green corrosion inhibitor for carbon steel in oil well acidizing environment," *Journal of Molecular Liquids*, vol. 264, pp. 515–525, 2018. doi: 10.1016/j.molliq.2018.05.058.

42. A. Singh, K. R. Ansari, I. H. Ali, Y. Lin, B. EL Ibrahimi, and L. Bazzi, "Combination of experimental, surface and computational insight into the corrosion inhibition of pyrimidine derivative onto Q235 steel in oilfield acidizing fluid under hydrodynamic condition," *Journal of Molecular Liquids*, vol. 353, p. 118825, 2022. doi: 10.1016/j.molliq.2022.118825.

43. Y. Wang et al., "Indolizine quaternary ammonium salt inhibitors part II: A reinvestigation of an old-fashioned strong acid corrosion inhibitor phenacyl quinolinium bromide

and its indolizine derivative," *New Journal of Chemistry*, vol. 42, no. 15, pp. 12977–12989, 2018. doi: 10.1039/C8NJ02505C.

44. Z. Yang et al., "Structure of a novel benzyl quinolinium chloride derivative and its effective corrosion inhibition in 15 wt.% hydrochloric acid," *Corrosion Science*, vol. 99, pp. 281–294, 2015. doi: 10.1016/j.corsci.2015.07.023.

45. L. B. Furtado et al., "Cleaner corrosion inhibitors using *Peumus boldus* Molina formulations in oil well acidizing fluids: Gravimetric, electrochemical and DFT studies," *Sustainable Chemistry and Pharmacy*, vol. 19, p. 100353, 2021. doi: 10.1016/j.scp.2020.100353.

46. J. Haque, K. R. Ansari, V. Srivastava, M. A. Quraishi, and I. B. Obot, "Pyrimidine derivatives as novel acidizing corrosion inhibitors for N80 steel useful for petroleum industry: A combined experimental and theoretical approach," *Journal of Industrial and Engineering Chemistry*, vol. 49, pp. 176–188, 2017. doi: 10.1016/j.jiec.2017.01.025.

47. K. R. Ansari, M. A. Quraishi, and A. Singh, "Chromenopyridin derivatives as environmentally benign corrosion inhibitors for N80 steel in 15% HCl," *Journal of the Association of Arab Universities for Basic and Applied Sciences*, vol. 22, no. 1, pp. 45–54, 2017. doi: 10.1016/j.jaubas.2015.11.003.

48. A. Singh et al., "Inhibition effect of newly synthesized benzoxanthones derivative on hydrogen evolution and Q235 steel corrosion in 15% HCl under hydrodynamic condition: Combination of experimental, surface and computational study," *International Journal of Hydrogen Energy*, vol. 46, no. 76, pp. 37995–38007, 2021. doi: 10.1016/j.ijhydene.2021.09.051.

49. A. Singh et al., "Understanding xanthone derivatives as novel and efficient corrosion inhibitors for P110 steel in acidizing fluid: Experimental and theoretical studies," *Journal of Physics and Chemistry of Solids*, vol. 172, p. 111064, 2023. doi: 10.1016/j.jpcs.2022.111064.

50. A. Singh, K. R. Ansari, M. A. Quraishi, H. Lgaz, and Y. Lin, "Synthesis and investigation of pyran derivatives as acidizing corrosion inhibitors for N80 steel in hydrochloric acid: Theoretical and experimental approaches," *Journal of Alloys and Compounds*, vol. 762, pp. 347–362, 2018. doi: 10.1016/j.jallcom.2018.05.236.

51. I. B. Obot, N. K. Ankah, A. A. Sorour, Z. M. Gasem, and K. Haruna, "8-Hydroxyquinoline as an alternative green and sustainable acidizing oilfield corrosion inhibitor," *Sustainable Materials and Technologies*, vol. 14, pp. 1–10, 2017. doi: 10.1016/j.susmat.2017.09.001.

52. H. M. Abd El-Lateef, A. M. Abu-Dief, and M. A. A. Mohamed, "Corrosion inhibition of carbon steel pipelines by some novel Schiff base compounds during acidizing treatment of oil wells studied by electrochemical and quantum chemical methods," *Journal of Molecular Structure*, vol. 1130, pp. 522–542, 2017. doi: 10.1016/j.molstruc.2016.10.078.

53. M. Salman, K. R. Ansari, V. Srivastava, D. S. Chauhan, J. Haque, and M. A. Quraishi, "Chromeno naphthyridines based heterocyclic compounds as novel acidizing corrosion inhibitors: Experimental, surface and computational study," *Journal of Molecular Liquids*, vol. 322, p. 114825, 2021. doi: 10.1016/j.molliq.2020.114825.

54. K. R. Ansari, M. A. Quraishi, and A. Singh, "Pyridine derivatives as corrosion inhibitors for N80 steel in 15% HCl: Electrochemical, surface and quantum chemical studies," *Measurement*, vol. 76, pp. 136–147, 2015. doi: 10.1016/j.measurement.2015.08.028.

55. K. R. Ansari and M. A. Quraishi, "Experimental and computational studies of naphthyridine derivatives as corrosion inhibitor for N80 steel in 15% hydrochloric acid," *Physica E: Low-dimensional Systems and Nanostructures*, vol. 69, pp. 322–331, 2015. doi: 10.1016/j.physe.2015.01.017.

56. X. Zhang, Y. Zheng, X. Wang, Y. Yan, and W. Wu, "Corrosion inhibition of N80 steel using novel diquaternary ammonium salts in 15% hydrochloric acid," *Industrial & Engineering Chemistry Research*, vol. 53, no. 37, pp. 14199–14207, 2014. doi: 10.1021/ie502405a.

57. E. Ituen, O. Akaranta, and A. James, "Influence of clomipramine-based blends on corrosion behaviour of J55 steel in simulated oilfield acidizing solution," *Journal of Bio- and Tribo-Corrosion*, vol. 3, no. 2, p. 23, 2017. doi: 10.1007/s40735-017-0082-3.

58. E. B. Ituen and J. E. Asuquo, "Inhibition of X80 steel corrosion in oilfield acidizing environment using 3-(2-chloro-5,6-dihydrobenzo[*b*][1]benzazepin-11-yl)-*N,N*-dimethylpropan-1-amine and its blends," *Journal of King Saud University*, vol. 31, no. 1, pp. 127–135, 2019. doi: 10.1016/j.jksus.2017.06.009.

59. S. Kumar, D. Sharma, P. Yadav, and M. Yadav, "Experimental and quantum chemical studies on corrosion inhibition effect of synthesized organic compounds on N80 steel in hydrochloric acid," *Industrial & Engineering Chemistry Research*, vol. 52, no. 39, pp. 14019–14029, 2013. doi: 10.1021/ie401308v.

17 Heterocycles as Corrosion Inhibitors in Harsh Conditions

Lipiar K. M. O. Goni[1], Ibrahim Y. Yaagoob[1], and Mohammad A. J. Mazumder[1,2]
[1]Chemistry Department, King Fahd University of Petroleum & Minerals, Dhahran, Saudi Arabia
[2]Interdisciplinary Research Center for Advanced Materials (IRC-AM), King Fahd University of Petroleum & Minerals, Dhahran, Saudi Arabia

17.1 INTRODUCTION

The phenomenon of corrosion that degrades metallic structures is considered a persistent global problem [1]. Technically, the simultaneous movement of mass and charge across a metal–solution interface is the definition of corrosion [2]. General or uniform corrosion has been identified as the most widespread form of corrosion that follows electrochemical reactions between the metals and the corrosive environments. It occurs on an entirely exposed metal surface and proceeds uniformly to damage the metals [3]. Usually, chemicals, such as acids (HCl, H_2SO_4, and HNO_3), moisture/water (H_2O), bases (NaOH, $NaHCO_3$, and $CaCO_3$), table salts (NaCl), gases (HCHO, NH_3, and sulfur-containing gases), aggressive metal polishes, etc., can cause metallic degradation upon contact. Any country in the world is found to spend somewhere in the range of 1–5% of its GNP on maintenance costs due to corrosion-related problems [4]. Corrosion is so pervasive that a study by the NACE said that the cost of corrosion worldwide in 2013 was 3.4% of the world's GDP [5]. Upstream oil and gas industries worldwide have been reported to spend US$1.3 billion annually as part of the corrosion cost, and 33% (US$463 million) of this cost is attributable to downhole corrosion and material issues [6].

Oil and gas industries use tubular steels, such as J55, N80, P110, Q235, carbon steel, etc., to construct downhole tubes, channeling pipelines, transmission pipelines, flow lines, etc. However, the acidizing process, which incorporates using highly concentrated (~15%) hydrochloric acid to dissolve carbonate rocks, takes a severe toll on the metallic structures due to the harsh corrosive nature of HCl [7]. Even though the oil and gas industries spend a lot to deal with this menace, economic losses are tough to stop. One of the most well-liked and economically practical approaches to dealing with this issue is corrosion inhibitors. Heterocyclic compounds are fundamental components of many natural products like amino acids, alkaloids, vitamins, nucleic acids, carbohydrates, etc., and are considered environmentally benign. From

DOI: 10.1201/9781003377016-17

the corrosion protection point of view, they are very efficient owing to the possession of heteroatoms (N, O, and S), polar functional groups ($-OH$, $-NH_2$, $-CN$, $-NO_2$, $-C=O$, $-COOH$), conjugated double bonds, and aromatic systems rich with π-electrons in their structures [1]. Therefore, this chapter will focus on various heterocycles employed in the oil and gas industries as acidizing corrosion inhibitors. However, a brief discussion on other acid treatment systems for oil-well acidizing has been made beforehand.

17.2 GENERAL BACKGROUND OF ACIDIZING

Oil-well stimulation, which in most circumstances entails adding acid to the reservoir to dissolve formation rocks and/or soluble components existing in the well, is necessary when flow line blockage causes a decline in oil production [8]. Oil and gas operators have utilized acid treatment (acidizing) to enhance well productivity for about 120 years. Acidizing injects acid into a well's geological formation or reservoir to dissolve the rocks that line the well's contours and enable oil and/or gas production. Any acidizing process aims to raise a well's productivity or injectivity. Acidizing is an ancient technique, preceding all other well-stimulation methods, including hydraulic fracturing, invented in the late 1940s. However, it is worth mentioning that the exploitation of the acidizing process was restrained to some extent because of the scarcity of effective corrosion inhibitors. With the advent of new, efficient acidizing corrosion inhibitors and improvement in the acidizing process itself, well-stimulation service-based industries kept growing. Acidizing is currently one of the most popular and effective methods accessible to the oil and gas sectors for stimulating (improving) well production. Acidizing is frequently done on new wells to increase their initial production and on older wells to increase productivity and recover as much energy as possible [9].

17.3 TYPES OF ACID TREATMENTS SYSTEMS

It is worth mentioning that two types of acids are used in the acidizing process: (i) mineral acids (HCl, HF, H_2SO_3, etc.) and (ii) organic acids ($HCOOH$, CH_3COOH, H_3NSO_3, and $C_2H_3ClO_2$) [10]. However, HCl is the most popular acid for use in the acidifying process. Also, mud acid, a mixture of HCl and HF, and organic acid, CH_3COOH (up to 10 wt%), are used sometimes. Acid treatment systems can be classified into three main types: acid washing, matrix acidizing, and fracture acidizing.

17.3.1 ACID WASHING

Tubular and wellbore cleaning is the primary objective of acid washing. This operation is usually done in carbonate formations to eliminate debris that retard flow in the well. However, this step is done at a low pressure using mild HCl (~7 wt%) and other additives to eradicate scale, debris, and rust, paving the way for oil and gas to flow easily into the well [8].

17.3.2 MATRIX ACIDIZING

The process of matrix acidizing involves pumping acid through a matrix, below the formation–fracture pressure, into a well to infiltrate rock pores, dissolving the mud and sediments and increasing the reservoir's innate pores. Usually, an acid concentration of 5–15 wt% and an acid volume of 15–200 gallons per foot of producing formation are used to restore or improve the wellbore's permeability (8–24 inches in radius) [8]. While HCl solution dissolves carbonate minerals, mud acid formulation is recommended to dissolve plugging minerals and many silicates.

17.3.3 FRACTURE ACIDIZING

Fracture acidizing, or fracking, is the most used acidizing technique to stimulate dolomite or limestone formations [9]. Fracking includes pouring acid at a pressure greater than the formation–fracturing pressure, in contrast to matrix acidizing. As a result, the fracture faces are randomly etched by the flowing acid without using a propping agent, making conductive channels that stay open even after the fracture closes. The recommended amount of HCl per foot of generating formation is 100–500 gallons [8].

17.4 HETEROCYCLES AS ACIDIZING CORROSION INHIBITORS

Acid fracturing is increasingly becoming challenging owing to advancements in drilling engineering, making the bottom-hole environment more complicated than ever. Even though the pickling liquid can increase the oil output in the oilfield development process, it can cause the metal tubing and other components to corrode, leading to a hike in operational costs. Using acidizing corrosion inhibitors is currently considered a prevalent and quite economical approach. However, it is expected that a corrosion inhibitor, especially in the case of an acidizing corrosion inhibitor, at quite a low concentration, can impart a reasonably good inhibition efficiency (IE) under the conditions of highly concentrated acid (~20%) and high temperature. These corrosion inhibitors are expected to be low-cost, nontoxic, and biodegradable. The following section will highlight a multitude of moderate to highly efficient, nontoxic heterocycles reported recently as acidizing corrosion inhibitors.

17.4.1 AZOLE-BASED HETEROCYCLES

Azole-based compounds are five-member aromatic rings containing N and at least one other N, O, or S atom as part of the ring. Azole-based compounds are incredibly soluble in harsh polar acid conditions due to the heteroatoms, conjugated double bonds, and aromatic electrons they contain, which also allows them to function very well as corrosion inhibitors [11]. Imidazoline-based corrosion inhibitors have a reputation for possessing nontoxicity attributes and being thermally very stable and highly efficient and are widely used for oilfield applications. Furthermore, the imidazole ring having three C and two N atoms is highly soluble in the aqueous media owing to possessing a very high polarity. Liu et al. [12] synthesized nano-SiO$_2$@

octadecylbisimidazoline quaternary ammonium salt (nano-SiO$_2$@OBQA) and studied its anticorrosive property toward protecting N80 steel sheet in 20% HCl at different temperatures. Nano-SiO$_2$@OBQA, at a concentration of 4%, reduced the corrosion rate by more than 95% at a remarkably high temperature of 180 °C. ΔG°_{ads} and K_{ads} were calculated to be –27.97 kJ mol^{-1} and 33.8983 L mol^{-1}, respectively. The inhibitor was adsorbed by following the Langmuir adsorption isotherm, as the ΔG°_{ads} value indicates via mixed physisorption and chemisorption. While the inorganic nanomaterials made nano-SiO$_2$@OBQA thermally very robust, a long hydrophobic alkyl chain helped repel the corrosive water molecules away from the metal surface. Additionally, π-electrons in the benzene ring, lone pair of electrons on N and O, and a C=N moiety in the bismidazoline ring helped it absorb strongly onto N80 steel sheets.

Benzimidazole, having an imidazole ring fused to a benzene ring, is another class of significant and widely used acid corrosion inhibitors. In addition to two coordinate covalent bond-forming nitrogen atoms in the imidazole ring, a fused benzene ring in benzimidazole compounds help them undergo superior adsorption onto the metal specimen and impart superb inhibition efficiency. Moreover, many natural products contain the benzimidazole moiety, and benzimidazole-based compounds have various pharmaceutical implications. Srivastava et al. [13] synthesized STBim as a corrosion inhibitor of carbon steel in 15% HCl. STBim was found to be environmentally benign with an IE of 98% at a temperature of 303 K, which showed an LD$_{50}$ value of 2000 mg kg^{-1}.

Moreover, the inhibitor maintained more than 95% IE even at 333 K. While EIS showed that the capacitive behavior got bettered with the inclusion of STBim in 15% HCl, PDP showed that it worked as an inhibitor of mixed types with a cathodic predominance, indicating that STBim blocked both the anodic dissolution of the metal and the cathodic evolution of molecular hydrogen. DFT simulation clarified that the inhibitor adsorbed onto the metal surface via the protonated form. It is worth mentioning that the protonated form of STBim can get adsorbed onto the cathodic sites via electrostatic interaction. In contrast, via a bridge-type linkage, the preadsorbed Cl- ions on the anodic sites attach the protonated form to the metal surface. Biotin was utilized by Xu et al. [14] to prevent the corrosion of mild steel in 15% HCl. Biotin, alternately known as vitamin H or B7, possesses a tetrahydroimidizalone ring fused with a tetrahydrothiophene ring and has many health applications. The drug's maximal IE was 97% at a 500 ppm concentration. Biotin was discovered to act as a mixed-type inhibitor and obey Langmuir adsorption isotherm. DFT study disclosed that biotin underwent adsorption onto a metal surface through the protonated form. SEM and AFM investigations validated the formation of biotin film on the metal surface. Some more recently reported imidazole- and benzimidazole-based acidizing corrosion inhibitors have been outlined in Table 17.1.

A compound (Py-OH) containing three heterocycles (pyrazole, pyran, and pyrimidine) fused with each other was synthesized and tested as a corrosion inhibitor of Q235 steel in 15% HCl under dynamic conditions [7]. The inhibitor was rendered nontoxic, considering the LD$_{50}$ value that came to be as high as 1468 mg kg^{-1}. Considering hydrodynamic conditions, the inhibitor was efficient enough to impart IE of 94.5% at a relatively low feeding dose of 200 mg L^{-1} at 308 K. The R_{ct} and

TABLE 17.1

Imidazole- and Benzimidazole-based Acidizing Corrosion Inhibitors

Inhibitor (Concentration)	Electrolyte (Metal) (Temperature)	IE (%)	Inhibitor Type	Mode of Adsorption (Isotherm)	Reference
CPBPM (10^{-3} M)	5.0 M HCl (MS) (30 °C)	98.6	Mixed	Chemical (Langmuir)	[15]
NBIDAB (404.37 μM)	15% HCl (304L SS) (25 °C)	97.0	Mixed	Physical (Langmuir)	[16]
1-Decyl-3-methylimidazolium chloride (400 ppm + 0.5 mM KI)	15% HCl (P110) (35 °C)	90.2	Mixed	Mixed (El-Awady)	[17]
2-Heptadecyl-1-[2-(octadecanoylamino)ethyl]-2-imidazoline (400 ppm + 1 mM KI)	15% HCl (CS) (25 °C)	90.0	Mixed	Chemical (N/A)	[18]
BINMPA (317.1 μM)	15% HCl (N80) (30 °C)	97.4	Mixed	Mixed (Langmuir)	[19]
Benzimidazole-based derivative (300 ppm + 2500 ppm KI + 100 ppm SDS + 2000 ppm NAC)	15% HCl (N80) (90 °C)	92.9	Mixed	Physical (Langmuir)	[20]

C_{dl} values, with those of the control, were increasing and decreasing, respectively, as found in the EIS study, revealing that raising the concentration of Py-OH helped lower the corrosion rate. While the PDP investigation demonstrated that Py-OH operated as a mixed-type inhibitor, as there was no discernible change in E_{corr}, β_a, or β_c concerning those of the control, it underwent pure physical adsorption following

Freundlich isotherm. A pyrazole-based compound, PZ-1, containing two pyrazole rings was synthesized by Singh et al. [21]. PZ-1 was highly efficient in imparting an excellent IE of 98.4% to N80 steel in 15% HCl. Additionally, molecular dynamics (MD) study found PZ-1 to bind to an N80 steel surface with a high binding energy ($E_{binding}$) of 631.22 kJ mol^{-1}. XPS, AFM, and SEM studies found a film of PZ-1 adsorbed onto the N80 steel surface. A pyridine- and pyrazole rings-containing compound, MMDPPM, synthesized by Yadav et al. [22], also imparted an excellent IE of 95.9% when the inhibitor was assessed for its efficacy among protecting mild steel in 15% HCl. Paul et al. [23], inspired by the biological properties of pyrazole- and carbohydrazide moiety-containing compounds, synthesized AMPC and studied its anticorrosive properties on mild steel corrosion in 15% HCl. AMPC showed a superb IE of 98.3% and retarded anodic and cathodic reaction rates efficiently. The inhibitor followed Langmuir's adsorption isotherm to form a monolayer on the metal surface.

Eid et al. [24] synthesized a series of tetrazole-based organoselenium compounds (Figure 17.1a), specifically TOS1, TOS2, and TOS3, for the corrosion protection of casings and well tubing during the well stimulation acidizing process. While all three tetrazoles imparted IE close to or more than 90%, TOS2 imparted a maximum IE of 94.6% toward J55 steel corrosion in 10% HCl. Additionally, TOS2 showed superiority when tested to prevent the growth of sulfate-reducing bacteria that causes microbially induced corrosion, making TOS2 useful for mitigating corrosion reasoned by infected stagnant waters in the annular spacing. In light of the numerous pharmacological properties shown by triazoles, Mehta et al. [25] designed and synthesized MTPT (Figure 17.1b) to alleviate the corrosion of mild steel in 15% HCl. MTPT, at 100 ppm concentration, conferred an IE of 94.3% to retard both anodic and cathodic reactions. The adsorption of MTPT onto a mild steel surface followed the Langmuir adsorption isotherm and involved both physical and chemical adsorption.

Compounds that contain both N and S have been reported as a superior corrosion inhibitor in comparison to those that include either N or S. As N and S are less electronegative than O, compounds containing both N and S heteroatoms would be anchoring more actively, through coordinate covalent bonds, with the vacant d-orbitals of the metal atoms in a particular electrolyte. Tiwari et al. [26] tested the IE of

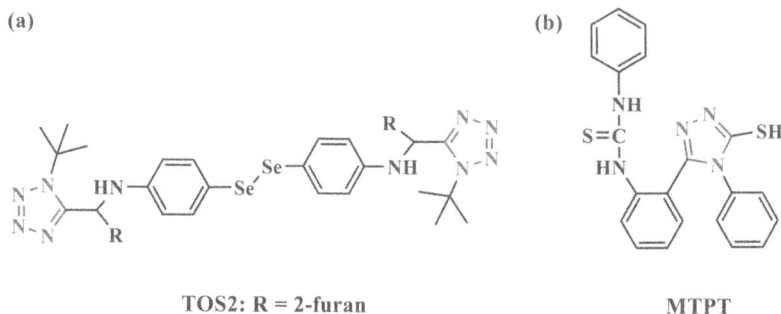

TOS2: R = 2-furan   MTPT

FIGURE 17.1 Tetrazole-based (a) and triazole-based (b) heterocyclic oil-well acidizing corrosion inhibitors.

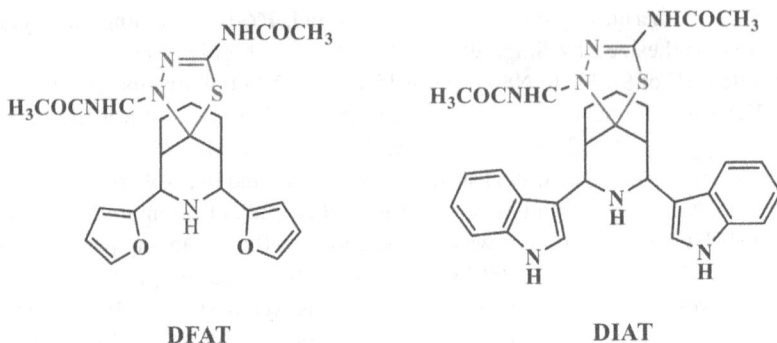

DFAT   **DIAT**

FIGURE 17.2 Thiadiazoline-based heterocycles for corrosion mitigation of N80 steel in acidizing conditions.

two thiadaizoline compounds (Figure 17.2), namely, DFAT and DIAT, for N80 CS corrosion in 15% HCl. Both inhibitors performed well at 303 K, with DFAT and DIAT imparting IEs of 96% and 97.1%, respectively, at a moderately small dose of 50 ppm. Both inhibitors exhibited mixed-type behavior and adhered to the Langmuir adsorption isotherm. DFT studies showed that both DFAT and DIAT got adsorbed onto N80 steel via their protonated forms. Basiony et al. [27] synthesized an ethoxylated aminoethyl thiazole, bis-ethoxylated (4-methyl thiazole-2-amine), and tested its IE for X60 steel corrosion in 15% HCl. The thiazole, at 10^{-3} M concentration, provided an IE of 95.5% at 25 °C. The inhibitor slowed the anodic dissolution of metal and cathodic evolution of H_2 molecules and followed the Langmuir adsorption isotherm. ΔG_{ads}° of -39.3 kJ mol^{-1} indicated that the thiazole was adsorbed spontaneously via mixed physiochemisorption. Some more thiazole- and thiadiazole-based acidizing heterocyclic corrosion inhibitors are listed in Table 17.2.

17.4.2 AZINE-BASED HETEROCYCLES

In heterocyclic chemistry, azines are aromatic six-membered ring compounds that contain one (pyridine) to six (hexazine) N atoms. Pyrazines are an essential class of inhibitors with widespread implications in this field. Furthermore, pyrazines have been deemed safe, following their usage in pharmaceutical and agro-based industries. In addition to being cost-effective and nontoxic, they, most importantly, offer superior IE because of having a planar structure that facilitates its easy adsorption, with an N atom and π-electron-rich aromatic ring helping the process. To illustrate, Obot et al.'s [36] work on an environmentally benign pyrazine derivative, 2,3-pyrazine dicarboxylic acid (1 wt%), has seen it imparting an IE of 90.0% at a very high temperature of 90 °C in the presence of 0.1% NaI and 0.01% glutathione additives. The PDP investigation showed that the pyrazine compound, which used an inhibitor of X60 steel corrosion in 15% HCl, was of mixed type with cathodic preference. In an approach to solving the problems of corrosion in acidizing conditions and waste graphite disposal, Haruna and Saleh [37] synthesized graphene oxide (GO) from waste graphite and grafted NAEP onto GO. The piperazine-grafted GO, NAEP-GO,

TABLE 17.2

Thiazole- and Thiadiazole-based Heterocyclic Corrosion Inhibitors for Corrosion Mitigation in Acidizing Conditions

Inhibitor (Concentration)	Electrolyte (Metal) (Temperature)	IE (%)	Inhibitor Type	Mode of Adsorption (Isotherm)	Reference
PPATPT (100 ppm)	15% HCl (MS) (30 °C)	97.7	Mixed	Mixed (Langmuir)	[25]
DPDHPNT (100 ppm)	15% HCl (MS) (30 °C)	96.0	Mixed	Physical (Langmuir)	[28]
BTHHP (500 ppm) R = Glucose	15% HCl (N80) (303 °C)	97.3	Mixed	Chemical (Langmuir)	[29]
BTDCDMT (50 ppm)	15% HCl (N80) (30 °C)	95.4	Mixed	Mixed (Langmuir)	[30]
NOPTAC (0.15%)	15% Lactic acid (N80) (90 °C)	97.6	Mixed	Mixed (Langmuir)	[31]
DHPPTEH (50 ppm)	15% HCl (N80) (30 °C)	98.0	Mixed	Mixed (Langmuir)	[32]

(Continued)

TABLE 17.2 (*Continued*)
Thiazole- and Thiadiazole-based Heterocyclic Corrosion Inhibitors for Corrosion Mitigation in Acidizing Conditions

Inhibitor (Concentration)	Electrolyte (Metal) (Temperature)	IE (%)	Inhibitor Type	Mode of Adsorption (Isotherm)	Reference
TMPTMTD (200 ppm)	15% HCl (MS) (60 °C)	97.1	Mixed	Chemical (Langmuir)	[33]
BHMPDTA (2×10^{-3} M)	15% HCl (CS) (50 °C)	98.3	Mixed	Mixed (Langmuir)	[34]
BONMTZA (600 ppm)	15% HCl (X60) (60 °C)	91.0%	Mixed	Mixed (Langmuir)	[35]

was used as a carbon steel corrosion inhibitor in 15% HCl. NAEP-GO provided an IE of 87% for a concentration of 25 ppm at room temperature. NAEP-GO was found to retard the cathodic reaction predominantly and followed Langmuir adsorption isotherm. In a bid to demonstrate the first-ever use of triazines as corrosion inhibitor of N80 steel corrosion in 15% HCl, Salman et al. [38] synthesized TZ and studied its anticorrosive behavior to find TZ imparting an IE of 93.2%, at a relatively high dose of 800 mg L^{-1} at 308 K. Surface morphological studies confirmed that TZ films formed on the N80 steel surface. It was found that TZ behaves as a mixed-type inhibitor and follows the Langmuir adsorption isotherm.

Pyridines are another very important class of inhibitors that show multifarious pharmacological and biological activities apart from being used as corrosion inhibitors. Farag et al. [39] synthesized two pyridine products, namely, Py I and Py II, and studied them as corrosion inhibitors of API CS in 6 M H_2SO_4. Both compounds containing aromatic rings, amino and methoxy groups, and N and O heteroatoms showed a little over 80% IE. On the other hand, Ansari et al.'s [40] work with ADP and AMP saw them imparting IE close to 90% toward N80 steel corrosion in 15% HCl at 200 mg L^{-1} inhibitor concentration, with ADP being a relatively better inhibitor. The existence of electron-donating —OH groups in the phenyl ring on ADP made it more efficient, as evidenced by a DFT study that showed that the energy gap (ΔE) between the HOMO and LUMO in the case of ADP is lower than that of AMP, meaning ADP is more reactive in case of getting adsorbed. Furthermore, El-Lateef et al.'s [41] work on a pyridine-based Schiff base, HA-2, yielded a fine IE of 94.1% when the inhibitor was tested for its efficacy in protecting carbon steel from corroding in 15% HCl at 60 °C. The existence of aromatic rings and −OH and −C=N groups paved the way for the Schiff base to impart a relatively high IE at such a high temperature.

Naphthyridines, containing two pyridine rings fused, are planar molecules containing multiple heteroatoms and many π-electrons. Polar substituents, such as −NH$_2$, −OH, −CN, etc., increase their solubility and inhibitive performance. Furthermore, it showed that they exhibit a variety of biological functions. Ansari and Quraishi [42] synthesized a series of napthyridines (Figure 17.3), namely, ANC-1, ANC-2, and ANC-3, and tested their IEs for N80 steel corrosion in 15% HCl. According to the PDP investigation, all the inhibitors were mixed-type and obeyed the Langmuir adsorption isotherm. ΔG_{ads}° values demonstrated that the adsorption process followed mixed physiochemisorption. The IEs, according to the WL test, were found to be on the order of ANC-1 (93.9%) > ANC-2 (91.5%) > ANC-3 (85.1%). The experimental findings pretty much followed the DFT study. The DFT-optimized HOMOs (Figure 17.3) show that electron density is mainly localized on pyridine and benzene rings in the case of ANC-1 and ANC-2, making those π-electrons more available for coordination. On the other hand, the electron concentration is mainly localized on N and O of the −NO$_2$ group, making ANC-3 less prone to adsorption onto the N80 steel surface. Three environmentally benign, following the measurement of LD$_{50}$ values, chromeno naphthyridines (Figure 17.4), namely, CN-1, CN-2, and CN-3, were synthesized and examined for their anticorrosive properties toward N80 steel corrosion in 15% HCl [43]. While all the inhibitors, at a dosage of 300 mg L^{-1}, imparted more

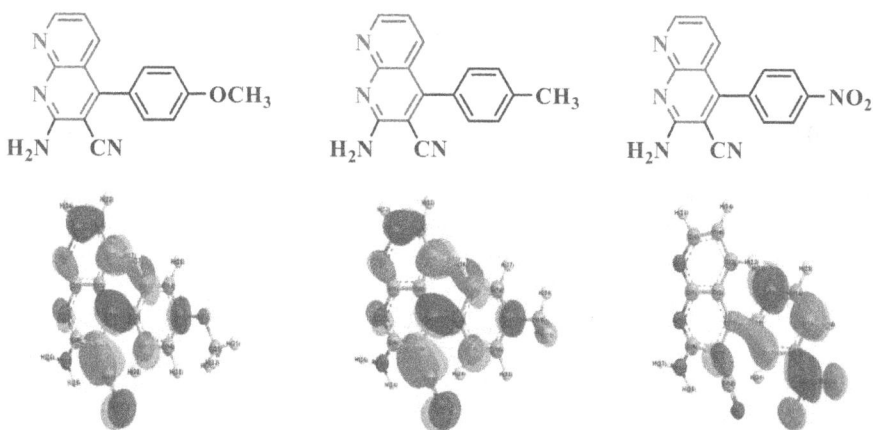

FIGURE 17.3 Molecular structures (*up*) and HOMOs (*down*) for ANC-1 (*left*), ANC-2 (*middle*), and ANC-3 (*right*).

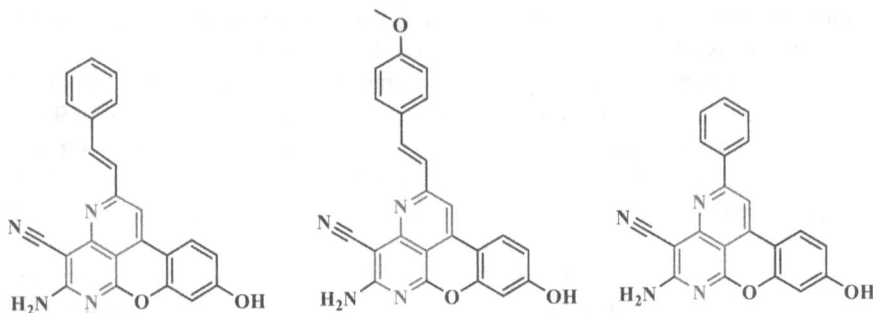

FIGURE 17.4   Chromeno naphthyridines CN-1 (*left*), CN-2 (*center*), and CN-3 (*right*) developed for the protection of N80 steel corrosion in 15% HCl.

than 95% IE at 35 °C, CN-1 showed a very impressive IE of 98.3%. CN-1 showing superior IE has been attributed to a long π-conjugation that has paved the way for CN-1 to chelate better with the empty d-orbitals of N80 steel. This has been further confirmed by the ΔE values that were lowest for CN-1. Additionally, ΔN, the proportion of electrons that were moved between the inhibitor and metal atoms, was found to be 0.148% for CN-1, higher than that of CN-2 and CN-3. MD study, furthermore, proved that the $E_{binding}$ for CN-1 (protonated form) was the highest (–870.1 kJ mol^{-1}) among all CN inhibitors.

Pyrimidines, a very important class of molecules, are widely present in nature, such as thiamine (vitamin B1), alloxan, and nucleotides. They have been reported to have shown a variety of biological functions, including antitumor, antiallergic, antifungal, antihypertensive, etc. [44]. In light of pyrimidines' excellent performance as corrosion inhibitors in light acid media [45, 46], there has been tremendous interest in their performance in acidizing conditions in recent years. Haque et al.'s [47] and Yadav et al.'s [48] works on PP and MPTS, both at a concentration of 250 ppm, saw them offering moderate IEs of 89.1% and 89.4% to N80 steel and MS in 15% HCl at 308 K and 333 K, respectively. However, Mehta et al.'s [44] work on POTC (250 ppm), Singh et al.'s [49] work on CP (200 ppm), and Singh et al.'s [50] work on PD (400 ppm, with 0.5 mM KI added) yielded an excellent IE of 96.9%, 96.4%, and 97.0% toward mild steel, N80 steel, and Q235 steel corrosion, respectively, in 15% HCl. Sarkar et al. [51] synthesized pyrimidine derivatives, MPA (Figure 17.5a), and CPA (Figure 17.5b) and tested their IEs for N80 steel corrosion in 15% HCl. Both inhibitors imparted well over 90% IE at a concentration of 250 μM L^{-1}, with MPA showing an amazingly high 99.5% IE. PDP study generated Tafel plots (Figure 17.5c and d) that showed that the shift of E_{ocp}, in comparison to the blank, is not >85 mV toward the anodic or cathodic direction, indicating that MPA and CPA both functioned as mixed-type inhibitors. Nyquist plots (Figure 17.5e and f) showed that the diameter of the capacitive loop increased with increasing inhibitor dosage, indicating that R_{ct} increased across the metal–solution interface. However, an increase in R_{ct} values is more pronounced in the case of MPA. Additionally, surface morphology studies aided by SEM, AFM, and

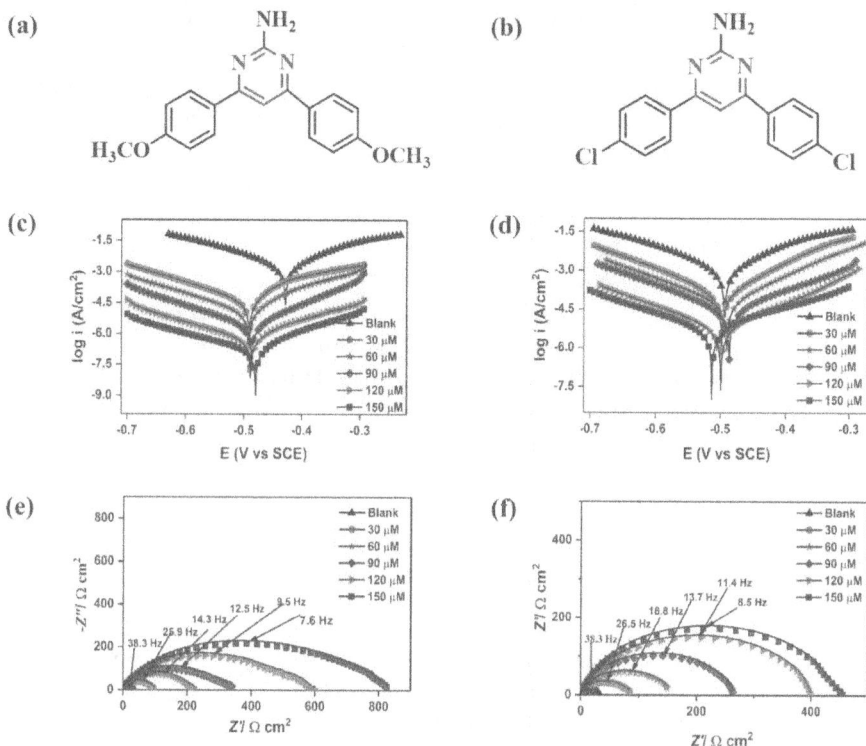

FIGURE 17.5 Molecular structures of MPA (a) and CPA (b); Tafel plots for MPA (c) and CPA (d); and Nyquist plots for MPA (e) and CPA (f) [51].

XPS proved the adsorption of inhibitors at the N80 steel surface. Furthermore, DFT and MD study-led findings agreed with the experimental results.

17.4.3 Quinoline- and Indolizine-based Heterocycles

Quinolines, double-ring structures with benzene and a pyridine ring joined at two nearby carbon atoms, have set numerous instances of acting as very effective corrosion inhibitors in acid media, especially 1 M HCl conditions [52–54]. Quinoline derivatives did not fall short of imparting good IEs in highly aggressive acidizing conditions. A dibenzyl amine-quinoline derivative (DEEQ) was synthesized and tested for its anticorrosive effect toward mild steel corrosion in 15% HCl [55]. DEEQ imparted an amazingly high IE of 99.0% and 95.4% at 60 °C and 90 °C, respectively. DEEQ underwent adsorption through mixed physi-chemisorption and followed Langmuir adsorption isotherm. Salman et al. [56] developed a quinoline derivative containing −CN, −OH, −NH₂ functionalities and an additional styrene moiety and studied its IE toward N80 steel corrosion in 15% HCl. The inhibitor AHQ imparted an excellent IE of 98.4% at 300 mg L⁻¹ concentration. AHQ was reportedly adsorbed via a mixed physi-chemisorption mechanism and followed Langmuir adsorption isotherm. Furthermore, the PDP study revealed that AHQ was a mixed-type inhibitor. A quaternary ammonium salt-based quinoline, benzyl quinolinium chloride derivative (BQD), was synthesized by Yang et al. [57]. BQD, at a concentration of 0.744 mM, containing multiple aromatic rings rich with π-electron and a quaternary N^+ ion, attached with N80 steel surface very firmly to impart an extremely high IE of 99.5%, measured via WL test done at 90 °C. ΔG_{ads}° value for BQD adsorption onto the N80 steel surface was measured to be more than −50 kJ mol^{-1}, indicating a powerful chemisorption mechanism taking place. The adsorption isotherm investigation also revealed that BQD obeyed Langmuir's adsorption isotherm. Some additional quinoline-based heterocyclic acidizing corrosion inhibitors are listed in Table 17.3.

Indolizines, uncommon isomers of indole with nitrogen located at the ring fusion position, have recently been used as acidizing corrosion inhibitors. A DiPaQBr derivative was synthesized and studied for its efficacy toward N80 steel

TABLE 17.3
Quinoline- and Indolizine-based Heterocyclic Acidizing Corrosion Inhibitors

Inhibitor (Concentration)	Electrolyte (Metal) (Temperature)	IE (%)	Inhibitor type	Mode of Adsorption (Isotherm)	Reference
OH 8-Hydroxyquinoline (0.4 wt% + 0.005 wt% KI)	15% HCl (X60) (25 °C)	83.3	Mixed	N/A* (N/A*)	[61]
Phenacyl quinolium chloride (100 mM)	4.410 M HCl (CS) (90 °C)	94.5	N/A	Mixed (Langmuir)	[62]
Phenacyl quinolium bromide (100 mM)	4.410 M HCl (CS) (90 °C)	97.3	N/A	Mixed (Langmuir)	[62]
N,N'-Octane-1,8-diyl-bisquinolium bromide (1.3 × 10⁻² M)	15% HCl (N80) (90 °C)	91.0	Mixed	Mixed (Langmuir)	[63]

(Continued)

TABLE 17.3 *(Continued)*
Quinoline- and Indolizine-based Heterocyclic Acidizing Corrosion Inhibitors

Inhibitor (Concentration)	Electrolyte (Metal) (Temperature)	IE (%)	Inhibitor type	Mode of Adsorption (Isotherm)	Reference
Di-BQC (1.0 wt%)	20% HCl (N80) (90 °C)	99.4	Mixed	N/A*	[64]
QM-DiBQC (1.0 wt%)	20% HCl (N80) (90 °C)	99.3	Mixed	N/A*	[64]

Note: *N/A: not available.

corrosion in 15% HCl [58]. DiPaQBr, at a 0.356 mM concentration, imparted a very high IE of 97.1% at 25 °C. DiPaQBr acted as a mixed-type inhibitor following Langmuir adsorption isotherm to undergo mixed physi-chemisorption. A Di-EAQBr derivative and another Di-BuQBr derivative were synthesized by Wang et al. [59] as corrosion inhibitors of N80 steel in 15% HCl. Di-EAQBr and Di-BuQBr, both at a concentration of 1.0 mM, imparted an IE of 93% and 97.7% at a very high temperature of 90 °C. Both inhibitors got adsorbed via a mixed physi-chemisorption mechanism following Langmuir adsorption isotherm. Yang et al. [60] synthesized and then used three indolizine quaternary ammonium salt inhibitors, Di-BQC, QM-DiBQC, and PyM-DiBQC, as corrosion inhibitors of N80 steel in 15% HCl. All inhibitors showed more than 95% IE, with Di-BQC imparting a maximum IE of 97% measured via the PDP method. The ecological toxicity and EC_{50} values for all the inhibitors were well above 2.0×10^4 mg L^{-1}, indicating that they are of very low toxicity or practically nontoxic. Some additional recently reported indolizine-based heterocyclic corrosion inhibitors are outlined in Table 17.3.

17.4.4 OTHER HETEROCYCLES

Chitosan (LD_{50}: 16,000 mg kg^{-1}) and piperanol (LD_{50}: 2700 mg kg^{-1}) were used, by Chauhan et al. [65], as reactants for the microwave-assisted synthesis of piperonal-chitosan Schiff base (Pip-Cht), the first-ever chemically functionalized chitosan used as a corrosion inhibitor in oil-well acidizing condition. Pip-Cht, at the expense of a very high inhibitor concentration of 600 mg L^{-1}, imparted a moderate IE of 85.2% to carbon steel in 15% HCl. Another environmentally

benign acidizing corrosion inhibitor, 5-hydroxy tryptophan, a naturally occurring amino acid, also imparted a moderate IE of 78% at a concentration of
10×10^{-5} M when it was used as a corrosion inhibitor of mild steel in 15% HCl
at 30 °C [66]. However, Yadav et al.'s [67] work on two amino acid-based compounds, OYAA and OYPA, used for the prevention of mild steel in 15% HCl
have seen OYAA and OYPA imparting an excellent IE of 93.2% and 97.9%,
respectively, at 50 ppm concentration at 30 °C. Rahimi et al. [68] synthesized
a corrosion inhibitor, FACI, based on bio-based furfuryl alcohol (Figure 17.6a)

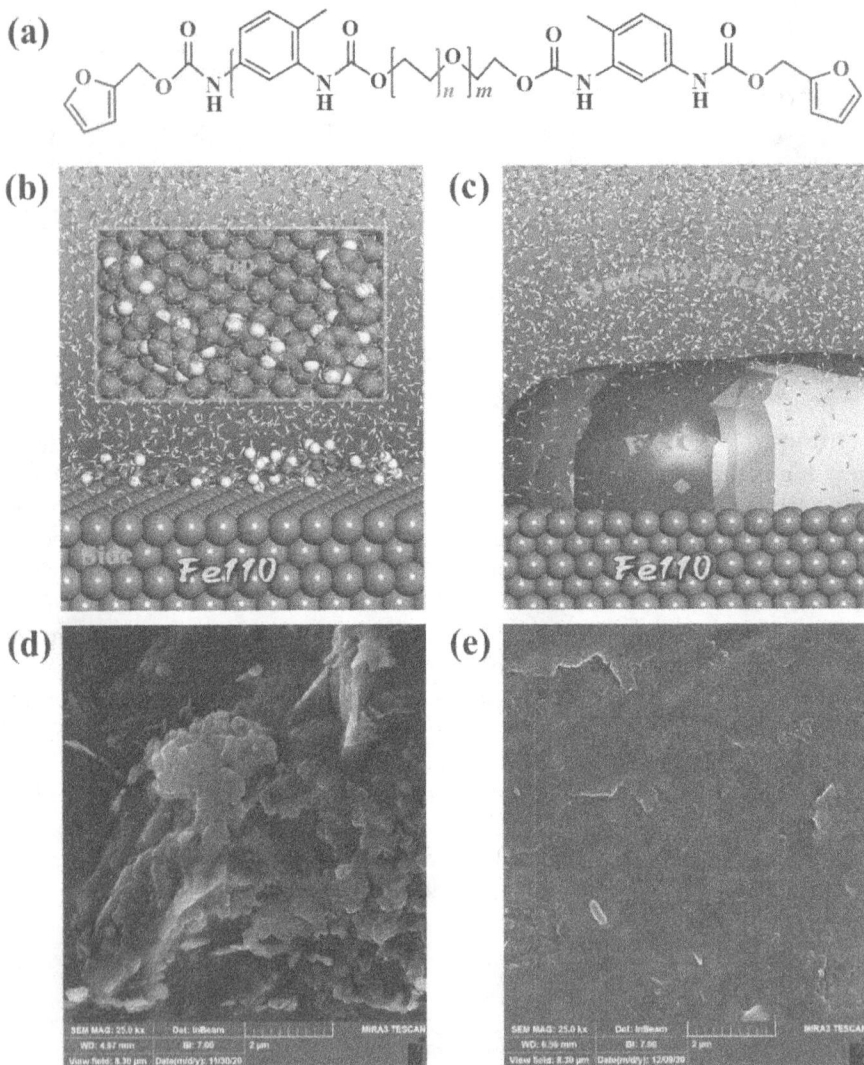

FIGURE 17.6   Molecular structure of FACI (a); equilibrium configuration (b) and density
field distribution (c) of FACI on Fe(110) substrate; and SEM images (resolution: 2 μm) of mild
steel sample treated without (d) and with (e) FACI [68].

that was examined as a corrosion inhibitor of MS in 15% HCl at various temperatures (20–80 °C). FACI, at 136×10^{-5} M concentration, yielded the best IE of 95.4% at 60 °C. Figure 17.6b, employing MD investigation, demonstrates the most stable configuration of FACI adsorbed onto Fe(110) surface. Figure 17.6c, representing the density field map of FACI on the iron substrate, shows the formation of a thick, hydrophobic film to protect the metal specimen. Figure 17.6d revealed the SEM image of the surface of a seriously damaged mild steel metal specimen treated with 15% HCl without any inhibitor, and Figure 17.6e depicts a much better, smoother surface representing the treatment of the same metal specimen treated with 15% HCl containing 136×10^{-5} M of FACI. Table 17.4 includes more heterocyclic compounds recently reported as acidizing corrosion inhibitors.

TABLE 17.4
Heterocyclic Compounds Developed as Oil-well Acidizing Corrosion Inhibitors

Inhibitor (Concentration)	Electrolyte (Metal) (Temperature)	IE (%)	Inhibitor Type	Mode of Adsorption (Isotherm)	Reference
MABPQAS (1.0 mM)	3 M HCl (CS) (50 °C)	96.3	N/A*	Mixed (Villamil)	[69]
4-Aminoantipyrine Schiff base derivative (4.0 mM)	15% HCl (SAE 1012 CS) (25 °C)	87.0	Mixed	Mixed (Langmuir)	[70]
4-[6-(1H-indol-3-yl)-2-sulfanylidene-2,3-dihydropyrimidin-4-yl]-2H-1-benzopyran-2-one [1.0 mM]	15% HCl (API X-65) (25 °C)	95.8	Mixed	Mixed (Langmuir)	[71]

(Continued)

TABLE 17.4 (*Continued*)

Heterocyclic Compounds Developed as Oil-well Acidizing Corrosion Inhibitors

Inhibitor (Concentration)	Electrolyte (Metal) (Temperature)	IE (%)	Inhibitor Type	Mode of Adsorption (Isotherm)	Reference
AMDMOTCC (300 ppm)	15% HCl (N80) (35 °C)	97.7	Mixed	Mixed (Langmuir)	[72]
HBADMPDHP (400 ppm)	15% HCl (P110 CS) (35 °C)	95.0	Mixed	Mixed (Langmuir)	[73]
DMTHBX (200 ppm)	15% HCl (Q235 CS) (35 °C)	92.3	Cathodic	N/A (Langmuir)	[74]
DPAMNDTI (200 ppm)	15% HCl (N80) (25 °C)	84.3	Mixed	Mixed (Langmuir)	[75]
MPTHA (400 ppm)	15% HCl (X80) (30 °C)	97.9	Mixed	Physical (Langmuir)	[76]

Note: *N/A = not available.

17.5 SUMMARY AND OUTLOOK

This chapter discusses different acidizing conditions used in oil and gas industries and many heterocyclic corrosion inhibitors used to tackle the menace of corrosion in highly concentrated HCl solutions. The focal points of this discussion are the acid solution, metallic components, inhibitors, and their efficiencies at optimum doses.

Many heterocyclic acidizing corrosion inhibitors developed to protect various steel sheets have been highlighted. Based on the discourse made so far, the following recommendations are offered:

1. Imidazole, pyrazole, thiazole, thiadiazole, pyrimidine, quinoline, and indolizine-based heterocycles are very efficient. In some cases, their IEs surpassed 99% [51, 55, 57, 64]. Therefore, these classes of heterocycles deserve more focus in future research.
2. Most acidizing inhibitors discussed have been tested below 50 °C. However, there are instances of inhibitors tested at 90 °C [20, 31, 36, 57, 59, 62–64] and 60 °C [13, 33, 35], and one inhibitor [12] produced high IE close to a temperature of more than 150 °C. Nevertheless, the acid concentration of ≥15 wt% and the acidizing temperature of ≥150 °C should be considered more for practical applications.
3. In parallel to developing and testing high-temperature and highly efficient synthetic inhibitors, the research community should also focus on researching bio-based green, nontoxic compounds as acidizing corrosion inhibitors. Modification of natural products with synthetic ones could be a good starting point.
4. Finally, an enormous effort should be made to understand the corrosion mechanism and the formation of corrosion products during the acidizing process at very high temperatures. Such kind of effort, eventually, will help us develop better models and corrosion inhibitors.

ACKNOWLEDGMENT

The authors gratefully acknowledge King Fahd University of Petroleum & Minerals (KFUPM) for funding this work through project # INAM 2112.

ABBREVIATIONS

ADP	2-amino-6-(2,4-dihydroxyphenyl)-4-(4-methoxyphenyl) nicotinonitrile
AFM	atomic force microscopy
AHQ	2-amino-7-hydroxy-4-styrylquinoline-3-carbonitrile
AMDMOTCC	2-amino-4-(4-methoxyphenyl)-7,7-dimethyl-5-oxo-5,6,7,8-tetrahydro-4H-chromene-3-carbonitrile
AMP	2-amino-4-(4-methoxyphenyl)-6-phenylnicotinonitrile
AMPC	(E)-5-amino-3-(4-methoxyphenyl)-N'-(1-(4-methoxyphenyl) ethylidene)-1H-pyrazole-4-carbohydrazide
ANC-1	2-amino-4-(4-methoxyphenyl)-1,8-naphthyridine-3-carbonitrile
ANC-2	2-amino-4-(4-methylphenyl)-1,8-naphthyridine-3-carbonitrile
ANC-3	2-amino-4-(3-nitrophenyl)-1,8-naphthyridine-3-carbonitrile
β_a	anodic slope
β_c	cathodic slope

BHMPDTA	2-((benzylidene)hydrazono)-4-methyl-5-(phenyldiazenyl) thiazol-3(2*H*)-amine
BINMPA	1-(1*H*-benzo[*d*]imidazole-2-yl)-*N*-((furan-2-yl) methylene)-2-phenylethanamine
BONMTZA	4-(2*H*-benzo[*e*][1, 3]oxazin-3(4*H*)-yl)-*N*-(5-methyl-1,3,4-thiadiazol-2-yl)benzenesulfonamide
BTDCDMT	3-(benzo[*d*]thiazol-2-yl)-2-(3,5-dichlorophenyl)-2,5-dimethylthiazolidin-4-one
BTHHP	6-(2-(benzo[*d*]thiazol-2-yl)hydrazono)-hexane-1,2,3,4,5-pentaol
CN-1	5-amino-9-hydroxy-2-(4-vinylphenyl)-1,11*b*-dihydrochromeno[4,3,2-de][1, 6]naphthyridine-4-carbonitrile
CN-2	5-amino-9-hydroxy-2-(4-methoxyphenyl)chromeno[4,3,2-de][1, 6]naphthyridine-4-carbonitrile
CN-3	5-amino-9-hydroxy-2-phenylchromeno[4,3,2-de][1, 6] naphthyridine-4-carbonitrile
CP	8,8-dimethyl-5-*p*-tolyl-8,9-dihydro-1*H*-chromenopyrimidine-2,4,6-(3*H*,5*H*,7*H*)-trione
CPA	4,6-bis(4-chlorophenyl)pyrimidin-2-amine
CPBPM	(2(-4(chloro phenyl-1*H*-benzo[*d*]imidazole)-1-yl) phenyl) methanone
C_{dl}	double-layer capacitance
CS	carbon steel
DHPPTEH	(13*E*)-2-(2,5-dihydro-2-phenylthiazol-5-yl)-1-(1-phenyl)-2-(1*H*-1,2,4-triazol-1-yl) ethylidene)hydrazine
DFAT	[2-4(difuran-2-yl-3-azabicyclo[3.3.1]nonan-9-yl)-5-spiro-4-acetyl-2-acetylamino-1,3,4-thiadiazoline]
DFT	density functional theory
DIAT	[2-4(di-1*H*-indole-3-yl-3-azabicyclo[3.3.1] nonan-9-yl)-5-spiro-acetyl-2-acetylimino-1,3,4-thiadiazoline]
Di-BQC	dimer indolizine derivative of benzyl quinoline chloride
Di-BuQB	*n*-butyl quinolinium bromide
Di-EAQBr	ethyl acetate quinolinium bromide
DiPaQBr	phenacyl quinolinium bromide
DMTHBX	9,9-dimethyl-12-*p*-tolyl-8,9,10,12-tetrahydrobenzo[*a*] xanthen-1-one
DPAMNDTI	1-diphenylaminomethyl-3-(1-*N*-dithiooxamide)iminoisatin
DPDHPNT	2-(3,5-diphenyl-4,5-dihydro-1*H*-pyrazol-1-yl)naphtho[2,3-*d*] thiazole
E_{corr}	corrosion potential
E_{ocp}	open-circuit potential
EIS	electrochemical impedance spectroscopy
HA-2	1-{(*Z*)-[(4,6-dimethylpyridin-2-yl)imino]methyl}naphthalen-2-ol
HBADMPDHP	4-((2-hydroxybenzylidene)amino)-1,5-dimethyl-2-phenyl-1, 2-dihydro-3*H*-pyrazol-3-one
HCHO	formaldehyde
HCl	hydrochloric acid

HOMO	highest occupied molecular orbital
ΔG°_{ads}	Gibbs free energy of adsorption
GDP	gross domestic product
GNP	gross national product
IE	inhibition efficiency
K_{ads}	equilibrium adsorption constant
LUMO	lowest unoccupied molecular orbital
MABPQAS	1,1′-((3-morpholinopropyl)azanediyl)-bis(propan-2-ol)-based quaternary ammonium salt
MMDPPM	5-(4-methoxyphenyl)-3-(4-methyl phenyl)-4,5-dihydro-1*H*-pyrazol-1-yl-(pyridine-4-yl)methanone
MPA	4,6-bis(4-methoxyphenyl) pyrimidin-2-amine
MPTHA	2-methyl-9-phenyl-1,2,3,4-tetrahydroacridine
MPTS	6′-(4-methoxyphenyl)-1′-phenyl-2′-thioxo-2′,3′-dihydro-1′*H*-spiro[indoline-3,4′-pyrimidine]-2-one
MS	mild steel
MTPT	1-(2-(5-mercapto-4-phenyl-4*H*-1,2,4-triazole-3-yl) phenyl)-3-phenylthiourea
NACE	National Association of Corrosion Engineers
NAEP	*N,N*′-bis-(2-aminoethyl)piperazine
NBIDAB	*N*-(6-(1*H*-benzo[*d*]imidazole-1-yl)hexyl)-*N,N*-dimethyldodecan-1-aminium bromide
NH_3	ammonia
NOPTAC	*N*-(3-oxo-3-phenylpro-pyl)thiazol-2-aminium chloride
OYAA	(Z)-2-(2-oxoindolin-3-ylideneamino)acetic acid
OYPA	2-(2-oxoindolin-3-ylideneamino)-3-phenyl propionic acid
PD	4-(3-methoxy-4-hydroxy-phenyl)-2-phenyl-1,4-dihydro-benzo[4, 5]imidazo[1,2-*a*]pyrimidine-3-carboxylicacid ethyl ester
PDP	potentiodynamic polarization
POTC	6,6′-(1,4-phenylene) bis(4-oxo-2-thioxo-1,2,3,4-tetrahydropyrimidine-5-carbonitrile)
PP	5-styryl-2,7-dithioxo 2,3,5,6,7,8-hexahydropyrimido[4,5-*d*] pyrimidin-4(1*H*)-one
PPATPT	1-phenyl-3-(2-(5-(phenylamino)-1,3,4-thiadiazol-2-yl)phenyl) thiourea
Py I	2-amino-6-(3,4-dimethoxyphenyl)-4-phenylnicotinonitrile
Py II	2-amino-4-(3,4-dimethoxyphenyl)-6-phenylnicotinonitrile
PyM-DiBQC	pyridyl-3-methene-dimer indolizine derivative of benzyl quinoline chloride
PZ-1	4,4′-((4-methoxyphenyl)methylene) bis(3-methyl-1-phenyl-1*H*-pyrazol-5-ol)
QM-DiBQC	quinolyl-3-methened-dimer indolizine derivative of benzyl quinoline chloride
R_{ct}	charge transfer resistance
SEM	scanning electron microscopy

STBim	2-styryl-1*H*-benzo[*d*]imidazole
TMPTMTD	5-[2-(3,4,5-trimethoxyphenyl)-6-(4-methoxylphenyl)-imidazo [2,1-*b*][1, 3, 4] thiadiazol-5-yl) methylidene]-1,3-thiazolidine-2,4-dione
TOS1	4,4′-diselanediylbis (*N*-(1-(1-(*tert*-butyl)-1*H*-tetrazol-5-yl)-2-methylpropyl) aniline)
TOS2	4,4′-diselane diylbis (*N*-((1-(*tert*-butyl)-1*H*-tetrazol-5-yl)(furan-2-yl) methyl) aniline)
TOS3	2-((4-(((1-(*tert*-butyl)-1*H*-tetrazol-5-yl) (*p*-tolyl)methyl)amino) phenyl)selanyl)-3-methylnaphthalene-1,4-dione
TZ	1,3,5-tris(4-methoxyphenyl)-1,3,5-triazine (TZ)
WL	weight loss
XPS	X-ray photoelectron spectroscopy

REFERENCES

1. Goni, L.K.M.O., Jafar Mazumder, M., Quraishi, A., and Mizanur Rahman, A. (2021) Bioinspired Heterocyclic Compounds as Corrosion Inhibitors: A Comprehensive Review. *Chem Asian J*, **16** (11), 1324–1364.
2. Mccafferty, E. (2010) *Introduction to Corrosion Science*, Springer, New York.
3. Goni, L.K.M.O., and Mazumder, M.A.J. (2019) Green Corrosion Inhibitors, in *Corrosion Inhibitors*, 1st ed., IntechOpen, London, pp. 77–95.
4. Aljeaban, N.A., Goni, L.K.M.O., Alharbi, B.G., Jafar, M.A., Ali, S.A., Chen, T., Quraishi, M.A., and Al-Muallem, H.A. (2020) Polymers Decorated with Functional Motifs for Mitigation of Steel Corrosion: An Overview. *Int J Polym Sci*, **2020**, 1–23.
5. Bowman, E., Thompson, N., Gl, D., Moghissi, O., Gould, M., and Payer, J. (2016) International Measures of Prevention, Application, and Economics of Corrosion Technologies Study.
6. Askari, M., Aliofkhazraei, M., Jafari, R., Hamghalam, P., and Hajizadeh, A. (2021) Downhole Corrosion Inhibitors for Oil and Gas Production: A Review. *Appl Surf Sci Adv*, **6**, 100128.
7. Caihong, Y., Singh, A., Ansari, K.R., Ali, I.H., and Kumar, R. (2022) Novel Nitrogen-based Heterocyclic Compound as Q235 Steel Corrosion Inhibitor in 15% HCl under the Dynamic Condition: A Detailed Experimental and Surface Analysis. *J Mol Liq*, **362**, 119720.
8. Solomon, M.M., Uzoma, I.E., Olugbuyiro, J.A.O., and Ademosun, O.T. (2022) A Censorious Appraisal of the Oil Well Acidizing Corrosion Inhibitors. *J Pet Sci Eng*, **215**, 110711.
9. Singh, A., and Quraishi, M.A. (2015) Acidizing Corrosion Inhibitors: A Review. *J Mater Environ Sci*, **6** (1), 224–235.
10. Ansari, K.R., Chauhan, D.S., Singh, A., Saji, V.S., and Quraishi, M.A. (2020) Corrosion Inhibitors for Acidizing Process in Oil and Gas Sectors, in *Corrosion Inhibitors in the Oil and Gas Industry*, 1st ed., Wiley-VCH Verlag GmbH, Weinheim, pp. 151–176.
11. El Ibrahimi, B., and Guo, L. (2021) Azole-Based Compounds as Corrosion Inhibitors for Metallic Materials, in *Azoles: Synthesis, Properties, Applications and Perspectives*, 1st ed., IntechOpen, London, pp. 85–112.

12. Liu, Y., Chen, L., Tang, Y., Zhang, X., and Qiu, Z. (2022) Synthesis and Characterization of Nano-SiO$_2$@octadecylbisimidazoline Quaternary Ammonium Salt Used as Acidizing Corrosion Inhibitor. *Rev Adv Mater Sci*, **61** (1), 186–194.
13. Srivastava, V., Salman, M., Chauhan, D.S., Abdel-Azeim, S., and Quraishi, M.A. (2021) (*E*)-2-Styryl-1*H*-benzo[*d*]imidazole as Novel Green Corrosion Inhibitor for Carbon Steel: Experimental and Computational Approach. *J Mol Liq*, **324**, 115010.
14. Xu, X., Singh, A., Sun, Z., Ansari, K.R., and Lin, Y. (2017) Theoretical, Thermodynamic and Electrochemical Analysis of Biotin Drug as an Impending Corrosion Inhibitor for Mild Steel in 15% Hydrochloric Acid. *R Soc Open Sci*, **4** (12), 170933.
15. Fergachi, O., Benhiba, F., Rbaa, M., Ouakki, M., Galai, M., Touir, R., Lakhrissi, B., Oudda, H., and Touhami, M.E. (2019) Corrosion Inhibition of Ordinary Steel in 5.0 M HCl Medium by Benzimidazole Derivatives: Electrochemical, UV-visible Spectrometry, and DFT Calculations. *J Bio Tribocorros*, **5** (1).
16. Odewunmi, N.A., Mazumder, A.J., and Ali, M. (2022) Tipping Effect of Tetra-Alkylammonium on the Potency of *N*-(6-(1*H*-benzo[*d*]imidazol-1-yl)hexyl)-*N*,*N*-dimethyldodecan-1-aminium Bromide (BIDAB) as Corrosion Inhibitor of Austenitic 304L Stainless Steel in Oil and Gas Acidization: Experimental and DFT Approach. *J Mol Liq*, **360**, 119431.
17. Singh, A., Ansari, K.R., Quraishi, M.A., and Banerjee, P. (2021) Corrosion Inhibition and Adsorption of Imidazolium-based Ionic Liquid over P110 Steel Surface in 15% HCl under Static and Dynamic Conditions: Experimental, Surface and Theoretical Analysis. *J Mol Liq*, **323**, 114608.
18. Solomon, M.M., Umoren, S.A., Quraishi, M.A., Tripathy, D.B., and Abai, E.J. (2020) Effect of Alkyl Chain Length, Flow, and Temperature on the Corrosion Inhibition of Carbon Steel in a Simulated Acidizing Environment by an Imidazoline-based Inhibitor. *J Pet Sci Eng*, **187**, 106801.
19. Yadav, M., Sarkar, T.K., and Purkait, T. (2015) Amino Acid Compounds as Eco-friendly Corrosion Inhibitor for N80 Steel in HCl Solution: Electrochemical and Theoretical Approaches. *J Mol Liq*, **212**, 731–738.
20. Yaocheng, Y., Caihong, Y., Singh, A., and Lin, Y. (2019) Electrochemical Study of Commercial and Synthesized Green Corrosion Inhibitors for N80 Steel in Acidic Liquid. *N J Chem*, **43** (40), 16058–16070.
21. Singh, A., Ansari, K.R., Quraishi, M.A., and Kaya, S. (2020) Theoretically and Experimentally Exploring the Corrosion Inhibition of N80 Steel by Pyrazol Derivatives in Simulated Acidizing Environment. *J Mol Struct*, **1206**, 127685.
22. Yadav, M., Sinha, R.R., Sarkar, T.K., and Tiwari, N. (2015) Corrosion Inhibition Effect of Pyrazole Derivatives on Mild Steel in Hydrochloric Acid Solution. *J Adhes Sci Technol*, **29** (16), 1690–1713.
23. Paul, P.K., Yadav, M., and Obot, I.B. (2020) Investigation on Corrosion Protection Behavior and Adsorption of Carbohydrazide-Pyrazole Compounds on Mild Steel in 15% HCl Solution: Electrochemical and Computational Approach. *J Mol Liq*, **314**, 113513.
24. Eid, A.M., Shaaban, S., and Shalabi, K. (2020) Tetrazole-based Organoselenium Bi-Functionalized Corrosion Inhibitors during Oil Well Acidizing: Experimental, Computational Studies, and SRB Bioassay. *J Mol Liq*, **298**, 111980.
25. Mehta, R.K., Yadav, M., and Obot, I.B. (2022) Electrochemical and Computational Investigation of Adsorption and Corrosion Inhibition Behavior of 2-Aminobenzo-hydrazide Derivatives at Mild Steel Surface in 15% HCl. *Mater Chem Phys*, **290**, 126666.

26. Tiwari, N., Mitra, R.K., and Yadav, M. (2021) Corrosion Protection of Petroleum Oil Well/Tubing Steel Using Thiadiazolines as Efficient Corrosion Inhibitor: Experimental and Theoretical Investigation. *Surf Interfaces*, **22**, 100770.

27. El Basiony, N.M., Elgendy, A., El-Tabey, A.E., Al-Sabagh, A.M., El-Hafez, A., El-raouf, G.M., and Migahed, M.A. (2020) Synthesis, Characterization, Experimental and Theoretical Calculations (DFT and MC) of Ethoxylated Aminothiazole as Inhibitor for X65 Steel Corrosion in Highly Aggressive Acidic Media. *J Mol Liq*, **297**, 111940.

28. Saraswat, V., and Yadav, M. (2020) Computational and Electrochemical Analysis on Quinoxalines as Corrosion Inhibitors for Mild Steel in Acidic Medium. *J Mol Liq*, **297**, 111883.

29. Shaw, P., Obot, I.B., and Yadav, M. (2019) Functionalized 2-Hydrazinobenzothiazole with Carbohydrates as a Corrosion Inhibitor: Electrochemical, XPS, DFT and Monte Carlo Simulation Studies. *Mater Chem Front*, **3** (5), 931–940.

30. Yadav, M., Sharma, D., and Kumar, S. (2015) Thiazole Derivatives as Efficient Corrosion Inhibitor for Oil-Well Tubular Steel in Hydrochloric Acid Solution. *Korean J Chem Eng*, **32** (5), 993–1000.

31. Zhang, X., Zhang, Y., Su, Y., Wang, X., and Lv, R. (2022) Synthesis and Corrosion Inhibition Performance of Mannich Bases on Mild Steel in Lactic Acid Media. *ACS Omega*, **7** (36), 32208–32224.

32. Yadav, M., Sharma, D., and Sarkar, T.K. (2015) Adsorption and Corrosion Inhibitive Properties of Synthesized Hydrazine Compounds on N80 Steel/Hydrochloric Acid Interface: Electrochemical and DFT Studies. *J Mol Liq*, **212**, 451–460.

33. Yadav, M., Behera, D., Kumar, S., and Yadav, P. (2015) Experimental and Quantum Chemical Studies on Corrosion Inhibition Performance of Thiazolidinedione Derivatives for Mild Steel in Hydrochloric Acid Solution. *Chem Eng Commun*, **202** (3), 303–315.

34. Abd El-Lateef, H.M., Sayed, A.R., and Shalabi, K. (2022) Studying the Effect of Two Isomer Forms Thiazole and Thiadiazine on the Inhibition of Acidic Chloride–Induced Steel Corrosion: Empirical and Computer Simulation Explorations. *J Mol Liq*, **356**, 119044.

35. Alamry, K.A., Hussein, M.A., Musa, A., Haruna, K., and Saleh, T.A. (2021) The Inhibition Performance of a Novel Benzenesulfonamide-based Benzoxazine Compound in the Corrosion of X60 Carbon Steel in an Acidizing Environment. *RSC Adv*, **11** (12), 7078–7095.

36. Obot, I.B., Umoren, S.A., and Ankah, N.K. (2019) Pyrazine Derivatives as Green Oil Field Corrosion Inhibitors for Steel. *J Mol Liq*, **277**, 749–761.

37. Haruna, K., and Saleh, T.A. (2021) N,N'-Bis-(2-aminoethyl)piperazine Functionalized Graphene Oxide (NAEP-GO) as an Effective Green Corrosion Inhibitor for Simulated Acidizing Environment. *J Environ Chem Eng*, **9** (1), 104967.

38. Salman, M., Ansari, K.R., Haque, J., Srivastava, V., Quraishi, M.A., and Mazumder, M.A.J. (2020) Ultrasound-assisted Synthesis of Substituted Triazines and Their Corrosion Inhibition Behavior on N80 Steel/Acid Interface. *J Heterocycl Chem*, **57** (5), 2157–2172.

39. Farag, A.A., Mohamed, E.A., Sayed, G.H., and Anwer, K.E. (2021) Experimental/computational Assessments of API Steel in 6 M H_2SO_4 Medium Containing Novel Pyridine Derivatives as Corrosion Inhibitors. *J Mol Liq*, **330**, 115705.

40. Ansari, K.R., Quraishi, M.A., and Singh, A. (2015) Pyridine Derivatives as Corrosion Inhibitors for N80 Steel in 15% HCl: Electrochemical, Surface and Quantum Chemical Studies. *Measurement (Lond)*, **76**, 136–147.

41. Abd El-Lateef, H.M., Abu-Dief, A.M., and Mohamed, M.A.A. (2017) Corrosion Inhibition of Carbon Steel Pipelines by Some Novel Schiff Base Compounds during Acidizing Treatment of Oil Wells Studied by Electrochemical and Quantum Chemical Methods. *J Mol Struct*, **1130**, 522–542.

42. Ansari, K.R., and Quraishi, M.A. (2015) Experimental and Computational Studies of Naphthyridine Derivatives as Corrosion Inhibitor for N80 Steel in 15% Hydrochloric Acid. *Physica E Low Dimens Syst Nanostruct*, **69**, 322–331.

43. Salman, M., Ansari, K.R., Srivastava, V., Chauhan, D.S., Haque, J., and Quraishi, M.A. (2021) Chromeno Naphthyridines Based Heterocyclic Compounds as Novel Acidizing Corrosion Inhibitors: Experimental, Surface and Computational Study. *J Mol Liq*, **322**, 114825.

44. Mehta, R.K., Gupta, S.K., and Yadav, M. (2022) Studies on Pyrimidine Derivative as Green Corrosion Inhibitor in Acidic Environment: Electrochemical and Computational Approach. *J Environ Chem Eng*, **10** (5), 108499.

45. Verma, C., Olasunkanmi, L.O., Ebenso, E.E., and Quraishi, M.A. (2018) Adsorption Characteristics of Green 5-Arylamino Methylene Pyrimidine-2,4,6-Triones on Mild Steel Surface in Acidic Medium: Experimental and Computational Approach. *Results Phys*, **8**, 657–670.

46. Anusuya, N., Saranya, J., Sounthari, P., Zarrouk, A., and Chitra, S. (2017) Corrosion Inhibition and Adsorption Behavior of Some Bis-Pyrimidine Derivatives on Mild Steel in Acidic Medium. *J Mol Liq*, **225**, 406–417.

47. Haque, J., Ansari, K.R., Srivastava, V., Quraishi, M.A., and Obot, I.B. (2017) Pyrimidine Derivatives as Novel Acidizing Corrosion Inhibitors for N80 Steel Useful for Petroleum Industry: A Combined Experimental and Theoretical Approach. *Journal of Industrial and Engineering Chemistry*, **49**, 176–188.

48. Yadav, M., Sinha, R.R., Kumar, S., and Sarkar, T.K. (2015) Corrosion Inhibition Effect of Spiropyrimidinethiones on Mild Steel in 15% HCl Solution: Insight from Electrochemical and Quantum Studies. *RSC Adv*, **5** (87), 70832–70848.

49. Singh, A., Ansari, K.R., Chauhan, D.S., Quraishi, M.A., and Kaya, S. (2020) Anti-corrosion Investigation of Pyrimidine Derivatives as Green and Sustainable Corrosion Inhibitor for N80 Steel in Highly Corrosive Environment: Experimental and AFM/XPS Study. *Sustain Chem Pharm*, **16**, 100257.

50. Singh, A., Ansari, K.R., Ali, I.H., Lin, Y., EL Ibrahimi, B., Bazzi, L. (2022) Combination of Experimental, Surface and Computational Insight into the Corrosion Inhibition of Pyrimidine Derivative onto Q235 Steel in Oilfield Acidizing Fluid under Hydrodynamic Condition. *J Mol Liq*, **353**, 118825.

51. Sarkar, T.K., Saraswat, V., Mitra, R.K., and Obot, I.B. (2021) Mitigation of Corrosion in Petroleum Oil well/tubing Steel Using Pyrimidines as Efficient Corrosion Inhibitor: Experimental and Theoretical Investigation. *Mater Today Commun*, **26**, 101862.

52. Ganapathi Sundaram, R., and Sundaravadivelu, M. (2018) Surface Protection of Mild Steel in Acidic Chloride Solution by 5-Nitro-8-hydroxy Quinoline. *Egypt J Pet*, **27** (1), 95–103.

53. Farag, A.A., and Noor El-Din, M.R. (2012) The Adsorption and Corrosion Inhibition of Some Nonionic Surfactants on API X65 Steel Surface in Hydrochloric Acid. *Corros Sci*, **64**, 174–183.

54. Achary, G., Sachin, H.P., Naik, Y.A., and Venkatesha, T. (2008) The Corrosion Inhibition of Mild Steel by 3-Formyl-8-hydroxy Quinoline in Hydrochloric Acid Medium. *Mater Chem Phys*, **107** (1), 44–50.

55. Li, Y., Wang, D., and Zhang, L. (2019) Experimental and Theoretical Research on a New Corrosion Inhibitor for Effective Oil and Gas Acidification. *RSC Adv*, **9** (45), 26464–26475.

56. Salman, M., Srivastava, V., Quraishi, M.A., Chauhan, D.S., Ansari, K.R., and Haque, J. (2021) Quinoline Carbonitriles as Novel Inhibitors for N80 Steel Corrosion in Oil-Well Acidizing: Experimental and Computational Insights. *Russ J Electrochem*, **57** (3), 228–244.

57. Yang, Z., Zhan, F., Pan, Y., LYu, Z., Han, C., Hu, Y.P., Ding, P., Gao, T., Zhou, X., and Jiang, Y. (2015) Structure of a Novel Benzyl Quinolinium Chloride Derivative and Its Effective Corrosion Inhibition in 15 wt.% Hydrochloric Acid. *Corros Sci*, **99**, 281–294.

58. Wang, Y., Yang, Z., Zhan, F., Lyu, Z., Han, C., Wang, X., Chen, W., Ding, M., Wang, R., and Jiang, Y. (2018) Indolizine Quaternary Ammonium Salt Inhibitors Part II: A Reinvestigation of an Old-fashioned Strong Acid Corrosion Inhibitor Phenacyl Quinolinium Bromide and Its Indolizine Derivative. *N J Chem*, **42** (15), 12977–12989.

59. Wang, Y., Yang, Z., Hu, H., Wu, J., and Finšgar, M. (2022) Indolizine Quaternary Ammonium Salt Inhibitors: The Inhibition and Anti-Corrosion Mechanism of New Dimer Derivatives from Ethyl Acetate Quinolinium Bromide and *n*-Butyl Quinolinium Bromide. *Colloids Surf A Physicochem Eng Asp*, **651**, 129649.

60. Yang, Z., Wang, Y., Zhan, F., Chen, W., Ding, M., Qian, C., Wang, R., and Hou, B. (2019) Indolizine Quaternary Ammonium Salt Inhibitors, Part III: Insights into the Highly Effective Low-toxicity Acid Corrosion Inhibitor – Synthesis and Protection Performance. *N J Chem*, **43** (47), 18461–18475.

61. Obot, I.B., Ankah, N.K., Sorour, A.A., Gasem, Z.M., and Haruna, K. (2017) 8-Hydroxyquinoline as an Alternative Green and Sustainable Acidizing Oilfield Corrosion Inhibitor. *Sustainable Materials Technologies*, **14**, 1–10.

62. Yang, Z., Qian, C., Chen, W., Ding, M., Wang, Y., Zhan, F., and Tahir, M.U. (2020) Synergistic Effect of the Bromide and Chloride Ion on the Inhibition of Quaternary Ammonium Salts in Haloid Acid, Corrosion Inhibition of Carbon Steel Measured by Weight Loss. *Colloids Interface Sci Commun*, **34**, 100228.

63. Zhang, X., Zheng, Y., Wang, X., Yan, Y., and Wu, W. (2014) Corrosion Inhibition of N80 Steel Using Novel Diquaternary Ammonium Salts in 15% Hydrochloric Acid. *Ind Eng Chem Res*, **53** (37), 14199–14207.

64. Wang, Y., Yang, Z., Wang, R., Chen, W., and Ding, M. (2019) High-efficiency Corrosion Inhibitor for Acidizing: Synthesis, Characterization and Anti-corrosion Performance of Novel Indolizine Derivative. SPE-193587-MS.

65. Chauhan, D.S., Mazumder, M.A.J., Quraishi, M.A., Ansari, K.R., and Suleiman, R.K. (2020) Microwave-assisted Synthesis of a New Piperonal-Chitosan Schiff Base as a Bio-Inspired Corrosion Inhibitor for Oil-Well Acidizing. *Int J Biol Macromol*, **158**, 231–243.

66. Ituen, E., Akaranta, O., and James, A. (2017) Electrochemical and Anticorrosion Properties of 5-Hydroxytryptophan on Mild Steel in a Simulated Well-Acidizing Fluid. *J Taibah Univ Sci*, **11** (5), 788–800.

67. Yadav, M., Gope, L., and Sarkar, T.K. (2016) Synthesized Amino Acid Compounds as Eco-friendly Corrosion Inhibitors for Mild Steel in Hydrochloric Acid Solution: Electrochemical and Quantum Studies. *Res Chem Intermed*, **42** (3), 2641–2660.

68. Rahimi, A., Abdouss, M., Farhadian, A., Guo, L., and Neshati, J. (2021) Development of a Novel Thermally Stable Inhibitor Based on Furfuryl Alcohol for Mild Steel Corrosion in a 15% HCl Medium for Acidizing Application. *Ind Eng Chem Res*, **60** (30), 11030–11044.

69. Farag, A.A., Abdallah, H.E., Badr, E.A., Mohamed, E.A., Ali, A.I., and El-Etre, A.Y. (2021) The Inhibition Performance of Morpholinium Derivatives on Corrosion Behavior of Carbon Steel in the Acidized Formation Water: Theoretical, Experimental and Biocidal Evaluations. *J Mol Liq*, **341**, 117348.

70. Gerengi, H., Cakmak, R., Dag, B., Solomon, M.M., Tuysuz, H., and Kaya, A. (2020) Synthesis and Anticorrosion Studies of 4-[(2-Nitroacetophenonylidene)-amino]-antipyrine on SAE 1012 Carbon Steel in 15 wt.% HCl Solution. *J Adhes Sci Technol*, **34** (22), 2448–2466.

71. Hashem, H.E., Farag, A.A., Mohamed, E.A., and Azmy, E.M. (2022) Experimental and Theoretical Assessment of Benzopyran Compounds as Inhibitors to Steel Corrosion in Aggressive Acid Solution. *J Mol Struct*, **1249**, 131641.

72. Singh, A., Ansari, K.R., Quraishi, M.A., Lgaz, H., and Lin, Y. (2018) Synthesis and Investigation of Pyran Derivatives as Acidizing Corrosion Inhibitors for N80 Steel in Hydrochloric Acid: Theoretical and Experimental Approaches. *J Alloys Compd*, **762**, 347–362.

73. Singh, A., Ansari, K.R., Quraishi, M.A., Kaya, S., and Guo, L. (2020) Aminoantipyrine Derivatives as a Novel Eco-friendly Corrosion Inhibitors for P110 Steel in Simulating Acidizing Environment: Experimental and Computational Studies. *J Nat Gas Sci Eng*, **83**, 103547.

74. Singh, A., Bedi, P., Ansari, K.R., Pramanik, T., Chaudhary, D., Santra, S., Alanazi, A.K., Das, S., Quraishi, M.A., Lin, Y., Kaya, S., and EL Ibrahimi, B. (2021) Inhibition Effect of Newly Synthesized Benzoxanthonesderivative on Hydrogen Evolution and Q235 Steel Corrosion in 15% HCl under Hydrodynamic Condition: Combination of Experimental, Surface and Computational Study. *Int J Hydrogen Energy*, **46** (76), 37995–38007.

75. Yadav, M., Sharma, U., and Yadav, P.N. (2013) Isatin Compounds as Corrosion Inhibitors for N80 Steel in 15% HCl. *Egypt J Pet*, **22** (3), 335–344.

76. Zhang, W., Li, H.J., Wang, M., Wang, L.J., Pan, Q., Ji, X., Qin, Y., and Wu, Y.C. (2019) Tetrahydroacridines as Corrosion Inhibitor for X80 Steel Corrosion in Simulated Acidic Oilfield Water. *J Mol Liq*, **293**, 111478.

18 Sustainable and Green Heterocycles Corrosion Inhibitors

Omar Dagdag[1], Rajesh Haldhar[2], Seong-Cheol Kim[2], Walid Daoudi[3], Elyor Berdimurodov[4], Ekemini D. Akpan[1] and Eno E. Ebenso[1]
[1]Centre for Materials Science, College of Science, Engineering and Technology, University of South Africa, Johannesburg, South Africa
[2]School of Chemical Engineering, Yeungnam University, Gyeongsan, Republic of Korea
[3]Laboratory of Molecular Chemistry, Materials and Environment (LCM2E), Department of Chemistry, Multidisciplinary Faculty of Nador, University Mohamed I, Nador, Morocco
[4]Faculty of Chemistry, National University of Uzbekistan, Tashkent, Uzbekistan

18.1 INTRODUCTION

Metallic resources are employed in several businesses for household and building purposes [1]. However, the interactions of the elements in their environment easily cause them to corrode [2]. Due to corrosion failure, there are a lot of incidents recorded across the world, which result in significant health, financial, and fatality losses. The NACE estimates that corrosion causes a 3.5% loss in global GDP [3]. As a result, several efforts have been undertaken to decrease corrosion damage. According to evidence, the previously employed corrosion monitoring techniques can reduce corrosion losses by 15–35% [3]. Utilizing chemical molecules is among the most efficient, well-liked, and cost-effective choices for lowering the risks and losses associated with corrosion [4]. At the boundary between metal and a harsh environment, organic compounds, particularly heterocyclic compounds with electron-rich centres, adsorb and create protective coatings that separate the metal surfaces from the electrolyte [5]. One of the most effective inhibitors is demonstrated to be the family of heterocyclic compounds' organic corrosion inhibitors. A cyclic organic molecule is called a heterocyclic compound if one or more carbon atoms are replaced by several other heteroatoms (P, O, S, N) [6]. Several types of research in recent years have employed heterocyclic compounds as corrosion protection compounds are reported pyridines [7], pyrimidines [8], imidazoles [9], triazoles [10], benzotriazoles [11], tetrazoles [12], indoles [13], thiazoles [14], oxadiazoles [15],

DOI: 10.1201/9781003377016-18

pyrazole [16], benzimidazoles [17], and macrocyclic compounds [18], and biomolecules as well as several amino acids [19], proteins, carbohydrates [20], etc. All of these kinds of molecules might therefore serve as corrosion prevention agents. Heterocyclic inhibitors rise with the presence of heteroatoms (N, S, P) [21–25] and π-electrons of the double bond and triple bond with the metal surfaces [26–28].

This chapter describes the collection of heterocyclic compounds as corrosion protection compounds. Diverse classes of sustainable and green heterocycles corrosion inhibitor compounds are often utilized for various metals and their alloys. This is the first complete chapter on sustainable and green heterocycles-based molecules as corrosion inhibitors.

18.2   SUSTAINABLE AND GREEN HETEROCYCLES CORROSION INHIBITORS

18.2.1   Azoles

Azoles are some of the most effective metals and alloys in electrolytic media. It has been discovered that the nitrogen and sulfur heteroatoms in the azole ring ordinate chemical bonds towards the metal surface.

18.2.1.1   Pyrazoles

Five-membered heterocyclic rings called "pyrazoles" have two neighbour ring nitrogen heteroatoms in their C_3N_2 rings. Pyrazoles' IE (%) can be explained by the adsorption of azole N's lone-pair electrons and the pyrazole ring's delocalized π-electrons [29]. The application of pyrazoles and their derivatives interact strongly with metal surfaces [30]. When two biologically active pyrazoles, DPP and 4-CP, were tested for MS in 1 M HCl, the chloro group produced a high IE of 94% [31]. Thiosemicarbazide and phenyl-substituted pyrazole carboxaldehyde analogues had a reaction product with an IE of 95.5% at 200 mg/L [32]. Figure 18.1 shows some important pyrazole derivatives used to prevent metal samples from corroding in various acid media and their corrosion IE (%).

18.2.1.2   Imidazoles

The positioning of the two nitrogen heteroatoms in the imidazole structure differs from that of the pyrazole, which has the skeletal formula C_3N_2. In contrast to pyrazoles, where the two nitrogen heteroatoms stand close, imidazoles' two nitrogen heteroatoms are divided by a carbon atom. It was noted that the nitrogen atoms of the pyrimidine and imidazole aromatic rings are mainly attributed to enhance corrosion inhibition. Imidazoles have significant interactions with the surface of metals because they are extremely polar, highly nucleophilic, and miscible in corrosive environments. They also act as major inhibitors when one or both of their derivatives, generated by decreasing one of the two double bonds (imidazoline), are reduced. These benzimidazole analogues were often used as chemicals to prevent corrosion. Fatty acid imidazolines are among the most popular in the oil industry. In an H_2SO_4 corrosive environment, non-toxic imidazoles were assessed for Cu corrosion [34]. It was shown that adding the methyl group and the phenyl ring improved

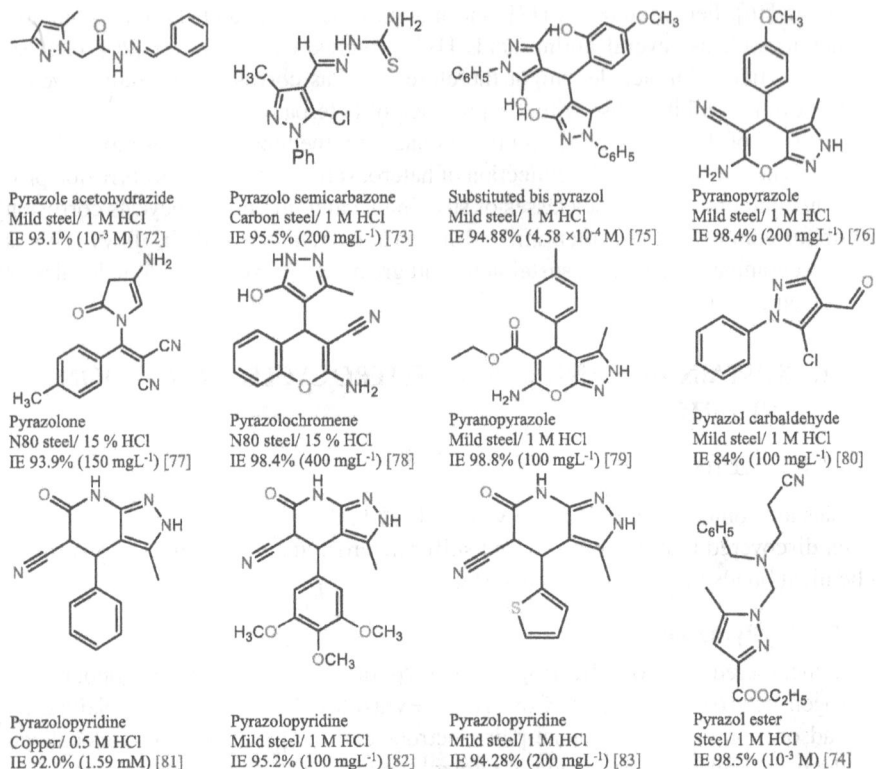

Pyrazole acetohydrazide	Pyrazolo semicarbazone	Substituted bis pyrazol	Pyranopyrazole
Mild steel/ 1 M HCl	Carbon steel/ 1 M HCl	Mild steel/ 1 M HCl	Mild steel/ 1 M HCl
IE 93.1% (10^{-3} M) [72]	IE 95.5% (200 mgL^{-1}) [73]	IE 94.88% (4.58 ×10^{-4} M) [75]	IE 98.4% (200 mgL^{-1}) [76]
Pyrazolone	Pyrazolochromene	Pyranopyrazole	Pyrazol carbaldehyde
N80 steel/ 15 % HCl	N80 steel/ 15 % HCl	Mild steel/ 1 M HCl	Mild steel/ 1 M HCl
IE 93.9% (150 mgL^{-1}) [77]	IE 98.4% (400 mgL^{-1}) [78]	IE 98.8% (100 mgL^{-1}) [79]	IE 84% (100 mgL^{-1}) [80]
Pyrazolopyridine	Pyrazolopyridine	Pyrazolopyridine	Pyrazol ester
Copper/ 0.5 M HCl	Mild steel/ 1 M HCl	Mild steel/ 1 M HCl	Steel/ 1 M HCl
IE 92.0% (1.59 mM) [81]	IE 95.2% (100 mgL^{-1}) [82]	IE 94.28% (200 mgL^{-1}) [83]	IE 98.5% (10^{-3} M) [74]

FIGURE 18.1 Structures of some important pyrazole and their corrosion IE (%). (Reprinted with permission from Ref. [33] © 2021 Elsevier Publications.)

corrosion inhibition compared to the bare imidazole. It has been demonstrated that the neutral form of imidazole adsorbs more effectively than the protonated version [35]. Figure 18.2 shows some important imidazole derivatives used to prevent metal samples from corroding in various acid media and their corrosion IE (%).

18.2.1.3   Thiazoles and Oxazoles

Similar to imidazoles in structure, thiazoles and oxazoles contain one of the nitrogen heteroatoms swapped out for a sulfur or oxygen heteroatom. Compared to corrosion inhibitors based on imidazoles, BPOX, an oxazole derivative, was claimed to have a 95% moderate efficacy for MS [36]. The IE (%) of the imidazole derivative was greater than that of the comparable oxazole at 1 mM. It was investigated that the oxazole analogues with various functional groups linked to the oxazole ring, including azidooxy, benzene sulfonate, and hydroxyl groups [37]. The IE (%) for the benzene sulfonate group was 94.5% at 10^{-3} M. Along with biocides and molybdates, 2-phenyl-4-methyl-4((tetrazol-5-yl)methyl)oxazoline was tested for MS at low temperatures [38]. Oxazole and molybdate worked together synergistically to produce IE = 94%, which fell to IE = 84% when a CTAB biocide was added. At 50 mg/L, poly(2-ethyl-2-oxazoline) has been reported to have an IE of 88.15% [39].

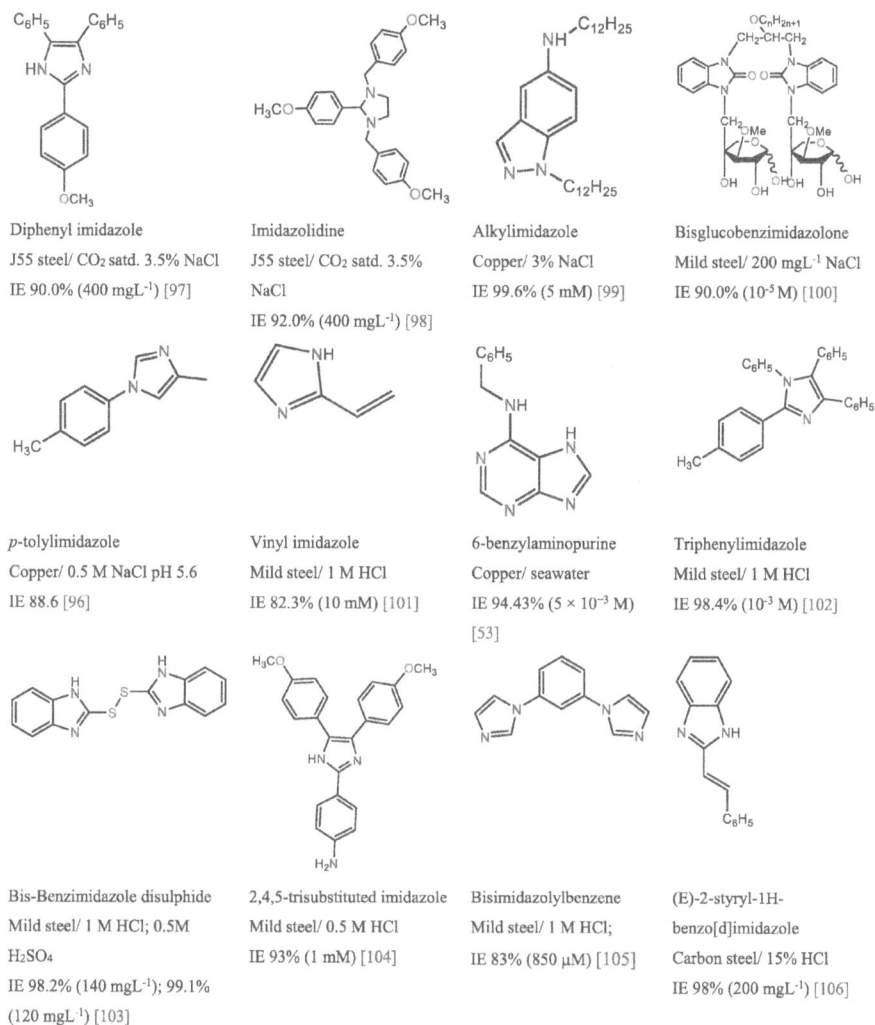

Diphenyl imidazole
J55 steel/ CO$_2$ satd. 3.5% NaCl
IE 90.0% (400 mgL^{-1}) [97]

Imidazolidine
J55 steel/ CO$_2$ satd. 3.5%
NaCl
IE 92.0% (400 mgL^{-1}) [98]

Alkylimidazole
Copper/ 3% NaCl
IE 99.6% (5 mM) [99]

Bisglucobenzimidazolone
Mild steel/ 200 mgL^{-1} NaCl
IE 90.0% (10^{-5} M) [100]

p-tolylimidazole
Copper/ 0.5 M NaCl pH 5.6
IE 88.6 [96]

Vinyl imidazole
Mild steel/ 1 M HCl
IE 82.3% (10 mM) [101]

6-benzylaminopurine
Copper/ seawater
IE 94.43% (5 × 10^{-3} M)
[53]

Triphenylimidazole
Mild steel/ 1 M HCl
IE 98.4% (10^{-3} M) [102]

Bis-Benzimidazole disulphide
Mild steel/ 1 M HCl; 0.5M
H$_2$SO$_4$
IE 98.2% (140 mgL^{-1}); 99.1%
(120 mgL^{-1}) [103]

2,4,5-trisubstituted imidazole
Mild steel/ 0.5 M HCl
IE 93% (1 mM) [104]

Bisimidazolylbenzene
Mild steel/ 1 M HCl;
IE 83% (850 μM) [105]

**(E)-2-styryl-1H-
benzo[d]imidazole**
Carbon steel/ 15% HCl
IE 98% (200 mgL^{-1}) [106]

FIGURE 18.2 Structures of some important imidazole and their corrosion IE (%). (Reprinted with permission from Ref. [33] © 2021 Elsevier Publications.)

There are recent studies on azole and thiazole compounds as corrosion inhibitors accessible [40]. Thiazoles TCA, MTT, and APT have effects upon the corrosion of MS in a corrosive medium containing 0.5 M H$_2$SO$_4$ [41]. For APT having *p*-tolyl group, a remarkable IE (%) of 98.10% has been found. Another research examined two thiazole derivatives [42]. The 2-amino-4-methyl-thiazole was well adsorbed by the methyl and amino groups, providing an IE (%) of 98% at 10.0 mM [43]. Common heterocyclic compounds function as mixed-type inhibitors, and the inhibitor's adsorption procedure includes donor–acceptor interactions that form a flat layer on the metal surface. It has been demonstrated that the presence of two amino groups facilitates the effective adsorption of (*S*)-4,5,6,7-tetrahydrobenzo[*d*]. Thiazole-2,6-diamine has effect on the corrosion of MS in hydrochloric solution [44]. At 320 mg/L, an efficiency of 89.71%

was achieved, whereas 2-((thiazole-2-ylimino)methyl)phenol exhibited an efficiency of 94% at 10 mM [45]. The order of thiazole, 2-aminothiazole, and 4-aminothiazole is in terms of IE (%) for 1,3-thiazole, and its amino analogues [46].

An imidazole–thiazole hydrochloride derivative was investigated on the corrosion of N80 steel in 0.5 M HCl corrosive media [47]. Figure 18.3 shows some important oxazole and thiazole derivatives used to prevent metal samples from corroding in various acid media and their corrosion IE (%).

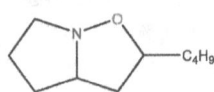

Pyrrolo-oxazole
Mild steel/ 0.5 M HCl
IE 95% (10^{-4} M) [124]

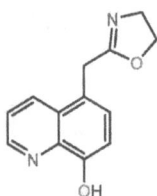

Oxazolylquinolin-ol
Mild steel/ 1 M HCl
IE 95.18% (10^{-3}M) [125]

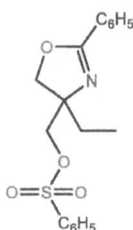

Oxazol benzenesulphonate
Mild steel/ 1 M HCl
IE 94.50% (10^{-3}M) [126]

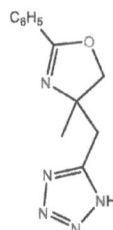

Tetrazolyloxazoline
Mild steel/ cooling water
IE 94.5% at (10^{-3} M) [127]

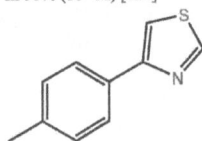

4-(p-tolyl)thiazole
Mild steel/ 0.5 M H_2SO_4
IE 98.1% (10^{-2} M) [129]

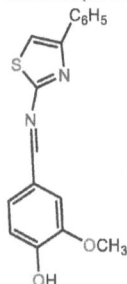

Vanillin-thiazole Schiff base
Mild steel/ 1 M HCl
IE 98.3% (300 mgL^{-1}) [130]

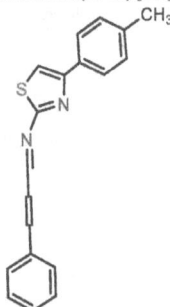

Cinnamaldehyde-thiazole Schiff base
Mild steel/ 20% formic acid, 20% acetic acid
IE 99.31% (300 mgL^{-1}), 98.72% (300 mgL^{-1}) [131]

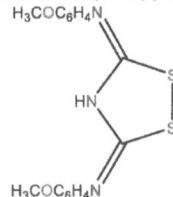

Dithiazolidine
Mild steel/ 1 M HCl
IE 99.5% (500 mgL^{-1}) [132]

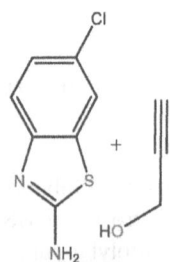

Aminochlorobenzothiazole + propargyl alcohol
Mild steel/ 15% HCl (boiling)
IE 99.3% (1000 + 2500 mgL^{-1}) [133]

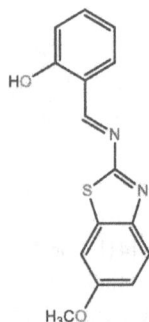

Aminobenzothiazole Schiff base
Mild steel/ 1 M HCl
IE 97.1% (500 mgL^{-1}) [134]

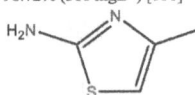

2-amino-4-methylthiazole
Mild steel/ 0.5 M HCl
IE 97% (10 mM) [135]

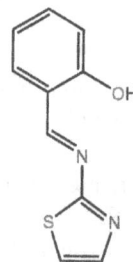

Thiazole-iminophenol
Mild steel/ 2 M HCl
IE 94% (10^{-2} M) [136]

FIGURE 18.3 Structures of some important oxazole and thiazole and their corrosion IE (%). (Reprinted with permission from Ref. [33] © 2021 Elsevier Publications.)

18.2.1.4 Oxadiazoles

Because of its symmetrical structure and good surface adsorption properties, 1,3,4-oxadiazole was used in most experiments. Bentiss et al. investigated novel categories of oxadiazole analogue on the corrosion of MS in corrosive solutions with concentrations of 0.5 M H_2SO_4 and 1 M HCl [48]. A compound with a 2-position hydroxyl group fared better. Later, the same team looked at 2,4-bis-1,3,4-oxadiazoles [49]. The presence of the nitro group had a detrimental impact, while the amino functional groups are attributed to a rise in the protection efficiency (IE > 97%). The corrosion inhibition analysis of 2,5-bis(n-methoxyphenyl)-1,3,4-oxadiazoles showed that the inhibitory behaviour was significantly influenced by the location of the methoxy functional additives [50]. The methoxy groups' location and presence of aromatic rings may promote the formation of hydrogen bonds, leading to increased adsorption.

An analysis of thiadiazole and oxadiazole analogues revealed that oxadiazole performed better in corrosive conditions containing H_2SO_4, whereas thiadiazole in media containing HCl [51]. The two investigated inhibitors weren't cytotoxic. The researchers also reported bio-oxadiazole analogues for MS in 1 M perchloric acid corrosive medium [52]. The best IE (%) was produced by the pyridine ring with nitrogen heteroatoms positioned in the p-position. High IE > 98% evaluations of 3 long fatty acid oxadiazoles have been performed [53]. 3,5-Bis(n-pyridyl)1,2,4-oxadiazoles were assessed for C38 steel in a corrosive medium containing 1 M HCl [54]. At 760 µM, a 1,3,4-oxadiazole derivative with a 2-ethylbenzimidazole moiety had an IE of 95.10% [55]. In 20% HCOOH, oxyadiazole compounds were assessed for MS [15]. The inhibitors provided a high IE of >99% at 500 mg/L. Figure 18.4 shows some important oxadiazole derivatives used to prevent metal samples from corroding in various acid media and their corrosion IE (%).

18.2.1.5 Thiadiazoles

Three hector bases with a thiadiazole moiety were the best inhibitors tested in Quraishi and his team's study, providing the best inhibition [56]. Thiadiazoles were tested for MS in the corrosive environment containing 1 M HCl [57], H_2SO_4 corrosive solution with 0.5 M concentration [58], and 20% corrosive media with H_2SO_4 [59]. The type of aromatic rings also influences the corrosion efficiency of drug-type anti-polarization agents for metal surfaces. The following research investigated four bisthiadiazoles [60]. It is also noted that the thiadiazole-based organic molecules effectively interacted with the metal surface by the interactions in which the covalent bonds are formed. In 0.5 M H_2SO_4 and 1 M HCl corrosive environments, 2,2'-benzothiazolyl disulfide was investigated for MS [61]. High IE > 97% was recorded in both acids at a relatively low dosage of 150 µM. In 0.5 M HCl acidic conditions, PDTT was suggested as a good anti-polarization agent for steel [62]. A complex surfaced from the protonation of the Schiff base. Investigations into another Schiff base revealed a high IE of 97% with 1 mM [63]. At high temperatures, the thiadiazole analogues DSTA and BTTA were suggested as good protective additive in the carbon dioxide saturated solution [64]. At 0.025 mM, DSTA performed better, providing 99.37% protection. In a 3 M H_2SO_4 aggressive environment, 2-amino-5-ethyl-1, 3, and 4-thiadiazoles were studied

Bisoxadiazole
Mild steel/ 1 M HCl
IE 98.2% (80 mgL^{-1})
[149]

Mercaptooxaidazole
Mild steel/ 15% HCl
(boiling)
IE 98.94% (500 mgL^{-1}) [154]

Bisoxadiazole
Mild steel/ 1 M HCl
IE 94.5% (10^{-4} M) [157]

Bisoxadiazoles
Mild steel/ 1 M HCl
IE 99.8% (60 mgL^{-1}) [157]

Bisoxadiazoles
Mild steel/ 0.5 M H$_2$SO$_4$
IE 95.1% (10^{-4} M) [152]

Oxadiazole Schiff base
Mild steel/ 1 M HCl
IE 90.2% (500 mgL^{-1}) [158]

Pyridyl-oxadiazole
Mild steel/ 1 M HClO$_4$
IE 91.6% (12×10^{-4} M)
[153]

Pyridyl-oxadiazole
C38 steel/ 1 M HCl
IE 97.43% (12×10^{-4} M)
[155]

Aminochlorobenzothiazole +
propargyl alcohol
Mild steel/ 15% HCl (boiling)
IE 99.3% (1000 + 2500 mgL^{-1})
[133]

Aminobenzothiazole Schiff
base
Mild steel/ 1 M HCl
IE 97.1% (500 mgL^{-1})
[134]

2-amino-4-methylthiazole
Mild steel/ 0.5 M HCl
IE 97% (10 mM) [135]

Thiazole-iminophenol
Mild steel/ 2 M HCl
IE 94% (10^{-2} M) [136]

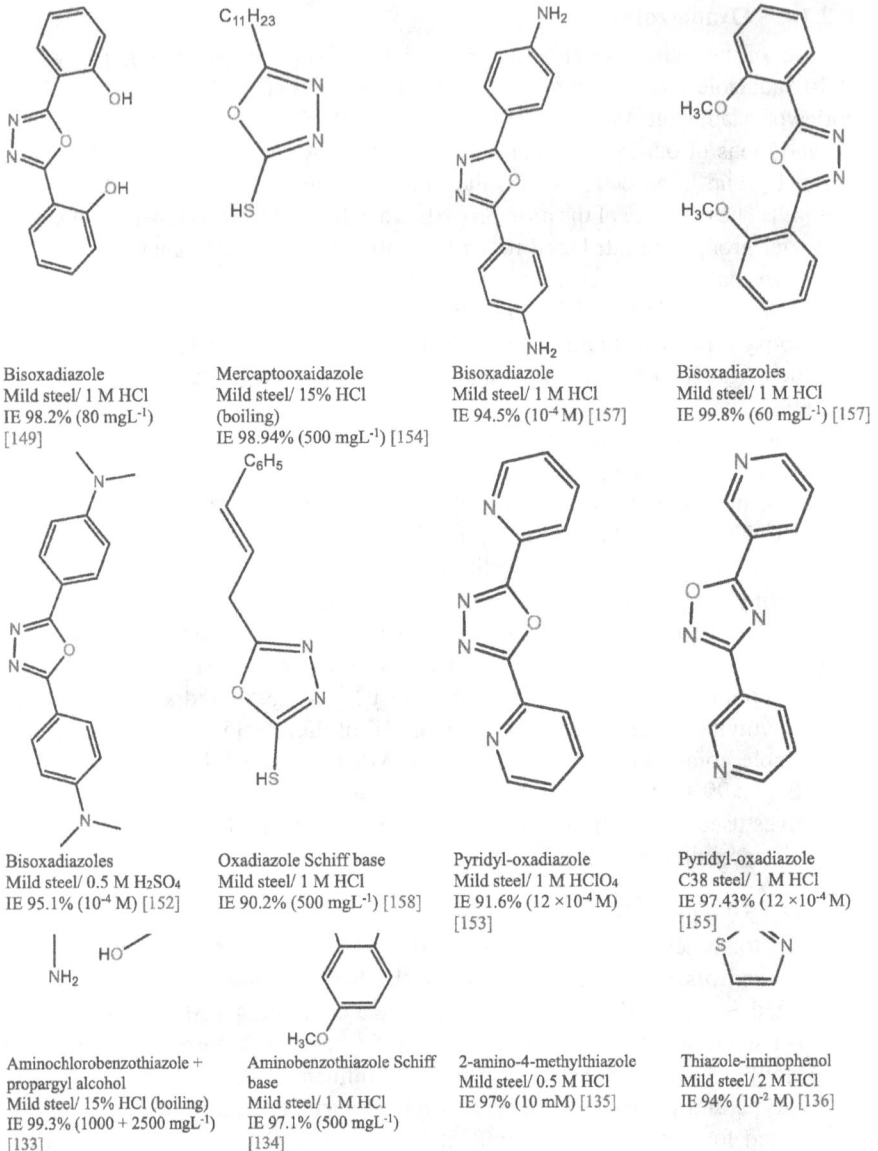

FIGURE 18.4 Structures of some important oxadiazoles and their corrosion IE (%). (Reprinted with permission from Ref. [33], © 2021 Elsevier Publications.)

for SS [65]. Figure 18.5 shows some important thiadiazole derivatives used to prevent metal samples from corroding in various acid media and their corrosion IE (%).

18.2.1.6 Tetrazoles

Tetrazoles are five-membered heterocycles with four heteroatoms of nitrogen. Three isomers, 5*H*-tetrazole, 2*H*-tetrazole, and 1*H*-tetrazole might exist depending on where the double bonds are placed. In 0.1 M HCl corrosive conditions, five tetrazole analogues

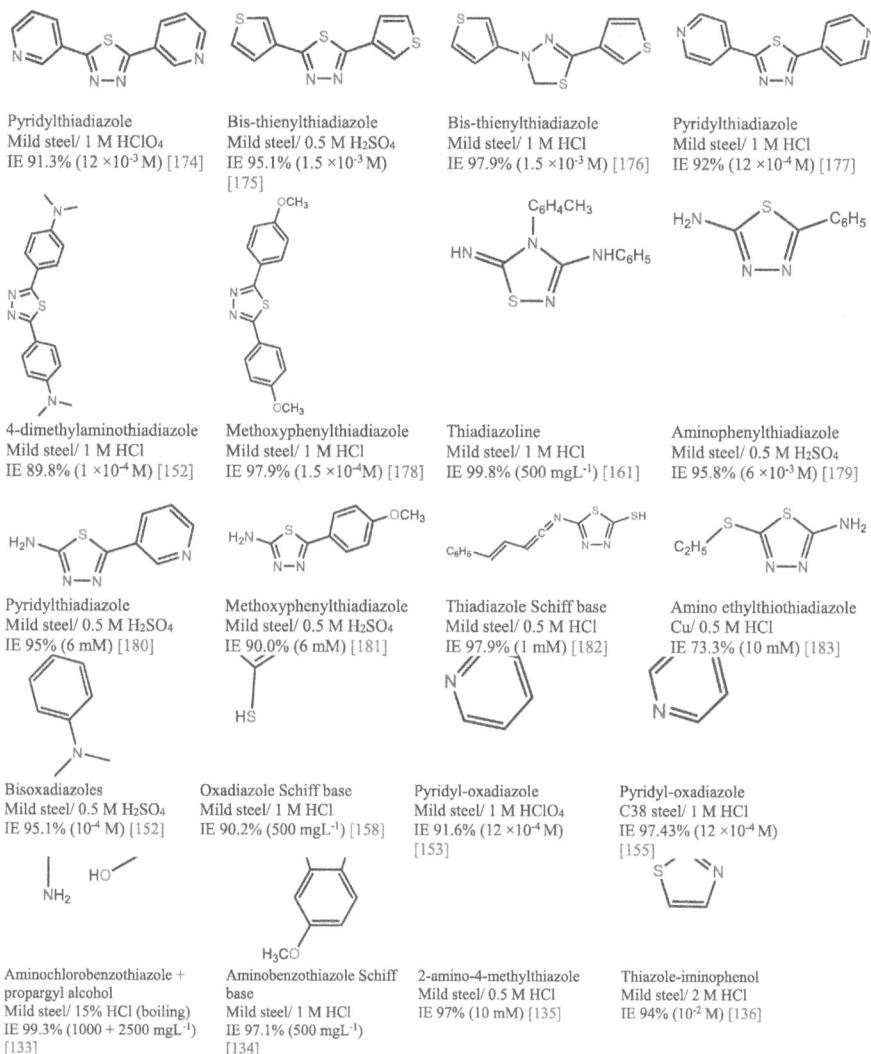

FIGURE 18.5 Structures of some important thiadiazole and their corrosion IE (%). (Reprinted with permission from Ref. [33], © 2021 Elsevier Publications.)

were tested for Cu [66]. IE > 98% was demonstrated for 5-phenyl tetrazole and 5-methyl-5-phenyl tetrazole. Several 5-substituted tetrazoles were studied by Verma et al. [67], with a maximum IE of 98.69% at 40 mg/L. For MS in 1 M HCl corrosive medium, tetrazole derivatives with IEs of 89% and 92.7% were observed [68]. These corrosion inhibitors were confirmed to be more effective for Cu in HNO_3-corrosive environments [69]. It was found that adding an amino group as a substituent enhanced the results when compared to 1H-tetrazole and that adding a phenyl ring (1-phenyl-1H-tetrazole-5-thiol) improved the results even more (97.01% IE at 10^{-2} M) [70]. The effects of Cu corrosion on 5-substituted tetrazoles with either pyridine or thiophene as substitutes were investigated [71]. When compared to pyridine, the thiophene derivative demonstrated better inhibition. Symmetrical tetrazole-based diselenides were created by El-Askalany and colleagues [72]. An excellent IE of 93.2% was attained at a low dosage of 50 m/L. The organoselenium functionalized tetrazole was tested for the steel in the aggressive acidic solutions, showing up to 94.6% IE (%) obtained [72]. Figure 18.6 shows some important triazole and tetrazole derivatives used to prevent metal samples from corroding in various acid media and their corrosion IE (%).

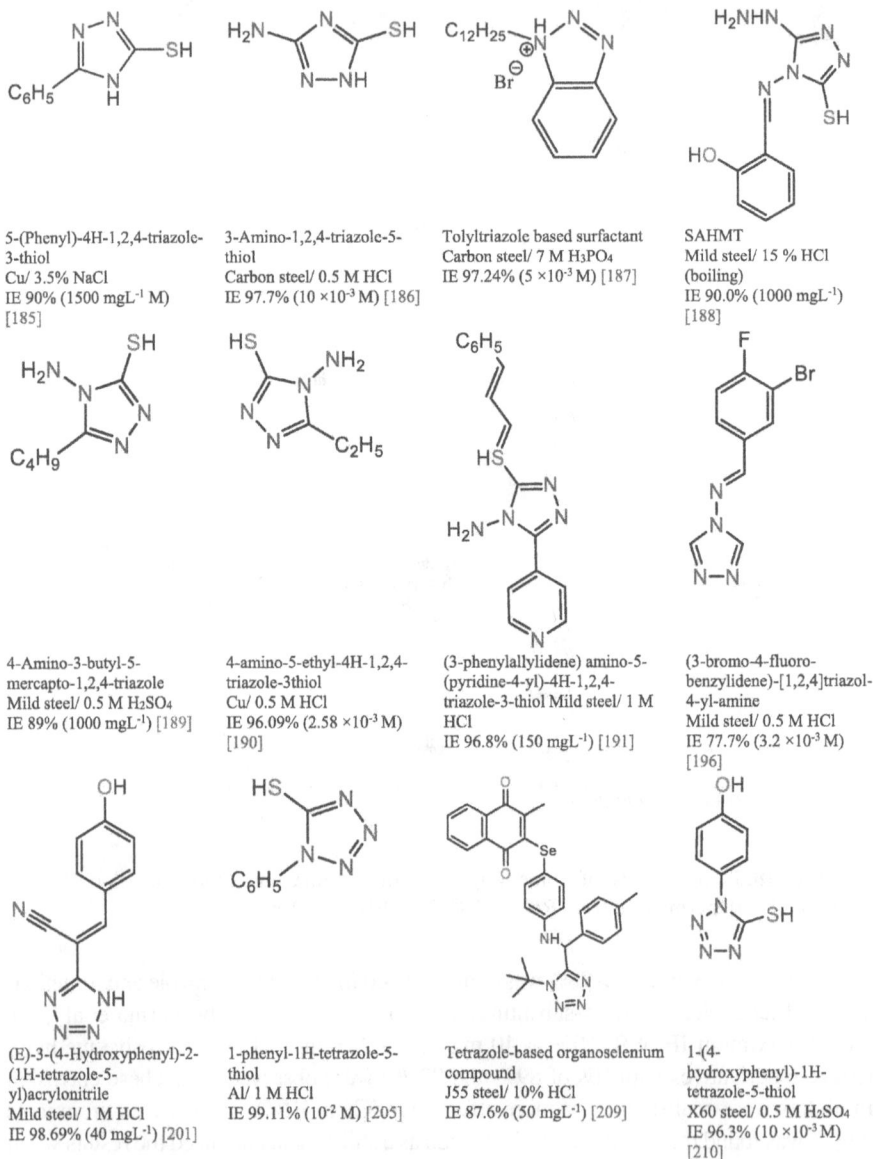

5-(Phenyl)-4H-1,2,4-triazole-3-thiol
Cu/ 3.5% NaCl
IE 90% (1500 mgL⁻¹ M)
[185]

3-Amino-1,2,4-triazole-5-thiol
Carbon steel/ 0.5 M HCl
IE 97.7% (10 ×10⁻³ M) [186]

Tolyltriazole based surfactant
Carbon steel/ 7 M H₃PO₄
IE 97.24% (5 ×10⁻³ M) [187]

SAHMT
Mild steel/ 15 % HCl
(boiling)
IE 90.0% (1000 mgL⁻¹)
[188]

4-Amino-3-butyl-5-mercapto-1,2,4-triazole
Mild steel/ 0.5 M H₂SO₄
IE 89% (1000 mgL⁻¹) [189]

4-amino-5-ethyl-4H-1,2,4-triazole-3thiol
Cu/ 0.5 M HCl
IE 96.09% (2.58 ×10⁻³ M)
[190]

(3-phenylallylidene) amino-5-(pyridine-4-yl)-4H-1,2,4-triazole-3-thiol Mild steel/ 1 M HCl
IE 96.8% (150 mgL⁻¹) [191]

(3-bromo-4-fluoro-benzylidene)-[1,2,4]triazol-4-yl-amine
Mild steel/ 0.5 M HCl
IE 77.7% (3.2 ×10⁻³ M)
[196]

(E)-3-(4-Hydroxyphenyl)-2-(1H-tetrazole-5-yl)acrylonitrile
Mild steel/ 1 M HCl
IE 98.69% (40 mgL⁻¹) [201]

1-phenyl-1H-tetrazole-5-thiol
Al/ 1 M HCl
IE 99.11% (10⁻² M) [205]

Tetrazole-based organoselenium compound
J55 steel/ 10% HCl
IE 87.6% (50 mgL⁻¹) [209]

1-(4-hydroxyphenyl)-1H-tetrazole-5-thiol
X60 steel/ 0.5 M H₂SO₄
IE 96.3% (10 ×10⁻³ M)
[210]

FIGURE 18.6 Structures of some important triazole and tetrazole and their corrosion IE (%). (Reprinted with permission from Ref. [33], © 2021 Elsevier Publications.)

18.2.2 Azines

18.2.2.1 Pyridines

Pyridine is a very important class of compounds widely used in pharmaceutical conception. Due to their presence in natural products and exceptional biological properties, the heterocyclic compounds have received the most significant research attention. These heterocyclic compounds rule the fields of biochemistry and medicinal chemistry, and they are becoming more significant in a variety of other fields [73]. At 200 mg L^{-1}, a remarkable IE of 95.7% was attained with a mixed effect. Cinnamaldehyde derivative had the highest performance in electron-rich regions [74]. The involvement of heteroatoms of nitrogen (PMPM, IE = 90%), oxygen (PMTM, IE = 92%), and sulfur (PMAM, IE = 95%) in various pyridyl Schiff bases was examined [75]. Utilizing WL, electrochemical, and morphological techniques, a Schiff base N,N'-(pyridine-2,6-diyl)bis(1-(4-methoxyphenyl)methanimine) has been assessed on J55 and N80 steels in 3.5% NaCl CO$_2$ saturated [76]. At a C_{inh} of 400 mg/L, high IE of 90% for N80 steel and 93% for J55 steel were observed. An ultrasound-assisted method was used to manufacture a number of pyrazolo[3,4 b]pyridines and test them for MS [77]. At 100 mg/L, the electron-donating methoxy group derivative had an IE of 95.2%. The 2-amino-3,5-dicarbonitrile-6-thiopyridines showed comparable outcomes (IE = 97.6%) [78], cinnamaldehyde moiety, and methoxy group-containing dihydropyridine derivative (IE=98.10% at 400 mg/L) [79] for MS in hydrochloric environment. Figure 18.7 shows some important pyridine derivatives used to prevent metal samples from corroding in various acid media and their corrosion IE (%).

18.2.2.2 Pyrimidines

Another part of the diazine category, pyrimidine has two nitrogen heteroatoms at positions 1 and 3. They make up a large portion of nucleic acids. The pyrimidine pharmacophore has considerable pharmacological value as an antimicrobial, veterinary, antibacterial, and cardiovascular medicine. It also plays a significant part in biological processes. Fluorouracil, Tegafur, Isethionate, Uramustine, and Piritrexim are a few of the well-known commercially accessible medications that are pyrimidine derivatives. It is among the most secure heterocyclic configurations because each nitrogen heteroatom has two lone pairs of electrons [80]. Pyrimidine derivatives work well to prevent rusting [81]. 2,6-Dimethylpyrimidine-2-amine and also its analogues were explored by Elewady et al. [82]. The IE (%) and C_{inh} have a direct relationship [83]. Figure 18.8 shows some important pyrimidine derivatives used to prevent metal samples from corroding in various acid media and their corrosion IE (%).

18.2.2.3 Triazines

Similar to benzene, triazine has three carbon atoms substituted by nitrogen heteroatoms [84]. One of the most researched corrosion inhibitors is 1,3,5-triazine (s-triazine) derivatives. Triazines can be used as complexation agents, insecticides, herbicides, fungicides, etc. To get excellent IE (%) with triazines, it is crucial to incorporate the right substituents. 6-Methyl-5-[m-nitro styryl]-3-mercapto-1,2,4-triazine has been investigated for MS in a 12% hydrochloric acidic medium at 50 °C

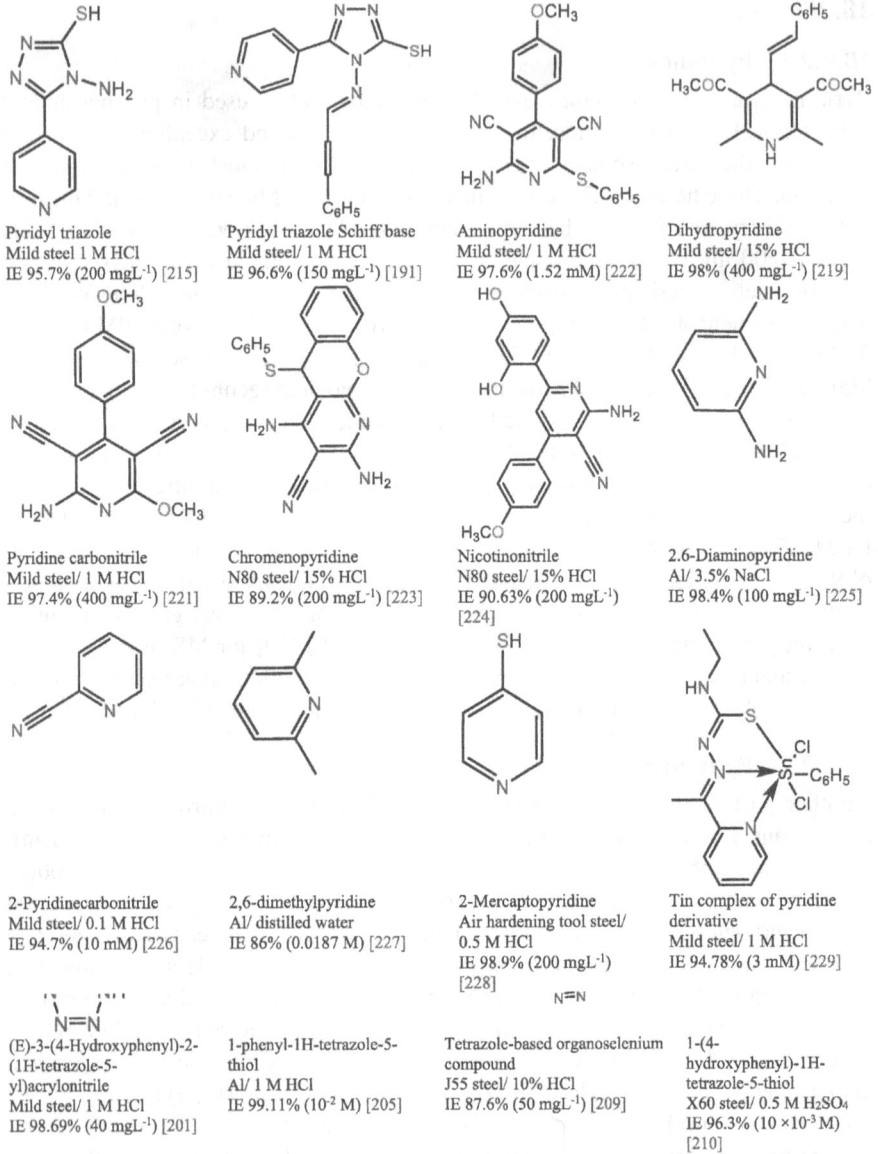

Pyridyl triazole
Mild steel 1 M HCl
IE 95.7% (200 mgL^{-1}) [215]

Pyridyl triazole Schiff base
Mild steel/ 1 M HCl
IE 96.6% (150 mgL^{-1}) [191]

Aminopyridine
Mild steel/ 1 M HCl
IE 97.6% (1.52 mM) [222]

Dihydropyridine
Mild steel/ 15% HCl
IE 98% (400 mgL^{-1}) [219]

Pyridine carbonitrile
Mild steel/ 1 M HCl
IE 97.4% (400 mgL^{-1}) [221]

Chromenopyridine
N80 steel/ 15% HCl
IE 89.2% (200 mgL^{-1}) [223]

Nicotinonitrile
N80 steel/ 15% HCl
IE 90.63% (200 mgL^{-1}) [224]

2.6-Diaminopyridine
Al/ 3.5% NaCl
IE 98.4% (100 mgL^{-1}) [225]

2-Pyridinecarbonitrile
Mild steel/ 0.1 M HCl
IE 94.7% (10 mM) [226]

2,6-dimethylpyridine
Al/ distilled water
IE 86% (0.0187 M) [227]

2-Mercaptopyridine
Air hardening tool steel/
0.5 M HCl
IE 98.9% (200 mgL^{-1}) [228]

Tin complex of pyridine derivative
Mild steel/ 1 M HCl
IE 94.78% (3 mM) [229]

(E)-3-(4-Hydroxyphenyl)-2-(1H-tetrazole-5-yl)acrylonitrile
Mild steel/ 1 M HCl
IE 98.69% (40 mgL^{-1}) [201]

1-phenyl-1H-tetrazole-5-thiol
Al/ 1 M HCl
IE 99.11% (10^{-2} M) [205]

Tetrazole-based organoselenium compound
J55 steel/ 10% HCl
IE 87.6% (50 mgL^{-1}) [209]

1-(4-hydroxyphenyl)-1H-tetrazole-5-thiol
X60 steel/ 0.5 M H$_2$SO$_4$
IE 96.3% (10 ×10^{-3} M) [210]

FIGURE 18.7   Structures of some important pyridine and their corrosion IE (%). (Reprinted with permission from Ref. [33]. © 2021 Elsevier Publications.)

[85]. According to research on the three triazines ABTDT, ATTDT, and AMTDT, the inhibitor ABTDT with the —CH$_2$Ph group offered the greatest IE (99.90% at 200 mg/L) [86]. The corrosion performance of triazines was enhanced by the modification of methoxy and hydroxyl functional groups, resulting in over 92.5% protection degrees [87]. Three Schiff base triazines, MHMMT, DHMMT, and HMMT, were

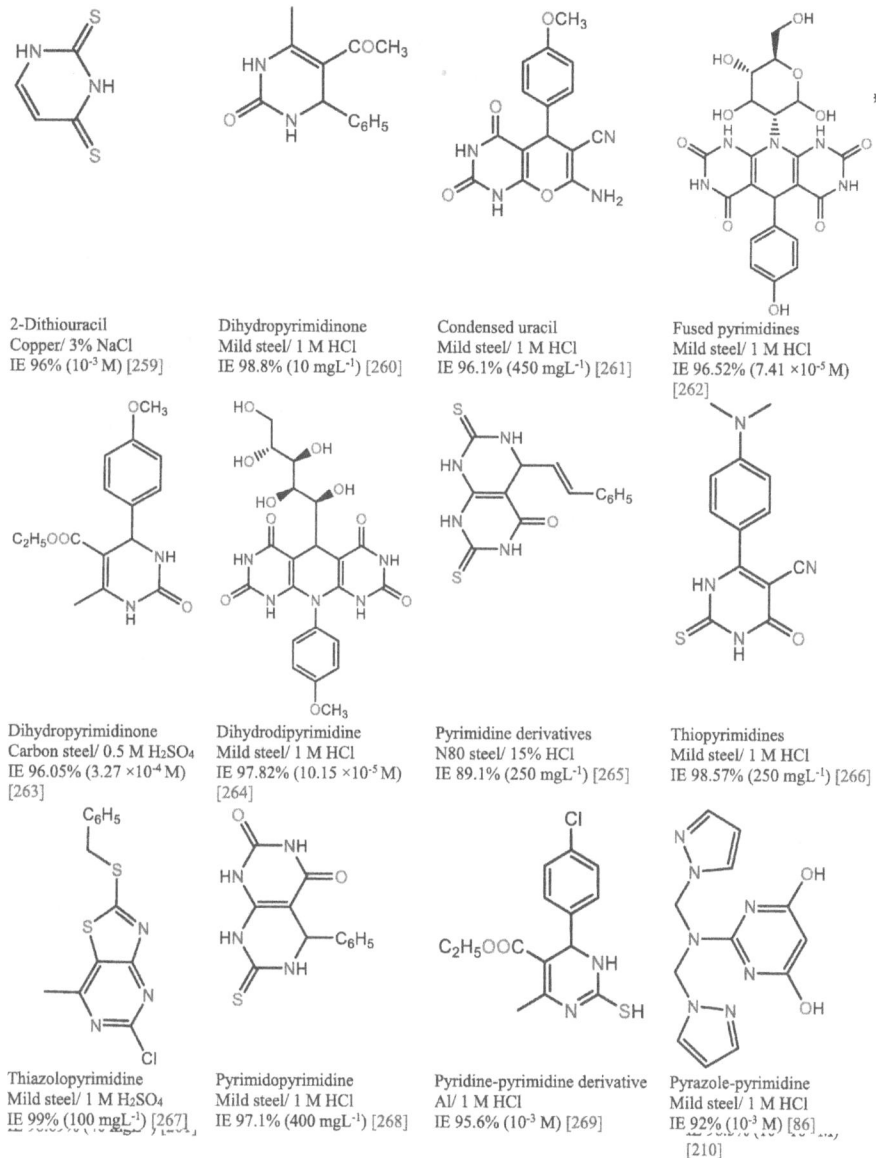

2-Dithiouracil
Copper/ 3% NaCl
IE 96% (10^{-3} M) [259]

Dihydropyrimidinone
Mild steel/ 1 M HCl
IE 98.8% (10 mgL^{-1}) [260]

Condensed uracil
Mild steel/ 1 M HCl
IE 96.1% (450 mgL^{-1}) [261]

Fused pyrimidines
Mild steel/ 1 M HCl
IE 96.52% (7.41 ×10^{-5} M)
[262]

Dihydropyrimidinone
Carbon steel/ 0.5 M H$_2$SO$_4$
IE 96.05% (3.27 ×10^{-4} M)
[263]

Dihydrodipyrimidine
Mild steel/ 1 M HCl
IE 97.82% (10.15 ×10^{-5} M)
[264]

Pyrimidine derivatives
N80 steel/ 15% HCl
IE 89.1% (250 mgL^{-1}) [265]

Thiopyrimidines
Mild steel/ 1 M HCl
IE 98.57% (250 mgL^{-1}) [266]

Thiazolopyrimidine
Mild steel/ 1 M H$_2$SO$_4$
IE 99% (100 mgL^{-1}) [267]

Pyrimidopyrimidine
Mild steel/ 1 M HCl
IE 97.1% (400 mgL^{-1}) [268]

Pyridine-pyrimidine derivative
Al/ 1 M HCl
IE 95.6% (10^{-3} M) [269]

Pyrazole-pyrimidine
Mild steel/ 1 M HCl
IE 92% (10^{-3} M) [86]
[210]

FIGURE 18.8 Structures of some important pyrimidine and their corrosion IE (%).
(Reprinted with permission from Ref. [33], © 2021 Elsevier Publications.)

investigated for MS corrosion in a 0.5 N H$_2$SO$_4$ corrosive environment. 2,4,6-Tris(n-carboxyalkylamino)1,3,5-triazine was tested for MS in corrosive media containing 1 M HCl. The findings demonstrated that a change from physical to chemical adsorption occurred with an extension in the alkyl chain [88]. Figure 18.9 shows some important triazine derivatives used to prevent metal samples from corroding in various acid media and their corrosion IE (%).

Nitro-mercaptotriazine
Tubing steel/ 12% HCl
IE 89.4% (375 mgL^{-1}) [279]

Dihydrotriazine-one
Mild steel/ 1 M HCl
IE 99.38% (200 mgL^{-1}) [280]

Triazinetrihexanoic acid
Mild steel/ 1 M HCl
IE 98.38% (5000 mgL^{-1}) [282]

Hexahydro(1,3,5)-triazine
Mild steel/ 1 M HCl
IE 97.07% (300 mgL^{-1}) [284]

Tris-pyridine-triazine
Mild steel/ 1 M HCl
IE 97.7% (10 mgL^{-1}) [294]

Triazine Schiff base
Mild steel/ 1 M HCl
IE 98.6% (80 mgL^{-1}) [286]

Trihydrizinotriazine
Steel/ 1 M HCl
IE 97.8% (25 mgL^{-1}) [290]

Triazine triamine
N80 steel/ 15% HCl
IE 93% (150 mgL^{-1}) [289]

Dihydrotriazine-one
Mild steel/ 1 M HCl
IE 94.28% (200 mgL^{-1}) [291]

Triazine Schiff base
Mild steel/ 2 N H$_2$SO$_4$
IE 95.1% (200 mgL^{-1}) [287]

Trispyridyltriazine
Tin/ 0.5 M HCl
IE 87% (10^{-3}M) [295]

Trithiocyanuric acid
Cu/ 3% NaCl
IE 95.3% (0.75 mM) [296]

Thiazolopyrimidine
Mild steel/ 1 M H$_2$SO$_4$
IE 99% (100 mgL^{-1}) [267]

Pyrimidopyrimidine
Mild steel/ 1 M HCl
IE 97.1% (400 mgL^{-1}) [268]

Pyridine-pyrimidine derivative
Al/ 1 M HCl
IE 95.6% (10^{-3} M) [269]

Pyrazole-pyrimidine
Mild steel/ 1 M HCl
IE 92% (10^{-3} M) [86] [210]

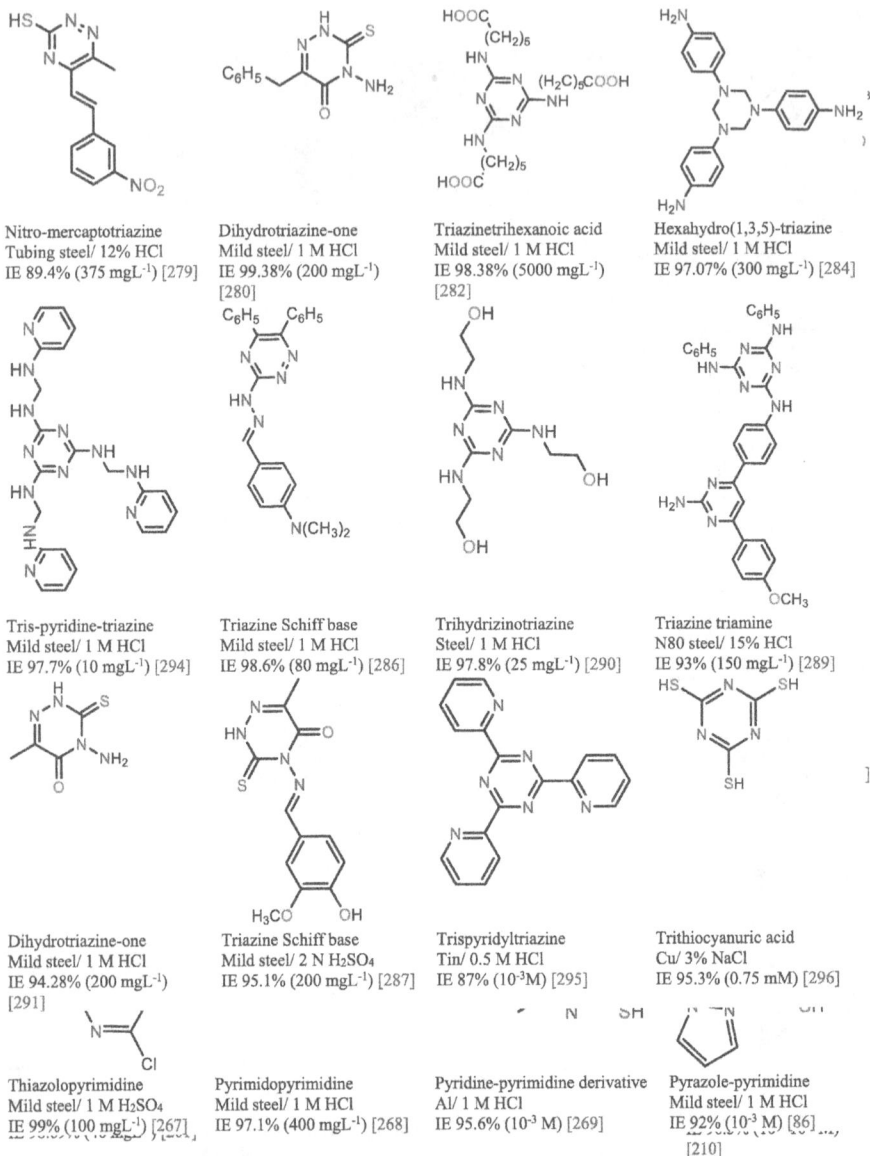

FIGURE 18.9 Structures of some important triazine and their corrosion IE (%). (Reprinted with permission from Ref. [33], © 2021 Elsevier Publications.)

18.2.3 QUINOLINES

Quinoline, an aromatic heterocycle, was initially produced by removing it from coal tar. They may be created in a step process in the lab using aniline. Quinolines are used to create a variety of colours, vitamins, and medications. The effectiveness of three quaternary ammonium quinolone phenolates as anti-polarization agents for Zn-Mg composition in batteries was studied through their synthesis [89]. One of the derivatives performed capacitively in the electrochemical investigations, whereas the other, which included nitrogen–oxygen and methyl substituents, displayed an inductive character. In a different investigation, quinaldic acid, quinaldine, and quinolone were assessed in corrosive environments using 0.5 M HCl [90]. Quinaldic acid's higher M_w allowed for more surface covering [91]. Deaerated 3 M HCl corrosive media were used to study quinoline and its phosphonium derivatives [92]. The thermodynamic characteristics suggested a chemical adsorption mechanism, while the inhibitors displayed a mixed activity. 8-Nitroquinoline and 8-aminoquinoline were investigated as Al alloy inhibitors in 3% NaCl corrosive environments [93]. Both inhibitors had anodic properties and worked by strongly hybridizing the sp- and p-orbitals of the active regions on the Al atoms in the inhibitor. Figure 18.10 shows some important quinoline and quinazoline derivatives used to prevent metal samples from corroding in various acid media and their corrosion IE (%).

18.2.4   HETEROCYCLIC PHARMACEUTICAL PRODUCTS

Drug molecules are seen as ecologically safe and green alternatives to the harmful inhibitors that are now used. Drugs typically have complicated structures with several centres with high electron densities, which makes it easier for them to bind to metal surfaces. As a result, various pharmacological chemicals were investigated and assessed as natural corrosion agents [94, 95]. Quinolones, tetracyclines, beta-lactam antibiotics, etc., are some of the several types examined [34]. It investigated how Donaxine and Gramine inhibited behaviour [96]. At an inhibitor dosage of 7.5 mM, the high IE (%) was 98%. The Langmuir isotherm was followed by surface adsorption. Amoxicillin, Penicillin G, and Ampicillin, three antibacterial medications, were examined for CS [35]. Another investigation examined the effects of the medication cephalothin on API 5L X52 steel in 1 M HCl acidic conditions [97]. At 600 mg/L, the inhibitor had a 92% efficiency rate. The well-known drug Telmisartan produced an excellent IE of 97.39% at 125 mg/L for MS in the corrosive environment containing 1 M HCl [98]. Figure 18.11 shows some important drug derivatives that prevent metal samples from corroding in various acid media and their corrosion IE(%).

18.2.5   HETEROCYCLIC AMINO ACIDS

Each amino acid has an amino or carboxyl functional group, making them all very soluble in water. Numerous studies about using amino acids as corrosion inhibitors have surfaced over time. Figure 18.12 shows some important heterocyclic amino acids used to prevent metal samples from corroding in various acid media and their corrosion IE (%).

18.2.6   MACROCYCLIC COMPOUNDS

In a carbon dioxide–saturated saline solution, an HPT was investigated as the macrocyclic corrosion inhibitor for J55 steel [99]. A dosage of 400 mg L^{-1} resulted in a high IE of 93%. The electrochemical and thermodynamic methods studied the macrocyclic effects of corrosion inhibition. When the corrosion inhibitor was present, SECM tests revealed that the steel substrate was protected to a significant amount. The same team used electrochemical and surface studies to assess the trilogy effects on corrosion

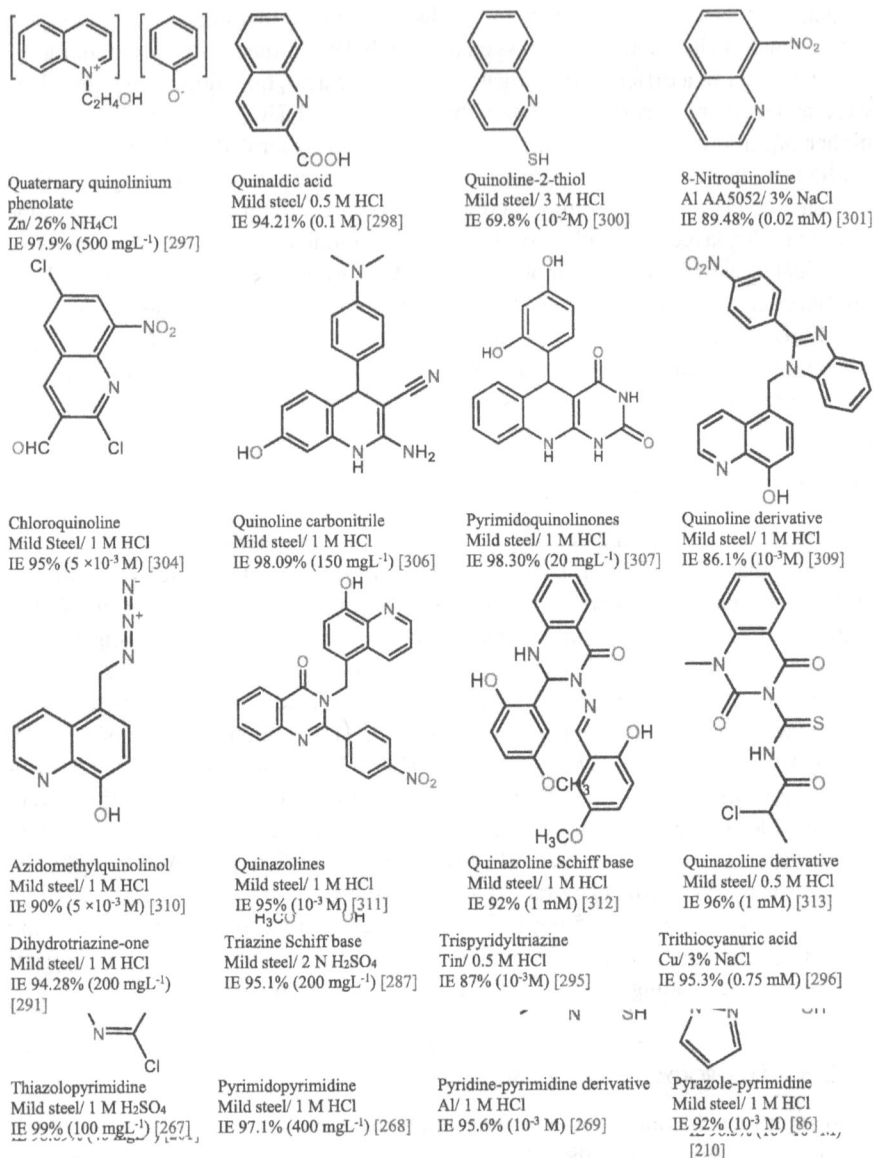

Quaternary quinolinium phenolate
Zn/ 26% NH₄Cl
IE 97.9% (500 mgL⁻¹) [297]

Quinaldic acid
Mild steel/ 0.5 M HCl
IE 94.21% (0.1 M) [298]

Quinoline-2-thiol
Mild steel/ 3 M HCl
IE 69.8% (10⁻²M) [300]

8-Nitroquinoline
Al AA5052/ 3% NaCl
IE 89.48% (0.02 mM) [301]

Chloroquinoline
Mild Steel/ 1 M HCl
IE 95% (5 ×10⁻³ M) [304]

Quinoline carbonitrile
Mild steel/ 1 M HCl
IE 98.09% (150 mgL⁻¹) [306]

Pyrimidoquinolinones
Mild steel/ 1 M HCl
IE 98.30% (20 mgL⁻¹) [307]

Quinoline derivative
Mild steel/ 1 M HCl
IE 86.1% (10⁻³M) [309]

Azidomethylquinolinol
Mild steel/ 1 M HCl
IE 90% (5 ×10⁻³ M) [310]

Quinazolines
Mild steel/ 1 M HCl
IE 95% (10⁻³ M) [311]

Quinazoline Schiff base
Mild steel/ 1 M HCl
IE 92% (1 mM) [312]

Quinazoline derivative
Mild steel/ 0.5 M HCl
IE 96% (1 mM) [313]

Dihydrotriazine-one
Mild steel/ 1 M HCl
IE 94.28% (200 mgL⁻¹)
[291]

Triazine Schiff base
Mild steel/ 2 N H₂SO₄
IE 95.1% (200 mgL⁻¹) [287]

Trispyridyltriazine
Tin/ 0.5 M HCl
IE 87% (10⁻³M) [295]

Trithiocyanuric acid
Cu/ 3% NaCl
IE 95.3% (0.75 mM) [296]

Thiazolopyrimidine
Mild steel/ 1 M H₂SO₄
IE 99% (100 mgL⁻¹) [267]

Pyrimidopyrimidine
Mild steel/ 1 M HCl
IE 97.1% (400 mgL⁻¹) [268]

Pyridine-pyrimidine derivative
Al/ 1 M HCl
IE 95.6% (10⁻³ M) [269]

Pyrazole-pyrimidine
Mild steel/ 1 M HCl
IE 92% (10⁻³ M) [86]
[210]

FIGURE 18.10 Structures of some important quinoline and quinazoline and their corrosion IE (%). (Reprinted with permission from Ref. [33], © 2021 Elsevier Publications.)

efficiency [100]. The water contact angle increased as the inhibitor was adsorbed to the metal surface at increasing quantities, demonstrating a development in the hydrophobicity of the metal surface. In a subsequent experiment, a Pd-porphyrin-based compound was investigated as a good anti-polarization agent in the carbon dioxide–saturated saline solution [101]. As a result, it provided an IE of 85% and the porphyrin PF-2 performed better (92%) due to the introduction of methoxy functional groups.

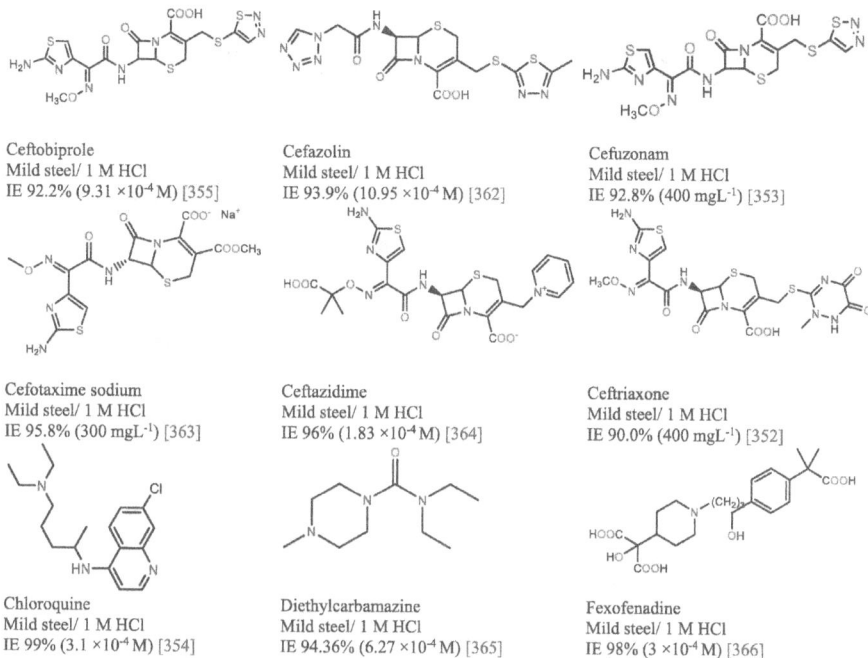

Ceftobiprole
Mild steel/ 1 M HCl
IE 92.2% (9.31 ×10⁻⁴ M) [355]

Cefazolin
Mild steel/ 1 M HCl
IE 93.9% (10.95 ×10⁻⁴ M) [362]

Cefuzonam
Mild steel/ 1 M HCl
IE 92.8% (400 mgL⁻¹) [353]

Cefotaxime sodium
Mild steel/ 1 M HCl
IE 95.8% (300 mgL⁻¹) [363]

Ceftazidime
Mild steel/ 1 M HCl
IE 96% (1.83 ×10⁻⁴ M) [364]

Ceftriaxone
Mild steel/ 1 M HCl
IE 90.0% (400 mgL⁻¹) [352]

Chloroquine
Mild steel/ 1 M HCl
IE 99% (3.1 ×10⁻⁴ M) [354]

Diethylcarbamazine
Mild steel/ 1 M HCl
IE 94.36% (6.27 ×10⁻⁴ M) [365]

Fexofenadine
Mild steel/ 1 M HCl
IE 98% (3 ×10⁻⁴ M) [366]

FIGURE 18.11 Structures of some different drugs and their corrosion IE (%). (Reprinted with permission from Ref. [33], © 2021 Elsevier Publications.)

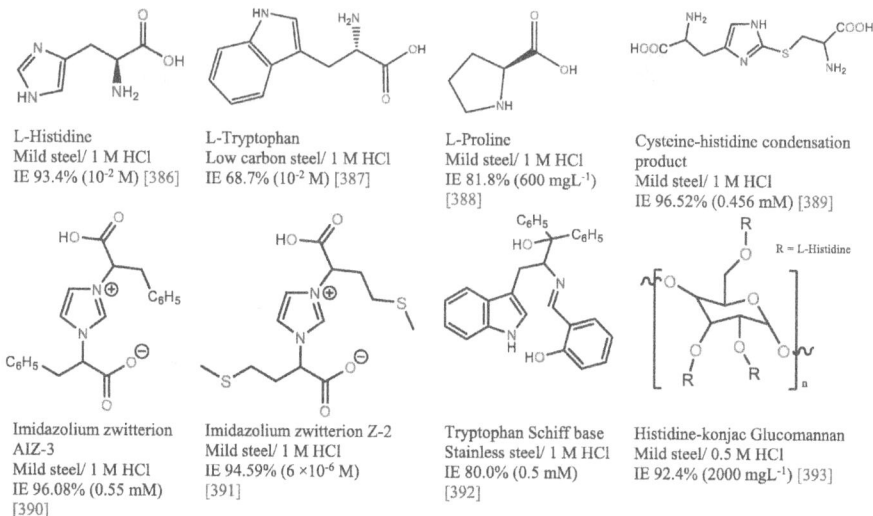

L-Histidine
Mild steel/ 1 M HCl
IE 93.4% (10⁻² M) [386]

L-Tryptophan
Low carbon steel/ 1 M HCl
IE 68.7% (10⁻² M) [387]

L-Proline
Mild steel/ 1 M HCl
IE 81.8% (600 mgL⁻¹) [388]

Cysteine-histidine condensation product
Mild steel/ 1 M HCl
IE 96.52% (0.456 mM) [389]

Imidazolium zwitterion AIZ-3
Mild steel/ 1 M HCl
IE 96.08% (0.55 mM) [390]

Imidazolium zwitterion Z-2
Mild steel/ 1 M HCl
IE 94.59% (6 ×10⁻⁶ M) [391]

Tryptophan Schiff base
Stainless steel/ 1 M HCl
IE 80.0% (0.5 mM) [392]

Histidine-konjac Glucomannan
Mild steel/ 0.5 M HCl
IE 92.4% (2000 mgL⁻¹) [393]

FIGURE 18.12 Structures of some important amino acids and their corrosion IE (%). (Reprinted with permission from Ref. [33], © 2021 Elsevier Publications.)

Figure 18.13 shows some important macrocyclic inhibitors that prevent metal samples from corroding in various acid media and their corrosion IE (%).

18.2.7   HETEROCYCLIC CARBOHYDRATES AND BIOPOLYMERS

The heterocyclic carbohydrates and biopolymers were recommended as good inhibitors in the carbon dioxide–saturated saline solution [103]. The glucose ring was modified

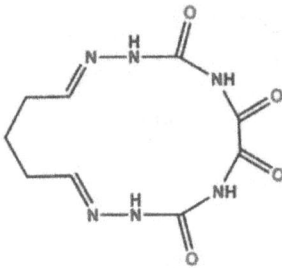

(1E,10E)-1,2,4,7,9,10-hexaazacyclopentadeca-10,15-diene-3,5,6,8-tetraone (IE 92%)

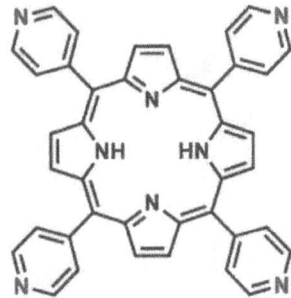

5,10,15,20-Tetra(4-pyridyl)-21H,23H-porphine (IE 92%)

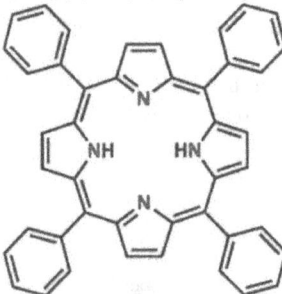

5,10,15,20-Tetraphenyl-21H,23H-porphine (IE 82%)

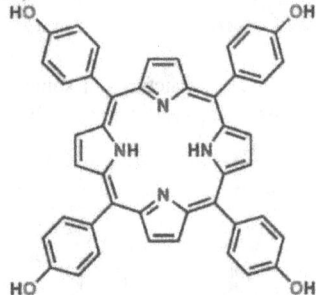

5,10,15,20-Tetrakis(4-hydroxyphenyl)-21H,23H-porphine (IE 84%)

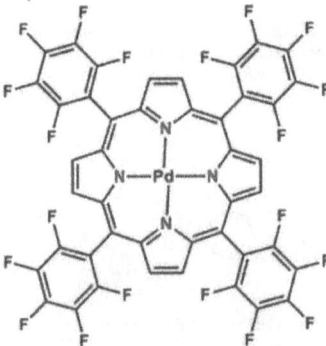

5,10,15,20-tetrakis(pentafluorophenyl)-21H,23Hporphyrin palladium(II) (IE 86%)

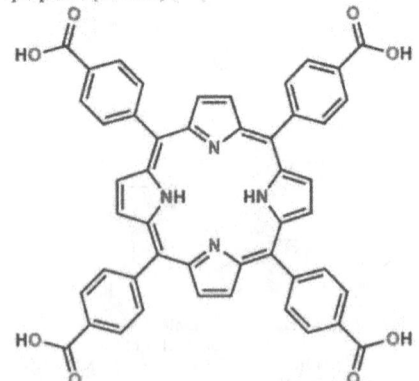

4,4',4",4'''-(porphyrin-5,10,15,20-tetrayl) tetrabenzoic acid (IE% 93%)

FIGURE 18.13   Structures of some important macrocyclic and their corrosion IE (%). (Reprinted with permission from Ref. [102], © 2021 Elsevier Publications.)

to the amino or carboxyl functional groups in these types of inhibitors. At a dosage of 2.27×10^{-4} M, the resulting product had good water solubility and a high IE of 91.82%. The steel has been examined at high temperatures for carbon dioxide–saturated saline-solution corrosion with chitosan-salicylaldehyde Schiff base [104], 95.2% of high IE at 150 mg/L. P110 steel was tested using two chitosan oligosaccharide analogues in a CO_2-saturated chloride environment [105]. To create the multifunctional corrosion inhibitor, benzaldehyde and propionaldehyde were added to the chitosan after it had been quaternized. The anti-corrosion agents were examined in both hydrodynamic and static circumstances, and at a Cinh of 100 mg/L, higher IE (%) for BHC was achieved in both static (IE = 88.59%) and hydrodynamic (IE = 84.68%) conditions. Figure 18.14 shows some important heterocyclic carbohydrates and biopolymers that prevent metal samples from corroding in various acid media and their corrosion IE (%).

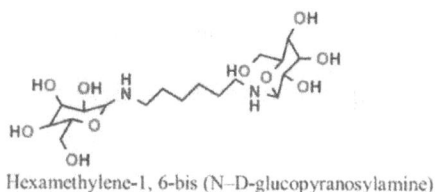

Hexamethylene-1, 6-bis (N–D-glucopyranosylamine)

API X60 steel/3.5 wt% NaCl saturated with CO_2

IE=91.82% (100 mg/L)

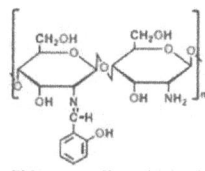

Chitosan-salicylaldehyde Schiff base

J55 Steel/3.5% NaCl saturated with carbon dioxide

IE= 95.2% (150 mg/L)

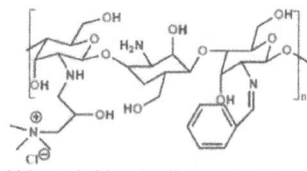

N-benzyl chitosan oligosaccharide quaternary ammonium salt (BHC)

P110 Steel/3.5% NaCl saturated with carbon dioxide

IE=93.35%(100mg/L)

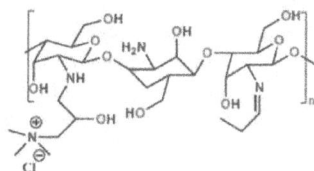

N-propyl chitosan oligosaccharide quaternary ammonium salt (PHC)

P110 Steel/3.5% NaCl saturated with carbon dioxide

IE=91.62%(100mg/L)

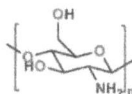

Chitosan

API 5L X60 Steel /3.5% NaCl saturated with carbon dioxide

IE= 55 %(100mg/L)

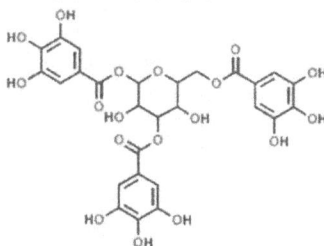

Tannic acid

API5L X60 steel/3.5 wt% NaCl saturated with CO_2

IE=90% (500 mg/L+2g/L KI)

5,10,15,20-tetrakis(pentafluorophenyl)-
21H,23Hporphyrin palladium(II) (IE 86%)

4,4',4'',4'''-(porphyrin-5,10,15,20-tetrayl)
tetrabenzoic acid (IE% 93%)

FIGURE 18.14 Structures of some important heterocyclic carbohydrates and biopolymers and their corrosion IE (%). (Reprinted with permission from Ref. [102], © 2021 Elsevier Publications.)

18.3   CONCLUSIONS

The basis of biomolecules, including carbs, hormones, amino acids, proteins, nucleic acids, and vitamins, are sustainable and green heterocycles corrosion inhibitors. In medicinal drugs, these types of green compounds may be modified. As this chapter covers, many of the aforementioned biomolecules, plant extracts, drugs, etc. have excellent solubility and the highest anti-corrosion efficiency. These compounds are highly biocompatible because they are of natural origin or were designed to suit the human body (for example, medications). Their release into the environment has minimal to no adverse effects. As a result, many current studies have been conducted to explore innovative eco-friendly heterocyclic organic inhibitors. The structure, characteristics, and IE (%) of many ecologically friendly heterocyclic corrosion inhibitors have been described in detail. Pharmaceutically active substances, carbohydrates, and amino acids are among the leading groups mentioned. As stated in this overview, each of them has advantages and disadvantages. A description of commonly used five- and six-membered heterocyclic compounds (fundamental components of medications, biomolecules, etc.) is also provided. A brief explanation of how Green Chemistry works to prevent rusting is given.

REFERENCES

1. N. Baddoo, "Stainless steel in construction: A review of research, applications, challenges and opportunities," *Journal of Constructional Steel Research*, vol. 64, pp. 1199–1206, 2008.
2. K. Suzumura and S.-I. Nakamura, "Environmental factors affecting corrosion of galvanized steel wires," *Journal of Materials in Civil Engineering*, vol. 16, pp. 1–7, 2004.
3. C. Verma, E. E. Ebenso, and M. Quraishi, "Ionic liquids as green and sustainable corrosion inhibitors for metals and alloys: An overview," *Journal of Molecular Liquids*, vol. 233, pp. 403–414, 2017.
4. L. T. Popoola, "Organic green corrosion inhibitors (OGCIs): A critical review," *Corrosion Reviews*, vol. 37, pp. 71–102, 2019.
5. S. K. Ahmed, W. B. Ali, and A. A. Khadom, "Synthesis and investigations of heterocyclic compounds as corrosion inhibitors for mild steel in hydrochloric acid," *International Journal of Industrial Chemistry*, vol. 10, pp. 159–173, 2019.
6. J. A. Joule, and K. Mills, *Heterocyclic Chemistry at a Glance*, John Wiley & Sons, Inc., 2012.
7. C. T. Ser, P. Žuvela, and M. W. Wong, "Prediction of corrosion inhibition efficiency of pyridines and quinolines on an iron surface using machine learning-powered quantitative structure–property relationships," *Applied Surface Science*, vol. 512, p. 145612, 2020.
8. K. Rasheeda, D. Vijaya, P. Krishnaprasad, and S. Samshuddin, "Pyrimidine derivatives as potential corrosion inhibitors for steel in acid medium: An overview," *International Journal of Corrosion and Scale Inhibition*, vol. 7, pp. 48–61, 2018.
9. M. B. P. Mihajlović, M. B. Radovanović, Ž. Z. Tasić, and M. M. Antonijević, "Imidazole based compounds as copper corrosion inhibitors in seawater," *Journal of Molecular Liquids*, vol. 225, pp. 127–136, 2017.
10. M. ElBelghiti, Y. Karzazi, A. Dafali, B. Hammouti, F. Bentiss, and I. Obot, et al., "Experimental, quantum chemical and Monte Carlo simulation studies of 3,5-disubstituted-4-amino-1, 2, 4-triazoles as corrosion inhibitors on mild steel in acidic medium," *Journal of Molecular Liquids*, vol. 218, pp. 281–293, 2016.
11. C. Gattinoni, and A. Michaelides, "Understanding corrosion inhibition with van der Waals DFT methods: The case of benzotriazole," *Faraday Discussions*, vol. 180, pp. 439–458, 2015.

12. J. Seetharaman, E. A. Reny, D. A. Johnson, K. B. Sawant, and V. Sivaswamy, *Tetrazole Based Corrosion Inhibitors*, Google Patents, 2021.
13. C. Verma, M. Quraishi, E. Ebenso, I. Obot, and A. El Assyry, "3-Amino alkylated indoles as corrosion inhibitors for mild steel in 1 M HCl: Experimental and theoretical studies," *Journal of Molecular Liquids*, vol. 219, pp. 647–660, 2016.
14. M. Quraishi, W. Khan, and M. Ajmal, "The influence of some condensation products of aminobenzothiazoles and salicylaldehyde on corrosion inhibition and hydrogen permeation in sulphuric acid solution," *Journal-Electrochemical Society of India*, vol. 46, pp. 133–138, 1997.
15. M. Quraishi and F. A. Ansari, "Fatty acid oxadiazoles as corrosion inhibitors for mild steel in formic acid," *Journal of Applied Electrochemistry*, vol. 36, pp. 309–314, 2006.
16. M. Yadav, R. R. Sinha, T. K. Sarkar, and N. Tiwari, "Corrosion inhibition effect of pyrazole derivatives on mild steel in hydrochloric acid solution," *Journal of Adhesion Science and Technology*, vol. 29, pp. 1690–1713, 2015.
17. I. Obot, A. Madhankumar, S. Umoren, and Z. Gasem, "Surface protection of mild steel using benzimidazole derivatives: Experimental and theoretical approach," *Journal of Adhesion Science and Technology*, vol. 29, pp. 2130–2152, 2015.
18. S. Hadisaputra, S. Hamdiani, M. A. Kurniawan, and N. Nuryono, "Influence of macrocyclic ring size on the corrosion inhibition efficiency of dibenzo crown ether: A density functional study," *Indonesian Journal of Chemistry*, vol. 17, pp. 431–438, 2017.
19. B. El Ibrahimi, A. Jmiai, L. Bazzi, and S. El Issami, "Amino acids and their derivatives as corrosion inhibitors for metals and alloys," *Arabian Journal of Chemistry*, vol. 13, pp. 740–771, 2020.
20. S. A. Umoren, and U. M. Eduok, "Application of carbohydrate polymers as corrosion inhibitors for metal substrates in different media: A review," *Carbohydrate Polymers*, vol. 140, pp. 314–341, 2016.
21. O. Dagdag, Z. Safi, R. Hsissou, H. Erramli, M. El Bouchti, and N. Wazzan, et al., "Epoxy pre-polymers as new and effective materials for corrosion inhibition of carbon steel in acidic medium: Computational and experimental studies," *Scientific Reports*, vol. 9, pp. 1–14, 2019.
22. N. Mechbal, M. Belghiti, N. Benzbiria, C.-H. Lai, Y. Kaddouri, and Y. Karzazi, et al., "Correlation between corrosion inhibition efficiency in sulfuric acid medium and the molecular structures of two newly eco-friendly pyrazole derivatives on iron oxide surface," *Journal of Molecular Liquids*, vol. 331, p. 115656, 2021.
23. S. A. Haladu, N. D. Mu'azu, S. A. Ali, A. M. Elsharif, N. A. Odewunmi, and H. M. Abd El-Lateef, "Inhibition of mild steel corrosion in 1 M H_2SO_4 by a Gemini surfactant 1,6-hexyldiyl-bis-(dimethyldodecylammonium bromide): ANN, RSM predictive modeling, quantum chemical and MD simulation studies," *Journal of Molecular Liquids*, vol. 350, p. 118533, 2022.
24. W. S. Abdrabo, B. Elgendy, K. A. Soliman, H. M. Abd El-Lateef, and A. H. Tantawy, "Synthesis, assessment and corrosion protection investigations of some novel peptidomimetic cationic surfactants: Empirical and theoretical insights," *Journal of Molecular Liquids*, vol. 315, p. 113672, 2020.
25. H. M. Abd El-Lateef, K. Shalabi, and A. H. Tantawy, "Corrosion inhibition of carbon steel in hydrochloric acid solution using newly synthesized urea-based cationic fluorosurfactants: Experimental and computational investigations," *New Journal of Chemistry*, vol. 44, pp. 17791–17814, 2020.
26. X. Ren, S. Xu, X. Gu, B. Tan, J. Hao, and L. Feng, et al., "Hyperbranched molecules having multiple functional groups as effective corrosion inhibitors for Al alloys in aqueous NaCl," *Journal of Colloid and Interface Science*, vol. 585, pp. 614–626, 2021.
27. B. Prakashaiah, D. V. Kumara, A. A. Pandith, A. N. Shetty, and B. A. Rani, "Corrosion inhibition of 2024-T3 aluminum alloy in 3.5% NaCl by thiosemicarbazone derivatives," *Corrosion Science*, vol. 136, pp. 326–338, 2018.

28. A. R. Sayed, and H. M. A. El-Lateef, "Thiocarbohydrazones based on adamantane and ferrocene as efficient corrosion inhibitors for hydrochloric acid pickling of C-steel," *Coatings*, vol. 10, p. 1068, 2020.

29. E. Geler and D. Azambuja, "Corrosion inhibition of copper in chloride solutions by pyrazole," *Corrosion Science*, vol. 42, pp. 631–643, 2000.

30. R. Vera, F. Bastidas, M. Villarroel, A. Oliva, A. Molinari, and D. Ramírez, et al., "Corrosion inhibition of copper in chloride media by 1,5-bis(4-dithiocarboxylate-1-dodecyl-5-hydroxy-3-methylpyrazolyl) pentane," *Corrosion Science*, vol. 50, pp. 729–736, 2008.

31. S. El Arrouji, K. Karrouchi, A. Berisha, K. I. Alaoui, I. Warad, and Z. Rais, et al., "New pyrazole derivatives as effective corrosion inhibitors on steel–electrolyte interface in 1 M HCl: Electrochemical, surface morphological (SEM) and computational analysis," *Colloids and Surfaces A: Physicochemical and Engineering Aspects*, vol. 604, p. 125325, 2020.

32. R. A. Hameed, H. Al-Shafey, A. A. Magd, and H. Shehata, "Pyrazole derivatives as corrosion inhibitor for C-steel in hydrochloric acid medium," *Journal of Materials and Environmental Science*, vol. 3, pp. 294–305, 2012.

33. M. A. Quraishi, D. S. Chauhan, and V. S. Saji, "Heterocyclic biomolecules as green corrosion inhibitors," *Journal of Molecular Liquids*, vol. 341, p. 117265, 2021.

34. G. Gece, "Drugs: A review of promising novel corrosion inhibitors," *Corrosion Science*, vol. 53, pp. 3873–3898, 2011.

35. G. Golestani, M. Shahidi, and D. Ghazanfari, "Electrochemical evaluation of antibacterial drugs as environment-friendly inhibitors for corrosion of carbon steel in HCl solution," *Applied Surface Science*, vol. 308, pp. 347–362, 2014.

36. G. Moretti, F. Guidi, and F. Fabris, "Corrosion inhibition of the mild steel in 0.5 M HCl by 2-butyl-hexahydropyrrolo[1, 2-*b*][1, 2]oxazole," *Corrosion Science*, vol. 76, pp. 206–218, 2013.

37. H. Rahmani, F. El-Hajjaji, A. El Hallaoui, M. Taleb, Z. Rais, and M. El Azzouzi, et al., "Experimental, quantum chemical studies of oxazole derivatives as corrosion inhibitors on mild steel in molar hydrochloric acid medium," *International Journal of Corrosion and Scale Inhibition*, vol. 7, pp. 509–527, 2018.

38. B. Labriti, N. Dkhireche, R. Touir, M. Ebn Touhami, M. Sfaira, and A. El Hallaoui, et al., "Synergism in mild steel corrosion and scale inhibition by a new oxazoline in synthetic cooling water," *Arabian Journal for Science and Engineering*, vol. 37, pp. 1293–1303, 2012.

39. Z. P. Mathew, K. Rajan, C. Augustine, B. Joseph, and S. John, "Corrosion inhibition of mild steel using poly(2-ethyl-2-oxazoline) in 0.1 M HCl solution," *Heliyon*, vol. 6, p. e05560, 2020.

40. B. El Ibrahimi and L. Guo, "Azole-based compounds as corrosion inhibitors for metallic materials," *Azoles: Synthesis, Properties, Applications and Perspectives*, 2020.

41. K. Khaled and M. A. Amin, "Corrosion monitoring of mild steel in sulphuric acid solutions in presence of some thiazole derivatives: Molecular dynamics, chemical and electrochemical studies," *Corrosion Science*, vol. 51, pp. 1964–1975, 2009.

42. A. Döner, R. Solmaz, M. Özcan, and G. Kardaş, "Experimental and theoretical studies of thiazoles as corrosion inhibitors for mild steel in sulphuric acid solution," *Corrosion Science*, vol. 53, pp. 2902–2913, 2011.

43. A. O. Yüce, B. D. Mert, G. Kardaş, and B. Yazıcı, "Electrochemical and quantum chemical studies of 2-amino-4-methyl-thiazole as corrosion inhibitor for mild steel in HCl solution," *Corrosion Science*, vol. 83, pp. 310–316, 2014.

44. A. Manivel, S. Ramkumar, J. J. Wu, A. M. Asiri, and S. Anandan, "Exploration of (*S*)-4,5,6,7-tetrahydrobenzo[*d*]thiazole-2,6-diamine as feasible corrosion inhibitor for mild steel in acidic media," *Journal of Environmental Chemical Engineering*, vol. 2, pp. 463–470, 2014.

45. N. Yilmaz, A. Fitoz, and K. C. Emregül, "A combined electrochemical and theoretical study into the effect of 2-((thiazole-2-ylimino) methyl) phenol as a corrosion inhibitor for mild steel in a highly acidic environment," *Corrosion Science*, vol. 111, pp. 110–120, 2016.

46. L. Guo, X. Ren, Y. Zhou, S. Xu, Y. Gong, and S. Zhang, "Theoretical evaluation of the corrosion inhibition performance of 1,3-thiazole and its amino derivatives," *Arabian Journal of Chemistry*, vol. 10, pp. 121–130, 2017.

47. A. A. Fadhil, A. A. Khadom, H. Liu, C. Fu, J. Wang, and N. A. Fadhil, et al., "(S)-6-Phenyl-2, 3, 5, 6-tetrahydroimidazo[2, 1-*b*] thiazole hydrochloride as corrosion inhibitor of steel in acidic solution: Gravimetrical, electrochemical, surface morphology and theoretical simulation," *Journal of Molecular Liquids*, vol. 276, pp. 503–518, 2019.

48. F. Bentiss, M. Traisnel, and M. Lagrenee, "The substituted 1,3,4-oxadiazoles: A new class of corrosion inhibitors of mild steel in acidic media," *Corrosion Science*, vol. 42, pp. 127–146, 2000.

49. M. Hegazy, H. Ahmed, and A. El-Tabei, "Investigation of the inhibitive effect of *p*-substituted 4-(*N,N,N*-dimethyldodecylammonium bromide) benzylidene-benzene-2-yl-amine on corrosion of carbon steel pipelines in acidic medium," *Corrosion Science*, vol. 53, pp. 671–678, 2011.

50. F. Bentiss, M. Traisnel, N. Chaibi, B. Mernari, H. Vezin, and M. Lagrenée, "2,5-Bis (*n*-methoxyphenyl)-1,3,4-oxadiazoles used as corrosion inhibitors in acidic media: Correlation between inhibition efficiency and chemical structure," *Corrosion Science*, vol. 44, pp. 2271–2289, 2002.

51. F. Bentiss, M. Traisnel, H. Vezin, H. Hildebrand, and M. Lagrenee, "2,5-bis(4-dimethylaminophenyl)-1,3,4-oxadiazole and 2,5-bis(4-dimethylaminophenyl)-1,3,4-thiadiazole as corrosion inhibitors for mild steel in acidic media," *Corrosion Science*, vol. 46, pp. 2781–2792, 2004.

52. M. Lebrini, F. Bentiss, H. Vezin, and M. Lagrenée, "Inhibiting effects of some oxadiazole derivatives on the corrosion of mild steel in perchloric acid solution," *Applied Surface Science*, vol. 252, pp. 950–958, 2005.

53. M. Quraishi and D. Jamal, "Corrosion inhibition by fatty acid oxadiazoles for oil well steel (N-80) and mild steel," *Materials Chemistry and Physics*, vol. 71, pp. 202–205, 2001.

54. M. Outirite, M. Lagrenée, M. Lebrini, M. Traisnel, C. Jama, and H. Vezin, et al., "ac Impedance, X-ray photoelectron spectroscopy and density functional theory studies of 3, 5-bis(*n*-pyridyl)-1,2,4-oxadiazoles as efficient corrosion inhibitors for carbon steel surface in hydrochloric acid solution," *Electrochimica Acta*, vol. 55, pp. 1670–1681, 2010.

55. P. R. Ammal, M. Prajila, and A. Joseph, "Effective inhibition of mild steel corrosion in hydrochloric acid using EBIMOT, a 1,3,4-oxadiazole derivative bearing a 2-ethylbenz-imidazole moiety: Electro analytical, computational and kinetic studies," *Egyptian Journal of Petroleum*, vol. 27, pp. 823–833, 2018.

56. M. Quraishi, and R. Sardar, "Hector bases: A new class of heterocyclic corrosion inhibitors for mild steel in acid solutions," *Journal of Applied Electrochemistry*, vol. 33, pp. 1163–1168, 2003.

57. M. Quraishi and S. Khan, "Thiadiazoles: A potential class of heterocyclic inhibitors for prevention of mild steel corrosion in hydrochloric acid solution," *Indian Journal of Chemical Technology*, vol. 12, pp. 576–581, 2005.

58. M. Quraishi and S. Khan, "Inhibition of mild steel corrosion in sulfuric acid solution by thiadiazoles," *Journal of Applied Electrochemistry*, vol. 36, pp. 539–544, 2006.

59. S. Khan and M. Quraishi, "Synergistic effect of potassium iodide on inhibitive performance of thiadiazoles during corrosion of mild steel in 20% sulfuric acid," *Arabian Journal for Science and Engineering*, vol. 35, pp. 71–81, 2010.

60. A. K. Singh and M. Quraishi, "The effect of some bis-thiadiazole derivatives on the corrosion of mild steel in hydrochloric acid," *Corrosion Science*, vol. 52, pp. 1373–1385, 2010.
61. A. K. Singh and M. Quraishi, "Effect of 2,2′-benzothiazolyl disulfide on the corrosion of mild steel in acid media," *Corrosion Science*, vol. 51, pp. 2752–2760, 2009.
62. R. Solmaz, G. Kardaş, M. Çulha, B. Yazıcı, and M. Erbil, "Investigation of adsorption and inhibitive effect of 2-mercaptothiazoline on corrosion of mild steel in hydrochloric acid media," *Electrochimica Acta*, vol. 53, pp. 5941–5952, 2008.
63. R. Solmaz, E. Altunbaş, and G. Kardaş, "Adsorption and corrosion inhibition effect of 2-((5-mercapto-1,3,4-thiadiazol-2-ylimino)methyl)phenol Schiff base on mild steel," *Materials Chemistry and Physics*, vol. 125, pp. 796–801, 2011.
64. L. D. Paolinelli, T. Pérez, and S. N. Simison, "The effect of pre-corrosion and steel microstructure on inhibitor performance in CO$_2$ corrosion," *Corrosion Science*, vol. 50, pp. 2456–2464, 2008.
65. R. T. Loto, C. A. Loto, A. P. Popoola, and T. Fedotova, "Inhibition effect of 2-amino-5-ethyl-1,3,4-thiadiazole on corrosion behaviour of austenitic stainless steel type 304 in dilute HCl solution," *Journal of Central South University*, vol. 23, pp. 258–268, 2016.
66. F. Zucchi, G. Trabanelli, and M. Fonsati, "Tetrazole derivatives as corrosion inhibitors for copper in chloride solutions," *Corrosion Science*, vol. 38, pp. 2019–2029, 1996.
67. C. Verma, M. Quraishi, and A. Singh, "5-Substituted 1*H*-tetrazoles as effective corrosion inhibitors for mild steel in 1 M hydrochloric acid," *Journal of Taibah University for Science*, vol. 10, pp. 718–733, 2016.
68. V. Dhayabaran, I. S. Lydia, J. P. Merlin, and P. Srirenganayaki, "Inhibition of corrosion of commercial mild steel in presence of tetrazole derivatives in acid medium," *Ionics*, vol. 10, pp. 123–125, 2004.
69. M. Mihit, R. Salghi, S. El Issami, L. Bazzi, B. Hammouti, and E. A. Addi, et al., "A study of tetrazoles derivatives as corrosion inhibitors of copper in nitric acid," *Pigment & Resin Technology*, vol. 35, pp. 151–157, 2006.
70. K. Khaled and M. Al-Qahtani, "The inhibitive effect of some tetrazole derivatives towards Al corrosion in acid solution: Chemical, electrochemical and theoretical studies," *Materials Chemistry and Physics*, vol. 113, pp. 150–158, 2009.
71. P. Liu, X. Fang, Y. Tang, C. Sun, and C. Yao, "Electrochemical and quantum chemical studies of 5-substituted tetrazoles as corrosion inhibitors for copper in aerated 0.5 M H$_2$SO$_4$ solution," *Materials Sciences and Applications*, vol. 2, p. 1268, 2011.
72. A. El-Askalany, S. Mostafa, K. Shalabi, A. Eid, and S. Shaaban, "Novel tetrazole-based symmetrical diselenides as corrosion inhibitors for N80 carbon steel in 1 M HCl solutions: Experimental and theoretical studies," *Journal of Molecular Liquids*, vol. 223, pp. 497–508, 2016.
73. K. Ansari, D. K. Yadav, E. E. Ebenso, and M. Quraishi, "Novel and effective pyridyl substituted 1,2,4-triazole as corrosion inhibitor for mild steel in acid solution," *International Journal of Electrochemical Science*, vol. 7, p. 4780, 2012.
74. K. Ansari, M. Quraishi, and A. Singh, "Schiff's base of pyridyl substituted triazoles as new and effective corrosion inhibitors for mild steel in hydrochloric acid solution," *Corrosion Science*, vol. 79, pp. 5–15, 2014.
75. A. Dutta, S. K. Saha, P. Banerjee, A. K. Patra, and D. Sukul, "Evaluating corrosion inhibition property of some Schiff bases for mild steel in 1 M HCl: Competitive effect of the heteroatom and stereochemical conformation of the molecule," *RSC Advances*, vol. 6, pp. 74833–74844, 2016.
76. A. Singh, K. Ansari, X. Xu, Z. Sun, A. Kumar, and Y. Lin, "An impending inhibitor useful for the oil and gas production industry: Weight loss, electrochemical, surface and quantum chemical calculation," *Scientific Reports*, vol. 7, pp. 1–17, 2017.

77. A. Dandia, S. Gupta, P. Singh, and M. Quraishi, "Ultrasound-assisted synthesis of pyrazolo[3, 4-*b*]pyridines as potential corrosion inhibitors for mild steel in 1.0 M HCl," *ACS Sustainable Chemistry & Engineering*, vol. 1, pp. 1303–1310, 2013.

78. M. A. Quraishi, "2-Amino-3,5-dicarbonitrile-6-thio-pyridines: New and effective corrosion inhibitors for mild steel in 1 M HCl," *Industrial & Engineering Chemistry Research*, vol. 53, pp. 2851–2859, 2014.

79. K. R. Ansari and M. A. Quraishi, "Experimental and quantum chemical evaluation of dihydropyridine derivative as environmental benign corrosion inhibitor for mild steel in 15% HCl," *Analytical & Bioanalytical Electrochemistry*, vol. 7, pp. 509–522, 2015.

80. V. Reznik, V. Akamsin, Y. P. Khodyrev, R. Galiakberov, Y. Y. Efremov, and L. Tiwari, "Mercaptopyrimidines as inhibitors of carbon dioxide corrosion of iron," *Corrosion Science*, vol. 50, pp. 392–403, 2008.

81. M. Abdallah, "Rhodanine azosulpha drugs as corrosion inhibitors for corrosion of 304 stainless steel in hydrochloric acid solution," *Corrosion Science*, vol. 44, pp. 717–728, 2002.

82. G. Elewady, "Pyrimidine derivatives as corrosion inhibitors for carbon-steel in 2 M hydrochloric acid solution," *International Journal of Electrochemical Science*, vol. 3, p. 1149, 2008.

83. M. Abdallah, E. Helal, and A. Fouda, "Aminopyrimidine derivatives as inhibitors for corrosion of 1018 carbon steel in nitric acid solution," *Corrosion Science*, vol. 48, pp. 1639–1654, 2006.

84. T. Eicher, S. Hauptmann, and A. Speicher, *The Chemistry of Heterocycles: Structures, Reactions, Synthesis, and Applications*, John Wiley & Sons, Inc., 2013.

85. M. Migahed and I. Nassar, "Corrosion inhibition of tubing steel during acidization of oil and gas wells," *Electrochimica Acta*, vol. 53, pp. 2877–2882, 2008.

86. S. John and A. Joseph, "Effective inhibition of mild steel corrosion in 1 M hydrochloric acid using substituted triazines: An experimental and theoretical study," *RSC Advances*, vol. 2, pp. 9944–9951, 2012.

87. A. Singh, K. Ansari, J. Haque, P. Dohare, H. Lgaz, and R. Salghi, et al., "Effect of electron donating functional groups on corrosion inhibition of mild steel in hydrochloric acid: Experimental and quantum chemical study," *Journal of the Taiwan Institute of Chemical Engineers*, vol. 82, pp. 233–251, 2018.

88. S. H. Yoo, Y. W. Kim, J. Shin, N. K. Kim, and J. S. Kim, "Effects of the chain length of tris(carboxyalkylamino)triazine on corrosion inhibition properties," *Bulletin of the Korean Chemical Society*, vol. 36, pp. 346–355, 2015.

89. D. Zhang, L. Li, L. Cao, N. Yang, and C. Huang, "Studies of corrosion inhibitors for zinc–manganese batteries: Quinoline quaternary ammonium phenolates," *Corrosion Science*, vol. 43, pp. 1627–1636, 2001.

90. E. E. Ebenso, I. B. Obot, and L. Murulana, "Quinoline and its derivatives as effective corrosion inhibitors for mild steel in acidic medium," *International Journal of Electrochemical Science*, vol. 5, pp. 1574–1586, 2010.

91. E. E. Ebenso, M. M. Kabanda, T. Arslan, M. Saracoglu, F. Kandemirli, and L. C. Murulana, et al., "Quantum chemical investigations on quinoline derivatives as effective corrosion inhibitors for mild steel in acidic medium," *International Journal of Electrochemical Science*, vol. 7, pp. 5643–5676, 2012.

92. M. Abdel-Aal and M. Morad, "Inhibiting effects of some quinolines and organic phosphonium compounds on corrosion of mild steel in 3 M HCl solution and their adsorption characteristics," *British Corrosion Journal*, vol. 36, pp. 253–260, 2001.

93. D. Wang, D. Yang, D. Zhang, K. Li, L. Gao, and T. Lin, "Electrochemical and DFT studies of quinoline derivatives on corrosion inhibition of AA5052 aluminium alloy in NaCl solution," *Applied Surface Science*, vol. 357, pp. 2176–2183, 2015.

94. A. Samide, B. Tutunaru, A. Dobrițescu, P. Ilea, A.-C. Vladu, and C. Tigae, "Electrochemical and theoretical study of metronidazole drug as inhibitor for copper corrosion in hydrochloric acid solution," *International Journal of Electrochemical Science*, vol. 11, p. e5534, 2016.

95. A. K. Singh and M. Quraishi, "Adsorption properties and inhibition of mild steel corrosion in hydrochloric acid solution by ceftobiprole," *Journal of Applied Electrochemistry*, vol. 41, pp. 7–18, 2011.

96. G. Quartarone, L. Ronchin, A. Vavasori, C. Tortato, and L. Bonaldo, "Inhibitive action of gramine towards corrosion of mild steel in deaerated 1.0 M hydrochloric acid solutions," *Corrosion Science*, vol. 64, pp. 82–89, 2012.

97. J. Aldana-Gonzalez, A. Espinoza-Vazquez, M. Romero-Romo, J. Uruchurtu-Chavarin, and M. Palomar-Pardave, "Electrochemical evaluation of cephalothin as corrosion inhibitor for API 5L X52 steel immersed in an acid medium," *Arabian Journal of Chemistry*, vol. 12, pp. 3244–3253, 2019.

98. C. Verma, M. Quraishi, and N. K. Gupta, "2-(4-{[4-Methyl-6-(1-Methyl-1H-1,3-benzodiazol-2-yl)-2-propyl-1H-1,3-benzodiazol-1-yl]methyl}phenyl) benzoic acid as green corrosion inhibitor for mild steel in 1 M hydrochloric acid," *Ain Shams Engineering Journal*, vol. 9, pp. 1225–1233, 2018.

99. A. Singh, Y. Lin, I. Obot, E. E. Ebenso, K. Ansari, and M. Quraishi, "Corrosion mitigation of J55 steel in 3.5% NaCl solution by a macrocyclic inhibitor," *Applied Surface Science*, vol. 356, pp. 341–347, 2015.

100. A. Singh, Y. Lin, K. Ansari, M. Quraishi, E. E. Ebenso, and S. Chen, et al., "Electrochemical and surface studies of some porphines as corrosion inhibitor for J55 steel in sweet corrosion environment," *Applied Surface Science*, vol. 359, pp. 331–339, 2015.

101. A. Singh, M. Talha, X. Xu, Z. Sun, and Y. Lin, "Heterocyclic corrosion inhibitors for J55 steel in a sweet corrosive medium," *ACS Omega*, vol. 2, pp. 8177–8186, 2017.

102. D. S. Chauhan, M. Quraishi, and A. Qurashi, "Recent trends in environmentally sustainable sweet corrosion inhibitors," *Journal of Molecular Liquids*, vol. 326, p. 115117, 2021.

103. I. Onyeachu, D. Chauhan, K. Ansari, I. Obot, M. Quraishi, and A. H. Alamri, "Hexamethylene-1, 6-bis(N-d-glucopyranosylamine) as a novel corrosion inhibitor for oil and gas industry: Electrochemical and computational analysis," *New Journal of Chemistry*, vol. 43, pp. 7282–7293, 2019.

104. K. Ansari, D. S. Chauhan, M. Quraishi, M. A. Mazumder, and A. Singh, "Chitosan Schiff base: An environmentally benign biological macromolecule as a new corrosion inhibitor for oil & gas industries," *International Journal of Biological Macromolecules*, vol. 144, pp. 305–315, 2020.

105. G. Cui, J. Guo, Y. Zhang, Q. Zhao, S. Fu, and T. Han, et al., "Chitosan oligosaccharide derivatives as green corrosion inhibitors for P110 steel in a carbon-dioxide-saturated chloride solution," *Carbohydrate Polymers*, vol. 203, pp. 386–395, 2019.

19 Modern Approaches of Sustainable Corrosion Inhibition Using Heterocyclic Compounds

Abhinay Thakur[1] and Ashish Kumar[2]
[1]Department of Chemistry, Faculty of
Technology and Science, Lovely Professional
University, Phagwara, Punjab, India
[2]NCE, Department of Science and Technology,
Government of Bihar, Bihar, India

19.1 INTRODUCTION

19.1.1 GENERAL CORROSION AND CORROSION INHIBITION

Corrosion is a natural process that occurs when metal materials react with their environment, leading to metal degradation and loss over time [1–5]. The mechanism of corrosion in metals can be described as a series of electrochemical reactions that occur when the metal interacts with an environment that is chemically different from the metal. These reactions lead in the formation of metal ions and corrosion products, which can cause the degradation of the metal over time. There are several different kinds of corrosion, including pitting corrosion, uniform corrosion, crevice corrosion, and galvanic corrosion. Uniform corrosion occurs when the entire substrate of the metal is subjected to a corrosive environment and corrosion products form evenly over the entire substrate. Pitting corrosion occurs when small, localized areas of the metal substrate become corroded, forming pits in the metal [6–8]. Crevice corrosion induces when a corrosive environment is trapped in small crevices on the metal substrate, leading to localized corrosion. When two distinct metals are in touch with one another in a corrosive environment, one of the metals functions as an anode and corrodes preferentially to the other metal. This is known as galvanic corrosion. This process can have serious consequences and result in incidents that can pose significant risks to people and the environment. One of the most well-known incidents of corrosion was the implosion of the Silver Bridge in West Virginia in 1967, which killed 46 people. The cause of the collapse was determined to be the corrosion of a critical support member, leading to its failure [9–12]. Another example of a corrosion-related incident was the Deepwater Horizon oil spill in 2010, which was caused by the failure of a blowout preventer due to corrosion and metal loss. The spill resulted in widespread environmental damage and economic losses [13–15].

DOI: 10.1201/9781003377016-19

Corrosion can also result in incidents in the transportation sector, such as train derailments caused by rail corrosion, and aviation incidents caused by corrosion in aircraft structures. In addition, corrosion can result in failures in power plants and other critical infrastructure, leading to outages and disruptions. The cost of corrosion is a major concern for industries and governments worldwide, with billions of dollars lost every year as a consequence of corrosion damage. According to estimates by the National Association of Corrosion Engineers (NACE), the cost of corrosion (COC) in the United States alone is estimated to be approximately $276 billion annually. In the gas and oil industry, corrosion can lead to reduced production and decreased efficiency, as well as increased maintenance costs [16, 17]. The COC in pipelines alone is predicted to be in the range of several billion dollars per year. In the transportation sector, corrosion can result in increased maintenance costs, reduced efficiency, and safety concerns. For example, the COC in the US transportation sector is predicted to be approximately $8 billion per year. In the construction industry, corrosion can result in reduced structural integrity, increased maintenance costs, and potential safety hazards [18, 19]. The COC in buildings and infrastructure is predicted to be several billion dollars per year. Many different types of metal materials can be affected by corrosion, with varying degrees of severity. The following are some of the most common metals affected by corrosion:

- *Iron:* Iron is amongst the most widely utilized metals in the world and is also one of the most susceptible to corrosion. Iron corrodes when it reacts with water and oxygen to form rust, a process that weakens the metal and can lead to structural failure.
- *Aluminium:* Al is widely used in the transportation and construction industries owing to its lightweight and high-strength properties. However, Al is susceptible to corrosion when exposed to certain environments, such as saltwater or industrial pollutants.
- *Copper:* Copper is widely used in electrical wiring, plumbing, and other applications due to its high conductivity and resistance to corrosion. However, copper can still corrode when exposed to certain conditions, such as acid rain or high humidity.
- *Stainless Steel:* SS is broadly utilized in a variety of applications due to its resistance to corrosion. However, SS can still corrode in certain environments, such as high temperatures and high chloride-containing environments.

Corrosion can occur as a result of several different factors, including exposure to moisture, high temperatures, high levels of pollutants, and improper maintenance. The rate and severity of corrosion can vary widely depending on the type of metal, the environment in which it is exposed, and other factors. The effects of corrosion can range from cosmetic damage to structural failure, relying on the severity and period of the corrosion process. Some of the most serious adverse effects of corrosion include reduced mechanical strength, decreased functionality, and increased risk of failure or collapse. To mitigate the adverse effects of corrosion, various techniques have been developed, including the use of corrosion inhibitors (CIs) [20–30].

CIs are substances that mitigates or stop the corrosion mechanism by forming a protective film on metal substrates. Traditional CIs, such as inorganic compounds and organic compounds, have been widely used for many years to inhibit corrosion in metal materials. However, these inhibitors have several drawbacks that have led to increased interest in green and environment-friendly alternatives. The following are some of the key drawbacks of traditional CIs:

- *Toxicity:* Many traditional CIs contain toxic compounds that can be harmful to human health and the environment.
- *Environmental Impact:* Traditional CIs can leach into the environment and cause damage to ecosystems, such as waterways and soil.
- *High Cost:* Traditional CIs can be expensive, especially those that require special handling and disposal due to their toxicity.
- *Limited Performance:* Some traditional CIs can be ineffective in certain environments, such as high temperature or high chloride-containing environments.
- *Compatibility Issues:* Some traditional CIs can cause compatibility issues with other materials, such as rubber and plastics, leading to reduced performance and increased maintenance costs.

However, in recent era, there has been an increasing demand in developing environment-friendly, or "green," CIs. Green inhibitors are derived from natural or renewable resources and are biodegradable, non-toxic, and safe for the environment. Plant extracts, amino acids, and carbohydrates are examples of green corrosion inhibitors (GCIs) that can effectively inhibit corrosion in metal materials [31–36]. Plant extracts, such as extracts from spices and herbs, contain compounds that can interfere with the corrosion process and slow the rate of corrosion. Amino acids, such as lysine and histidine, have also been shown to be effective CIs owing to their capability to develop protective layers on metal substrates. In addition, carbohydrates, such as sugars and starches, have been shown to effectively inhibit corrosion by producing a protective layer on metal substrates and reducing the reaction of metal with the environment. For instance, in an experiment, as a CI for CS in a 1.0 M HCl media at various temperatures, Mahfoud et al. [37] researched the methylene dichloride extract of *Lamium flexuosum*. The outcome showed the studied extract's effective inhibition, which improved with rising dosage and declined as temperature raised. At 900 ppm, this compound was an effective mixed-type CIs, with an efficacy value of 83.50%. It was discovered that the extract's adsorption over the CS substrate followed the Freundlich isotherm in a physisorption manner. AFM study of the steel's substrate topology reveals that the acquired average variation was 40.42 nm in the devoid of the extract, but extensive troughs and spikes with a mean roughness of 451.11 nm were seen in the acidic medium. The inclusion of the studied extract (LFMDE), though, resulted in significantly less corrosion impacts with tiny spikes at 90.91 nm, showing that the binding of a micro-fine layer over the CS substrate inhibited the invasion by an HCL media.

Similarly, Ravi Deyab et al. [38] investigated *Equisetum arvense* aerial part (EAAP) extract as CIs for steel-based components in multi-stage flash (MSF) sections during acid treatment. EAAP extract occupied nearly 97% of the CS substrate in 2 M HCl media as contrasted to the sample subjected to blank media, and the CR was decreased to 00.58 ± 0.02 μg cm^{-2} h^{-1} at 300 mg L^{-1}. On the substrate of CS, EAAP extract typically has a mixed effect on the cathodic and anodic regions. Quantum chemical estimations relied on the DFT model were carried out to identify the molecular functionality of the substances under discussion. As a consequence, Figure 19.1 shows the associated chemical characteristics of four components. The HOMO orbital often reflects a molecule's capability to donate electrons, whereas the LUMO orbital usually reflects a molecule's capability to accept electrons. It was discovered that the caftaric acid's LUMO and HOMO electron cloud is virtually exactly situated at the aromatic ring unit. This condition can be compared to other elements. This proves that these active adsorption spots were capable of exchanging electrons for metals and creating covalent links. In addition, the blue and red regions of the electrostatic potential (ESP) maps, which show electrophilic and nucleophilic character, correspondingly, the hydroxyl units are primarily where the red colour is concentrated. This suggests that while binding at the steel substrate, the red regions were essentially the primary active adsorption locations.

When the system's energy and temperature levels were equal, the dynamics phase was finished, and the overall system attained stability. Figure 19.2 shows the lower energy adsorption geometries of four components adsorbed over the substrate of Fe(1 1 0). To maximize substrate distribution and interaction and to create a strong bond between the substrate and adsorbate states, the inhibitor molecules were virtually flatly adsorbed on the substrate platform. The mean adsorption energy of the established equilibria geometries served as the basis for the calculation of the adsorption energies in this study. The estimated E_{ads} values for

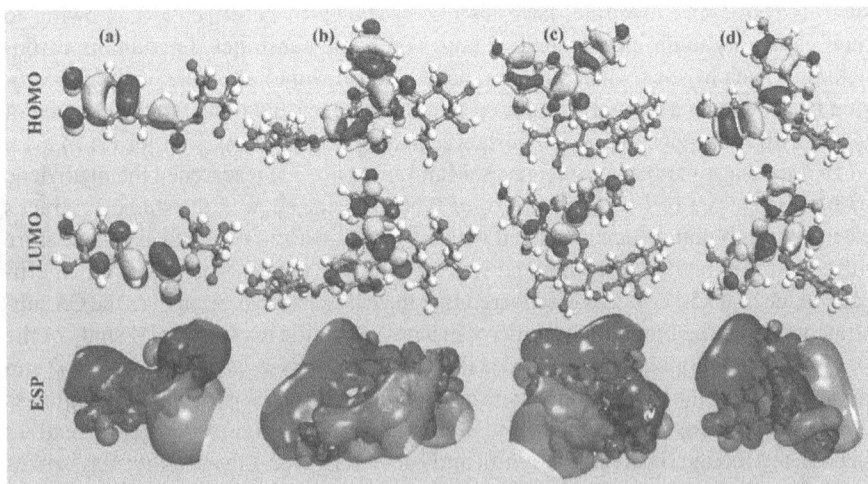

FIGURE 19.1 Representation of the HOMOs, LUMOs, and ESP for the constituents under investigation: (a) caftaric acid, (b) kaempferol-3,7-di-O-glucoside, (c) kaempferol-3-O-rutinoside, and (d) isoquercetin. (Adapted with permission from Ref. [38] under CCBY 4.0.)

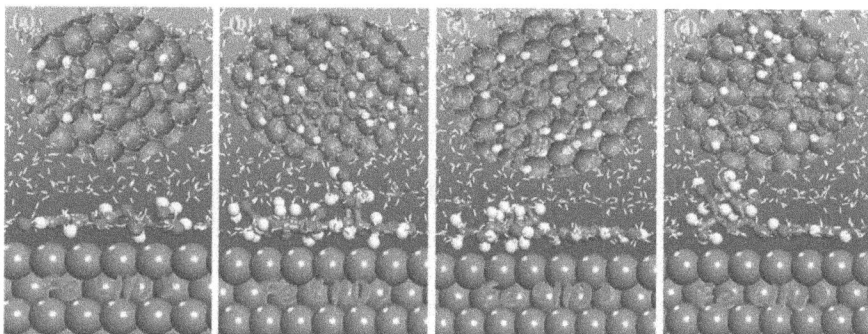

FIGURE 19.2 Views from the top and side of the four components' most persistent adsorption arrangements on the substrate of Fe(1 1 0): (a) caftaric acid, (b) kaempferol-3,7-di-O-glucoside, (c) kaempferol-3-O-rutinoside, and (d) isoquercetin. (Adapted with permission from Ref. [38] under CCBY 4.0.)

kaempferol-3,7-di-O-glucoside, caffeic acid, isoquercetin, and kaempferol-3-O-rutinoside tend to be 1107.3, 637.1, 810.6, and 1162.9 kJ mol^{-1}, correspondingly. We may observe that the sorption energies are negative, and we may therefore anticipate impulsive adsorption. In principle, the firmer the engagement among the inhibitors and metallic substrate, the larger the absolute value of E_{ads}. It is clear that kaempferol-3-O-rutinoside and kaempferol-3,7-di-O-glucoside possess greater absolute values of E_{ads} than isoquercetin and caftaric acid, suggesting that these might be key players in the mechanism of corrosion protection.

Similarly, by utilizing mass spectrometry and gas chromatography (GC) in conjunction, Zaher et al. [39] assessed the chemical components of the methanolic extract of *Ammi visnaga umbels* (AVU) and utilized for inhibiting corrosion of CS in 1.0 mol L^{-1} HCl utilizing electrochemical reaction methods. Binapacryl (4.32%), Khellin (1.97%), Edulisin III (72.88%), and Visnagin (1.65%) were among the 46 compounds that were discovered, accounting for 89.89% of the AVU extract's entire chemical components. The examined extract effectively prevents CS from corroding in 1.0 mol L^{-1} HCl media, according to various methodologies. The inhibiting effect of AVU extract attained an IE% of 84% at a minimal concentration of 700 ppm. Polarization studies revealed that the examined inhibitor functioned as a blended inhibitor, mitigating both cathodic and anodic corrosion processes. The EIS analyses show that the double layer dropped and the polarization resistance improved when AVU extract was added to HCl media. A shielded CS substrate was visible in the inhibited media on SEM pictures. The study's findings additionally supported the finding that employing plant-based CIs to preserve metals and alloys had considerable advantages.

As observed, the use of GCIs has several benefits over traditional inhibitors, including reduced toxicity and environmental impact, lower cost, and improved performance. In addition, green inhibitors are often derived from renewable resources and are biodegradable, making them a more sustainable and environment-friendly option for corrosion inhibition. GCIs have been shown to be highly effective in controlling corrosion, particularly in applications such as pipelines, storage tanks, and marine structures. In addition to their effectiveness, green inhibitors also offer several other benefits, such as improved safety, reduced environmental impact, and lower costs compared to traditional inhibitors.

19.2   GREEN CORROSION INHIBITORS

Green corrosion inhibitors are substances that are utilized to prevent the corrosion of metal substrates but do so in an environment-friendly manner. They are considered "green" because they are non-toxic, biodegradable, and have minimal impact on the environment. Corrosion is a natural process that occurs when metal reacts with its environment, leading to the gradual degradation of the metal's properties and, eventually, its structural integrity. This can result in costly and dangerous failures of metal structures, such as pipelines, bridges, and buildings [40–42]. To prevent corrosion, inhibitors can be added to the environment in which the metal is exposed. These inhibitors alter the corrosion reaction in such a way as to reduce or stop it altogether. They work by forming a protective layer over the metal substrate that slows down the reaction rate, or by changing the environment itself, making it less corrosive. Traditional CIs, such as chromates and phosphates, have been used for decades. However, these substances have been shown to have negative impacts on the environment, such as toxicity and biodegradability issues, and are being phased out in many countries. This has emerged the finding for alternative, environment-friendly inhibitors [43, 44]. GCIs include organic compounds, such as plant extracts, and inorganic compounds, such as nitrates and silicates.

Organic inhibitors are often attained from natural sources, such as plants, and are biodegradable. They work by adsorbing onto the metal substrate and developing a barrier that slows down the corrosion reaction. Inorganic inhibitors, on the other hand, work by changing the environment around the metal, making it less corrosive. One of the most commonly used green inhibitors is citric acid. This organic inhibitor is derived from citrus fruits and is non-toxic, biodegradable, and has a low environmental impact. Citric acid works by forming a protective layer over the metal substrate that slows down the corrosion reaction. It is often used in aqueous environments, such as boilers and cooling systems, where it is added to the water to prevent corrosion. Another popular green inhibitor is vinegar, which is a solution of acetic acid in water. Vinegar is commonly used in household cleaning and is an effective CIs, especially in acidic environments. The acetic acid in vinegar acts as a mild acid, slowing down the corrosion reaction and developing a protective layer on the metal substrate. Inorganic inhibitors, such as nitrates and silicates, are also commonly used as green inhibitors [45–48].

Nitrates are effective in neutral to slightly acidic environments and work by changing the environment around the metal, making it less corrosive. Silicates, on the other hand, are effective in neutral to alkaline environments and work by forming a protective film on the metal substrate. Recently, several studies have demonstrated the effective utilization of natural sources as CIs for preventing metallic corrosion such as methanolic extract of *Spilanthes uliginosa* leaves [49] revealed IE% of 90% at 500 ppm conc., aqueous extract of *Thevetia peruviana* [50] showed maximum

IE% of 90.3% at 300 ppm, *Pennisetum purpureum* extract [51] demonstrated highest IE% of 82.6% at 5000 ppm, *Gongronemena latifolium* extract [52] exhibited 91.24% of IE% at 1000 ppm, *Caulerpa prolifera* extract [53] showed a maximum IE% of 94.33% at 1.0 g L^{-1}, *Chromolaena odorata* extract [54] attained an optimum IE% of 92.39% at 4.0 g L^{-1}, *Chinese Yam Peel* extracts (CYPE) [55] exhibited 96.33% of IE% using 900 mg L^{-1} CYPE in artificial seawater at 298 K, and *Sugarcane purple rind* ethanolic extract (SPRE) [56] revealed a maximum IE% of 96.2% for CS having 800 ppm SPRE at ambient temperature. These compounds primarily contain polar groups like $-OH$, $-NH_2$, $-SH$, and $-COOH$ or several heteroatoms, including O, S, or N. Additionally, the molecular mass of inhibitor molecules with the aid of a number of polar groups and heteroatoms is a critical factor in determining their excellent corrosion prevention characteristics. Thus, one of the main objectives of corrosion researchers and the scientific community is the production of revolutionary, environment-friendly CIs generated from natural extract.

In conclusion, GCIs are essential for preventing the corrosion of metal substrates in an environment-friendly manner. They offer a safer, more sustainable alternative to traditional inhibitors, which have been shown to have negative impacts on the environment. The use of green inhibitors is becoming increasingly popular and is expected to continue to grow as awareness of their benefits increases and as regulations continue to phase out the use of traditional inhibitors.

19.2.1 SYNTHETIC CORROSION INHIBITORS

Synthetic CIs are chemicals that are added to a system (such as metal or oil) to prevent or reduce corrosion. They are designed to protect metal substrates from various types of corrosion and are widely used in industries, including oil and gas, water treatment, marine, and construction. Synthetic CIs work by adsorbing onto the metal substrate and developing a protective covering that prevents corrosive agents from reaching the metal. This protective layer is made up of a molecular film that provides a shield between the metal and the corrosive environment, effectively reducing the rate of corrosion [57, 58]. The film is stable and durable, providing long-lasting protection even in harsh and corrosive environments. There are several distinctive kinds of synthetic inhibitors, including nitrites, phosphates, and imidazolines. Nitrite inhibitors are commonly used in oil and gas pipelines to protect against corrosion caused by oxygen. Phosphate inhibitors are used to prevent corrosion in water treatment systems, as well as in oil and gas pipelines. Imidazoline inhibitors are generally utilized in marine environments, as well as in cooling systems and pipelines, to protect against corrosion caused by acids and salts.

In order to identify the role of its active components in a 1 M HCl media, Kaban et al. [59] investigated the anti-corrosion activity of aqueous smoke from rice husk ash. In this research, EIS, PDP, and deep learning techniques were used to evaluate the technique established to evaluate, analyse, and predict the innovative kind of GCIs for C1018. The inhibitor was assumed to be a mixed-type inhibitor to attain the best prevention of 80 ppm at 323 K, achieving up to 99% IE%, according to the corrosion experiment findings. With a decreased skewness value at 0.5190 nm on the coated MS, the AFM outcomes indicate a flatter substrate. The artificial neural network (ANN) successfully models the correlation in both the EIS and PDP as well as the emergence of the passivation film on the coated MS with lesser overfitting on the impeded metal, a highly accurate prognostication of 81.08%, and a reduced C_R of 0.6001. The experimental and predicted inhibitor adsorption outcome match very adequately. The research could be applied as a blueprint to create GCI relying on empirical and artificial intelligence (AI) methods in the future.

Similarly, Uzoma et al. [60] used the weight loss (WL), EIS, and PDP methods to investigate aspartame (ASP), a naturally occurring conventionally produced substance, as a CIs for T95 CS in a 15 wt% HCl media at 60, 70, 80, and 90 °C. It was discovered that in the investigated circumstances, ASP has a corrosion-inhibiting impact. Temperature rises resulted in an improvement in inhibition effectiveness. Following four hours of exposure in 2000 ppm ASP, the WL technique showed an inhibitory efficiency of 86%. According to PDP findings, in the tested circumstances, ASP performed as a mixed-type CIs. The findings from the EDX and SEM have all verified that the inhibitory capacity of ASP was caused by sorption over the steel substrate. Thus, ASP is a potential active ingredient for the creation of CI.

For the inaugural moment, Zhao et al. [61] created two carboxymethyl chitosan derivatives and utilized them as safe, effective, and environment-friendly CIs. Synthesized compounds were effective CIs for the carbon dioxide corrosion of P110 steel in 3.6 wt% NaCl media at 80 °C, according to EIS and WL experiments. CAHC and CHC had the highest IE% of 93.95% and 87.97%, correspondingly, at a dosage of 100 mg L^{-1}. These CI successfully lessen the destruction of the corrosion media to the metal substrate, as demonstrated by corrosion specimens, the AFM and SEM analyses. Quantum chemical simulations also support the finding that CAHC exhibited a greater inhibitory capability than CHC amongst the two synthesized chitosan derivatives. These two synthetic chitosan derivatives, which are in accordance with the goal for clean, less hazardous, and ecological sustainability, successfully inhibited CO_2 corrosion. They offer significant promise as innovative and powerful CI.

Furthermore, Chen et al. [62] conducted an approach to incorporate biomass basis molecules into the N_1-(2-aminoethyl)-N_2-(2-(2-(furan-2-yl)-4,5-dihydro-1H-imidazol-1-yl) ethyl) ethane-1,2-diamine (NNED) CIs synthesis. Utilizing EIS and WL techniques, the inhibitive capability was examined. Although the dosage was just 5 mg L^{-1}, the results showed that NNED has outstanding Q235 MS inhibition ability in 1 M HCl vigorous media. In fact, the inhibition effectiveness was achieved by up to 90%. Additionally, according to the thermodynamic characteristics, the sorption of NNED upon metal substrates follows the Langmuir adsorption isotherm. Both physisorption, which is related to electrostatic contact, and chemisorption, which is due to covalent bonds, were used in the development of adsorption films, but the latter performs a more prominent influence. Additionally, PDP tests show that NNED functions as a mixed-type inhibitor and predominately blocks the electrochemical process of metallic breakdown in the anode portion. Furthermore, the bonding intensity of the substrate of the metal enhanced substantially upon the inclusion of the NNED inhibitor, which firmly illustrates the construction of the NNED layer over the metallic substrate.

The pathway/process of synthetic corrosion inhibition is relied on the adsorption of the inhibitor molecules onto the metal substrate, creating a protective coating that prevents corrosive agents from reaching the metal. The mechanism can be divided into three main stages:

- *Adsorption of Inhibitory Molecules:* The inhibitory molecules are attracted to the metal substrate due to their polar nature and the presence of electrostatic forces. They adsorb onto the metal substrate and form a thin, continuous film.
- *Formation of the Protective Layer:* The adsorbed inhibitor molecules interact with the metal substrate and each other, developing a stable and dense protective layer that reduces the rate of corrosion. The protective layer behaves as a barrier between the environment and metal, reducing the concentration of corrosive agents and limiting the reaction between the metal and the environment.
- *Inhibition of Corrosion Reaction:* The protective covering created by the adsorbed inhibitor molecules inhibits the corrosion reaction by reducing the concentration of corrosive agents and preventing them from reaching the metal substrate. The protective layer also reduces the reaction rate by altering the corrosion reaction kinetics, making it more difficult for corrosion to occur.

The effectiveness of synthetic CIs depends on several factors such as the type of inhibitor, the type of corrosion, the environment, and the concentration of the inhibitor. Synthetic inhibitors can protect against distinctive types of corrosion, such as pitting corrosion, uniform corrosion, and crevice corrosion, by changing the corrosion reaction kinetics and reducing the concentration of corrosive agents. It is crucial to point that the mechanism of synthetic corrosion inhibition is not completely understood, and there is still ongoing research in this area. However, the adsorption-based mechanism described above provides a basic understanding of how synthetic inhibitors work to prevent metal substrates from corrosion. One of the main advantages of synthetic inhibitors is their versatility. They can be tailored to protect against specific types of corrosion, such as rust or pitting, and could be utilized in a wide range of industrial applications, including pipelines, storage tanks, and marine environments. Synthetic inhibitors are also effective in corrosive environments that are too harsh for other types of inhibitors, such as organic inhibitors. Synthetic inhibitors are easy to use and can be added to a system in a variety of ways, including through direct injection, continuous dosing, or by blending with other chemicals. They are also environment-friendly, as they are biodegradable and do not contain harmful substances that can cause environmental harm. However, synthetic inhibitors also have some limitations. One of the main disadvantages is that they can be expensive compared to other types of inhibitors, such as organic inhibitors. Additionally, synthetic inhibitors can be less effective in certain environments, such as those that are highly corrosive or have extreme temperature variations. Another potential limitation is that synthetic inhibitors can cause compatibility issues with other chemicals that are present in a system. For example, synthetic inhibitors can react with other chemicals and form deposits, leading to reduced efficacy. They can also cause fouling and scaling, which can impact the performance of equipment and systems.

19.2.2 GREEN CORROSION INHIBITORS

Green corrosion inhibitors are alternative, environment-friendly, and biodegradable substances that mitigate or stop the corrosion process. They are mainly derived from renewable sources, including plant extracts, natural extracts, amino acids, and carbohydrates [63–66]. Green corrosion inhibitors can be categorized into several categories based on their source materials:

- Plant Extracts–based CIs
  Plant extracts–based CIs are attained from distinctive plant species, including leaves, stems, roots, and fruits. The active compounds in plant extracts that inhibit corrosion are mainly phenols, flavonoids, tannins, terpenoids, and alkaloids.
- Natural Extracts–based CIs
  Natural extracts–based CIs are attained from distinctive natural sources, such as seaweed, microorganisms, and minerals. The active compounds in natural extracts that inhibit corrosion are mainly polysaccharides, proteins, and minerals.
- Amino Acids–based CIs
  Amino acids–based CIs are derived from various sources, such as plant proteins and waste products. The active compounds in amino acids that inhibit corrosion are mainly amino acids and peptides.
- Carbohydrates-based CIs
  Carbohydrates-based CIs are derived from various sources, such as plant polysaccharides and waste products. The active compounds in carbohydrates that inhibit corrosion are mainly polysaccharides, such as cellulose, starch, and pectin.

GCIs form a protective layer on the substrate of the metal. This protective layer acts as a shield between the metal and the corrosive environment, slowing down or preventing the corrosion process. The mechanism of action of GCIs can be categorized into three main categories:

- Adsorption
  The mechanism of adsorption involves the physical or chemical adsorption of the inhibitor molecule onto the metal substrate. This creates a protective layer on the substrate of the metal that prevents the corrosion reaction from taking place.
- Chemical Reactions
  The mechanism of chemical reactions involves the interaction of the inhibitor molecule with the metal substrate, forming a chemical bond. This bond creates a defensive film on the substrate of the metal that reduces or prevents the corrosion process.
- Passivation
  The mechanism of passivation involves the development of an oxide layer on the substrate of the metal that prevents the corrosion reaction from taking place. The inhibitor molecule stimulates the formation of this oxide film, providing a protective covering towards corrosion.

Common sources of plant extracts used for GCIs include materials from leaves, stems, roots, and fruits of various plants, such as cinnamon, clove, basil, rosemary, thyme, and lemon. Natural extracts used for GCIs can come from a variety of sources, including seaweed, fungi, and bacteria. Common amino acids used for GCIs include lysine, histidine, and arginine. Carbohydrates used for GCIs can come from various sources such as sugar, starch, and cellulose. For example, glucose and fructose are commonly used as CIs owing to their capability to chelate with metal ions and form stable complexes.

Advantages of GCIs:

- *Environment-friendly:* GCIs are biodegradable, non-toxic, and do not contain harmful chemicals, making them safe for the environment.
- *Effective Protection:* GCIs are effective in preventing corrosion in various metal substrates and offer long-term protection.
- *Cost-effective:* In comparison to traditional CIs, GCIs are cost-effective as they do not require special disposal methods and are biodegradable.
- *Wide Range of Applications:* GCIs could be used in a broad array of applications, including the oil and gas industry, marine industries, and more.

According to a report by Markets and Markets, the global GCIs market was valued at US$2.64 billion in 2018 and is projected to reach US$4.13 billion by 2023, growing at a CAGR of 9% from 2018 to 2023. The report also mentions that the increasing demand for eco-friendly and biodegradable products, along with stringent environmental regulations, is driving the growth of the GCI's market. The use of GCIs is particularly high in the marine industries due to the need for environment-friendly products in this sector. In the following section, we will be discussing the utilization of plant extracts, amino acids, and carbohydrates along with their derivatives as CIs for several metals and alloys.

19.3 SUSTAINABLE APPROACHES TO GREEN CORROSION INHIBITORS

19.3.1 UTILIZATION OF NATURAL PRODUCTS

19.3.1.1 Utilization of Plant Extracts

Plant extracts have been shown to have the potential as CIs for metal substrates. These extracts contain active compounds that can prevent the oxidation and corrosion of metal substrates by developing a protective covering on the substrate or by reacting with corrosive agents to reduce their effectiveness. Some commonly used plant extracts for corrosion inhibition include clove oil, eucalyptus oil, thyme oil, cinnamon oil, and rosemary oil. The effectiveness of these plant extracts as CIs relies on various factors, including the type of metal, the corrosive environment, and the concentration of the extract used. Additionally, some plant extracts may also have biodegradable properties, making them environment-friendly alternatives to traditional CIs. The following are the advantages of using plant extracts as CIs:

- *Environment-friendly:* Plant extracts are biodegradable and have low toxicity, making them an environment-friendly alternative to traditional CIs.
- *Wide Availability:* Plant extracts are readily available and can be easily sourced from natural resources.
- *Cost-effective:* Compared to traditional CIs, plant extracts are relatively inexpensive and accessible.
- *Ease of Application:* Plant extracts can be easily applied to metal substrates and do not require special handling or storage conditions.
- *Versatility:* Plant extracts can be used in various corrosive environments and can be effective against different types of corrosion.
- *Improved Performance:* Some plant extracts have been found to have superior corrosion inhibition performance compared to traditional inhibitors.
- *Green Chemistry:* The utilization of plant extracts as CIs promotes the principles of green chemistry and sustainable development.

In an experiment, using a solvent extraction procedure, Chapagain et al. [67] was able to extract the alkaloid from *Rhynchostylis retusa* (*RR*), which was then verified using biochemical and FTIR spectroscopy testing. In a medium of 1.0 M H_2SO_4, extracted alkaloids were evaluated as environment-friendly CIs for MS. The WL and EIS techniques were used to examine the IE% of RR alkaloid extracts. Alkaloids possess nitrogen in their composition, and there is a possibility that oxygen could also be present as a functional component. Figure 19.3a and b shows *RR* with alkaloids in its rhizome, leaflets, and flowers. The maximal IE% in the WL analysis was 87.51% in a 1000 ppm media at a 6-hour soaking duration, according to the findings. The findings of PDP showed that the extracted alkaloids worked as a blended type of inhibitor. The specimen was submerged for 6 hours, and the IE determined using the

FIGURE 19.3   (a) *Rhynchostylis retusa.* (b) Rhizome, leaflet, and flower. (Adapted with permission from Ref. [67] under CCBY 4.0.)

PDP method was 93.24%. According to the research on the impacts of temperature, *RR* was only effective below 35 °C. Therefore, in an acidic media under 35 °C, alkaloids of *RR* could be effectively recovered and utilized as CI for MS.

Similarly, alkaloids were effectively isolated by Thapa et al. [68] from the stem of *Solanum xanthocarpum*, and they were identified using qualitative chemical analyses and spectroscopy examinations and used as MS's GCI. The WL and EIS measuring techniques were utilized to evaluate the inhibitor's potency. The maximal IE% of 93.14% was attained based on the WL assessment. The inhibitor might function up to a temperature of 58 °C, according to the temperature impact experiment. According to the EIS analysis, the alkaloids act as a mixed-type GCI and efficiently prevent up to 98.14% of corrosion. An analysis of the kinetic variables shows that the inhibitor might act as a shield to prevent corrosion over an MS substrate. According to the results of the analysis of the thermodynamic variables, the mechanism was endothermic and instantaneous. The results show that *S. xanthocarpum*'s alkaloids could be highly effective, environmentally sustainable, and potentially GCI towards MS corrosion. By using the WL technique, Rubaye et al. [69] investigated the *Cherry Sticks* as a GCI. At temperatures of 30 °C, 40 °C, 50 °C, and 60 °C, the effect of temperature on consuming behaviour in the presence of an inhibitor at a dosage of 50×10^{-2} g L^{-1} was investigated. The findings show that under all tested circumstances, the green inhibitor CSCI (Cherry Sticks CIs) ensures the inhibitive effects on the deterioration of MS. The eco-friendly IE% increased as dosage improves, reaching an optimum of 89.5% at 50×10^{-2} g L^{-1}, and its sorption over MS substrate follows Langmuir isotherm.

Morales et al. [70] revealed the fabrication and utilization of eco-benign CI obtained from plant waste. An environmentally sustainable CIs was made using natural gums like latex and xanthan gum from the "lechero" plant (*Euphorbia laurifolia*), as well as pectin recovered from King mandarins (*Citrus nobilis* L.) and Tahiti limes (*Citrus latifolia*). A 23 factorial model was used to discover the best temperature, pH, and duration permutations for pectin isolation and purification, and the results were assessed based on the pectin output. At 85 °C, pH = 1, and 2 hours, the maximum pectin extraction outputs (38.10% and 41.20% from lime and King mandarin, correspondingly) were obtained. Because more inhibitory molecules were able to bind to the metallic substrate as a result of the introduction of xanthan gum to pectin (formulation % pectin-50% xanthan gum), the corrosion inhibitory effectiveness increased from 29.20% to 78.21% at 400 ppm.

Kamran et al. [71] researched the economical emulsion polymerization process for the production of eco-friendly *Prunus domestica* gum grafted polyaniline (PDG-g-PANI) composite, which was used as an excellent anti-corrosion substance for SS and MS in corrosive environments. A bilayered arrangement comprising a primary permeable coating of PANI covered by a fibrous coating of PDG was visible in SEM pictures. PDG-g-PANI was shown to dissolve in ethanol, acetone, propanol, chloroform, butanol, *N*-methyl-2-pyrrolidone, dimethyl sulfoxide, and a combination of chloroform and propanol in the solubility experiment. The corrosion-resistant function of the composite on MS and SS in 3.5% NaCl and 1 M H_2SO_4 media was investigated using the WL, EIS, and PDP analyses. In a 3.5% NaCl media, the PDG-g-PANI-coated MS had a corrosion prevention effectiveness of 96% as contrasted to

43% and 86% for uncoated PDG and PANI, respectively. The PDG-g-PANI-coated SS demonstrated a corrosion IE% of 98%. Additionally, PDG-g-PANI-coated SS and MS exhibit 96.6% and 99% corrosion prevention, respectively, in 1 M H_2SO_4 media. According to WL measurements, PDG-g-PANI covering may shield MS up to 93% for 14 days in salt media, whereas 97% of its ability to prevent corrosion was maintained for two months outside.

19.3.2 UTILIZATION OF AMINO ACIDS

Amino acids can be utilized in corrosion mitigation as CIs. The basic mechanism of inhibition involves the adsorption of amino acid molecules on the metal substrate, producing a protective film that prevents further corrosion. The inhibitive properties of amino acids could be attributed to their functional groups, including carboxyl, amine, and hydroxyl groups. They are effective CIs for various metals in distinctive environments, including acidic, neutral, and alkaline solutions. However, the efficacy of amino acids as CIs is dependent on factors such as concentration, pH, temperature, and the presence of other ions. In an experiment, using FT-IR and Raman (RS) spectroscopic techniques, Swiech et al. [72] investigated the structural alterations of tryptophan (Trp) upon the damaged 316 L SS substrate achieved under regulated modelled inflammatory circumstances. Utilizing the PDP methodology in phosphate-buffered saline (PBS) media acidified to pH 3.0 at 37 °C in the absence and presence of 10^{-2} M Trp, with varying soaking periods, the corrosion behavior and preventive effectiveness of the studied specimens were assessed (2 and 24 hours). The protonated nitrogen atom and the amine group of the indole ring that takes a mostly tilted configuration were the primary mechanisms by which the amino acid was deposited over the degraded SS substrate. It was inferred that a more hydrophobic condition was present based on the discernible variations in the Fermi doublet's intensity after Trp was adsorbed onto the damaged SS substrate. Following 2 hours as opposed to 24 hours of exposition could lead to an enhancement in corrosion resistance. The electrochemical analyses support this claim, showing that following only a brief soaking, Trp, functioning as a mixed-type inhibitor, significantly increases its IE%.

Similarly, using Monte Carlo simulations (MCS) modelling, Kasprzhitskii et al. [73] assessed the inhibiting impact of L-amino acids (AAs) having various side chain lengths on Fe(1 0 0) substrates. The adsorption behavior of AAs upon the Fe substrate was described quantitatively and qualitatively. According to studies, L-amino acid adsorption energy increases in absolute terms with side chain lengthening and was also influenced by the existence of heteroatoms. The equilibria low-energy configurations of L-amino acids deposited over the Fe(1 0 0) substrate in the gaseous state that was discovered using the MCS are shown in Figure 19.4. As a consequence of adsorption, it could be observed that all of the inhibitor molecules are orientated orthogonal to the metallic substrate. This suggests increased L-amino acid molecular layer durability, as seen by the involvement of more L-amino acid atomic centres engaging with the Fe substrate. In this instance, the side chain amino acids' lengthening would result in a greater attraction for the Fe substrate. This suggests a greater capacity to adequately shield the Fe substrate from the impacts of

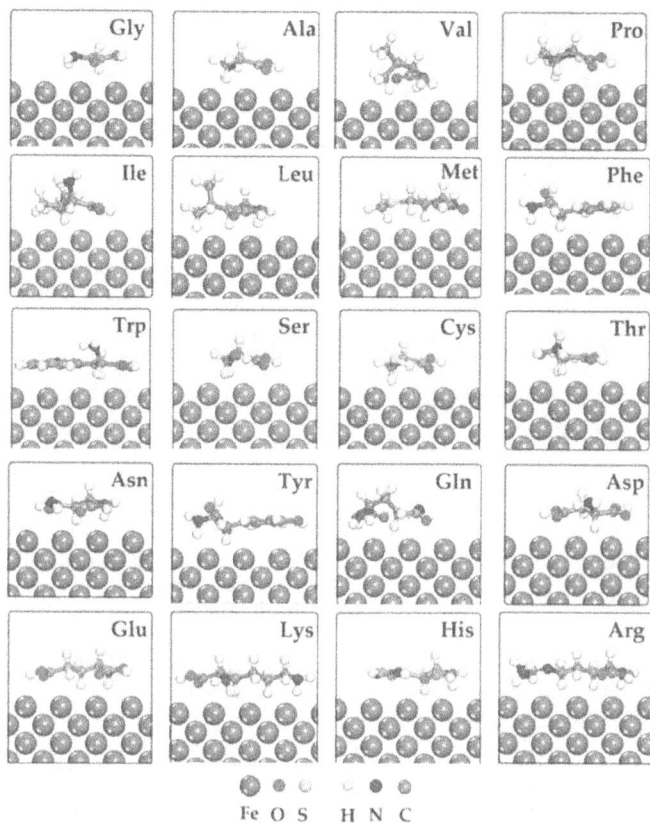

FIGURE 19.4 The side view of the L-amino acid adsorption geometries with the lowest energy predicted by the MCS on the substrate of Fe(1 0 0). (Adapted with permission from Ref. [73] under CCBY 4.0.)

harsh surroundings. In this instance, the kind of side chain significantly impacts the adsorption layer's shape and the type of interactions with the Fe substrate. According to the side chain categorization, the maximal absolute value of the adsorption energy AAs over the Fe substrate grows in the following order: Trp (nonpolar), Glu (acidic), Gln (polar), and Arg (basic).

In order to investigate the effects of the diethyl(4-methylphenyl)-N-(phenyl) aminomethylphosphonate (AP) compounds on the CS corrosion in 0.5M H_2SO_4 media at room temperature, Thoume et al. [74] synthesized, described, and examined them. Electrochemical methods were used to assess the inhibitory efficiency. PDP curves illustrated that the AP inhibitor behaves as a mixed inhibitor in 0.5 M H_2SO_4 with a predominately anodic effect. With an increment in inhibitor dosage, the IE% rises. At a dosage of 1.14×10^{-3} M, the IE% approaches a value of 91%. The synergism, which was caused by a cooperative process involving the anion iodide and the cation AP^+, causes the inhibitory ability of AP to enhance with the presence of KI. The Langmuir isotherm governs the sorption of AP over the CS substrate. The ΔGo_{ads} favour mixed adsorption. The predicted inhibition effectiveness is well correlated across empirical and theoretical research. The heterogeneous areas that the oxygen and nitrogen, along the aromatic rings in the composition, are the primary centres of adsorption are shown by computed areas on the MESP. With a molecular plan that is almost perpendicular to the Fe substrate, the molecule AP can accumulate over Fe(1 0 0). In 0.5 M H_2SO_4, the diethyl(4-methylphenyl)-N-(phenyl) aminomethylphosphonate effectively inhibits CS corrosion.

With outstanding quantities, Mazumder et al. [75] produced zwitterionic monomers, N,N'-diallylamino propanephosphonate, and leftover N,N'-diallyl-L-methionine hydrochloride. These monomers were used to create the co-cyclopoly-mers 5–7 and zwitterionic homo in an aquatic mixture with the help of the initiator 2,2′-azobis(2-methylpropionamidine)dihydrochloride. CIs 5–7 were reported to have maximal IE% of 85.2%, 83.3%, and 99.5%, correspondingly, at 313 K and 4.50×10^{-4} mol L^{-1}. There was a clear correlation between the IE% determined by WL, EIS, and PDP analyses. The Langmuir adsorption isotherm was discovered to be the perfectly suited among the several adsorption isotherms that were investigated. According to EIS studies, the zwitterionic copolymer 7 behaves as a blended sort of inhibitor when subjected to anodic regulation. Similarly, using EIS, PDP, and SEM techniques, Haque et al. [76] created and utilized an innovative CIs called 2-amino-3-((4-((S)-2-amino-2-carboxyethyl)-1H-imidazol-2-yl)thio)propionic acid (AIPA) for MS in 1 M HCl. The outcomes of the aforementioned techniques demonstrate how effective AIPA is as an inhibitor. At 0.456 mM L^{-1} (125 ppm; 125 mg L^{-1}), it provides a maxi-mal level of IE% of 96.52% and prevents corrosion by adhering to the MS substrate. AIPA functions as a mixed-type inhibitor having a significant cathodic tendency, according to the PDP analysis. In addition to AIPA, MS substrate topology was enhanced according to AFM and SEM investigations.

Advantages of using amino acids as CIs include the following:

- *Environment-friendly:* Amino acids are biodegradable and do not pose a threat to the environment, unlike some chemical inhibitors that can be toxic and harmful to aquatic life.
- *Effective at Low Concentrations:* Amino acids can be effective inhibitors even at low concentrations, making them cost-effective and efficient.
- *Versatility:* Amino acids can be used in various environments, including acidic, neutral, and alkaline solutions, making them suitable for a wide range of applications.
- *Selectivity:* Different amino acids can have different inhibiting properties, making it possible to select the most appropriate amino acid for a particular application.
- *Reduced Metal Wastage:* By inhibiting corrosion, the use of amino acids can reduce metal wastage, saving resources and costs associated with metal replacement.
- *Improved Metal Performance:* The protective film formed by adsorbed amino acid molecules can improve the performance of the metal by reduc-ing the risk of corrosion and extending its service life.

19.3.3 UTILIZATION OF CARBOHYDRATES

Carbohydrates have been found to have the potential to mitigate corrosion in some metal substrates. This is owing to their capability to form a protective covering on the metal substrate that acts as a barrier against corrosive agents. The most commonly used carbohydrates for this purpose are saccharides (sugars) and polysaccharides (complex sugars). For example, fructose has been shown to be effective in mitigating the corrosion of Al alloys in seawater. There are a few examples of the use of carbohydrates as CIs:

- *Fructose as a CIs for Al Alloys in Seawater:* Research has shown that fructose is effective in reducing the C_R of Al alloys in seawater. In a study, the addition of fructose to seawater was found to develop a protective covering on the Al substrate, which behaved as a shield to prevent further corrosion. The results illustrated that the C_R of the Al alloy was reduced by up to 80% in the presence of fructose.
- *Starch as a CIs for MS in Acidic Solutions:* Starch is a polysaccharide that has been found to be an excellent CIs for MS in acidic solutions. In a study, MS coupons were immersed in an acidic solution containing starch, and the C_R was measured over time. The results showed that the C_R was effectively mitigated in the presence of starch, suggesting that it is effective in mitigating the corrosion of MS in acidic media.
- *Dextrin as a CIs for CS in Neutral Environments:* Dextrin is a complex sugar that has been found to be effective in reducing the C_R of CS in neutral environments. In a study, CS coupons were immersed in a neutral solution containing dextrin, and the C_R was measured over time. The findings elaborated that the C_R was significantly reduced in the presence of dextrin, demonstrating its potential as a CIs for CS in neutral environments.

It is important to note that the effectiveness of carbohydrates as CIs can depend on various factors, such as the type of metal, the corrosive environment, and the specific type of carbohydrate used. Further research is needed to fully understand the mechanisms by which carbohydrates can inhibit corrosion and optimize their use as CIs.

In an experiment, a new chitosan oligosaccharide macromolecule using a glucose moiety (COS-g-Glu) was created by Rbaa et al. [77] and employed as a non-hazardous, long-lasting, and recyclable MS CI in an acid media (1.0 M HCl). According to EIS testing, the IE% rises with elevating inhibitor concentration and achieves an ideal level of 97% at 10^{-3} M. The adsorption of COS-g-Glu was validated by thermodynamic variables and steel substrate characterization techniques. The outcomes of the MD and DFT computations were consistent with the outcomes of the experiments. On the premise of polyurethane science, Farhadian et al. [78] developed an easy-to-use strategy to improve the inhibitory performance of natural polymers using hydroxyethyl cellulose (HEC) as a carbohydrate paradigm. In this experiment, WL, EIS, and PDP were utilized to evaluate the effectiveness of chemically modified hydroxyethylcellulose (CHEC) in preventing MS corrosion. All electrochemical tests showed that even at elevated temperatures, adding just 1% of polyurethane pre-polymer to the CHEC framework significantly increased its acidic solution IE%. At 80 °C, CHEC had a maximal inhibitory efficacy of 93% as a mixed-type inhibitor after adhering to the MS substrate. The shape of the MS substrate in the context of CHEC further supported the additive's preventive effect, and the XPS data amply demonstrated CHEC adsorption on the MS substrate. The electronic configuration of CHEC and its interconnections with the metallic substrate was also clarified at the molecular level by simulations using MDS and DFT. These results show that the polyurethane prepolymer technique is a novel and successful strategy for improving the anti-corrosion efficacy of natural polymer-based CI in harsh acidic conditions at extreme temperatures.

For MS in an acidic media, Rbaa et al. [79] documented the synthesis and corrosion prevention properties of chitosan (CH) and its 5-chloromethyl-8-hydroxyquinoline derivative (CH-HQ). Since the maximum IE% for CH-HQ and CH were 93% and 78%, correspondingly, it was found that CH-HQ had a higher protective efficacy than CH. CH-HQ and CH were mixed-type CIs over the examined temperature range (298 K ± 1 to 328 K ± 1), according to PDP experiments. Similarly, thiocarbohydrazide and chitosan were cross-linked by Mouaden et al. [80] using a simple one-step synthesis to produce thiocarbohydrazide-chitosan (TC-Cht), which was initially tested as a SS CIs in a 3.5% NaCl media. A thorough EIS and PDP were conducted, and the results demonstrated that the TC-Cht demonstrates mixed-type behavior with a predominately cathodic essence and behaves by sorption over the metallic substrate. The Langmuir isotherm was observed in the sorption of TC-Cht molecules over the SS substrate. At a dosage of 500 mg L^{-1}, the TC-Cht demonstrated a significant IE% of >94%.

Utilizing inulin, a carbohydrate polymer, as an eco-friendly inhibitor, B.P. and Rao [81] controlled acid corrosion of 6061 Al alloy and 6061 Al-15% (v) SiC(P) composite material (Al-CM). Inulin concentrations ranged from 0.2 g L^{-1} to 1.0 g L^{-1}, and experiments were performed at temperatures between 303 K and 323 K. The research revealed that the effectiveness of the inhibitor's reduction rose with higher inulin dosages and reduced with higher temperatures. By physical sorption on both alloy and Al-CM, inulin adhered to the Langmuir adsorption isotherm on metallic substrate. Findings from the EIS approach and the PDP technique were in excellent accordance with one another.

19.4   USE OF SEMINATURAL MATERIALS

19.4.1   Utilization of Amino Acids Derivatives

Amino acid derivatives have been used for corrosion mitigation due to their unique chemical structures and ability to adsorb onto metal substrates. When these compounds adsorb onto metal substrates, they form a protective layer that inhibits corrosion by acting as a barrier between the metal and the corrosive environment. In an experiment, through the conjunction of quantum chemistry (QM) computation, MDS, and MCS, Xu et al. [82] probed the effectiveness of glutamine (Glu), arginine (Arg), and asparagine (Asp) in hindering the corrosion of Al in NaCl media as well as the inhibitory activity process at the atomic scale to expose the connection between chemical composition, structure, and effectiveness. According to MC findings, the adsorption energies for Glu, Arg, and Asp in aquatic media are 69.22 kcal mol^{-1}, 210.39 kcal mol^{-1}, and 66.88 kcal mol^{-1}, correspondingly. The outcomes demonstrated that a significant absolute value of adsorption energy corresponds to a significant IE%. For Glu, Arg, and Asp, correspondingly, MDS reveals that the comparative content function of the peak range of Cl$^-$ was 18.96, 17.45, and 20.06. For Glu, Arg, and Asp, the inhibitor film's corresponding Cl$^-$ diffusion coefficients are 8.01×10^{-8} cm^2 s^{-1}, 5.18×10^{-8} cm^2 s^{-1}, and 1.22×10^{-7} cm^2 s^{-1}. The effectiveness of corrosion inhibition is dependent on both the adsorption energy and the capacity to prevent the dispersion and agglomeration of corrosion components. Theoretically computed inhibitory ability is Arg > Glu > Asp, which is well in line with experimental findings. This chapter serves as a theoretical guide for subsequent investigations into CIs of amino acids.

Zhang et al. [83] created unique N-doped carbon dots (N-CDs) using L-serine and citric acid. The N-CDs include a significant number of unsaturated bonds and polar moieties, according to the findings of XPS and FTIR. Integrating electrochemical analysis was used to analyse the copper-specific 0D nanomaterial's ability to suppress corrosion. The reduction efficacy of the N-CDs was determined to be as excellent as 98.5% average following being submerged for 24 hours, according to the findings, and they engaged with Cu substrate by physical and chemical adsorption. Additionally, the associated anti-corrosion process was thoroughly investigated and explained. The aim of this effort was to investigate effective and environmentally sustainable CI compounds for metal treatment.

In CO_2-saturated formation water, Zhang et al. [84] produced two amino acid derivatives (BPT and MPT) that were highly effective and environmentally benign CIs for N80 CS. According to theoretical estimations using the GFN-xTB technique, the chemisorption of BPT and MPT particles occurs primarily as a result of the bonding of S atoms to the substrate of Fe. The deposited MPT and BPT coatings successfully prevent the propagation of aggressive ions, hence preventing the corrosion of steel, according to the MDS for the propagation of aggressive ions. According to EIS testing, BPT outperforms MPT in terms of inhibitory effectiveness. It is discovered that BPT and MPT both have exceptional inhibitory effects (inhibition effectiveness of 99.44% for BPT and 99.26% for MPT), that is much greater than all those inhibitors of amino acid derivatives that have previously been documented. The conjunction of the enhanced electron-releasing capacity of benzyl and the wettability of BTP might explain the greater inhibitory efficacy and greater persistence of BPT.

In order to test for their capability to prevent corrosion on MS in 1 M HCl, Satpati et al. [85] generated three Schiff bases consisting of cinnamaldehyde and three distinct amino acids: tryptophan (CTSB), glycine (CGSB), and histidine (CHSB). WL and EIS experiments show that the CGSB was the least efficient of the three inhibitors, whereas the CTSB was the outstanding inhibitor. The efficiency of the Schiff bases was examined at 60 °C for 6 hours after susceptibility to HCl. At a consistent temperature of 30 °C, the effects of different exposure times – up to 96 hours – were also studied. Under these severe circumstances, all three inhibitors provide a notable amount of CI effectiveness. These mixed-type CI's chemisorption over metal was determined by their thermodynamic adsorption and kinetic characteristics.

Similarly, phenylalanine (P1) and aspartic acid (P2) were two organic substances that Oubaaqa et al. [86] investigated for their ability to prevent MS corrosion in a 1 M HCl media. The ability of two amino acids (P1 and P2) to prevent one another grew as their quantity improved, peaking at an ideal value of 89% and 87% for P2 and P1, correspondingly. Their adsorption process was in line with the Langmuir isotherm. The two substances function as blended inhibitors, according to PDP studies. According to UV-visible, the two compounds when combined result in less ferric ion disintegration in acidic environments. Photographs taken using SEM reveal that both inhibitors were thoroughly deposited on the MS.

In order to produce a water-soluble cyclo terpolymer 6 having a 1:1 M proportion of sulfide and sulfoxide clusters as a consequence of oxygen transmission from dimethyl sulfoxide (DMSO), Goni et al. [87] undertook varying N,N-diallyl methionine ethyl ester hydrochloride 5 copolymerization to SO_2 using the Butler cyclopolymerization protocol in DMSO. When H_2O_2 was used to oxidize half of the sulfide groups in 6, it produced the polymers sulfone and sulfoxide. Using a viscometrical method, the media characteristics of CI were identified. The content of the polymers improved along with the corrosion efficiency. The maximal inhibitory efficacy of copolymer compounds 6–8 was found to be 92%, 97%, and 95%, correspondingly, at a polymer dosage of 175 µM. The synthetic polymer molecules functioned as inhibitors of diverse types. By using both physisorption and chemisorption, polymer compound 7 adhered to the Temkin, Langmuir, and Freundlich isotherms. Analysis using SEM and XPS showed that the deposited polymers created a fine coating over the meal and inhibited additional chemical attacks.

By consolidating amino acids, glyoxal, and formaldehyde, Srivastava et al. [88] created three novel amino acid–based CIs: 2-(3-(carboxymethyl)-1H-imidazol-3-ium-1-yl)acetate (AIZ-1), 2-(3-(1-carboxyethyl)-1H-imidazol-3-ium-1-yl)propanoate (AIZ-2), and 2-(3-(1-carboxy-2-phenylethyl)-1H-imidazol-3-ium-1-yl)-3-phenylpropanoate (AIZ-3). The effectiveness of manufactured inhibitors at inhibiting corrosion was investigated by employing the EIS and PDP techniques. AIZ-3 demonstrated the highest IE% of the tested inhibitors, attaining 96.08% at a dose as minimal as 0.55 mM (200 ppm). AIZ-1 functions as a cathodic inhibitor, according to the findings of PDP research, whereas AIZ-3 and AIZ-2 function as blended kind of inhibitors. The findings of the EIS investigations demonstrated that the polarization resistance improved and the C_{dl} reduced with the addition of inhibitors owing to the inhibitors' deposition on the MS. The Langmuir adsorption isotherm was observed when AIZs were adsorbed over the MS.

The mechanism of amino acid derivatives in corrosion mitigation can be explained through various theories, including the adsorption theory, the pH adjustment theory, and the passivating film theory:

- *Adsorption Theory*: According to this theory, amino acid derivatives adsorb onto metal substrates through physical/chemical adsorption, or a combination of both. This adsorption forms a protective covering on the metal substrate that prevents the transfer of electrons between the metal and the corrosive environment.

- *pH Adjustment Theory*: This theory states that amino acid derivatives can adjust the pH of the environment near the metal substrate, making it less corrosive. This can be achieved through the release of basic or acidic species from the amino acid derivatives.
- *Passivating Film Theory*: This theory suggests that the amino acid derivatives might develop a passivating film on the metal substrate that prevents corrosion by acting as a barrier between the metal and the corrosive environment. The film can be composed of a complex network of organic molecules and metal ions that provides long-lasting protection.

Advantages of using amino acid derivatives as CIs include the following:

- *Eco-friendliness*: Amino acid derivatives are biodegradable and non-toxic, making them a safer and more environment-friendly alternative to traditional CIs.
- *Effectiveness*: Studies have shown that amino acid derivatives are effective in inhibiting corrosion in a wide range of environments and for different metals, including steel, Al, and copper.
- *Versatility*: Amino acid derivatives can be used in different forms, including as solutions, suspensions, or coatings, making them a versatile choice for corrosion protection.

Examples of amino acid derivatives used as CIs include the following:

- *Glycine*: Glycine has been shown to be an effective inhibitor for the corrosion of Al and its alloys in acidic solutions.
- L-*Histidine*: L-Histidine has been revealed to significantly inhibit the corrosion of MS in acidic and neutral environments.
- L-*Cysteine*: L-Cysteine has been revealed to be an effective CI for the corrosion of Al in an alkaline solution.

19.4.2 Utilization of Carbohydrate Derivatives

Carbohydrate derivatives such as glucose, fructose, and saccharides have been found to have anti-corrosion properties and have been used as CIs [89–92]. The use of carbohydrate derivatives as CIs is based on the adsorption of these molecules over the metal substrate, producing a protective covering. This film blocks the access of corrosive species, including oxygen and water to the metal, thereby reducing the C_R.

The mechanism of corrosion inhibition by carbohydrate derivatives varies depending on the type of carbohydrate and the metal. However, some common mechanisms include the following:

- *Adsorption*: The polar groups of the carbohydrate derivatives adsorb onto the metal substrate, creating a shield that prevents the access of corrosive species.
- *pH Adjustment*: The acidic or basic nature of the carbohydrate derivatives can alter the pH of the surrounding environment, which impacts the C_R.

- *Passivation*: The carbohydrate derivatives can react with the metallic substrate to develop a passive film that slows down the corrosion process.
- *Electrostatic Repulsion*: The negatively charged groups of the carbohydrate derivatives can repel positively charged corrosion products, reducing their rate of formation.

In an experiment, in 1 M HCl, Mobin et al. [93] tested the polysaccharide of *Plantago ovata* for its ability to stop the corrosion of CS. The extremely convoluted polysaccharide arabinosyl (galaturonic acid) rhamnosylxylan (AX), that compensates for the granules of the *Plantago*, was primarily in charge of preventing the corrosion of CS. The WL analyses demonstrated AX's effectiveness as a CS CIs in 1 M HCl. As AX content and media temperature rose, the effectiveness of the CI improved, pointing to a critical function for chemical adsorption. At 1000 ppm AX in 1 M HCl, the AX's inhibition effectiveness was 93.54%. The Langmuir adsorption isotherm governed the adsorption of AX over CS substrates. The mixed and endothermic adsorption processes of adsorption both were associated with a reduction in entropy. According to EIS tests, AX acted as a blended kind of inhibitor, forming a protective coating on the top of the CS to shield it from the harsh HCl media. Owing to the adhesion of a fine coating of AX molecules on the metallic specimen, CS was shielded from the corrosive HCl media.

Hexamethylene-1,6-bis(N-D-glucopyranosylamine) (HGA), an amine-functionalized glucose, was produced and studied by Onyeachu et al. [94]. By using EIS, PDP, and SEM, the corrosion prevention behavior of HGA upon API X60 steel in 3.5 wt% NaCl swamped by CO_2 was assessed. The Langmuir isotherm was followed by the sorption of HGA over the metal. The PDP findings showed that the HGA molecule acted as a blended-class inhibitor by slowing anodic and cathodic electrochemical mechanisms. At a dosage of 2.27×10^{-4} M, the greatest corrosion IE% was attained at 91.82% (100 ppm). Following adsorption over the metal, the inhibitor adopts a horizontal configuration after being protonated in the environment. According to quantum chemical computations and protonated MCS, the atoms of nitrogen and oxygen operate as reactive centres for adsorption.

Verma et al. [95] developed dihydropyrido-[2,3-d:6,5-d]-dipyrimidine-2,4,6,8($1H,3H,5H,7H$)-tetraone (GPHs) as derivatives of D-glucose and examined as CI for MS in 1 M HCl media by employing WL, EIS, and MCS techniques. GPH-3 was inhibited more effectively than GPH-2 and GPH-1. The outcomes also demonstrated that inhibitor compounds having (−OH, −OCH3) substituents which release electrons perform more effectively than the progenitor molecule with no substituents. MCS was utilized to better comprehend how inhibitor compounds and metallic substrates interacted. The data attained aid in predicting the GPH molecules' highly reliable adsorption locations upon the metallic substrate. Figure 19.5 displays the top perspectives of the density distributions of the optimum persistent lower energy arrangement for the adsorption of inhibitor molecules onto Fe(1 1 0) substrate. The significant negative value of the E_{ads} showed that all of the inhibitor compounds under investigation had a high tendency to adsorb. The molecules over the metallic substrate were encouraged to absorb by the aquatic media. The concentration of the GPH molecules adsorbing on the substrate of Fe(1 1 0) is shown in Figure 19.5.

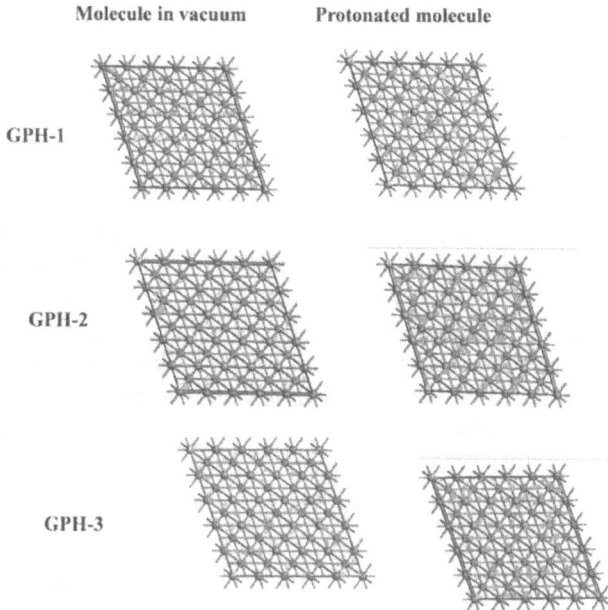

FIGURE 19.5 Top perspectives of the density dispersion of the least stable lower energy arrangement for the sorption of CI molecules on the substrate of Fe(1 1 0) attained by MCS. (Adapted with permission from Ref. [95 under CCBY 4.0.)

It is clear that the inhibitors do not favour any particular locations over the Fe substrate. The isodensity forms uniformly blanket the substrate that is being shown. It is particularly noticeable for protonated compounds. Because there aren't any atomic ledges and the atoms were packed closely together on the substrate under study, this uniform dispersion results. In accordance with the empirical sequence of IE%, the estimates of E_{ads} for the neutral states of the investigated molecules are as follows: GPH-3 (11.88 kcal mol) > GPH-2 (10.65 kcal mol) > GPH-1 (8.35 kcal mol). The patterns for inhibitor compounds in their aqueous and protonated forms are nearly identical, with a few minor differences.

Sayed et al. [96] developed thiocarbohydrazones through the reaction of 2-adamantanone (Ad-Th)/2-acetylferrocene (Fe-Th) with carbonothioic dihydrazide. The compounds were used as efficient CIs for CS pickling in HCl. The analysis showed that Fe-Th provided a better level of inhibition than Ad-Th with an IE% of 93.6% (Ad-Th) and 97.9% (Fe-Th) achieved at 200 ppm. The sorption of Fe-Th or Ad-Th additions resembled the Langmuir isotherm, exhibiting chemisorption predominating over physisorption. EIS tests showed that the CIs improved the capacitive behavior. The PDP investigations revealed that the inhibitors displayed mixed-type behavior. Experiments using SEM and FTIR highlight the presence of an organic inhibitor coating protecting the CS substrate. This study offers very important implications for producing innovative thiocarbohydrazone inhibitors exhibiting great protective effectiveness.

Advantages:

- Non-toxic and environment-friendly
- Low cost
- Easy availability
- Inhibits corrosion effectively

There have been numerous studies and research reports that show the effectiveness of carbohydrate derivatives as CIs. In one study, it was found that the inclusion of glucose to an acidic solution significantly reduced the C_R of MS. Another study found that a fructose solution was effective in inhibiting the corrosion of Al in an aqueous environment. It is important to note that the IE% of carbohydrate derivatives can be impacted by variables, including concentration, temperature, and the type of metal being protected.

19.5 USE OF SUSTAINABLE TACTICS

The use of inhibitors has long been recognized as an effective way to reduce the corrosion process and extend the service life of metal components. However, conventional CIs, including organic inhibitors and inorganic inhibitors, often have disadvantages such as high cost, toxicity, and environmental concerns. In recent years, there has been a growing interest in the use of green inhibitors, which are derived from natural sources and are considered environment-friendly. Green inhibitors, such as plant extracts, essential oils, and organic acids, have been revealed to possess excellent corrosion inhibition characteristics, but their efficacy is often limited. The concept of synergism, which refers to the combination of two or more agents to produce a combined effect that is greater than the sum of their individual effects, has been implemented to the field of corrosion inhibition to enhance the performance of green inhibitors. Synergistic combinations of green inhibitors with other corrosion inhibitors can improve the performance of the inhibitors by increasing their adsorption on the metal substrate, altering the corrosion process, or providing a protective film that prevents corrosion [97–100].

One of the ways in which synergism can be achieved is by increasing the adsorption of inhibitors over the metallic substrate. Adsorption is the process by which inhibitor molecules adhere to the metal substrate and form a protective film. The efficacy of an inhibitor is closely related to its adsorption on the metal substrate, and increasing the adsorption can enhance its performance. Synergistic combinations of green inhibitors with other inhibitors can increase the adsorption of the inhibitors by forming a stronger bond between the inhibitors and the metal substrate, reducing the inhibition potential, or increasing the stability of the protective film. Another way in which synergism can be achieved is by altering the corrosion process. Corrosion is a complex process that involves multiple steps, and the introduction of inhibitors can alter the kinetics of the reaction and slow down the C_R. Synergistic combinations of green inhibitors with other inhibitors can alter the corrosion mechanism by restricting the reaction sites, reducing the availability of corrosive species, or changing the potential of the metal substrate.

Finally, synergistic combinations of green inhibitors with other inhibitors can provide a protective film that prevents corrosion. A protective film can be formed by the adsorption of CIs upon the metal substrate, and the combined effect of multiple inhibitors can produce a more stable and durable film that provides better protection against corrosion. The protective film can also have a mechanical effect by reducing the exposure of the metal substrate to the corrosive environment, and a chemical effect by altering the reactivity of the metal substrate. The use of synergism effects for increasing the corrosion IE% of green inhibitors is a promising approach for improving the performance of environment-friendly corrosion inhibitors.

By combining green inhibitors with other inhibitors, the performance of the inhibitors can be enhanced, and the CR can be slowed down. However, it is important to note that the synergistic effect depends on the specific combination of inhibitors and the system being studied, and further research is needed to fully understand the underlying mechanisms and optimize the performance of synergistic combinations of inhibitors. Examples of synergistic combinations of green inhibitors with other inhibitors to enhance their corrosion IE% include the following:

- *Essential Oils and Organic Inhibitors:* Essential oils, such as clove oil and eucalyptus oil, have been shown to offer excellent corrosion inhibition attributes, but their efficacy is often limited. Synergistic combinations of essential oils with organic inhibitors, such as benzotriazole and imidazoline, have been reported to improve the efficacy of the inhibitors. For example, a study found that the combination of clove oil with benzotriazole showed a synergistic effect on the corrosion inhibition of MS in 1 M HCl solution, and the C_R was significantly reduced compared to the use of each inhibitor alone.
- *Plant Extracts and Inorganic Inhibitors:* Plant extracts, such as tannins, have been revealed to offer remarkable corrosion inhibition characteristics, but their efficacy is often limited. Synergistic combinations of plant extracts with inorganic inhibitors, such as zinc oxide and cerium oxide, have been reported to improve the performance of the inhibitors. For example, a study found that the combination of tannins with cerium oxide showed a synergistic effect on the inhibition of corrosion of Al in an acidic solution, and the C_R was significantly reduced compared to the use of each inhibitor alone.
- *Organic Acids and Inorganic Inhibitors:* Organic acids, including acetic acid and citric acid, have been shown to offer excellent corrosion inhibition properties, but their efficacy is often limited. Synergistic combinations of organic acids with inorganic inhibitors, such as magnesium oxide and zinc oxide, have been reported to improve the performance of the inhibitors. For example, a study found that the combination of acetic acid with magnesium oxide showed a synergistic effect on the corrosion inhibition of CS in an acidic solution, and the C_R was significantly reduced compared to the use of each inhibitor alone.

In an experiment, the synergism impact of 1-acetyl-3-thiosemicarbazide (AST) and I⁻ ions on the corrosion of C1018 CS in 1 M HCl media was explained empirically by Alamri et al. [101]. According to the EIS findings, AST solely prevented C1018 CS from corroding when exposed to acid. At 750 ppm of AST, the IE% increased to the highest dose of 72.27%. When 5 mM KI was added to 250 ppm of AST, the IE rate increased to 81.64%. According to the findings of the dispersion and protonated phase predictions made by the ACD/LABS Percepta software, AST was extremely solvable in aqueous acidic media and roughly 95% of it resides in the neutral state in 1 M HCl (pH = 0). The active reactivity locations of AST were predicted using MCS and DFT computations. The minimal adsorption energy and arrangement of AST individually and AST + I$^-$ on/Fe(1 1 0)/water junction were also calculated. DFT was additionally used to assess the active regions in charge of AST's adsorption over C1018 CS. It was expected that at pH = 0, about 0.2% of AST exists in the neutral phase and 0.5% in the protonated state predicated on the assessment of the protonation phase of AST. Understanding the correlation between the molecular constitution of AST and its capacity to prevent C1018 corrosion in 1 M HCl required the use of DFT computations. Figures 19.6a and b show the chemical orbitals, Figure 19.6c shows the Fukui function for AST and ASTH, and Figure 19.6c shows the optimized configurations. Figure 19.6 makes it evident that both the AST (neutral) state and the ASTH state (protonated state) of the molecules possess HOMO and LUMO orbitals that were shown to extend across the whole molecule. For the S4 atom, though, additional electrons were apparent. This suggests that the S atom, combined with the contributions of the N1, N2, N5, and O7 atoms, seems expected to constitute the principal active region. When a molecule interacting with the substrate of C1018 steel, these active regions were in charge of providing electron densities to the metallic substrate and receiving electrons from the metallic substrate.

By using *olive leaf extract* (OLE) as a reductant, Umoren et al. [102] created the chitosan-copper oxide (CHT-CuO) nanocomposite *in situ*. The OLE-mediated CHT–CuO nanocomposite was tested as a CI for X60 CS in a 5 wt% HCl media and contained distinctive dosage of chitosan (0.5, 1.0, and 2.0 g). The effect of KI inclusion on the nanocomposites' capability to prevent corrosion was investigated. The greatest IE% was attained at the optimal concentration of 0.5% of nanocomposites, with the corrosion inhibiting impact being seen to rise with increment in concentration. According to impedance studies, the sequence of corrosion prevention efficiency was CHT1.0–CuO (90.35%) > CHT0.5–CuO (90.16%) > CHT2.0–CuO (89.52%) nanocomposite. Additionally, it was discovered that IE% increased with temperature increases from 25 °C to 40 °C before declining with temperature increases from 50 °C to 60 °C. The PDP data imply that an active region-inhibiting process was used by the nanocomposites to prevent corrosion of X60 CS both on their own and in conjunction with KI. The calculated synergism factor (S1) was determined to be below 1 with coefficients of 0.74, 0.89, and 0.75 for CHT1.0–CuO, CHT0.5–CuO, and CHT2.0–CuO nanocomposites, correspondingly, at 60 °C. Additionally, the IE was enhanced by KI inclusion when the temperature increased from 25 °C to 60 °C. The production of a defensive layer, which may be attributable to the nanocomposites adhering to the CS substrate, was confirmed by the findings of the substrate investigation.

Using a 3.5 wt% NaCl solution, Ahangar et al. [103] investigated the synergistic function of the sodium lignosulfonate–zinc acetate (SLZA) mixture and its subsequent impact on the corrosion protection of MS. According to the EIS experiments'

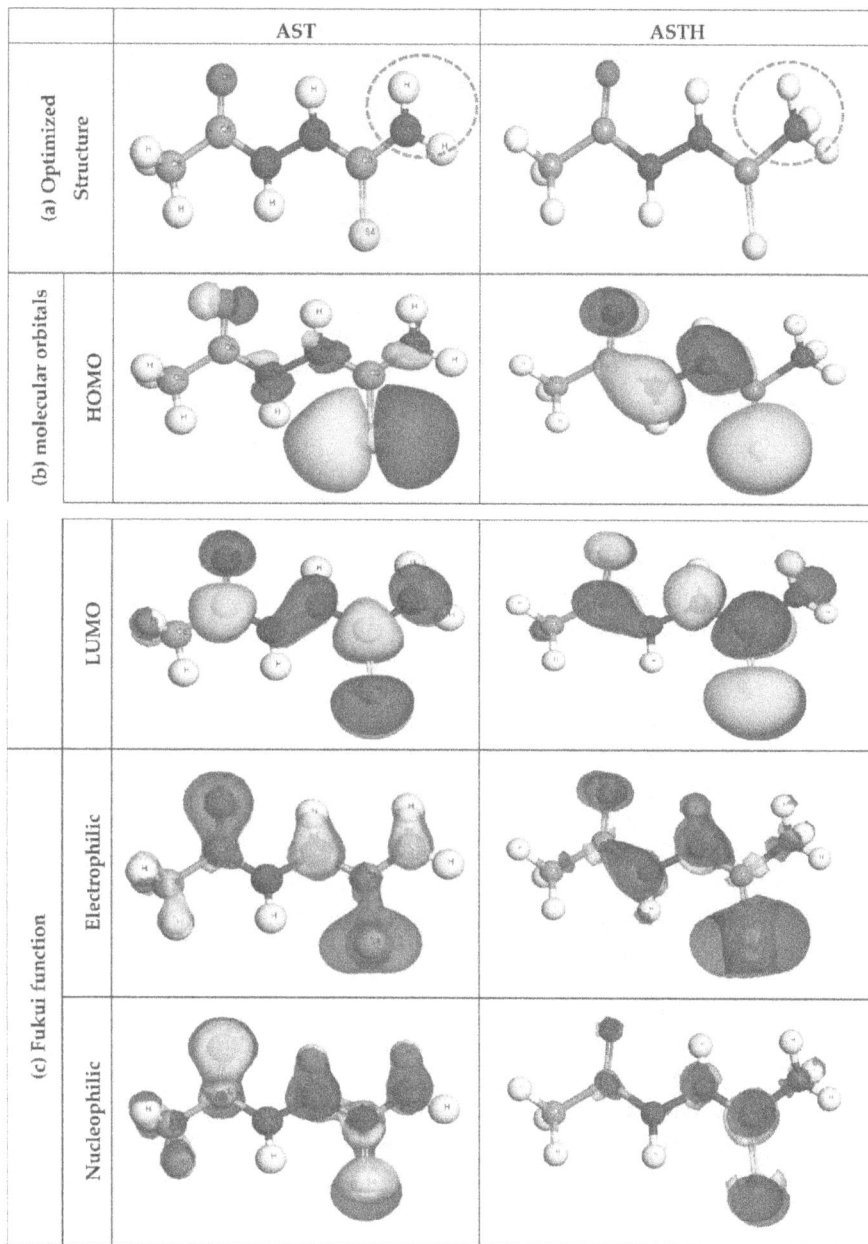

FIGURE 19.6 (a) Optimized configuration, (b) FMOs and (c) AST and ASTH Fukui function by utilizing GGA/BLYP/DNP set. (Adapted with permission from Ref. [101] under CCBY 4.0.)

findings, sodium lignosulfonate (SL) and zinc acetate (ZA) work together to synergistically reduce MS corrosion (ZA). In the addition of ZA and SL, the synergism value was around 9, and the overall MS resistance was greater than 400 k cm^2. Additionally, the estimated IE% for the SLZA system remained 96%. It is incredibly uncommon to find a corrosion-protecting system in the literaxture that performs as well as this one. Moreover, through the outcomes of XPS and SEM–EDS investigations, the primary corrosion protection process in the vicinity of the SLZA complex was hypothesized to be the co-deposition of substrate protective layer consisting of zinc-containing components and SL-based complex.

These above-stated studies demonstrate the potential of synergistic combinations of green inhibitors with other inhibitors to enhance their corrosion IE%. The data reported in these studies includes measurements of C_R, IE%, substrate analysis, and electrochemical methods. The results of these studies indicate that synergistic combinations of green inhibitors with other inhibitors can effectively slow down the C_R and extend the service life of metal components. However, it is important to note that the synergistic effect depends on the specific combination of inhibitors and the system being studied, and further research is needed to fully understand the underlying mechanisms and to optimize the performance of synergistic combinations of inhibitors.

Moreover, the efficacy of green inhibitors is often limited by their rapid degradation and loss of inhibiting properties. This can be due to exposure to environmental factors such as humidity, temperature, and UV light, which can cause the extract to evaporate or deteriorate. This limits the duration of their protective effect and reduces their overall efficacy. Self-healing materials have acquired significant interest in recent years owing to their ability to repair themselves in response to damage. This self-repair capability is based on the presence of microencapsulated healing agents, which are triggered by mechanical damage or environmental factors. When a self-healing material is damaged, the microcapsules are broken, releasing the healing agents that then react to form a new protective layer. This self-healing process can be repeated multiple times, leading to increased durability and lifespan [104–108]. The use of self-healing effects for increasing the corrosion IE% of green extracts has the potential to overcome the limitations of traditional green inhibitors. By incorporating self-healing materials into the extract-based coating, the continuous release of the extract can be ensured, leading to a longer-lasting protective effect. Additionally, the self-healing properties of the material can provide additional protective properties, such as physical barrier formation, which can further enhance the corrosion IE% of the green extract.

The self-healing effect can be achieved by using different methods, including microencapsulation, entrapment, and embedding. In microencapsulation, the green extract is encapsulated in a thin film, which is then incorporated into the coating. This film acts as a barrier, preventing the extract from evaporating or deteriorating. When the coating is damaged, the film is broken, releasing the green extract, which can then inhibit corrosion. In entrapment, the green extract is embedded in a polymer matrix, which acts as a barrier to prevent evaporation and deterioration. When the coating is damaged, the polymer matrix is broken, releasing the green extract, which can then inhibit corrosion. Embedding involves incorporating the green extract directly into the coating. This method can be used to create a self-healing coating by using a reactive species in the coating that can react with the extract to form a protective layer. Because of their permeable nature, which offers a high initial reservoir for CIs, plasma electrolytic oxidation (PEO) coatings, for example, could be employed as containers for inhibitors. Inhibitors are typically included into PEO pores during the coating procedure. This is because processing parameters, including strong local heat production and the development of high-energy arc discharges, are prone to change the inhibitor's molecular composition or cause it to oxidize, specifically in the context of organic inhibitors. Figure 19.7 provides a comprehensive representation of the corrosion-resistant processes offered by PEO covering with integrated inhibitors [109].

The use of self-healing effects for increasing the corrosion IE% of green extracts has been demonstrated in several studies. For example, a study by Zhang et al. (2020) found that a self-healing extract-based coating made from microcapsules containing green inhibitors showed significantly improved corrosion inhibition compared to a traditional extract-based coating. It could be stated that the use of self-healing effects for increasing the corrosion IE% of green extracts has the potential to overcome the limitations of traditional green inhibitors. This approach can ensure a longer-lasting protective effect and provide additional protective properties, which can further enhance corrosion IE%. In an experiment, Jamshidnejad et al. [110] researched self-healing coverings made of urethane and epoxy that contained OLE as a CIs. In urethane-shelled pellets, the healing compounds PDES and HOPDMS were utilized. In order to significantly improve the coating's ability to defend against corrosion, OLE was also included in the healing pellets. The effectiveness of the OLE as CI for steel in chloride media was demonstrated using polarization and EIS experiments, with an IE% of roughly 80% for ethanol-extracted inhibitor at an optimum concentration of 300 ppm. Concurrent agglomeration and ultrasound-assisted were used to create submicrometer capsules carrying inhibitor and healing compounds. Eventually, the self-healing properties of covering comprising 3%, 4%, and 5% of capsules were investigated. It was discovered that as the coating's capsule concentration increased, corrosion protection enhanced to the point where in a coating having 5 wt% of capsules, there was no sign of deterioration in the damaged region.

Arukalam et al. [111] researched into the abilities of *Lupinus arboreus* (LA) gum extract to function as a corrosion inhibitor and self-healing reagent in epoxy-based coatings. To identify the chemical components of the gum extract, a phytochemical

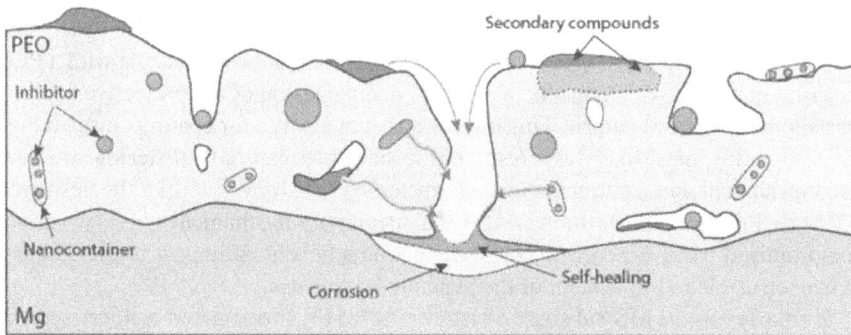

FIGURE 19.7 Depiction of the corrosion prevention process offered by PEO covering with inhibitors showing the self-healing mechanism. (Adapted with permission from Ref. [109] under CCBY 4.0.)

analysis was conducted. Both melamine-formaldehyde (MF) and urea- and melamine-formaldehyde (UF) were used to encase the *Lupinus arboreus* gum extract, and the microcapsules were analysed using XRD and FTIR methods. Remarkably, epoxy-(MF + LA) covering had greater repairing capability than epoxy-(UF + LA) covering, as demonstrated by the utilization of SEM. This supports the blocking effectiveness findings that showed that the epoxy-(MF + LA) covering had quite consistent barrier properties compared to the epoxy-(UF + LA) covering, whose efficacy initially was good but subsequently declined over the course of submerging. The findings indicate that a simple epoxy covering would not self-heal, and it could be because LA was not present. There was healing activity seen with regard to the epoxy coverings comprising the LA healing reagent. The preceding actions were seen the healing capability performing. Microcapsules comprising LA gum, which was formed of polymerizable monomeric components, were disseminated in the covering matrices during coating formation. Knife scribing caused mechanical degradation to the covering, which led to substrate flaws that caused the microcapsules to burst. The monomeric constituents of the LA gum extract were released when the microcapsules autonomously ruptured. The LA gum extract soaked into the substrate's cracks and, when combined with water, formed a polymeric layer that had the impact of shielding the substrate. Consequently, the substrate's cracks were repaired, and the coating's barrier effectiveness was recovered.

Gnedenkov et al. [112] researched self-healing coatings made of novel polycaprolactone to enhance the resistance to corrosion of Mg and related alloys. By examining the EIS and PDP slopes, the degree of corrosion prevention of coverings produced on the Mg substrate using an additional technique was determined (Figure 19.8). There were two temporal variables included in the θ–f dependence graphs for all kinds of investigated coverings (Figure 19.8a). Comparing the composite coverings with benzotriazole content to samples coated with PEO, the assessment of how the corrosion rate changed after prolonged exposition to 0.9% NaCl media revealed a drop in $|Z|/f = 0.1$ Hz (Figure 19.8b). The characteristics of this substance that were formerly described perhaps underlie this. Throughout the operation of covering insemination, heterogeneity of the substratum and the produced oxide film may result in a reduction in the resistance of the generated substrate layer to corrosion. For AT-Mg coated with the PEO technique, the inhibitor dosage utilized in this investigation was inadequate to improve corrosion resistance. The association of benzotriazole with a PEO covering matrices, which results in a partial disintegration of the protective coating, was another potential culprit. This impact does not emerge for coatings formed over MA8 Mg alloy, most likely as a result of the substrate materials' differing structure and mechanical and electrochemical characteristics. However, unlike the destroyed AT-Mg + PEO specimen, the AT-Mg + CC specimen's mechanical stability was not compromised when exposed to NaCl, indicating a beneficial impact of the addition of benzotriazole to the content of the generated coverings.

Similarly, on an MS substrate, Sanaei et al. [113] investigated a novel type of corrosion-inhibitory pigment predicated on zinc acetate-*Cichorium intybus* L. leaf extract (ZnA-CIL.L) in 3.5 wt% NaCl media. Several corrosion-inhibiting compounds, such as caffeic acid, flavone, and chicoric acid, which were packed with electron-rich groups and had a significant capacity for exchanging their lone pairs to

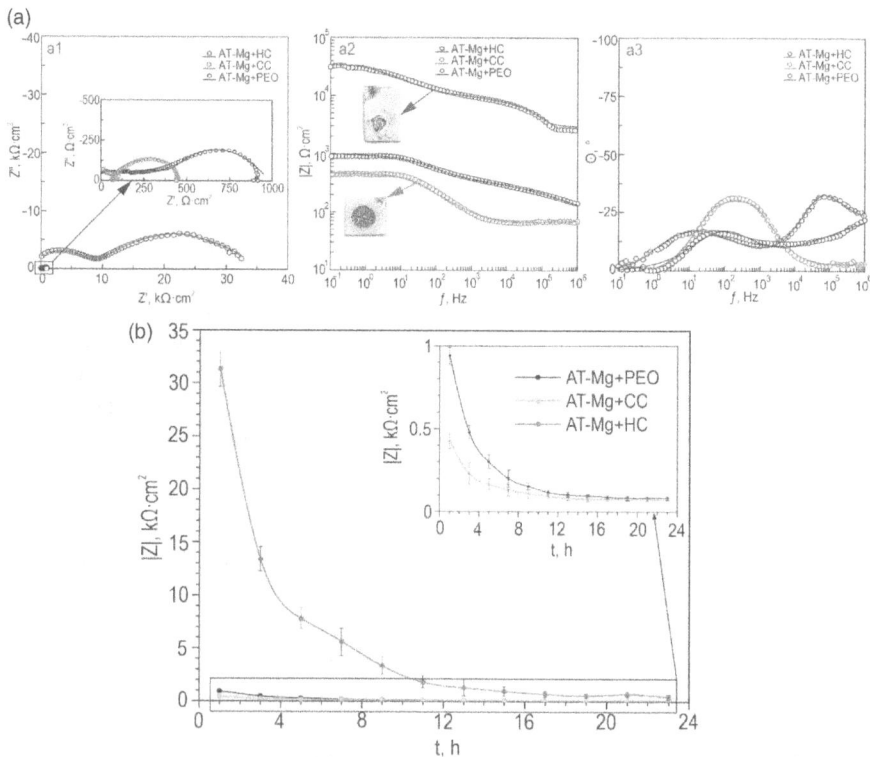

FIGURE 19.8 Impedance spectrum shown as Nyquist (a1) and Bode graphs (a2,a3), acquired upon AT-Mg specimens with different kinds of conceived coatings were exposed to 0.9% NaCl for 1 hour. Optical pictures of the specimens following 24-hour dipping in NaCl media shown in (a), along with kinetics of transformation in the magnitude of impedance modulus, evaluated at lower frequencies ($|Z|f = 0.1$ Hz) (b). (Adapted with permission from Ref. [112] under CCBY 4.0.)

Zn^{2+} cations that had vacant 3D orbitals, were present in the CIL.L extract. EIS and PDP examination findings proved that the ZnA-CIL.L hybrid pigment, in its extract state in saline media, significantly delay the corrosion of the steel specimen. Both the cathodic and anodic processes were significantly repressed and reduced in the addition of ZnA-CIL.L extract owing to the combinatorial impact among Zn2p cations and CIL.L. By chelating the CIL.L molecules using Zn/Fe cations and depositing the zinc cations as zinc hydroxide upon the cathodic region, the hybrid green corrosion inhibitive pigment was able to achieve the desired inhibition capability.

19.6 CONCLUSION AND FUTURE DIRECTIONS

In conclusion, the utilization of heterocyclic compounds as corrosion inhibitors has become a promising approach in sustainable corrosion inhibition due to their high efficiency, low toxicity, and environment-friendly properties. This chapter reviewed the modern approaches to sustainable corrosion inhibition using heterocyclic compounds, including various types of heterocyclic compounds and their mechanisms of corrosion inhibition. The effectiveness of these compounds as corrosion inhibitors has been studied on different metals, and the results show their potential as alternative corrosion inhibitors. Green inhibitors are eco-benign, less hazardous, biodegradable, and have minimal impact on the environment. Synthetic inhibitors provide long-lasting protection against various types of corrosion and are widely used in industries, including oil and gas, water treatment, marine, and construction. The mechanism of inhibition involves the adsorption of the inhibitor molecules onto the metallic substrate, creating a protective film that prevents corrosive agents from reaching the metal. In conclusion, the use of synergistic combinations of green inhibitors with other inhibitors has shown the potential in enhancing the corrosion IE% of metal components. Similarly, essential oils, plant extracts, and organic acids are examples of green inhibitors that have been combined with other inhibitors to produce a greater inhibiting effect. The synergistic effect is dependent on the specific combination of inhibitors and the system being studied.

Despite the potential of green inhibitors, their limited efficacy due to rapid degradation and loss of inhibiting properties remains a challenge that needs to be addressed. The effectiveness of both types of inhibitors depends on various factors, such as the type of inhibitor, the type of corrosion, the environment, and the concentration of the inhibitor. The use of GCIs is expected to continue to grow in the future, while synthetic inhibitors will continue to exhibit an important part in industries where long-lasting protection against corrosion is necessary. This chapter provides valuable insights for future research and development in the field of corrosion inhibition and sustainable materials. It is expected that the use of heterocyclic compounds will continue to grow in the future, leading to more sustainable and environment-friendly corrosion inhibition solutions. Further research is needed to fully understand the underlying mechanisms and to optimize the performance of synergistic combinations of inhibitors.

REFERENCES

1. S. GK, J.M. Jacob, A. Raj, Synergistic effect of salts on the corrosion inhibitive action of plant extract: A review, J. Adhes. Sci. Technol. 35 (2021) 133–163. https://doi.org/10.1080/01694243.2020.1797336.
2. N.R.J. Hynes, R.M. Selvaraj, T. Mohamed, A.M. Mukesh, K. Olfa, M.P. Nikolova, *Aerva lanata* flowers extract as green corrosion inhibitor of low-carbon steel in HCl solution: An in vitro study, Chem. Pap. 75 (2021) 1165–1174. https://doi.org/10.1007/s11696-020-01361-5.
3. C.C. Aralu, H.O. Chukwuemeka-Okorie, K.G. Akpomie, Inhibition and adsorption potentials of mild steel corrosion using methanol extract of *Gongronema latifoliuim*, Appl. Water Sci. 11 (2021) 1–7. https://doi.org/10.1007/s13201-020-01351-8.
4. A.E.A.S. Fouda, E.S. El-Gharkawy, H. Ramadan, A. El-Hossiany, Corrosion resistance of mild steel in hydrochloric acid solutions by *Clinopodium acinos* as a green inhibitor, Biointerface Res. Appl. Chem. 11 (2021) 9786–9803. https://doi.org/10.33263/BRIAC112.97869803.
5. A.A.S. Begum, R.M.A. Vahith, V. Kotra, M.R. Shaik, A. Abdelgawad, E.M. Awwad, M. Khan, *Spilanthes acmella* leaves extract for corrosion inhibition in acid medium, Coatings. 11 (2021) 1–24. https://doi.org/10.3390/coatings11010106.

6. V. Bashir, P. Sharma, N. Dhaundiyal, A. Shafi Kumar, *Gymneme sylvestre* as a green corrosion inhibitor for aluminum in an acidic medium, Port. Electrochim. Acta. 39 (2021) 199–212. https://doi.org/10.4152/pea.2021390304.

7. A. Mazkour, S. El Hajjaji, N. Labjar, E.M. Lotfi, M. El Mahi, Investigation of corrosion protection of austenitic stainless steel in 5.5 M polluted phosphoric acid using 5-azidomethyl-7-morpholinomethyl-8-hydroxyquinoline as an ecofriendly inhibitor, Int. J. Corros. 2021 (2021). https://doi.org/10.1155/2021/6666811.

8. A. Saxena, J. Kumar, Phytochemical screening, metal-binding studies and applications of floral extract of *Sonchus oleraceus* as a corrosion inhibitor, J. Bio- Tribo-Corrosion. (2020) 1–10. https://doi.org/10.1007/s40735-020-00349-8.

9. S.A. Alazzam, M.M. Sharqi, A.F. Almehemdi, Allelochemicals analysis of *Rumex vesicarius* L. and *Zygophyllum coccineum* L., and their effect on seed germination and seedling growth of wheat, *Triticum aestivum* L., IOP Conf. Ser. Earth Environ. Sci. 761 (2021). https://doi.org/10.1088/1755-1315/761/1/012077.

10. A. Berrissoul, E. Loukili, N. Mechbal, F. Benhiba, A. Guenbour, B. Dikici, A. Zarrouk, A. Dafali, Anticorrosion effect of a green sustainable inhibitor on mild steel in hydrochloric acid, J. Colloid Interface Sci. 580 (2020) 740–752. https://doi.org/10.1016/j.jcis.2020.07.073.

11. J.K. Odusote, T.B. Asafa, J.G. Oseni, A.A. Adeleke, A.A. Adediran, R.A. Yahya, J.M. Abdul, S.A. Adedayo, Inhibition efficiency of gold nanoparticles on corrosion of mild steel, stainless steel and aluminium in 1 M HCl solution, Mater. Today Proc. 38 (2021) 578–583. https://doi.org/10.1016/j.matpr.2020.02.984.

12. H. Fathima, M. Pais, P. Rao, Anticorrosion performance of biopolymer pectin on 6061 aluminium alloy: Electrochemical, spectral and theoretical approach, J. Mol. Struct. 1243 (2021) 130775. https://doi.org/10.1016/j.molstruc.2021.130775.

13. K.J. Al-Sallami, Impact of *Conyza bonariensis* extract on the corrosion protection of carbon steel in 2 M HCl solution, Int. J. Electrochem. Sci. 16 (2021) 210929. https://doi.org/10.20964/2021.09.40.

14. Y. Lekbach, F. Bennouna, S. El Abed, M. Balouiri, M. El Azzouzi, A. Aouniti, S. Ibnsouda Koraichi, Green corrosion inhibition and adsorption behaviour of *Cistus ladanifer* extract on 304L stainless steel in hydrochloric acid solution, Arab. J. Sci. Eng. 46 (2021) 103–113. https://doi.org/10.1007/s13369-020-04791-1.

15. K. Mahalakshmi, Corrosion inhibition of mild steel by ethanol extracts of *Acacia nilotica* in hydrochloric acid media, Malaya J. Mat, S (2020) 2082–2084.

16. D.S. Chauhan, M.A. Quraishi, W.B.W. Nik, V. Srivastava, Triazines as a potential class of corrosion inhibitors: Present scenario, challenges and future perspectives, J. Mol. Liq. 321 (2021) 114747. https://doi.org/10.1016/j.molliq.2020.114747.

17. D.S. Chauhan, C. Verma, M.A. Quraishi, Molecular structural aspects of organic corrosion inhibitors: Experimental and computational insights, J. Mol. Struct. 1227 (2021) 129374. https://doi.org/10.1016/j.molstruc.2020.129374.

18. M. Barbouchi, B. Benzidia, A. Aouidate, A. Ghaleb, M. El Idrissi, M. Choukrad, Theoretical modeling and experimental studies of terebinth extracts as green corrosion inhibitor for iron in 3% NaCl medium, J. King Saud Univ. - Sci. 32 (2020) 2995–3004. https://doi.org/10.1016/j.jksus.2020.08.004.

19. D.K. Verma, A. Al Fantazi, C. Verma, F. Khan, A. Asatkar, C.M. Hussain, E.E. Ebenso, Experimental and computational studies on hydroxamic acids as environmental friendly chelating corrosion inhibitors for mild steel in aqueous acidic medium, J. Mol. Liq. 314 (2020) 113651. https://doi.org/10.1016/j.molliq.2020.113651.

20. A. Thakur, A. Kumar, S. Kaya, R. Marzouki, F. Zhang, L. Guo, Recent advancements in surface modification, characterization and functionalization for enhancing the biocompatibility and corrosion resistance of biomedical implants, Coatings. 12 (2022) 1459. https://doi.org/10.3390/coatings12101459.

21. S. Bashir, A. Thakur, H. Lgaz, I.-M. Chung, A. Kumar, Corrosion inhibition performance of acarbose on mild steel corrosion in acidic medium: An experimental and computational study, Arab. J. Sci. Eng. 45 (2020) 4773–4783. https://doi.org/10.1007/s13369-020-04514-6.

22. A. Thakur, S. Sharma, R. Ganjoo, H. Assad, A. Kumar, Anti-corrosive potential of the sustainable corrosion inhibitors based on biomass waste: A review on preceding and perspective research, J. Phys. Conf. Ser. 2267 (2022) 012079. https://doi.org/10.1088/1742-6596/2267/1/012079.

23. S. Bashir, A. Thakur, H. Lgaz, I.M. Chung, A. Kumar, Corrosion inhibition efficiency of bronopol on aluminium in 0.5 M HCl solution: Insights from experimental and quantum chemical studies, Surf. Interfaces. 20 (2020) 100542. https://doi.org/10.1016/j.surfin.2020.100542.

24. S. Bashir, H. Lgaz, I.M. Chung, A. Kumar, Effective green corrosion inhibition of aluminium using analgin in acidic medium: An experimental and theoretical study, Chem. Eng. Commun. 208 (2020) 1–10. https://doi.org/10.1080/00986445.2020.1752680.

25. S. Bashir, A. Thakur, H. Lgaz, I.-M. Chung, A. Kumar, Computational and experimental studies on phenylephrine as anti-corrosion substance of mild steel in acidic medium, J. Mol. Liq. 293 (2019) 111539. https://doi.org/10.1016/j.molliq.2019.111539.

26. A. Thakur, S. Kaya, A. Kumar, Recent innovations in nano container-based self-healing coatings in the construction industry, Curr. Nanosci. 18 (2021) 203–216. https://doi.org/10.2174/1573413717666210216120741.

27. C. Dhonchak, N. Agnihotri, Computational insights in the spectrophotometrically 4H-chromen-4-one complex using DFT method, Biointerface Res. Appl. Chem. 13 (2023) 357.

28. A. Thakur, K. SAVAŞ, A. Kumar, Recent trends in the characterization and application progress of nano-modified coatings in corrosion mitigation of metals and alloys, Appl. Sci. 13 (2023) 730. https://doi.org/10.3390/app13020730.

29. A. Thakur, S. Kaya, A.S. Abousalem, A. Kumar, Experimental, DFT and MC simulation analysis of Vicia sativa weed aerial extract as sustainable and eco-benign corrosion inhibitor for mild steel in acidic environment, Sustain. Chem. Pharm. 29 (2022) 100785. https://doi.org/10.1016/j.scp.2022.100785.

30. A. Thakur, A. Kumar, Recent advances on rapid detection and remediation of environmental pollutants utilizing nanomaterials-based (bio)sensors, Sci. Total Environ. 834 (2022) 155219. https://doi.org/10.1016/j.scitotenv.2022.155219.

31. A. Thakur, A. Kumar, S. Sharma, R. Ganjoo, H. Assad, Materials today : Proceedings computational and experimental studies on the efficiency of Sonchus arvensis as green corrosion inhibitor for mild steel in 0. 5 M HCl solution, Mater. Today Proc. 66 (2022) 609–621. https://doi.org/10.1016/j.matpr.2022.06.479.

32. A. Thakur, A. Kumar, S. Kaya, D.V.N. Vo, A. Sharma, Suppressing inhibitory compounds by nanomaterials for highly efficient biofuel production: A review, Fuel. 312 (2022) 122934. https://doi.org/10.1016/j.fuel.2021.122934.

33. A. Thakur, A. Kumar, Sustainable inhibitors for corrosion mitigation in aggressive corrosive media: A comprehensive study, J. Bio- Tribo-Corrosion. 7 (2021) 1–48. https://doi.org/10.1007/s40735-021-00501-y.

34. A. Thakur, S. Kaya, A.S. Abousalem, S. Sharma, R. Ganjoo, H. Assad, A. Kumar, Computational and experimental studies on the corrosion inhibition performance of an aerial extract of Cnicus benedictus weed on the acidic corrosion of mild steel, Process. Saf. Environ. Prot. 161 (2022) 801–818. https://doi.org/10.1016/j.psep.2022.03.082.

35. G. Parveen, S. Bashir, A. Thakur, S.K. Saha, P. Banerjee, A. Kumar, Experimental and computational studies of imidazolium based ionic liquid 1-methyl- 3-propylimidazolium iodide on mild steel corrosion in acidic solution experimental and computational studies of imidazolium based ionic liquid 1-methyl- 3-propylimidazolium, Mater. Res. Express. 7 (2020) 016510. https://doi.org/10.1088/2053-1591/ab5c6a.

36. A. Thakur, A. Kumar, Potential of weeds extract as a green corrosion inhibitor on mild steel: A review, Think India J. 22 (2019) 3226–3240.
37. H. Mahfoud, N. Rouag, S. Boudiba, M. Benahmed, K. Morakchi, S. Akkal, Mathematical and electrochemical investigation of *Lamium flexuosum* extract as effective corrosion inhibitor for CS in acidic solution using multidimensional minimization program system, Arab. J. Sci. Eng. (2022). https://doi.org/10.1007/s13369-021-06546-y.
38. M.A. Deyab, Q. Mohsen, L. Guo, Theoretical, chemical, and electrochemical studies of *Equisetum arvense* extract as an impactful inhibitor of steel corrosion in 2 M HCl electrolyte, Sci. Rep. 12 (2022) 1–14. https://doi.org/10.1038/s41598-022-06215-6.
39. A. Zaher, R. Aslam, H.S. Lee, A. Khafouri, M. Boufellous, A. Alrashdi, Y. El aoufir, H. Lgaz, M. Ouhssine, A combined computational & electrochemical exploration of the *Ammi visnaga* L. extract as a green corrosion inhibitor for carbon steel in HCl solution, Arab. J. Chem. 15 (2022) 103573. https://doi.org/10.1016/j.arabjc.2021.103573.
40. A.V. Gambo, Corrosion inhibition performance of *Acacia nolitica* pod extract on aluminium alloy in 1 M alkaline solution, FUDMA J. Sci. 4 (2020) 154–161. https://doi.org/10.33003/fjs-2020-0403-284.
41. Y. Wang, X. Zhang, G. Ma, X. Du, N. Shaheen, H. Mao, Recognition of weeds at asparagus fields using multi-feature fusion and backpropagation neural network, Int. J. Agric. Biol. Eng. 14 (2021) 190–198. https://doi.org/10.25165/j.ijabe.20211404.6135.
42. H.F. Wali, S.S. Bahar, A study of natural *Rosmarinus* corrosion inhibitor for zinc in HCl solution, J. Phys. Conf. Ser. 1973 (2021) 012126. https://doi.org/10.1088/1742-6596/1973/1/012126.
43. M. Razizadeh, M. Mahdavian, B. Ramezanzadeh, E. Alibakhshi, S. Jamali, Synthesis of hybrid organic–inorganic inhibitive pigment based on basil extract and zinc cation for application in protective construction coatings, Constr. Build. Mater. 287 (2021) 123034. https://doi.org/10.1016/j.conbuildmat.2021.123034.
44. B.T. Ogunyemi, D.F. Latona, A.A. Ayinde, I.A. Adejoro, Theoretical investigation to corrosion inhibition efficiency of some chloroquine derivatives using density functional theory, Adv. J. Chem. A. 3 (2020) 485–492. https://doi.org/10.33945/sami/ajca.2020.4.10.
45. A.E.S. Fouda, A. Motaal, A.S. Ahmed, H.B. Sallam, Corrosion protection of carbon steel in 2 M HCl using *Aizoon canariense* extract, Biointerface Res. Appl. Chem. 12 (2021) 230–243. https://doi.org/10.33263/briac121.230243.
46. R. Rosliza, A. Nurashimah, The effectiveness of *Musa paradisiaca* as green inhibitor for mild steel in marine corrosion, J. Phys. Conf. Ser. 1874 (2021) 012073. https://doi.org/10.1088/1742-6596/1874/1/012073.
47. H. Kumar, V. Yadav, S. Anu, N. Kr. Saha Kang, Adsorption and inhibition mechanism of efficient and environment friendly corrosion inhibitor for mild steel: Experimental and theoretical study, J. Mol. Liq. 338 (2021) 116634. https://doi.org/10.1016/j.molliq.2021.116634.
48. A.T.J. Rani, A. Thomas, A. Joseph, Inhibition of mild steel corrosion in HCl using aqueous and alcoholic extracts of *Crotalaria pallida*: A combination of experimental, simulation and theoretical studies, J. Mol. Liq. 334 (2021) 116515. https://doi.org/10.1016/j.molliq.2021.116515.
49. S.S. Durodola, A.S. Adekunle, L.O. Olasunkanmi, J.A.O. Oyekunle, Inhibition of mild steel corrosion in acidic medium by extract of *Spilanthes uliginosa* leaves, Electroanalysis. 32 (2020) 2693–2702. https://doi.org/10.1002/elan.202060227.
50. A.S. Fouda, H.E. Megahed, N. Fouad, N.M. Elbahrawi, Corrosion inhibition of carbon steel in 1 M hydrochloric acid solution by aqueous extract of *Thevetia peruviana*, J. Bio- Tribo-Corrosion. 2 (2016) 1–13. https://doi.org/10.1007/s40735-016-0046-z.
51. E.B. Ituen, A.O. James, O. Akaranta, Elephant grass biomass extract as corrosion inhibitor for mild steel in acidic medium, J. Mater. Environ. Sci. 8 (2017) 1498–1507.

52. C.O. Akalezi, C.E. Ogukwe, E.A. Ejele, E.E. Oguzie, Corrosion inhibition properties of *Gongronema latifollium* extract in acidic media, Int. J. Corros. Scale Inhib. 5 (2016) 232–247. https://doi.org/10.17675/2305-6894-2016-5-3-4.
53. M. Ramdani, H. Elmsellem, N. Elkhiati, B. Haloui, A. Aouniti, M. Ramdani, Z. Ghazi, A. Chetouani, B. Hammouti, *Caulerpa prolifera* green algae using as eco-friendly corrosion inhibitor for mild steel in 1 M HCl media, Der Pharma Chem. 7 (2015) 67–76.
54. F.O. Nwosu, M.M. Muzakir, Thermodynamic and adsorption studies of corrosion inhibition of mild steel using lignin from siam weed (*Chromolaena odorata*) in acid medium, J. Mater. Environ. Sci. 7 (2016) 1663–1673.
55. D. Li, X. Zhao, Z. Liu, H. Liu, B. Fan, B. Yang, X. Zheng, W. Li, H. Zou, Synergetic anticorrosion mechanism of main constituents in Chinese yam Peel for copper in artificial seawater, ACS Omega. 6 (2021) 29965–29981. https://doi.org/10.1021/acsomega.1c04500.
56. S. Meng, Z. Liu, X. Zhao, B. Fan, H. Liu, M. Guo, H. Hao, Efficient corrosion inhibition by sugarcane purple rind extract for carbon steel in HCl solution: Mechanism analyses by experimental and in silico insights, RSC Adv. 11 (2021) 31693–31711. https://doi.org/10.1039/d1ra04976c.
57. A.A. Paul, B.O. Kelechukwu, S.M. Uchenna, Adsorption and thermodynamic studies of the corrosion inhibition effect of *Rosmarinus officinalis* L. leaves on aluminium alloy in 0.25 M HCl and effect of an external magnetic field, Int. J. Phys. Sci. 16 (2021) 79–95. https://doi.org/10.5897/ijps2021.4945.
58. S.C. Ikpeseni, H.I. Owamah, K. Owebor, E.S. Ameh, S.O. Sada, E. Otuaro, Corrosion inhibition efficiency, adsorption and thermodynamic studies of *Ocimum gratissimum* on carbon steel in 2 M sodium chloride solution, J. Bio- Tribo-Corrosion. 7 (2021) 1–14. https://doi.org/10.1007/s40735-021-00505-8.
59. A.P.S. Kaban, J.W. Soedarsono, W. Mayangsari, M.S. Anwar, A. Maksum, A. Ridhova, R. Riastuti, Insight on corrosion prevention of C1018 in 1.0 M hydrochloric acid using liquid smoke of rice husk ash: Electrochemical, surface analysis, and deep learning studies, Coatings. 13 (2023) 1–18. https://doi.org/10.3390/coatings13010136.
60. I.E. Uzoma, M.M. Solomon, R.T. Loto, S.A. Umoren, Aspartame as a green and effective corrosion inhibitor for T95 carbon steel in 15 wt.% HCl solution, Sustainability. 14 (2022). https://doi.org/10.3390/su14116500.
61. Q. Zhao, J. Guo, G. Cui, T. Han, Y. Wu, Chitosan derivatives as green corrosion inhibitors for P110 steel in a carbon dioxide environment, Colloids Surf. B: Biointerfaces. 194 (2020) 111150. https://doi.org/10.1016/j.colsurfb.2020.111150.
62. Z. Chen, A.A. Fadhil, T. Chen, A.A. Khadom, C. Fu, N.A. Fadhil, Green synthesis of corrosion inhibitor with biomass platform molecule: Gravimetrical, electrochemical, morphological, and theoretical investigations, J. Mol. Liq. 332 (2021) 115852. https://doi.org/10.1016/j.molliq.2021.115852.
63. O.K. Abiola, N.C. Oforka, E.E. Ebenso, N.M. Nwinuka, Eco-friendly corrosion inhibitors: The inhibitive action of *Delonix regia* extract for the corrosion of aluminium in acidic media, Anti-Corrosion Methods Mater. 54 (2007) 219–224. https://doi.org/10.1108/00035590710762357.
64. M.H. Shahini, M. Ramezanzadeh, G. Bahlakeh, B. Ramezanzadeh, Superior inhibition action of the Mish Gush (MG) leaves extract toward mild steel corrosion in HCl solution: Theoretical and electrochemical studies, J. Mol. Liq. 332 (2021) 115876. https://doi.org/10.1016/j.molliq.2021.115876.
65. S. Aourabi, M. Driouch, M. Sfaira, F. Mahjoubi, B. Hammouti, C. Verma, E.E. Ebenso, L. Guo, Phenolic fraction of *Ammi visnaga* extract as environmentally friendly antioxidant and corrosion inhibitor for mild steel in acidic medium, J. Mol. Liq. 323 (2021) 114950. https://doi.org/10.1016/j.molliq.2020.114950.

66. A. Salmasifar, M. Edraki, E. Alibakhshi, B. Ramezanzadeh, G. Bahlakeh, Combined electrochemical/surface investigations and computer modeling of the aquatic artichoke extract molecules corrosion inhibition properties on the mild steel surface immersed in the acidic medium, J. Mol. Liq. 327 (2021) 114856. https://doi.org/10.1016/j. molliq.2020.114856.

67. A. Chapagain, D. Acharya, A.K. Das, K. Chhetri, H.B. Oli, A.P. Yadav, Alkaloid of *Rhynchostylis retusa* as green inhibitor for mild steel corrosion in 1 M H_2SO_4 solution, Electrochem. 3 (2022) 211–224. https://doi.org/10.3390/electrochem3020013.

68. O. Thapa, J.T. Magar, H.B. Oli, A. Rajaure, D. Nepali, Alkaloids of *Solanum xanthocarpum* stem as green inhibitor for mild steel corrosion in one molar sulphuric acid solution, Electrochem. 3 (2022) 820–842. https://doi.org/10.3390/electrochem3040054.

69. I. Rubaye, A.A. Abdulwahid, S.B. Al-Baghdadi, A.A. Al-Amiery, A.A.H. Kadhum, A.B. Mohamad, Cheery sticks plant extract as a green corrosion inhibitor complemented with LC-EIS/MS spectroscopy, Int. J. Electrochem. Sci. 10 (2015) 8200–8209.

70. J. Núñez-Morales, L.I. Jaramillo, P.J. Espinoza-Montero, V.E. Sánchez-Moreno, Evaluation of adding natural gum to pectin extracted from *Ecuadorian citrus* peels as an eco-friendly corrosion inhibitor for carbon steel, Molecules. 27 (2022). https://doi. org/10.3390/molecules27072111.

71. M. Kamran, A.H.A. Shah, G. Rahman, S. Bilal, Potential impacts of *Prunus domestica* based natural gum on physicochemical properties of polyaniline for corrosion inhibition of mild and stainless steel, Polymers (Basel). 14 (2022). https://doi.org/10.3390/ polym14153116.

72. D. Święch, G. Palumbo, N. Piergies, E. Pięta, A. Szkudlarek, C. Paluszkiewicz, Spectroscopic investigations of 316l stainless steel under simulated inflammatory conditions for implant applications: The effect of tryptophan as corrosion inhibitor/ hydrophobicity marker, Coatings. 11 (2021). https://doi.org/10.3390/coatings11091097.

73. A. Kasprzhitskii, G. Lazorenko, T. Nazdracheva, V. Yavna, Comparative computational study of l-amino acids as green corrosion inhibitors for mild steel, Computation. 9 (2021) 1–11. https://doi.org/10.3390/computation9010001.

74. A. Thoume, A. Elmakssoudi, D.B. Left, N. Benzbiria, F. Benhiba, M. Dakir, M. Zahouily, A. Zarrouk, M. Azzi, M. Zertoubi, Amino acid structure analog as a corrosion inhibitor of carbon steel in 0.5 M H_2SO_4: Electrochemical, synergistic effect and theoretical studies, Chem. Data Collect. 30 (2020) 100586. https://doi.org/10.1016/j. cdc.2020.100586.

75. M.A. Jafar Mazumder, New, amino acid based zwitterionic polymers as promising corrosion inhibitors of mild steel in 1 M HCl, Coatings. 9 (2019) 675. https://doi. org/10.3390/coatings9100675.

76. J. Haque, V. Srivastava, C. Verma, M.A. Quraishi, Experimental and quantum chemical analysis of 2-amino-3-((4-((S)-2-amino-2-carboxyethyl)-1H-imidazol-2-yl)thio) propionic acid as new and green corrosion inhibitor for mild steel in 1 M hydrochloric acid solution, J. Mol. Liq. 225 (2017) 848–855. https://doi.org/10.1016/ j.molliq.2016.11.011.

77. M. Rbaa, F. Benhiba, R. Hssisou, Y. Lakhrissi, B. Lakhrissi, M.E. Touhami, I. Warad, A. Zarrouk, Green synthesis of novel carbohydrate polymer chitosan oligosaccharide grafted on D-glucose derivative as bio-based corrosion inhibitor, J. Mol. Liq. 322 (2021) 114549. https://doi.org/10.1016/j.molliq.2020.114549.

78. A. Farhadian, S. Assar Kashani, A. Rahimi, E.E. Oguzie, A.A. Javidparvar, S.C. Nwanonenyi, S. Yousefzadeh, M.R. Nabid, Modified hydroxyethyl cellulose as a highly efficient eco-friendly inhibitor for suppression of mild steel corrosion in a 15% HCl solution at elevated temperatures, J. Mol. Liq. 338 (2021) 116607. https://doi.org/10.1016/j. molliq.2021.116607.

79. M. Rbaa, M. Fardioui, C. Verma, A.S. Abousalem, M. Galai, E.E. Ebenso, T. Guedira, B. Lakhrissi, I. Warad, A. Zarrouk, 8-Hydroxyquinoline based chitosan derived carbohydrate polymer as biodegradable and sustainable acid corrosion inhibitor for mild steel: Experimental and computational analyses, Int. J. Biol. Macromol. 155 (2020) 645–655. https://doi.org/10.1016/j.ijbiomac.2020.03.200.

80. K.E.L. Mouaden, D.S. Chauhan, M.A. Quraishi, L. Bazzi, Thiocarbohydrazide-crosslinked chitosan as a bioinspired corrosion inhibitor for protection of stainless steel in 3.5% NaCl, Sustain. Chem. Pharm. 15 (2020) 100213. https://doi.org/10.1016/j.scp.2020.100213.

81. C. B.P., P. Rao, Carbohydrate biopolymer for corrosion control of 6061 Al-Alloy and 6061Aluminum-15%(v) SiC(P) composite: Green approach, Carbohydr. Polym. 168 (2017) 337–345. https://doi.org/10.1016/j.carbpol.2017.03.098.

82. X.T. Xu, H.W. Xu, W. Li, Y. Wang, X.Y. Zhang, A combined quantum chemical, molecular dynamics and Monto Carlo study of three amino acids as corrosion inhibitors for aluminum in NaCl solution, J. Mol. Liq. 345 (2022). https://doi.org/10.1016/j.molliq.2021.117010.

83. Y. Zhang, S. Zhang, B. Tan, L. Guo, H. Li, Solvothermal synthesis of functionalized carbon dots from amino acid as an eco-friendly corrosion inhibitor for copper in sulfuric acid solution, J. Colloid Interface Sci. 604 (2021) 1–14. https://doi.org/10.1016/j.jcis.2021.07.034.

84. Q.H. Zhang, B.S. Hou, Y.Y. Li, Y. Lei, X. Wang, H.F. Liu, G.A. Zhang, Two amino acid derivatives as high efficient green inhibitors for the corrosion of carbon steel in CO_2-saturated formation water, Corros. Sci. 189 (2021). https://doi.org/10.1016/j.corsci.2021.109596.

85. S. Satpati, A. Suhasaria, S. Ghosal, A. Saha, S. Dey, D. Sukul, Amino acid and cinnamaldehyde conjugated Schiff bases as proficient corrosion inhibitors for mild steel in 1 M HCl at higher temperature and prolonged exposure: Detailed electrochemical, adsorption and theoretical study, J. Mol. Liq. 324 (2021) 115077. https://doi.org/10.1016/j.molliq.2020.115077.

86. M. Oubaaqa, M. Ouakki, M. Rbaa, A.S. Abousalem, M. Maatallah, F. Benhiba, A. Jarid, M. Ebn Touhami, A. Zarrouk, Insight into the corrosion inhibition of new amino-acids as efficient inhibitors for mild steel in HCl solution: Experimental studies and theoretical calculations, J. Mol. Liq. 334 (2021) 116520. https://doi.org/10.1016/j.molliq.2021.116520.

87. L.K.M.O. Goni, M.A.J. Mazumder, S.A. Ali, M.K. Nazal, H.A. Al-Muallem, Biogenic amino acid methionine-based corrosion inhibitors of mild steel in acidic media, Int. J. Miner. Metall. Mater. 26 (2019) 467–482. https://doi.org/10.1007/s12613-019-1754-4.

88. V. Srivastava, J. Haque, C. Verma, P. Singh, H. Lgaz, R. Salghi, M.A. Quraishi, Amino acid based imidazolium zwitterions as novel and green corrosion inhibitors for mild steel: Experimental, DFT and MD studies, J. Mol. Liq. 244 (2017) 340–352. https://doi.org/10.1016/j.molliq.2017.08.049.

89. C. Verma, M.A. Quraishi, Chelation capability of chitosan and chitosan derivatives: Recent developments in sustainable corrosion inhibition and metal decontamination applications, Curr. Res. Green Sustain. Chem. 4 (2021) 100184. https://doi.org/10.1016/j.crgsc.2021.100184.

90. A. Sánchez-Eleuterio, C. Mendoza-Merlos, R. Corona Sánchez, A.M. Navarrete-López, A. Martínez Jiménez, E. Ramírez-Domínguez, L. Lomas Romero, R. Orozco Cruz, A. Espinoza Vázquez, G.E. Negrón-Silva, Experimental and theoretical studies on acid corrosion inhibition of API 5L X70 steel with novel 1-*N*-α-D-Glucopyranosyl-1*H*-1,2,3-triazole xanthines, Molecules. 28 (2023). https://doi.org/10.3390/molecules28010460.

91. B.A. Al Jahdaly, M.F. Elsadek, B.M. Ahmed, M.F. Farahat, M.M. Taher, A.M. Khalil, Outstanding graphene quantum dots from carbon source for biomedical and corrosion inhibition applications: A review, Sustainability. 13 (2021) 1–33. https://doi.org/10.3390/su13042127.

92. C. Verma, M.A. Quraishi, Carbohydrate polymer–metal nanocomposites as advanced anticorrosive materials: A perspective, Int. J. Corros. Scale Inhib. 11 (2022) 507–523. https://doi.org/10.17675/2305-6894-2022-11-2-3.

93. M. Mobin, M. Rizvi, Polysaccharide from *Plantago* as a green corrosion inhibitor for carbon steel in 1 M HCl solution, Carbohydr. Polym. 160 (2017) 172–183. https://doi.org/10.1016/j.carbpol.2016.12.056.

94. I.B. Onyeachu, D.S. Chauhan, K.R. Ansari, I.B. Obot, M.A. Quraishi, A.H. Alamri, Hexamethylene-1,6-bis(N-D-glucopyranosylamine) as a novel corrosion inhibitor for oil and gas industry: Electrochemical and computational analysis, New J. Chem. 43 (2019) 7282–7293. https://doi.org/10.1039/c9nj00023b.

95. C. Verma, M.A. Quraishi, K. Kluza, M. Makowska-Janusik, L.O. Olasunkanmi, E.E. Ebenso, Corrosion inhibition of mild steel in 1 M HCl by D-glucose derivatives of dihydropyrido [2,3-d:6,5-D′] dipyrimidine-2,4,6,8($1H,3H,5H,7H$)-tetraone, Sci. Rep. 7 (2017) 1–17. https://doi.org/10.1038/srep44432.

96. A.R. Sayed, H.M. Abd El-Lateef, Thiocarbohydrazones based on adamantane and ferrocene as efficient corrosion inhibitors for hydrochloric acid pickling of c-steel, Coatings. 10 (2020) 1–20. https://doi.org/10.3390/coatings10111068.

97. O. Benali, H. Benmehdi, O. Hasnaoui, C. Selles, R. Salghi, Green corrosion inhibitor: Inhibitive action of tannin extract of *Chamaerops humilis* plant for the corrosion of mild steel in 0.5M H_2SO_4, J. Mater. Environ. Sci, 4 (2013) 127–138.

98. M. Benahmed, I. Selatnia, N. Djeddi, S. Akkal, H. Laouer, Adsorption and corrosion inhibition properties of butanolic extract of *Elaeoselinum thapsioides* and its synergistic effect with *Reutera lutea* (Desf.) Maires (Apiaceae) on A283 carbon steel in hydrochloric acid solution, Chem. Africa. 3 (2020) 251–261. https://doi.org/10.1007/s42250-019-00093-8.

99. X. Li, S. Deng, Synergistic inhibition effect of walnut green husk extract and potassium iodide on the corrosion of cold rolled steel in trichloroacetic acid solution, J. Mater. Res. Technol. 9 (2020) 15604–15620. https://doi.org/10.1016/j.jmrt.2020.11.018.

100. A.S. Fouda, K. Shalabi, M.S. Shaaban, Synergistic effect of potassium iodide on corrosion inhibition of carbon steel by *Achillea santolina* extract in hydrochloric acid solution, J. Bio-Tribo-Corrosion. 5 (2019). https://doi.org/10.1007/s40735-019-0260-6.

101. A.H. Alamri, Experimental and theoretical insights into the synergistic effect of iodide ions and 1-acetyl-3-thiosemicarbazide on the corrosion protection of c1018 carbon steel in 1 M HCl, Materials (Basel). 13 (2020) 1–17. https://doi.org/10.3390/ma13215013.

102. S. Umoren, D. Kavaz, S.A. Umoren, Corrosion inhibition evaluation of chitosan–CuO nanocomposite for carbon steel in 5% HCl solution and effect of KI addition, Sustainability. 14 (2022). https://doi.org/10.3390/su14137981.

103. M. Ahangar, M. Izadi, T. Shahrabi, I. Mohammadi, The synergistic effect of zinc acetate on the protective behavior of sodium lignosulfonate for corrosion prevention of mild steel in 3.5 wt% NaCl electrolyte: Surface and electrochemical studies, J. Mol. Liq. 314 (2020) 113617. https://doi.org/10.1016/j.molliq.2020.113617.

104. B. Tan, J. He, S. Zhang, C. Xu, S. Chen, H. Liu, W. Li, Insight into anti-corrosion nature of betel leaves water extracts as the novel and eco-friendly inhibitors, J. Colloid Interface Sci. 585 (2021) 287–301. https://doi.org/10.1016/j.jcis.2020.11.059.

105. A. Dehghani, G. Bahlakeh, B. Ramezanzadeh, Construction of a sustainable/controlled-release nano-container of non-toxic corrosion inhibitors for the water-based siliconized film: Estimating the host-guest interactions/desorption of inclusion complexes of cerium acetylacetonate (CeA) with beta-cyclodextrin (β-CD) via detailed electronic/atomic-scale computer modeling and experimental methods, J. Hazard. Mater. 399 (2020) 123046. https://doi.org/10.1016/j.jhazmat.2020.123046.

106. J. Hu, Y. Zhu, J. Hang, Z. Zhang, Y. Ma, H. Huang, Q. Yu, J. Wei, The effect of organic core–shell corrosion inhibitors on corrosion performance of the reinforcement in simulated concrete pore solution, Constr. Build. Mater. 267 (2021) 121011. https://doi.org/10.1016/j.conbuildmat.2020.121011.

107. E. Aquino-Torres, R.L. Camacho-Mendoza, E. Gutierrez, J.A. Rodriguez, L. Feria, P. Thangarasu, J. Cruz-Borbolla, The influence of iodide in corrosion inhibition by organic compounds on carbon steel: Theoretical and experimental studies, Appl. Surf. Sci. 514 (2020) 145928. https://doi.org/10.1016/j.apsusc.2020.145928.

108. M.T. Majd, S. Asaldoust, G. Bahlakeh, B. Ramezanzadeh, M. Ramezanzadeh, Green method of carbon steel effective corrosion mitigation in 1 M HCl medium protected by *Primula vulgaris* flower aqueous extract via experimental, atomic-level MC/MD simulation and electronic-level DFT theoretical elucidation, J. Mol. Liq. 284 (2019) 658–674. https://doi.org/10.1016/j.molliq.2019.04.037.

109. B. Vaghefinazari, E. Wierzbicka, P. Visser, R. Posner, E. Matykina, M. Mohedano, C. Blawert, M.L. Zheludkevich, S.V. Lamaka, Chromate-free corrosion protection strategies for magnesium alloys: A review. Part III: Corrosion inhibitors and combining them with other protection strategies, Materials (Basel). 15 (2022) 8489. https://doi.org/10.3390/ma15238489.

110. Z. Jamshidnejad, A. Afshar, M.A. RazmjooKhollari, Synthesis of self-healing smart epoxy and polyurethane coating by encapsulation of olive leaf extract as corrosion inhibitor, Int. J. Electrochem. Sci. 13 (2018) 12278–12293. https://doi.org/10.20964/2018.12.83.

111. I.O. Arukalam, I.O. Madu, E.Y. Ishidi, High performance characteristics of *Lupinus arboreus* gum extract as self-healing and corrosion inhibition agent in epoxy-based coating, Prog. Org. Coatings. 151 (2021) 106095. https://doi.org/10.1016/j.porgcoat.2020.106095.

112. A.S. Gnedenkov, S.L. Sinebryukhov, V.S. Filonina, A.Y. Ustinov, S.V. Sukhoverkhov, S.V. Gnedenkov, New polycaprolactone-containing self-healing coating design for enhance corrosion resistance of the magnesium and its alloys, Polymers (Basel). 15 (2023). https://doi.org/10.3390/polym15010202.

113. Z. Sanaei, T. Shahrabi, B. Ramezanzadeh, Synthesis and characterization of an effective green corrosion inhibitive hybrid pigment based on zinc acetate-*Cichorium intybus* L. leaves extract (ZnA-CIL.L): Electrochemical investigations on the synergistic corrosion inhibition of mild steel in aqueous chloride solutions, Dye. Pigment. 139 (2017) 218–232. https://doi.org/10.1016/j.dyepig.2016.12.002.

Index

For Product Safety Concerns and Information please contact our EU
representative GPSR@taylorandfrancis.com
Taylor & Francis Verlag GmbH, Kaufingerstraße 24, 80331 München, Germany

www.ingramcontent.com/pod-product-compliance
Lightning Source LLC
Chambersburg PA
CBHW060749220326
41598CB00022B/2384